酪氨酸酶抑制剂的合成及在食品中的应用

刘进兵　著

中国纺织出版社

内容提要

酪氨酸酶是与酶促褐变相关的重要酶，酶促褐变是影响食品行业发展的一个重要问题，不仅影响果蔬价值，也降低了其内在品质。为了防止果蔬褐变，延长果蔬货架期，酪氨酸酶抑制剂的研究已经成为了当前研究的热点。目前市场上并无酪氨酸酶抑制剂的专著出现，本书结合作者多年研究酪氨酸酶抑制剂的成果，阐述化学合成酪氨酸酶抑制剂研究进展及其在食品中的应用。本书可以作为从事食品工业相关技术人员的工具书和大专院校相关专业的教学参考书。

图书在版编目(CIP)数据

酪氨酸酶抑制剂的合成及在食品中的应用/刘进兵著．—北京：中国纺织出版社，2016．12（2022.9 重印）
ISBN 978－7－5180－3180－1

Ⅰ．①酪… Ⅱ．①刘… Ⅲ．①酪氨酸酶—抑制剂—研究 Ⅵ．①Q554

中国版本图书馆 CIP 数据核字(2016)第 322205 号

责任编辑：国帅　　责任设计：品欣排版　　责任印制：王艳丽

中国纺织出版社出版发行
地址：北京市朝阳区百子湾东里 A407 号楼　邮政编码：100124
销售电话：010—67004422　传真：010—87155801
http://www.c-textilep.com
E-mail:faxing@c-textilep.com
中国纺织出版社天猫旗舰店
官方微博 http://weibo.com/2119887771
佳兴达印刷（天津）有限公司印刷　各地新华书店经销
2016 年 12 月第 1 版　2022 年 9 月第 2 次印刷
开本：710×1000　1/16　印张：26.25
字数：418 千字　　定价：89.00 元

前言

众所周知，酪氨酸酶是生物合成黑色素的关键酶，包括决定哺乳类动物皮肤和头发的颜色。不同的皮肤病，像黄黑斑、老年斑都是因为表皮色素过度积累的结果。黑色素恶性肿瘤，特别是威胁生命的皮肤肿瘤，引起人们的极大关注。黑色素通过一系列氧化反应形成，包括酪氨酸酶参与的氧化酪氨酸的反应。在食品工业当中，酪氨酸酶是一种非常重要的酶，在控制蔬菜和水果的质量和经济价值中，酪氨酸酶催化酚类化合物的氧化来形成相应的醌和引起水果和蔬菜的酶促褐变。另外，醌类化合物同带有氨基和巯基官能团的蛋白质反应，生产不可逆的褐变色素，形成令人不愉快的颜色和气味。因此，高活性的酪氨酸酶抑制的发展在农业和食品工业领域有很大的需求。酪氨酸酶在昆虫的生长和防御作用方面发挥着重要的作用。它参与到昆虫体内黑色素元的生成，伤口的愈合、蜕皮等。另外，酪氨酸酶抑制剂已经成为医药和化妆品中的重要组成部分。因此，酪氨酸酶抑制剂的研究对医药、农业、食品、化妆品等行业有重要意义。

目前国内还没有一部关于酪氨酸酶抑制剂的专著，本书作者一直在从事酪氨酸酶抑制剂的研究，基于作者的研究成果，借鉴国内外同行关于酪氨酸酶抑制剂的研究特色编写了这本书。本书在内容选取上力求反映酪氨酸酶抑制剂的最新成果和前沿发展动态。本书的内容安排是根据官能团而定，共分五章，第 1 章介绍酪氨酶和酪氨酸酶抑制剂概况，第 2 ~4 章分别介绍不同种类的酪氨酸酶抑制剂，同时附上了作者最近的研究成果，第 5 章介绍酪氨酸酶抑制剂在食品领域的应用。

本书由刘进兵编著，第 5 章包含了硕士生陈玲娟、唐君源毕业论文部分内容，第 2 ~4 章包含了编者最近的研究成果和陈昌红硕士论文部分内容。本书在编写过程中得到了很多前辈、同行和朋友的帮助，得到了邵阳学院和中国纺织出版社的大力支持，在此，作者表示衷心感谢！

由于水平有限，时间仓促，书中难免有不足甚至谬误之处，特别是在内容选取和表述上都还存在许多不足，诚恳希望读者给予批评指正。

目 录

第1章　酪氨酸酶与酪氨酸酶抑制剂

众所周知,酪氨酸酶是生物合成黑色素的关键酶,包括决定哺乳类动物皮肤和头发的颜色。不同的皮肤病,像黄黑斑、老年斑都是因为表皮色素过度积累的结果。另外,酪氨酸酶引起褐变导致营养降低,质量变差和经济损失。当前,传统的技术不足以防止酪氨酸酶的作用,鼓励大家寻找新的有效的酪氨酸酶抑制剂。

黑色素是一种分布最广泛的的色素之一,在细菌、真菌、植物、动物中都有发现。它是不同酚类的生物高聚物,有着复杂的结构,颜色从黄色到黑色间变化。哺乳动物的皮肤和毛发的颜色是由许多因素共同决定的,其中最重要的因素是色素的含量、种类和分布。黑色素充当着保护皮肤的角色,吸收紫外线,消除活性氧。各种各样的皮肤病导致皮肤色素的过度积累,影响皮肤的美容。黑色素恶性肿瘤,特别是威胁生命的皮肤肿瘤,引起人们的极大关注。黑色素通过一系列氧化反应形成,包括酪氨酸酶参与的氧化酪氨酸的反应。

酪氨酸酶是一种含铜离子酶,催化黑色素合成的两步关键反应:酪氨酸的羟基化和将多巴氧化成多巴醌的反应。酪氨酸酶在植物和动物中广泛存在,并参与到黑色素的形成之中。在食品工业当中,酪氨酸酶是一种非常重要的酶,在控制蔬菜和水果的质量和经济价值中[1],酪氨酸酶催化酚类化合物的氧化来形成相应的醌和引起水果和蔬菜的酶促褐变。另外,醌类化合物同带有氨基和巯基官能团的蛋白质反应,生产不可逆的褐变色素,形成令人不愉快的颜色和气味。这种醌和蛋白质的反应降低了蛋白质的可消化性和必要氨基酸的利用率,包括赖氨酸和半胱氨酸。因此,高活性的酪氨酸酶抑制的发展在农业和食品工业领域有很大的需求。

酪氨酸酶在昆虫的生长和防御作用方面发挥着重要的作用。它参与到昆虫体内黑色素元的生成,伤口的愈合、蜕皮等[2]。酪氨酸酶抑制剂的发展在控制虫害方面已经变成一种积极的可选择的方式。另外,酪氨酸酶抑制剂已经变成医药和化妆品中的重要组成部分,它用于防止黑色素过度积累[3]。

1.1 酪氨酸酶的生化反应特征和反应机理

1.1.1 酪氨酸酶的分类和活性

酪氨酸酶是研究最充分多铜加氧酶之一，它含有两个铜离子的活性中心酶。酪氨酸酶催化单酚的羟基化和二酚到领苯醌。后者的反应速度比前者快很多，因此，这酪氨酸的羟基化到L－多巴被认为是反应的决速步骤。分类研究证明了被纳入到酚的底物氧分子来源于氧气[4]。

最有特征的酪氨酸酶是来源于链霉菌、脉孢菌和蘑菇。前面两种是单体蛋白，最后一种是两种不同亚基的四聚体。酪氨酸酶已经从许多的植物和动物资源中被分离，纯化，但是只有很少一部分已经明确了特征。不像真菌的酪氨酸酶，人酪氨酸是膜束缚的糖蛋白（13%糖类）[5]。这些酶中的大部分基因已经测序。在人酪氨酸酶基因中各种突变都与褪色缺失的眼皮肤白化病有关。在真菌和脊椎动物中，酪氨酸酶催化酪氨酸是黑色素形成基本步骤。在植物当中，酪氨酸酶催化的底物是不同的酚类物质。当病原体和昆虫的组织受伤时，酪氨酸酶在褐变途径中氧化酚类物质对伤口进行保护。[4]

1.1.2 酪氨酸酶催化机理

酪氨酸酶单酚酶活性的机理与它的三种形态紧密相关。三种亚型分别是还原形态、氧化形态和脱氧形态。在单酚酶的反应周期中，单酚酶只能同氧化形态反应和束缚一个氧化形式铜离子的轴向位置。靠束缚过氧化物穿过一个三角形双锥体的中间体重排实现单酚的羟基化。这样产生一个邻苯二酚分子，邻苯二酚被氧化成领苯二醌，结果在脱氧形态形成中进一步束缚氧分子。[5]

领苯二酚能与酪氨酸酶的静止形态反应，形成结合的邻苯二酚。在二酚酶周期中，氧化和还原两种形态形式同领苯二酚反应，氧化成为邻苯二醌。然而，单酚能同二酚竞争束缚静止亚型的活性位点，抑制二酚的减少。比较单酚底物和二酚底物的动力学常数，在苯环上较大体积取代物显著降低单酚酶的活性，但不是二酚酶的活性[6]。这些暗示了当单酚底对羟基化物要求轴向赤道的重排，邻苯二酚底物不需要经过重排，在铜离子位置简单的电子转移就能完成羟基化。

稳定状态途径的动力学研究表明酪氨酸酶催化单酚比苯邻苯二酚的效率低。单酚酶的活性有一个明显的延迟时间的特征[7]。这个延迟时间取决于这些因素,像底物和酶的浓度和氢供体的参与。在动力学的研究中,延迟时间是要求静止态通过还原剂被卷入到脱氧态,这段时间出现少量氧化态通常地伴随有静止态亚型的活动。在还原剂的出现认为是辅助因子,特别是邻苯二酚衍生物像L-DOPA和(+)-儿茶酸,酪氨酸酶是活跃的,并且这延迟时间被缩短或消失。L-DOPA在非常低的浓度是最有效的消除延迟时间的化合物。

1.1.3 在哺乳动物中黑色素生物合成的途径

哺乳动物的黑色素细胞能能产生两种类型的黑色素:真黑色素和褐黑色素。真黑色素的颜色是黑色或褐色,褐黑色素的颜色是红色或黄色。野生型刺鼠毛发变色就是在毛囊黑色素细胞中这两种黑色素的转变的结果。几十年来,黑色素生物合成已经被研究和表征,特别是那些所需要的真黑色素。[4]

褐黑色素的合成的开始阶段是通过酪氨酸酶催化L-DOPA到与它相应的多巴醌,再到多巴色素。初期的多巴醌能经过两种不同的反应类型:分子内的1,4-加成到这苯环上,或与水加成反应。一个邻多巴醌侧链的氨基首先经历分子内的1,4-加成到苯环上,这导致它催化成白色多巴素。白色多巴色素迅速通过另一个邻多巴醌参与,被氧化成多巴色素。水加成到醌上形成三羟基酚。2,4,5-三羟基苯丙氨酸被化学法氧化成间羟基多巴醌通过另外一个邻位多巴醌[8]。这间羟基多巴醌参与,通过一系列缓慢反应成为多巴色素,这是开始阶段的最终产物。

最后阶段是用化学和酶促反应来表示,它们发生在多巴色素生产后造成真黑素的合成。它从缓慢的化学的多巴色素的去羰酸基生成5,6-二羟基吲哚开始,并且它的底物氧化成吲哚醌。多巴色素可以被多巴色素互变异构酶催化变成5,6-二羟基吲哚-2-羧基酸。5,6-二羟基吲哚-2-羧基酸进一步被氧化,通过邻位多巴醌来形成吲哚,5,6-二醌羧酸的氧化还原反应。吲哚,5,6-二醌羧酸存在三种互变异构的形式,包括醌亚胺和相应的高反应活性的醌甲氧化物。5,6-二羟基吲哚和5,6-二羟基吲哚-2-羧酸的衍生的黑色素的功能是不同的,前者是黑色,羊毛状的,然而后者是黄棕褐色,高分散的状态[9]。

在褐黑素生成时,像谷胱甘肽和半胱氨酸等硫氢基化合物的硫醇基亲核地进攻邻位多巴醌通过酪氨酸酶酶促地催化来生成半胱氨酸多巴或谷胱甘肽多巴。这些硫醇基能够加到苯环的不同位置,尽管这5-位是有利的位置。半胱氨酸多巴或谷胱甘肽多巴的取代环合作用和聚合在一些列不典型的反应中,导致褐黑素和表面色素的产生。这真黑素和褐黑素之间的相互作用引起混合型黑色素的不均匀分布。

1.1.4 植源性食物的酶促褐变

植源性食物和饮料,在刷洗,削皮和压碎等操作时,破坏细胞结构之后,酪氨酸酶和它们的多酚物被混合,在有氧的条件下发生酶促褐变。酶促褐变的基本步骤是领苯二酚的氧化,像 L - DOPA 到相应的醌,醌进一步氧化变成褐色的黑色素。邻位醌是很强的亲电体,它能遭到亲核体进攻,如水、其它的酚、氨基酸、多肽和蛋白质等,生成迈克尔加成产物。这些酶促褐变能够通过半胱氨酸和抗坏血酸捕获邻位中间体来阻止[10]。

绿原酸、植源性食物的主要酚类化合物,也被酪氨酸酶氧化成高反应活性的邻位醌中间体。邻位醌中间体能同赖氨酸的氨基,蛋氨酸的 SCH_3 和色氨酸的吲哚环发生亲核加成和聚合反应,这就是所谓的褐变和绿变反应。这些转变毁坏了重要的氨基酸,降低了消化吸收度和营养质量,也可能导致毒性化合物的形成。

1.2 酪氨酸酶抑制剂

通过避开紫外线的照射,酪氨酸酶的抑制,黑色素细胞新陈代谢和分裂的抑制或黑色素消融等方法可以抑制黑色素的生成。色素沉着过度紊乱标准典型的治疗方法,像治疗黑斑和炎症后的色素沉着过度的方法,包括用对苯二酚,通过维生素 A 和酪氨酸酶抑制剂的使用来增白。

1.2.1 从植物中分离提取得到的黄酮类酪氨酸酶抑制剂

植物多酚通常被归类为不同基团的具有酚功能的混合物。这些化合物是高等植物的次级代谢产物,具有许多的生物活性。黄酮类化合物是发现量最大,研究最充分的植物酚的一类,它是苯并吡喃酮衍生物,由吡喃环和酚环组成。已经被鉴别的超过 4000 种黄酮广泛分布在植物的花、皮、叶子、种子里面。根据三碳键结构的氧化程度和 B 环的连接位置等特点,黄酮类化合物可分为:黄烷醇、黄酮、黄酮醇、异黄酮、黄烷酮醇、查耳酮、橙酮等,它们有不同的羟基、甲氧基、糖苷基在不同的位置,并在 A,B 环间发生共轭。Y. - J. Kim[10]总结了从植物中分离出来的黄酮类酪氨酸酶抑制剂,见图 1 - 1。

最近,又报道了一些黄酮类酪氨酸酶抑制剂[11-13]。Carole Dubois[11]等研究

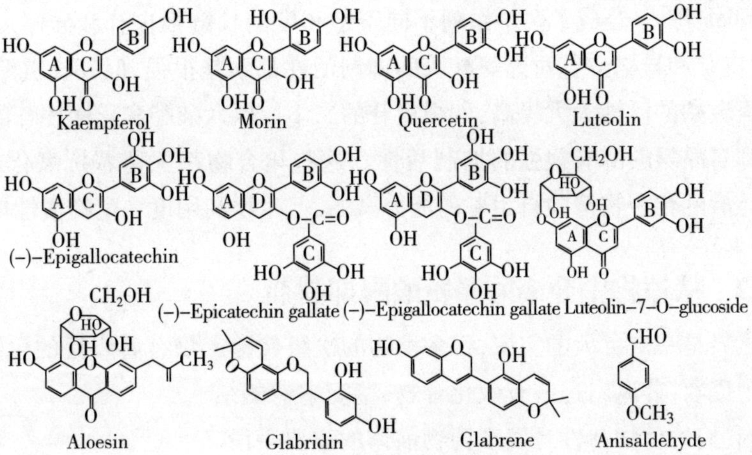

Kaempferol
Morin
Quercetin
Luteolin
(-)-Epigallocatechin
(-)-Epicatechin gallate
(-)-Epigallocatechin gallate
Luteolin-7-O-glucoside
Aloesin
Glabridin
Glabrene
Anisaldehyde

图　1-1

图　1-2

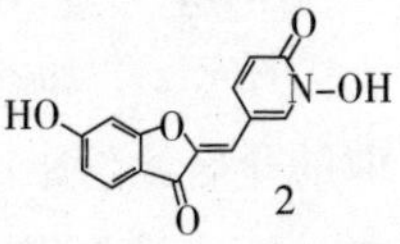

橙酮抑制酪氨酸酶活性，发现羟基在 A 环的对位，B 环没有取代，没有抑制作用，而在 B 环邻间对位上分别被羟基取代，具有提高酪氨酸酶催化活性的作用。将 B 环换成氧化 N-2-羟基吡啶后的杂环橙酮（结构式如图 1-2）表现出很强的酪氨酸酶抑制活性，$IC_{50}=1.5\mu mol/L$，其抑制表现为混合抑制的特征。说明了酪氨酸酶不止一个活性位点催化氧化橙酮，并且杂环酮和酪氨酸酶双铜的螯合，是发挥其抑制酪氨酸酶活性的重要原因。

木菠萝的木材曾经用作抗炎、抗氧化、抗衰老剂，于是 Nhan Trung Nguye 等[12]用甲醇溶液从中分离出了 4 个黄酮类化合物和 12 个查耳酮了化合物（图 1-3），它们中的大部分对酪氨酸酶的抑制作用，比现在化妆品中的增白剂曲酸强，特别是莫拉查耳酮 12，其 $IC_{50}=0.013\mu mol/L$。

	R_1	R_2
1	H	OH
2	H	ρ-coumarate
3	H	ferulate
4	OH	OH

	R_1	R_2	R_3
5	H	OH	H
6	H	ferulate	H
7	H	H	sinapate
8	H	H	OH
9	H	H	ρ-coumarate
10	H	H	ferulate
11	H	ρ-coumarate	H
12	OH		H

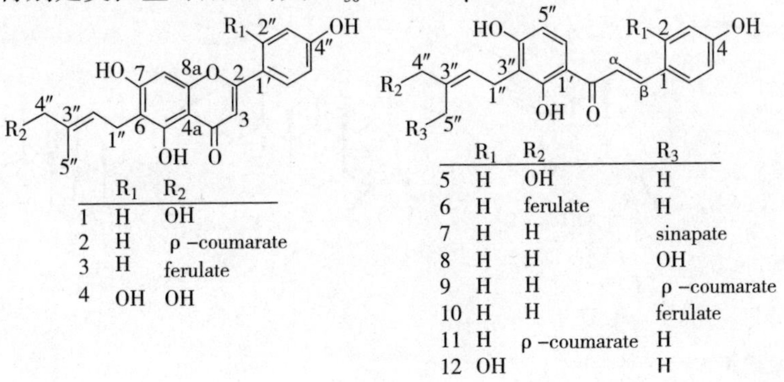

图　1-3

Ziaullah 等[13]合成了两个系列不同不饱和度的长链根皮苷和异槲皮苷酯衍生物,根皮苷和异槲皮苷对酪氨酸酶的抑制活性都不是很强,但其长链酯化以后抑制酪氨酸酶的活性大大提高,如根皮苷的二十二碳六酸酯和异槲皮苷的 α - 亚麻酸酯都对酪氨酸酶有很强的抑制活性。这些化合物都保留着抗氧化的能力,虽然酯化后的化合物清除自由基的能力减弱,但生物利用度和生物活性增强。

1.2.2 从植物中分离的酪氨酸酶抑制剂

自然界是一个巨大的宝库,很多植物的次级代谢产物有着广泛的生物活性。从药材中提取、分离出这些有效化合物,鉴定它们的结构,确定它们的生物活性是发现药物前体的重要手段。近两年,许多研究者从常见的植物和药材中,分离,提取,报道了许多结构新颖的酪氨酸酶抑制剂。

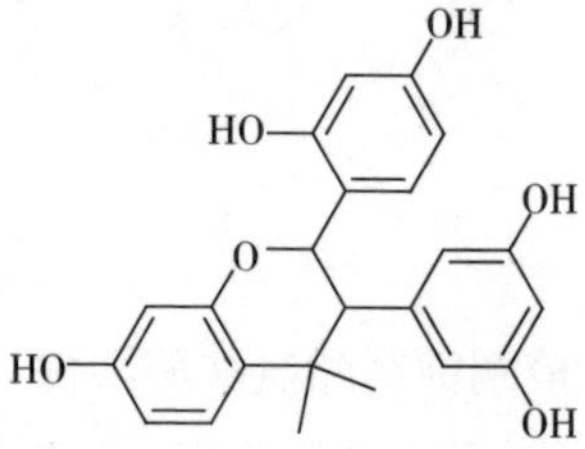

Horng - Huey Ko 等[14]从木菠萝的木心中分离提取的化合物(图 1 - 4),具有酪氨酸酶抑制活性,其抑制活性比维生素 A 和熊果苷之间,还具有抗 ABTS + 和 O_2 - 的作用。

图 1 - 4

蒲包草具有抵御食草动物和昆虫的作用,因为在这草种中有天然特殊的生物杀虫混合物。Evelyn Muñoz 等[15]用乙酸乙酯和正己烷从蒲包草中提取,分离出阿松香三稀 2、熊果酸 3α - 董尼酮 6、羽扇醇 4、谷甾醇 5、毛蕊花苷 8、马替诺皂苷 9 等具有抑制酪氨酸酶催化 L - 多巴合成黑色素的活性,它们抑制酪氨酸酶的 IC_{50} 在 10.0 到 200PPM 之间,结构见图 1 - 5。这些次级代谢产物对昆虫的作用各不相同,它们在杀灭方面的机理还要深入的检测和研究。

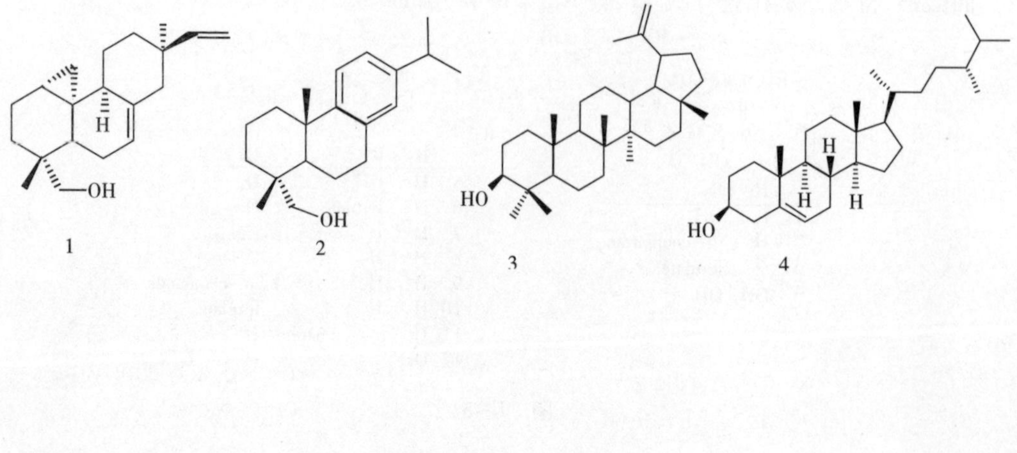

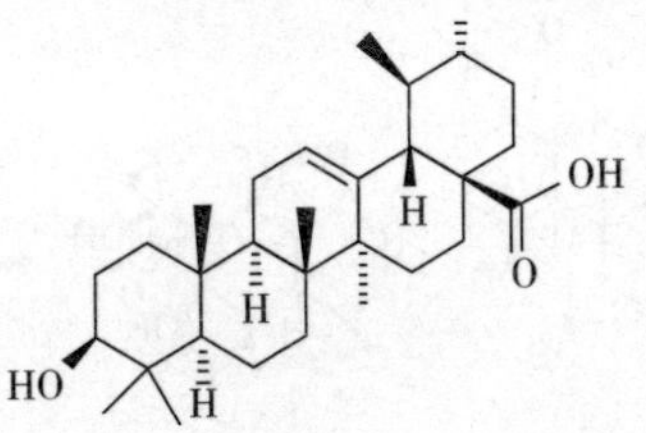

5

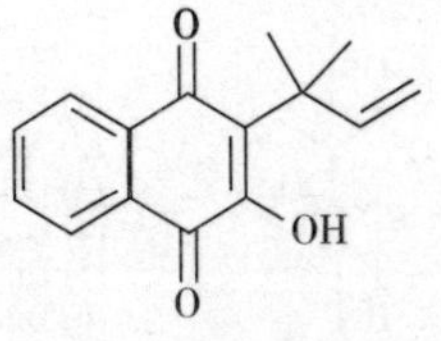

6

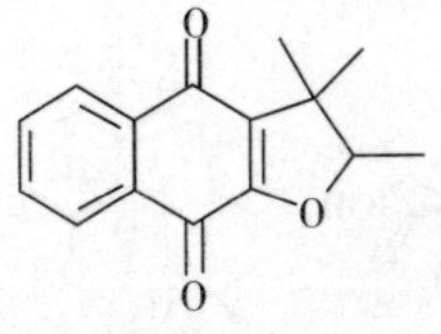

7

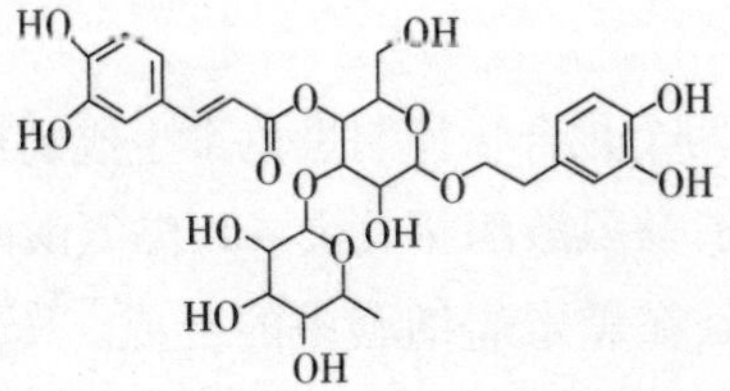

8

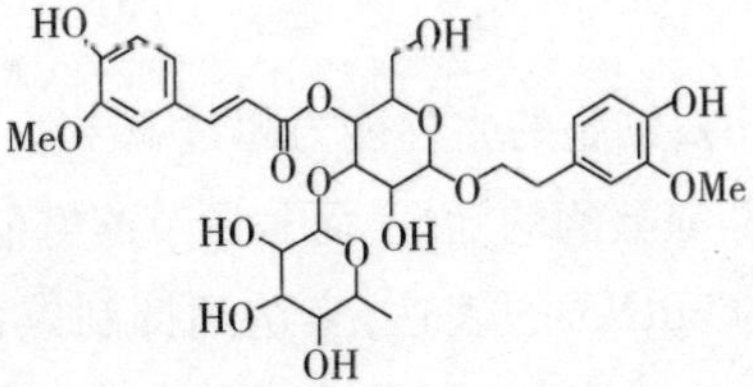

9

图 1-5

Xiaofang Wu 等[16]从芦荟中提取的芦荟苦素对酪氨酸酶有比较强抑制作用，其 IC_{50} 为 108.62μg/mL，其结构如图 1-6。

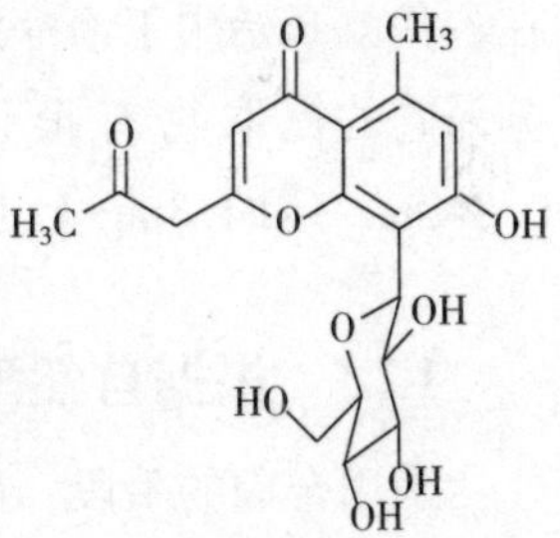

图 1-6

HirokiSatooka 等[17]从冬凌草山茶中分离出了4个贝壳杉稀二萜化合物结构（见图1-7），它们有很强的抑制黑色素生产的作用，但对小鼠 B16-F10 黑素瘤细胞表现出很强的细胞毒性。冬凌草素和 nodosin 对黑素瘤细胞毒性的 IC_{50} 分别为 1.1μmol/L 和 1.3μmol/L，当其浓度为 4.0μmol/L 和 8μmol/L 时，差不多全部黑素瘤细胞被杀死。Dihydronosin 对 B16-F10 的细胞毒性大大降低，而对黑色合成有比较强的抑制作用，所以 dihydronodosin 是一个很有潜力的黑色合成的抑制剂。

Seikou Nakamura 等[18]从睡莲的花芽和叶子的甲醇容易对 B16 黑素瘤细胞合成黑色素有抑制作用，其 IC_{50} =35.7μg/mL，并从中分离出 11 种苯基异喹啉生物碱。N-甲氧基-阿朴啡莲碱 2 和 N-甲基阿婆碱 5 抑制黑色素合成的 IC_{50} 分别为 15.8μmol/L 和 14.5μmol/L。他们对黑素瘤细胞的细胞毒性不是很强。在

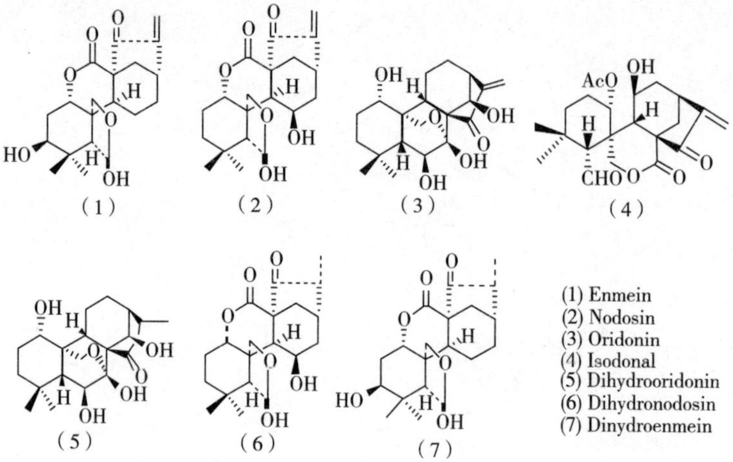

(1) Enmein
(2) Nodosin
(3) Oridonin
(4) Isodonal
(5) Dihydrooridonin
(6) Dihydronodosin
(7) Dinydroenmein

图 1-7

30μmol/L 时，细胞的生存能力超过 86%。这两个化合物抑制黑色素生成的机理不是抑制酪氨酸酶的活性，而是调节酪氨酸酶、酪氨酸酶相关蛋-1、酪氨酸酶相关蛋-2 的 mRNA 的表达，来达到抑制黑色素在黑素瘤细胞中的合成。Haroon Khan 等[19]从黄精茎中分离出了两个长链的酯，正十五羧酸丙酯和正十五羧酸甘油酯。它们是有效的酪氨酸酶抑制剂，IC_{50}分别为 22.34 和 9.45μmol/L。白芪在古代就开始用于治疗疾病，Fusheng Jiang 等[20]用氯仿提取白芪根中的有效成分。它的提取物有很强的清除自由基、抑制酪氨酸酶的能力，IC_{50} 分别为 0.848mg/L和4.3mg/L，比熊果苷强。

1.2.3 根据自然酪氨酸酶的结构合成的酪氨酸酶抑制剂

采用合成的手段，寻找药物，讨论它的构效关系是最基本的方法之一。研究者根据作用靶点的结构特征，作用机理和现有的活性结构，设计并合成新型的酪氨酸酶抑制剂。像根据白藜芦醇设计的偶氮白藜芦醇等具有很强的酪氨酸酶抑制活性。

Lyubomir Georgiev[21]合成了一些列含苯乙烯基和羟基肉桂酰胺的化合物，骨架见图 1-8，酪氨酸酶抑制活性检测，发现羟基或甲氧基在苯环上对位对酪氨酸酶的活性有很大的影响，增加一个取代在同一个苯环的另一个位置，明显减弱对酪氨酸酶的抑制作用。

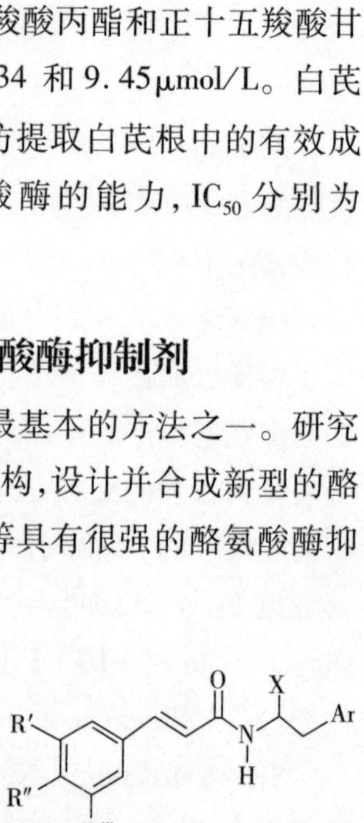

图 1-8

Young Mi Ha 等[22]模仿白藜芦醇的结构，设计合成了一些列1,6－二取代苯基－1,3,5－己三烯化合物。在苯环的对位被羟基取代得到的化合物1对酪氨酸酶有很强的抑制作用,IC_{50}为1.63μmol/L,比曲酸的抑制作用强30倍。但再在苯环的邻位置引入羟基、甲氧基、乙氧基就会减弱酪氨酸酶的抑制活性。或将对位的羟基甲基化,同样大大减弱其抑制酪氨酸酶的活性。而2,4－位同时被羟基取代2和2－位被羟基,3－位被甲氧基取代3后的化合物,对酪氨酸酶的抑制活性略有减弱,但都比曲酸的强。它们的抑制类型都属于竞争型抑制。

Sung Jin BAE等[23]同样模仿白藜芦醇的结构,合成了偶氮－白藜芦醇,得到了三个有效的酪氨酸酶抑制剂。抑制效果最好的IC_{50}为17.85μmol/L,结构如图1－9。它们不仅对B16F10细胞没有细胞毒性,还对酪氨酸酶和黑色素的合成都有抑制作用。抑制动力学分析,这些化合物是竞争型抑制剂。分子对接研究表明这些化合物与与酪氨酸酶之间氢结合作用比曲酸同酪氨酸酶的作用还高。曲酸只同酪氨酸酶的Asp262残基有氢键作用,而偶氮－白藜芦醇可以预料到与酪氨酸酶的Glu258,His261和Asp276残基有氢键作用。药效团模型表明这些化合物有一个氢键供体和三个氢键受体。偶氮－白藜芦醇类似物可考虑为很好的抗黑色素合成的治疗剂。

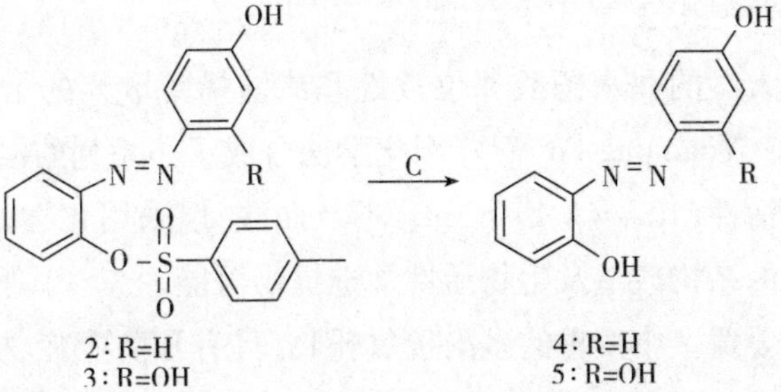

2: R=H
3: R=OH
4: R=H
5: R=OH

图 1－9

Yu Min Song等[24]也合成了一系列偶氮－白藜芦醇和偶氮－氧化白藜芦醇,但其中对酪氨酸酶抑制最好的化合物,IC_{50}＝36.28μmol/L,接近天然的白藜芦醇的抑制活性。Sung Jin Bae等[25]也得到了一系列偶氮－白藜芦醇衍生物,4－甲氧基或4－羟基取代的化合物表现出比较好的抑制活性,其中E－4－羟基苯亚胺－1',2'－二羟基苯甲烯,对蘑菇酪氨酸酶的抑制最好,IC_{50}为17.22μmol/L,其抑制类型属于非竞争型抑制。

Larissa Lavorato Lima等[26]由白藜芦醇的结构,设计合成了氮杂芪类似物。氮杂芪两个苯环的邻位引入羟基后的化合物,对酪氨酸酶的抑制活性最强,IC_{50}

$=65.67\mu g/mL$,比天然的化合物白藜芦醇($IC_{50}=87.1\mu g/mL$)要好。

RameshL 等[27]合成了一些列2,5-二取代-1,3,4-恶二唑化合物类酪氨酸酶抑制剂。通过抑制活性检测,3D QSAR 和分子对接研究发现2,5-二取代-1,3,4-恶二唑化合物的正电基团对酪氨酸酶的抑制起决定性的作用。并预测较小体积和更多正电荷基团的1,3,4-恶二唑化合物对酪氨酸酶具有更好的抑制作用。

芦笋常用作药物,作为利尿剂和净化剂,Alessandro Venditti 等[28]简化芦笋的活性物质芦笋酸的结构,得到二氢芦笋酸(图1-10)。他有很强的清除自由基和抑制酪氨酸酶的能力。其抑制酪氨酸酶的 IC_{50} 为 0.13mmol/L,抑制类型属于竞争型抑制。虽然其抑制酪氨酸酶的能力没有曲酸强,但它们的抑制机理不完全一样。由于其含有两个巯基,增加它在食品工业、农业、化妆品行业和药品中的用途。

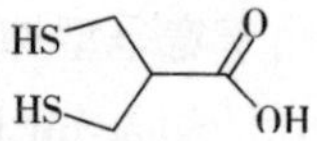

HS HS O OH

图 1-10

Dong-Fang Li 等[29]根据曲酸的结构,设计合成了一系列羟吡啶酮-L-苯基丙氨酸类酪氨酸酶抑制剂。其中最有效的抑制剂对酪氨酸酶单酚酶和双酚酶的抑制活性的 IC_{50} 分别为 12.6 和 $4\mu mol/L$。它属于混合抑制类型,既能结合自由的酪氨酸酶,又能结合底物作用的酪氨酸酶,而且它的细胞毒性比曲酸还低。

银杏酚酸是天然的在水杨酸邻位被烷基或链烯基取代的衍生物,在不同的植物中广泛存在。Yuanqing Fu 等[30]用化学法合成了一系列的银杏酚酸,虽然它的抗酪氨酸酶的活性($IC_{50}=2.8mg/mL$)不及曲酸,但采用化学的方法合成银杏酚酸为银杏酚酸的结构改造及生物活性筛选提供可能。

姜黄素是从姜黄茎中发现的多酚抗氧化物,具有预防癌症、抗炎、抗氧化、抗菌活性。它已经被批准用于化妆品色素和抗氧化剂,还能够治疗一些皮肤病。它的同类物四氢姜黄素已经被推荐用于化妆品中的增亮剂。Yongfu Jiang 等[31]也合成了一系列姜黄素类似物(图1-11)。在他们之中, 如下图1,2,3,4,其抑制酪氨酸酶的 IC_{50} 分别为 $1.74\mu mol/L$,$7.78\mu mol/L$,$16.74\mu mol/L$,$2.78\mu mol/L$,都比正丁基间苯二酚和曲酸都强。4-羟基官能团在酪氨酸酶的抑制活性中发挥着关键的作用。2 和 4 做小鼠急毒性试验表明化合物在 1200mg/Kg 没有毒性。双倒数作图法的动力学分析表明1,3 属于混合竞争型类型,而2,4 属于竞争型抑制类型。

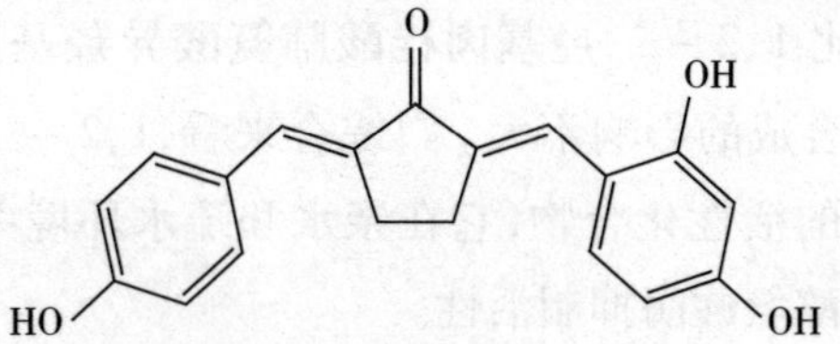

1

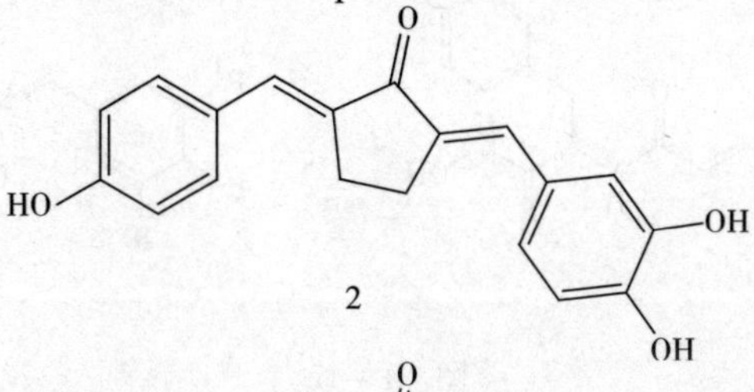

2

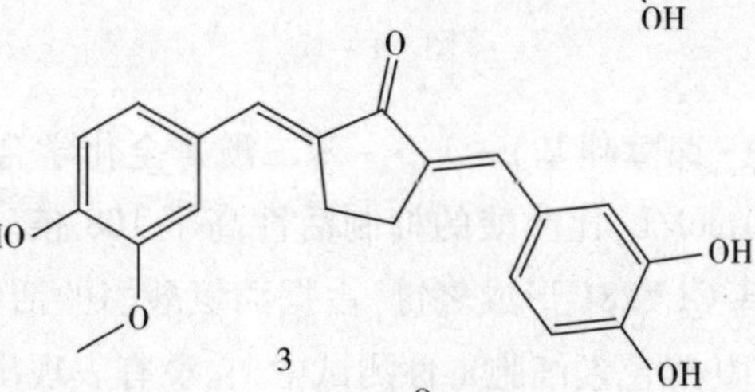

3

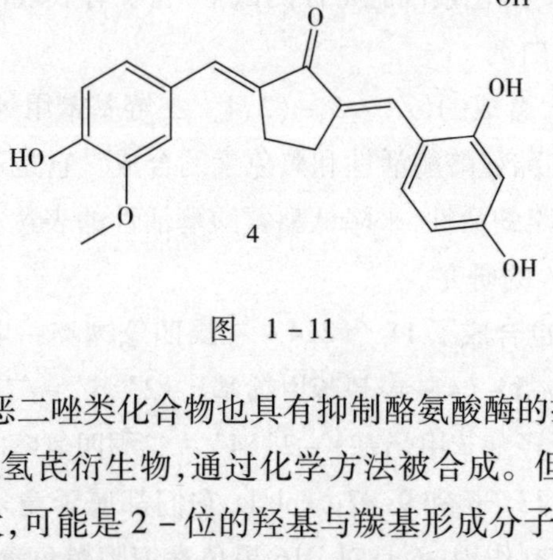

4

图　1－11

呋喃－1,3,4－恶二唑类化合物也具有抑制酪氨酸酶的抑制作用。[32]

脱氧安息香和二氢芪衍生物,通过化学方法被合成。但脱氧安息香对酪氨酸酶的抑制活性不大,可能是 2－位的羟基与羰基形成分子内的氢键造成的;而二氢芪就表现出很好的酪氨酸酶抑制活性,其中 2,4,4'－三羟基－二氢芪的活性最好,$IC_{50}=8.44\mu mol/L$。2,4,4'－三羟基－二氢芪分子对接模型研究表明 2－位的羟基 Asn260 残基形成氢键,4－位的羟基与 Val1283 残基作用,4'－的羟基与 His244 残基作用,其中 4'－羟基的作用,对酪氨酸酶的活性起到了关键的作用。这为进一步设计酪氨酸酶抑制提供了依据。[33]

羟基肉桂酸衍生物和苯丙氨酸异羟基肟(图 1－12),脯氨酸异羟基肟形成的化合物,既有抗氧化的作用,又有抑制酪氨酸酶的活性,而且它们的抗氧化活性比肉桂酸好。1,2－二羟基肉桂酸苯丙氨酸异羟基肟氨抑制酪氨酸酶的活性

(IC_{50}为4.9μmol/L)比1,2－二羟基肉桂酸脯氨酸异羟基肟氨要好一点,但对Mel－Ab细胞黑色素合成的影响不大。[34]综合来看,1,2－二羟基肉桂酸脯氨酸异羟基肟氨是有希望的活性化合物,它在亲水和疏水环境中都能表现出较好的抗氧化活性和较好的酪氨酸酶抑制活性。

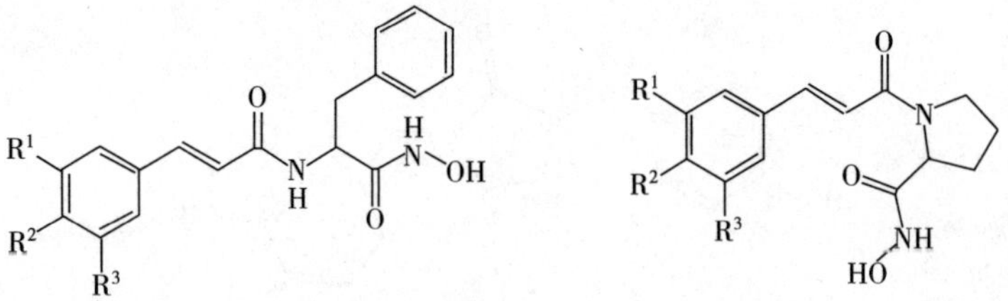

图 1－12

4－(6,7－二氢－茚噻唑基)－1,3－苯二酚是全化学合成的酪氨酸酶抑制剂[35],其IC_{50}=0.5μmol/L,比曲酸的抑制活性高了108倍。它在1,3－位的羟基同酪氨酸酶的残基GLY281形成氢键,占据酪氨酸酶的活性位置,抑制酪氨酸酶的活性。在B16F10黑色素细胞毒性测试中,它没有表现出细胞毒性。说明它是一个很有潜力的增白剂。

So Hee Kim等[36]合成的(Z)－5－(2,4－二羟基苯甲烯基)－2',4'－二酮四氢噻唑1,能够抑制酪氨酸酶活性和黑色素的合成。它是通过清除一氧化碳和抑制诱导NO信号通路的活性,来降低酪氨酸酶活性的表达。对于它用于治疗剂和增白剂,还需要深入的研究。

Young Mi Ha[37]也合成了11个2,4－二酮四氢噻唑－取代苯甲亚基衍生物(图1－13)其中(Z)－5－(4－羟基苯甲烯基)－2',4'－二酮四氢噻唑2和(Z)－5－(3,甲氧基,4－羟基苯甲烯基)－2',4'－二酮四氢噻唑3对酪氨酸酶的抑制活性的IC_{50}分别为13.36和9.87μmol/L,它们都属于竞争型抑制剂。不仅对酪氨酸酶有很强的抑制作用,而且对B16黑色素细胞黑色素的生成有抑制作用。分子对接研究表明它们对土豆二酚氧化酶的活性位点的束缚作用很强。说明它们有比较强的褪色能力,且没有明显的细胞毒性。

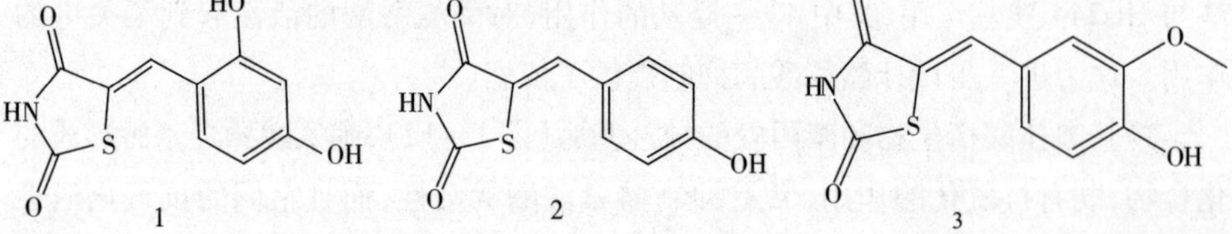

图 1－13

芪是一类天然的化合物,在植物中扮演着植物抗毒素的角色,抵抗真菌病原体。像白藜芦醇、康普立停等活性化合物都属于芪类。Mitko Miliovsky 等[38]用一锅法合成了 6 个顺式多羟基芪。其中最具前景的活性化合物是 E－2－(1－羧基－2－苯亚甲基)－4,5－二羟基安息香酸(图 1－14),它有很强的清除 DPPH 自由基的能力($IC_{50} \leqslant 10\mu mol/L$),抑制 Fusarium gramine arum 真菌(0.17 μmol/mL 抑制 89%),对酪氨酸酶的抑制的 IC_{50} 为 23μm。

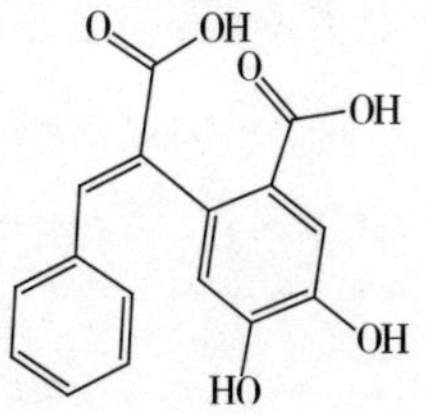

图　1－14

Babasaheb P. Bandgar 等[39]合成了咔唑衍生物(图 1－15)。其中大部分具有很好的抑制黄嘌呤氧化酶(IC_{50} =4.3－5.6μmol/L)和抑制酪氨酸酶的活性。7a 对酪氨酸酶的活性最好,分别用酪氨酸和 L－多巴作为底物,其抑制的 IC_{50} 值分别为 14.01μmol/L 和 15.15μmol/L,它同 L－多巴竞争酪氨酸酶的活性位点,其 Ki 值为 11.8μmol/L。老鼠黑素瘤细胞的抑制黑色素和细胞毒性的测试中,在 20μg/mL 浓度下,7a 和 7g 抑制了 42.5% 和 40.5% 的黑色素的合成,没有表现出明显的细胞毒性。

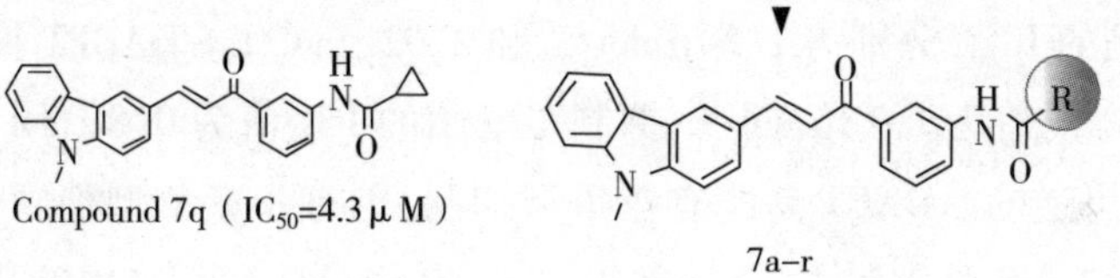

Compound 7q (IC_{50}=4.3 μM)　　7a–r

图　1－15

Shrikant S. Gawande 等[40]合成了一些列具有噻唑啉酮骨架的酪氨酸酶抑制剂,部分化合物相比于酪氨酸酶是有效的酪氨酸酶抑制剂。其中,1 对酪氨酸酶的抑制活性最高,IC_{50} = 34.12μmol/L。1 的抑制是可逆的竞争型抑制。从构效关系分析,氯苯基部分在酪氨酸酶的抑制活性中发挥着重要的作用。原因可能是氯原子绑定的酪氨酸酶的活性位置,使酪氨酸酶失去它的催化能力。

Zhixuan Zhou 等[41]合成了 3,5－二芳基－4,5－二氢－1H－吡唑类衍生物,他们中的大部分对酪氨酸酶都有很强的抑制作用。其中,抑制活性最好的是 1－3－(2,4－二羟基苯基)－5－(3,－羟基－4－甲氧基苯基)－4,5－二氢－乙酰基吡唑(图 1－16),其抑制酪氨酸酶的 IC_{50} =0.301μmol/L。动力学研究,它属于竞争型抑制剂。

Hooshang Hamidian 等[42]合成了 6 个包含偶氮 5(4H)－恶唑酮类化合物,它们对酪氨酸酶抑制的 IC_{50} 在 1.44μmol/L 和 4.33μmol/L 之间,远远比曲酸的抑

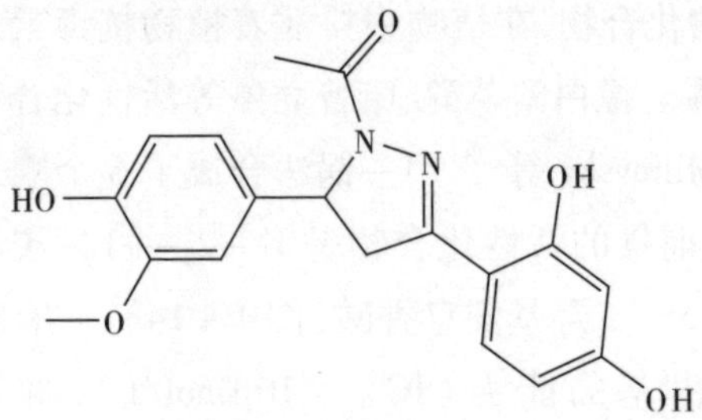

图 1-16

制活性要好，但他们对 A B16F10 的毒性都比曲酸大。所以，对这类化合物还需要进一步的设计、研究。构效关系表明恶唑酮环上 2 位和 4 位取代的芳基决定酪氨酸酶的活性。

苯甲醛缩氨基硫脲衍生物已经报道了有效的酪氨酸酶抑制剂。[43-44] Mei-Hua Yang 等[45]也合成了两个对酪氨酸酶有很强抑制作用的氨苯硫脲衍生物，对二甲氨基苯甲醛缩氨基硫脲希夫碱（DABT）和对二甲氨基苯甲醛缩氨基苯基硫脲希夫碱（DABPT），见图 1-17。DABT 对酪氨酸酶单酚酶和双酚酶都有抑制作用，其中抑制的 IC_{50}，分别为 1.54μmol/L 和 2.02μmol/L。DABPT 抑制酪氨酸酶单酚酶的 IC_{50} 的值为 1.78μmol/L，抑制二酚酶的 IC_{50} 值为 0.8μmol/L，它们的抑制机理都是可逆的。DABT 是既有竞争型抑制，也有非竞争型抑制，属于混合型抑制，其 KI 和 KIS 分别为 1.77μmol/L 和 6.49mmol/L；而 DABPT 是非竞争型抑制，其 KI = KIS = 0.77μmol/L。DABT 和 DABPT 表现出了比熊果苷和曲酸更强的抑制能力，说明氨基硫脲结构是设计酪氨酸酶抑制剂很好的选择。

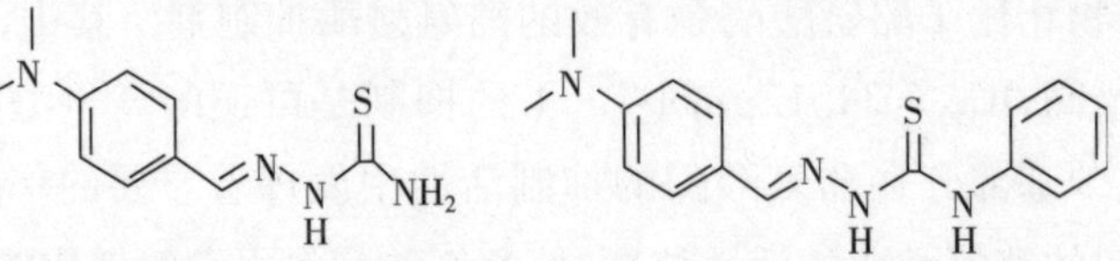

图 1-17

1.2.4 已有应用的化合物，作为酪氨酸酶抑制剂的研究

有效霉素 A[46] 是链霉菌的次级代谢产物，在稻子生长过程中用作生物杀虫剂，还报道了它具有抑制葡萄糖苷酶和海藻糖酶的作用。Zhi-Jiang Wang 发现效霉素 A 对酪氨酸酶具有有效的抑制作用（IC_{50} = 19.23mmol/L），并研究了它抑制酪氨酸酶的动力学和分子模型（图 1-18）。有效霉素 A 对酪氨酸酶的抑制是

可逆的，有效霉素 A 和酪氨酸酶非共价键分子间的相互作用降低酪氨酸酶的活性。它既能同自由的酪氨酸酶结合，又能同已经于底物发生作用的酪氨酸酶结合。通过分子对接模型，有效霉素 A 直接与酪氨酸酶活性位点的几个残基发生作用，包括 HIS85，HIS244，GLU256，HIS259 和 ASN260。

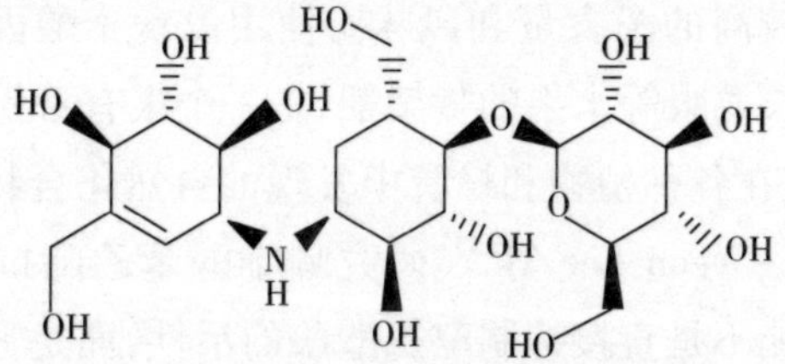

Chemical structure of validamycin A

图　1－18

呋喃甲醇常用作糊剂、电极材料、抗酸腐蚀材料等；糠醛来自于农产业和森林业的废弃物，用作火箭原料和溶剂等；呋喃甲酸已经用于药物中间体、杀菌剂等。最近报道，他们还是有效的酪氨酸酶抑制剂，其抑制能力糠醛 > 呋喃甲酸 > 呋喃甲醇，都属于可逆的抑制剂。糠醛和呋喃甲醇是混合抑制类型，呋喃甲酸是非竞争型[47]。

乙酰水杨酸也是[48]一个已经有广泛用途的化合物，如抗炎、头痛治疗、预防肿瘤，中风等。Zhi－Jiang　Wang 等报道了它对酪氨酸酶也有抑制作用（IC_{50} = 28.35mmol/L），抑制类型属于混合抑制，其 Ki = 11.778 ± 2.01mmol/L。当乙酰水杨酸的浓度为 200mmol/L 时，酪氨酸酶基本上完全被抑制。

4－丁雷锁辛[49－50]是一个对局部着色过度治疗有效的酪氨酸酶抑制剂。L. Kolbe 等在黑种人皮肤细胞模型中，发现熊果苷和曲酸对黑色素合成的抑制率不高，它们的 IC_{50} 分别 > 5000 μmol/L 和 > 400 μmol/L，而 4－丁雷锁辛抑制黑色素生成的 IC_{50} 值为 13.5μmol/L。使用含 4－丁雷锁辛的产品涂抹前臂的老年斑每天 2 次，8 周内老年斑的外观明显改善。4－丁雷锁辛不仅对老年斑，还对黑斑和其它表面的着色过度有很好的效果。

甲醛取代的吡啶，[51]是常见的药物中间体，化工原料。最近报道了他们对酪氨酸酶有抑制作用。它们抑制能力 3－甲醛吡啶 > 2－甲醛吡啶 > 4－甲醛吡啶，抑制二酚酶的 IC_{50} 值，分别为 0.0225mmol/L，1.3mmol/L，5mmol/L。他们都是混合型酪氨酸酶抑制剂。

没食子酸丙酯[52]是在脱水食品、脂肪、油中使用最广泛的酚类抗氧化剂之一。它还具有清除自由基,抗褐变的能力。Yi－Fen Lin 等报道了没食子酸丙酯具有抑制酪氨酸酶活性的能力,其抑制酪氨酸酶的 IC_{50}值为 0.685mmol/L。动力学分析,没食子酸丙酯是可逆的混合型抑制剂,抑制常数 KIS 和 KI 分别为 2.135 和 0.66mmol/L。没食子酸丙酯处理采摘的龙眼,发现延迟了龙眼外壳褐变的时间和程度,保持龙眼较高的酚含量和风味。使用没食子酸丙酯作为抗氧化剂和酪氨酸酶抑制剂,延长龙眼等水果的货架期,是一个很有希望的选择。

咖啡酸苯乙酯是在各种植物和蜂胶中发现的自然化合物,具有抗炎、免疫调节和细胞毒性剂。Ji－Yeon Lee 等[53]研究咖啡酸苯乙酯抗黑色素合成的过程中,发现咖啡酸苯乙酯不是直接抑制酪氨酸酶的活性,而是通过抑制黑色素酶的表达,如酪氨酸酶、酪氨酸酶相关蛋白－1、酪氨酸酶相关蛋白－2,来有效地减少了 a－MSH 刺激黑色素的合成。另一方面,咖啡酸乙酯对酪酸酶调节黑色素生成的转录因子,转录因子(MITF)的表达和核转运没有影响。咖啡酸苯乙酸酯对糖原合成酶激酶－3β 和 Akt 对 MITF 的激活和表达也没有影响。鉴于 MITF 在黑色素原合成中的重要角色,咖啡因苯乙酯对酪氨酸酶启动子的抑制,可能代表一种新型的治疗色素过度积累病的治疗方法。

1.3 抑制剂的重要性和应用

1.3.1 农业和食品领域

在食品储藏、销售过程中,酶催化有机胺、氨基酸、多肽和蛋白质与醌类的反应导致变质叫酶促褐变。重要氨基酸的破坏,消化吸收度的降低,蛋白水解酶和糖分解酶的抑制都会导致营养价值减少。抗营养物质和有毒化合物的生成可能进一步减少营养价值和出现可能的食品安全问题。因此,很有必要找到不同方法来阻止酪氨酸酶引起的酶促褐变。当前常规的避免褐变的技术是高压热汤的方法,食品侵入 80～90℃的液体当中 10～12min 或通过加压流动的蒸汽。这些传统的过程不可避免的造成重量和营养质量的损失。微波加热已经被提议为替代传统方法灭酶活的方法,但微波加热过程中在食品样品中产生温度梯度,对灭活程度不易控制。温度过热的地方,酶失去了活力,温度不够的地方,酶的活力还在。因此,严格控制微波加热对酶的灭活要样品均匀和绝对的温度控制。

已经报道了很多能够干扰酪氨酸酶中间的反应或减少二酚到醌抑制酶促褐变的化合物。然而,事实上报道的化合物受限于异味、变味、毒性和经济性,很少能用于食品中抑制酶促褐变。亚硫酸盐已经广泛用于防止酶促褐变在农业和海鲜产品中。由于安全的考虑，亚硫酸添加剂在一些食品中已经禁止了。在食品工业中，大部分替代物是维生素 C 和柠檬酸，但他们都不及亚硫酸有效，因为维生素 C 在减少酪氨酸酶形成的醌的过程中很快就被消耗掉了。最近，4－己基间苯二酚由于安全，用于防止虾的黑变和鱼肉和水果片的褐变。

1.3.2　化妆品和药品领域

酪氨酸酶抑制剂在化妆品和药品行业中的使用是很重要的，因为他们有效的防治色素沉着病。酪氨酸酶抑制剂可以减少黑色素的合成和用于化妆品中退去雀斑等。酪氨酸酶和酪氨酸酶抑制剂也可能发展成治疗色素相关的疾病的治疗剂，像白化病和斑驳病。已经报道了很多酪氨酸酶抑制剂，但由于不同的安全考虑只有很少的几个抑制剂用于皮肤增白。

在皮肤增白剂之中，1，4－对苯二酚是最广泛被使用的增白剂之一。但它能引起突变，所以很多国家已经禁止 1，4－对苯二酚在化妆品中使用。所以迫切需要找到可替代的退色剂。

当前，熊果苷和芦荟苦素由于无毒，对酪氨酸酶有很强的抑制作用，而作为增白剂用于化妆品中。熊果苷和芦荟苦素对酪氨酸酶的抑制，前者是非竞争型的抑制剂，后者是竞争型抑制剂。两个同时使用，具有协同作用，增强多酪氨酸酶的抑制能力；从而用作试剂时，可以降低试剂的使用量。

酪氨酸酶催化氧化，导致植食性食物和饮料发生褐变，造成关键氨基酸的破坏、消化吸收度和营养质量降低、形成有毒物质。为了解决这些不适宜的褐变，需要找到有效的酪氨酸酶抑制剂。另外，在黑色素积累过度相关的皮肤病的治疗、化妆品的增白剂、晒斑的褪色中，酪氨酸酶抑制剂扮演着重要的角色。然而，当前只有很少几个抗黑色素剂，在商业上能够应用，如曲酸和熊果苷。大部分的酪氨酸酶抑制剂，由于高细胞毒性，在水和有氧条件下的低稳定性等限制了应用。

酪氨酸酶抑制剂的安全问题是首先被考虑的，特别是用在食品和化妆品中的产品。来自于自然资源的抑制剂具有较大的潜力应用到食品当中，但有报道来一些天然的抑制剂具有很强的诱变活性，如一些顺式烯烃结构的化合物。

新酪氨酸酶抑制剂的发现和特征描述对他们在提高食品质量和营养价值，控制昆虫类害虫，防治色素过度积累和其它人体与黑色素相关的健康问题等方面的应用是很有益的。另外，理解酪氨酸酶激活剂和抑制剂在找到新颖的抑制剂和更多的变色过程规律的方法是很重要的。为了得到更好的抑制剂,天然和合成的不同类型的化合物已经被充分研究。然而,总的来说,酪氨酸酶抑制剂的研究,需要去确认他们的结构和活性,并且提高他们的安全性和不同应用的有效性。

参考文献

[1]Friedman M. Food browning and its prevention: anoverview[J]. J. Agric. Food Chem, 1996, 44, 631 -653.

[2] Sugumaran M. Molecular mechanisms for mammalianmelanogenesis comparison with insect cuticular sclerotiza - tion[J]. FEBS Lett., 1991, 293, 4 -10.

[3]邹先伟, 蒋志胜. 植物源酪氨酸酶抑制剂研究进展[J]. 中草药,2004, 35,702 -705.

[4]陈清西, 宋康康. 酪氨酸酶的研究进展[J]. 厦门大学学报,2006,45: 731 -737.

[5] Nishioka K. Particulate tyrosinase of human malignant melanoma. Solubilization, purification following trypsin treat - ment and characterization[J]. Eur. J. Biochem, 1987,85, 137 -146.

[6] Wilcox D E, Porras A G, Hwang Y T, et al. Substrate analogue binding to the coupled binuclear copper active site in tyrosinase[J]. J. Am. Chem. Soc., 1985, 107, 4015 -4027.

[7]Fenoll L G, Rodríguez - López J N, García - Sevilla F, et al. Analysis and interpretation of the action mechanism of mushroom tyrosinase on monophenols and diphenols generating highly unstable o - quinones[J]. Biochim. Biophys. Acta, 2001, 1548, 1 -22.

[8]Brun A, Rosset R. Ètude electrochimique del' oxy - dation dela dihydroxy - 3,4 phénylalanine (Dopa): mécanisme d' oxydation descatéchols enmilieuacide[J]. J. Electroanal. Chem., 1974,49, 287 -300.

[9]Aroca P, García - Borrón J C, Solano F, et al. Regulation of mammalian melanogenesis I: partial pu - rification and characterization of a dopachrome converting factor: dopachrome tautomerase[J]. Biochim. Biophys. Acta, 1990,

1035, 266 – 275.

[10] Kim Y J, Uyama, H. Tyrosinase inhibitors from natural and synthetic sources: structure, inhibition mechanism and perspective for the future[J]. Mol. Life Sci., 2005, 62, 1707 – 1723.

[11] CaroleD, Romain H, Maylis O, et al. Versatile Effects of Aurone Structure on Mushroom Tyrosinase Activity [J]. ChemBioChem, 2012, 13, 559 – 565.

[12] Nhan T N, Mai H K N, Hai X N, et al. Tyrosinase Inhibitors from the Wood of Artocarpus heterophyllus[J] J. Nat. Prod., 2012, 75, 1951 – 1955.

[13] Ziaullah K S, Bhullar S N, Warnakulasuriya, H P. et al. Biocatalytic synthesis, structural elucidation, antioxidant capacity and tyrosinase inhibition activity of long chain fatty acid acylated derivatives of phloridzin and isoquercitrin [J]. Bioorganic & Medicinal Chemistry, 2013, 21, 684 – 692.

[14] Ko H H, Jin Y J, Lu T M, et al. A Novel Monoterpene Stilbene Adduct with a 4, 4 – Dimethyl – 2, 3 – diphenylchromane Skeleton from Artocarpus xanthocarpus[J]. Chemistry biodiversity, 2013, 10, 1269 – 1275.

[15] Evelyn M, Jose G A, Julio A, Kubo I, et al. Tyrosinase Inhibitors from Calceolaria integrifolia s. l.: Calceolaria talcana Aerial Parts[J]. J. Agric. Food Chem., 2013, 6, 4336 – 4343.

[16] Wu X, Yin S, Zhong J, et al. Mushroom tyrosinase inhibitors fromAloe barbadensis Miller[J]. Fitoterapia, 2012, 83, 1706 – 1711.

[17] Satooka H, Isobe T, Nitoda T, et al. Melanogenesis inhibitors from Rabdosia japonica [J]. Phytomedicine, 2012, 19, 1016 – 1023.

[18] Seikou N, Souichi N, Genzo T, et al. Alkaloid constituents from flower buds and leaves of sacred lotus (Nelumbo nucifera, Nymphaeaceae) with melanogenesis inhibitory activity in B16 melanoma cells [J]. Bioorganic & Medicinal Chemistry, 2013, 21, 779 – 787.

[19] Haroon K, Muhammad S, Murad A K, et al. Isolation of long – chain esters from the rhizome of Polygonatum verticillatum by potent tyrosinase inhibition [J]. Med. Chem. Res., 2013, 22, 2088 – 2092.

[20] Jiang F, Li W, Huang Y, et al. Antioxidant, Antityrosinase and Antitumor Activity Comparison: The Potential Utilization of Fibrous Root Part of

Bletilla striata(Thunb.) Reichb. f. [J]. Plos One,2013,8,1 - 11.

[21]Lyubomir G, Maya C, Iskra T, et al. Anti - tyrosinase, antioxidant and antimicrobial activities of hydroxycinnamoylamides[J]. Med. Chem. Res., 2013, 22,4173 - 4182.

[22]Young M H, Hye J L, Park D, et al. Molecular Docking Studies of (1 E, 3 E,5 E) - 1,6 - Bis(substituted phenyl) - hexa - 1,3,5 - triene and 1,4 - Bis (substituted trans - styryl)benzene Analogs as Novel Tyrosinase Inhibitors[J]. Biol. Pharm. Bull., 2013,36, 55 - 65.

[23] Sung J B, Ha Y M, Kim J A, et al. A Novel Synthesized Tyrosinase Inhibitor: (E) - 2 - ((2, 4 - Dihydroxyphenyl) diazenyl) phenyl 4 - Methylbenzenesulfonate as an Azo - Resveratrol Analog[J]. Biosci. Biotechnol. Biochem., 2013, 77, 65 - 72.

[24] Song Y M, Ha Y M, Kim J A, et al. Synthesis of novel azo - resveratrol, azo - oxyresveratrol and their derivatives as potent tyrosinase inhibitors [J]. Bioorganic & Medicinal Chemistry Letters, 2012, 22, 7451 - 7455.

[25] Bae S J, Ha Y M, Park Y J, et al. Design, synthesis, and evaluation of (E) - N - subs tituted benzylidene e aniline derivatives as tyrosinase inhibitors[J]. European Journal of Medicinal Chemistry, 2012, 57, 383 - 390.

[26] Larissa LL, Rebeca M L, Annelisa F S, et al. Azastilbene analogs as tyrosinase inhibitors: new molecules with depigmenting potential[J]. The Scientiic World Journal, 2013,23:356 - 363.

[27] Sawant R L, Lanke P D, Wadekar J B. Tyrosinase inhibitory activity, 3DQSAR, and molecular docking study of 2,5 - disubstituted - 1,3,4 - oxadiazoles [J]. Journal of Chemistry, 2013,25,157 - 164.

[28]Alessandro V, Manuela M, Serrilli A M, et al. Dihydroasparagusic acid: antioxidant and tyrosinase inhibitory activities and improved synthesis[J]. J. Agric. Food Chem., 2013, 61, 6848 - 6855.

[29] Li D F, Hu P P, Liu M S, et al. Design and synthesis of hydroxypyridinone - L - phenylalanine conjugates as potential tyrosinase inhibitors [J]. J. Agric. Food Chem., 2013, 61,6597 - 6603.

[30] Fu Q, Hong S, Li D, et al. Novel chemical synthesis of Ginkgolic Acid (13:0) and evaluation of its tyrosinase inhibitory activity[J]. J. Agric. Food

Chem. , 2013, 61, 5347 -5352.

[31] Jiang Y, Du Z, Xue G, et al. Synthesis and biological evaluation of unsymmetrical curcumin analogues as tyrosinase inhibitors[J]. Molecules, 2013, 18, 3948 -3961.

[32] Mohamed M E S, Seham Y H, Huda E A, et al. Synthesis and bioassay of a new class of furanyl -1,3,4 - oxadiazole derivatives[J]. Molecules, 2013, 18, 8550 -8562.

[33] Argyro V, Grigoris Z, Eliza C, et al. Design, synthesis and molecular simulation studies of dihydrostilbene derivatives as potent tyrosinase inhibitors [J]. Bioorganic & Medicinal Chemistry Letters, 2012,22, 5523 -5526.

[34] Kwak S Y, Yang J K, Choi H R, et al. Synthesis and dual biological effects of hydroxycinnamoyl phenylalanyl/prolyl hydroxamic acid derivatives as tyrosinase inhibitor and antioxidant [J]. Bioorganic & Medicinal Chemistry Letters, 2013,23, 1136 -1142.

[35] Park J W, Ha Y M, Moon K, et al. Denovo tyrosinase inhibitor: 4 -(6, 7 - Dihydro - 5 H - indeno[5, 6 - d] thiazol - 2 - yl) benzene - 1, 3 - diol (MHY1556) [J]. Bioorganic & Medicinal Chemistry Letters, 2013, 23, 4172 -4176.

[36] Kim S H, Choi Y J, Moon K M, et al. The inhibitory effect of a synthetic compound, (Z) - 5 - (2, 4 - dihydroxybenzylidene) thiazolidine - 2, 4 - dione (MHY498), on nitric oxide - induced melanogenesis[J]. Bioorganic & Medicinal Chemistry Letters, 2013,23,4332 -4335.

[37] Ha Y M, Park Y J, Kim J A, et al. Design and synthesis of 5 - (substituted benzylidene) thiazolidine - 2, 4 - dione derivatives as novel tyrosinase inhibitors[J]. European Journal of Medicinal Chemistry, 2012, 49, 245 -252.

[38] Mitko M, Ivan S, Yavor M, et al. A novel one - pot synthesis and preliminary biolo gical activity evaluation of cis - restricted polyhydroxy stilbenes incorporating protocatechuic acid and cinnamic acid fragments[J]. European Journal of Medicinal Chemistry, 2013,66,185 -192.

[39] Babasaheb P B, Laxman K A, Hemant V C, et al. Synthesis, biological evaluation, and molecular docking of N - {3 - [3 - (9 - methyl - 9 H - carbazol - 3 - yl) - acryloyl] - phenyl} - benzamide/amide derivatives as xanthine oxidase and

tyrosinase inhibitors[J]. Bioorganic & Medicinal Chemistry, 2012,20,5649 - 5657.

[40] Shrikant S G, Suchita C W, Babasaheb P B, et al. Synthesis of new heterocyclic hybrids based on pyrazole and thiazolidinone scaffolds as potent inhibitors of tyrosinase[J]. Bioorganic & Medicinal Chemistry, 2013,21,2772 - 2777.

[41] Zhou Z, Zhuo J, Yan S, et al. Design and synthesis of 3,5 - diaryl - 4,5 - dihydro - 1H - pyrazoles as new tyrosinase inhibitors[J]. Bioorganic & Medicinal Chemistry, 2013, 21,2156 - 2162.

[42] Hooshang H, Roya T, Samieh F, et al. Synthesis of novel azo compounds containing 5(4H) - oxazolone ring as potent tyrosinase inhibitors[J]. Bioorganic & Medicinal Chemistry, 2013,21, 2088 - 2092.

[43] Zhu Y J, Song K K, Li Z C, et al. Antityrosinase and antimicrobial activities of transcinnamaldehyde thiosemicarbazone[J], J. Agric. Food Chem., 2009,57,5518 - 5523.

[44] Li Z C, Chen L H, Yu X J, et al. Inhibition kinetics of chlorobenzaldehyde thiosemicarbazones on mushroom tyrosinase[J], J. Agric. Food Chem., 2010, 58,13537 - 13540.

[45] Yang M H, Chen C M, Hu Y H, et al. Inhibitory kinetics of DABT and DABPT as novel tyrosinase inhibitors[J]. Journal of Bioscience and Bioengineering, 2013,115,514 - 517.

[46] Wang Z J, Ji S, Si Y X, et al. The effect of validamycin A on tyrosinase: Inhibition kinetics and computational simulation [J]. International Journal of Biological Macromolecules, 2013,55,15 - 23.

[47] Chai W M, Liu X, Hu Y H, et al. Antityrosinase and antimicrobial activities of furfuryl alcohol, furfural and furoic acid[J]. International Journal of Biological Macromolecules, 2013,57,151 - 155.

[48] Wang Z J, Lee J, Si Y X, et al. Toward the inhibitory effect of acetylsalicylic acid on tyrosinase: Integrating kinetics studies and computational simulations[J]. Process Biochemistry, 2013,48,260 - 266.

[49] Kolbe L, Mann T, Gerwat W, et al. 4 - n - butylresorcinol, a highly effective tyrosinase inhibitor for the topical treatment of hyperpigmentation[J]. Jeadv, 2013, 27 (Suppl. 1), 19 - 23.

[50] Huh S Y, Shin J W, Na J I, et al. The efficacy and safety of 4 - n -

butylresorcinol 0. 1% cream for the treatment of melasma：a randomized controlled splitface trial[J]. Ann Dermatol, 2010,22,21 –25.

[51] Punam B, Sujeetkumar J. Effect of substitution on the inhibition of mushroom tyrosinase by pyridine carboxaldehydes[J]. Progress in Reaction Kinetics and Mechanism, 2013, 38,75 –85.

[52] Lin Y F, Hu Y H, Lin H T, et al. Inhibitory effects of propyl gallate on tyrosinase and its application in controlling pericarp browning of harvested longan fruits[J]. J. Agric. Food Chem., 2013, 61,2889 –2895.

[53] Lee J Y, Choi H J, Chung T, et al. Caffeic acid phenethyl ester inhibits alpha – melanocyte stimulating hormone – induced melanin synthesis through suppressing transactivation activity of microphthalmia – associated transcription factor [J]. J. Nat. Prod., 2013, 76,1399 –1405.

第2章　醛酮及其衍生物酪氨酸酶抑制剂

醛酮及其衍生物具有广泛的生物活性,其中包括抗菌、消炎、抗肿瘤、酶抑制活性等,醛酮及其衍生物作为酪氨酸酶抑制剂,可以来源于化学合成及天然产物中。

2.1　醛类酪氨酸酶抑制剂

Kubo 于 1999 年报道了肉桂醛类化合物具有较好的酪氨酸酶抑制活性[1]。

从植物中可以提取一些具有酪氨酸酶抑制活性的醛类化合物,如茴香醛[2]、枯茗醛[3]等,1998 年 Kubo 等人从孜然中提取了孜然醛(4－异丙基苯甲醛)及其衍生物,并证明这类化合物具有对酪氨酸酶的抑制活性[4]。

对天然提取的醛类化合物进行结构改造,研究其抑制酪氨酸酶的构效关系,结果表明,在芳香醛化合物的对位引入推电子基团将使化合物对酪氨酸酶的抑制强度明显增强,并且随着推电子基团的推电子能力增强,其抑制强度也增大[5]。

2014 年赵连梅等人从中药木鳖子中提取了几种单体醛类化合物[6],其中以对羟基肉桂醛(PHD)单体对黑色素瘤细胞 B16 的抑制活性最好(图 2－1)。以 10、20、40μmol/L 对羟基肉桂醛作用 B16 细胞,设空白对照及福司克林阳性对照组。磺酰罗丹明法检测对羟基肉桂醛对 B16 细胞生长的抑制率,平板克隆集落形成实验检测细胞的克隆形成能力,Giemsa 染色观察细胞形态,比色法检测细胞中黑色素含量和酪氨酸酶活性,划痕愈合实验检测细胞的迁移能力,Western blotting 检测 Try、酪氨酸酶相关蛋白 1(tyrosinase protein, Trp1)及 MAPK 信号转导通路中 p－P38、p－JNK 和 p－ERK1/2 的表达。结果表明:对羟基肉桂醛对黑素瘤 B16 细胞具有明显的增殖抑制作用。得出结论:对羟基肉桂醛通过上调 MAPK 信号通路中 p－P38 和 p－JNK 的活性而对小鼠黑素瘤 B16 细胞产生明显的增殖抑制,并在形态和功能上诱导黑素瘤细胞的分化的结论。

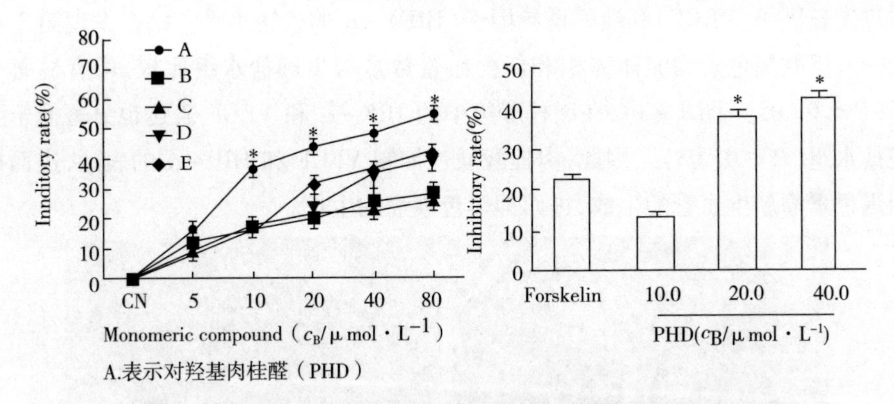

A.表示对羟基肉桂醛（PHD）

图 2－1　对羟基肉桂醛对 B16 细胞增殖的抑制作用及增殖影响

崔毅等人于 2015 年发现三种 alfa 取得肉桂醛对酪氨酸酶具有较好的抑制作用[7]，这三种化合物是 alfa 溴代肉桂醛、alfa 氯代肉桂醛和 alfa 甲基取代肉桂醛（图 2－2）。通过抑制活性测试后的结果表明：这三个化合物对单酚酶和二酚酶都具有抑制作用，为可逆抑制剂。从半抑制率来看，对于单酚酶，alfa 溴代肉桂醛、alfa 氯代肉桂醛和 alfa 甲基取代肉桂醛的 IC_{50} 依次为 0.075，0.140 和 0.440mmol/L，对于二酚酶的 IC_{50} 依次为 0.049，0.110 和 0.450mmol/L，并且进一步研究的抑制类型和抑制常数。采用紫外扫描、荧光焠灭、分子对接等方法对这类化合物抑制酪氨酸酶的分子机制进行了研究，结果表明：这类化合物能够减少喹啉的形成，是静态焠灭剂。通过分子对接分析发现，该类化合物并不是和活性中心铜离子发生作用，而是和活性中心氨基酸残基发生作用。这类化合物可以作为进一步开展酪氨酸酶抑制剂研究的先导化合物。

O H Br　O H C1　O H CH_3

图 2－2　三种肉桂醛衍生物化学结构式

2014 年李佳青选用肉桂醛为目标化合物，考察其对黑色素瘤小鼠的抑制作用及抑制机制[8]。取 20 只 Balb /c null 雌性裸鼠建立黑色素瘤模型，分为生理盐水组和肉桂醛组，每组 10 只，第 2 周开始每鼠分别注射 3 mL 生理盐水（生理盐水组）和 3 mL、10 μg /mL 肉桂醛药物溶液（肉桂醛组），比较两组小鼠瘤体体积差异，并通过免疫组化法观察两组之间的新生血管数量，免疫印迹法检测血管内皮生长因子(VEGF)和缺氧诱导因子(HIF)-α的表达水平。结果发现第3周肉桂醛组的黑色素瘤瘤体体积和新生血管数量与生理盐水组比较,均有显著下降($P < 0.05$)(图2-3),在肉桂醛组中的HIF-α和VEGF表达也显著低于生理盐水组($P < 0.05$)。因此,肉桂醛通过抑制VEGF和HIF-α的表达,进而抑制黑色素瘤新生血管的生成,进而抑黑色素瘤的生长。

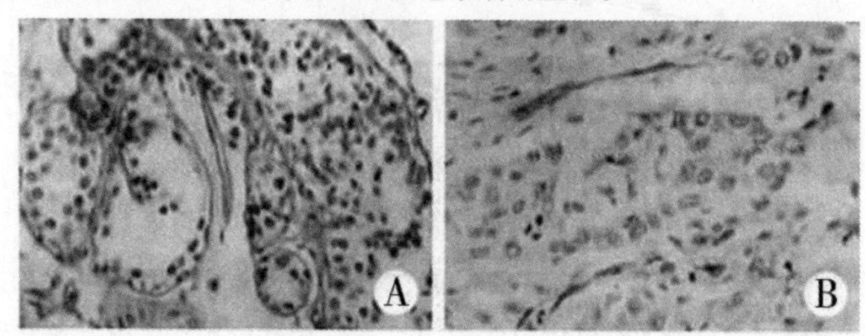

图2-3　两组黑色素瘤小鼠肿瘤新生血管比较(HE×200)

注:A.生理盐水组新生血管,细胞核染色呈蓝色;B.肉桂醛组新生血管,明显少于生理盐水组

2013印度学者Punam Bandyopadhyay等人考察了吡啶醛类化合物对酪氨酸酶的抑制活性[9],结果表明:这类化合物是酪氨酸酶可逆抑制剂。2-吡啶甲醛、3-吡啶甲醛、4-吡啶甲醛(图2-4)对蘑菇酪氨酸酶单酚酶和二酚酶都有抑制活性,其抑制活性顺序为:3-吡啶甲醛>2-吡啶甲醛>4-吡啶甲醛,其IC_{50}值分别为0.0225mmol/L、1.3mmol/L和5mmol/L。抑制机理分析表明:这些醛类化合物不仅能够和蛋白质氨基酸残基形成席夫碱,而且能够与酶形成氢键及发生亲核反应,这些因素在抑制酪氨酸酶的过程中起着重要的作用(图2-5)。在抑制单酚酶的活性时,延迟期并未拖长,但稳定期减少。抑制动力学结果表明,这类化合物都是酪氨酸酶的混合抑制剂。

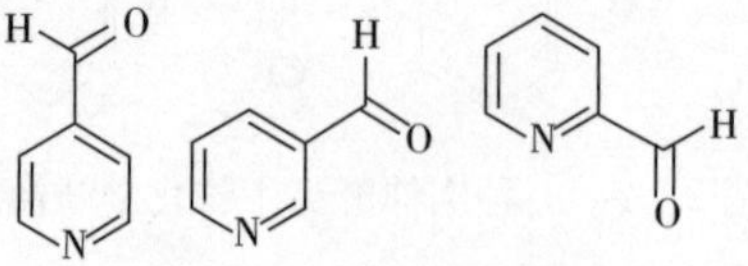

图2-4　吡啶醛化学结构式

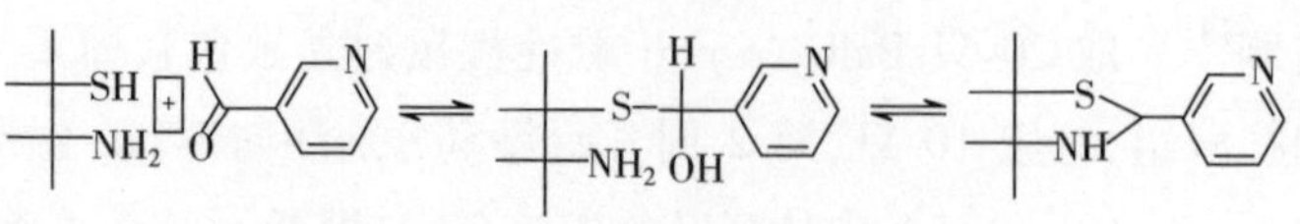

图2-5　与氨基酸残基的可能反应历程

2.2　酮类酪氨酸酶抑制剂

2016 年 Zong－Ping Zheng 等人以 2,4－二羟基苯甲醛为原料[10,11]，分别和 2－羟基苯乙酮及 2,4－二羟基苯乙酮反应，采用一釜法合成了 1,3,5－三苯基－1,5－戊二酮、黄酮衍生物、三苯甲醇、查尔酮（图 2－6）。该反应以聚乙二醇 400 为溶剂，以硼酸为催化剂，通过羟醛缩合、麦克加成、F－C 反应得到目标产物。目标化合物的酪氨酸酶抑制活性测试表明，这类化合物的抑制活性都强于曲酸，更重要的是首次证明了 1,3,5－三苯基－1,5－戊二酮、三苯甲醇具有酪氨酸酶抑制活性。同时对抗褐变性能进行了测试，结果表明，该类化合物于室温下对新切的藕片具有较好的抗褐变能力，同样 1,3,5－三苯基－1,5－戊二酮抗褐变性能最好。

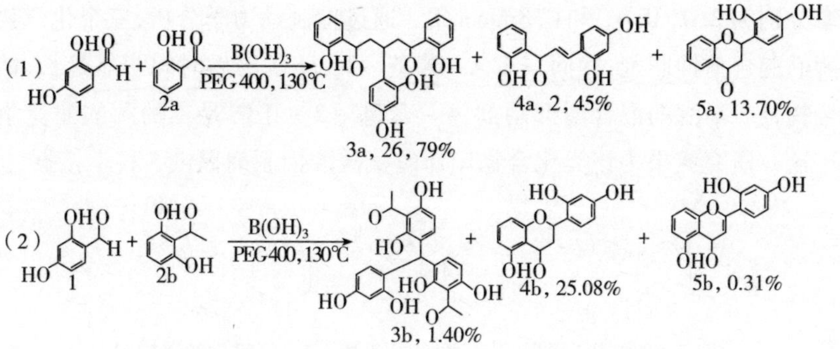

图 2－6　一釜法合成 1,3,5－三苯基－1,5－戊二酮、黄酮衍生物、三苯甲醇

2015 年韩国学者 Vivek K. Bajpai 等人从水杉中分离得到了一种二萜紫杉酮（图 2－7），并证实该化合物具有一定的酪氨酸酶抑制活性（图 2－8），具有潜在的抑制黑色素合成的能力，该发现表明该化合物可以用于食品工业和医药行业的皮肤增白剂[12]。

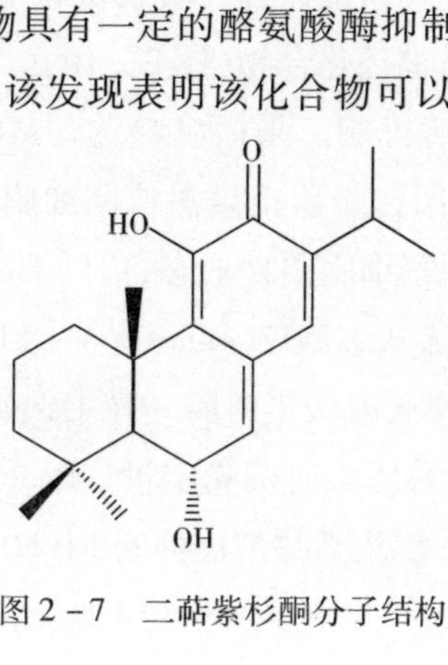

图 2－7　二萜紫杉酮分子结构

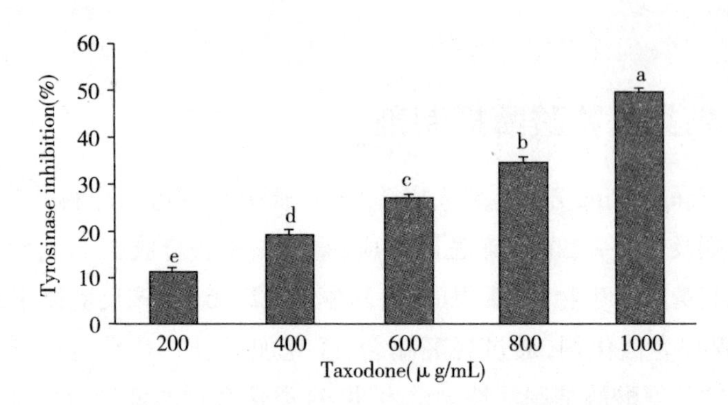

图 2 - 8　二萜紫杉酮的酪氨酸酶抑制活性

2015 年 Xuan Liu 等人首次对苯乙烯酮、苯丁酮、4 - 苯基 - 2 - 丁醇(图 2 - 9)的酪氨酸酶抑制活性进行了测试[13]，结果表明，这三类化合物是酪氨酸酶的可逆抑制剂，对于单酚酶的半抑制率分别为 1.5，2.8，和 1.1mmol/L，对于二酚酶的半抑制率分别为 2.0，0.6，和 0.8mmol/L。通过抑制动力学分析，三个化合物都是单酚酶的混合型抑制剂，然而只有 4 - 苯基 - 2 - 丁醇滞后时间明显延长，两个酮类化合物是二酚酶的混合型抑制剂，4 - 苯基 - 2 - 丁醇是二酚酶的非竞争型抑制剂。这一研究成果为该类化合物用作酪氨酸酶抑制剂提供了科学依据。

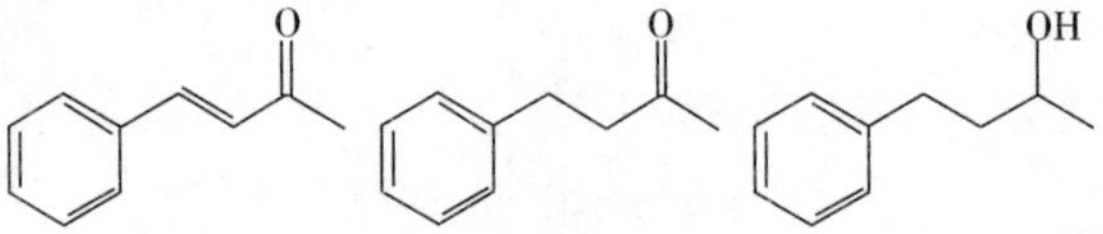

图 2 - 9　苯乙烯酮、苯丁酮、4 - 苯基 - 2 - 丁醇化学结构式

2013 年石河子大学药学院杨帆课题组在发酵、纯化紫杉醇工艺的同时从该菌的菌丝体中提取分离出一种新的单体化合物维康醇[14]，维康醇从化学结构来看是一种含有酮羰基的芳香稠环化合物(图 2 - 10)。作者探讨维康醇诱导小鼠黑色素瘤 B16F0 细胞凋亡的机制。采用噻唑蓝(MTT)法检测维康醇对小鼠 B16F0 细胞增殖的影响；采用台盼蓝拒染法检测细胞致死率；采用吖啶橙/溴乙啶(AO/EB)荧光染色法观察细胞凋亡形态；借助 Hoechst 33258 染色观察药物处理后细胞形态的变化；用流式细胞仪 Annexin V - FITC/PI 检测细胞凋亡率；用 Caspase - 9 /3 试剂盒检测半胱氨酸蛋白酶 - 9(Caspase - 9)和半胱氨酸蛋白酶 - 3(Caspase - 3)的活性；采用实时荧光定量(Real - Time PCR)法检测 Bax、Bcl - 2 基因表达水平。结果表明：维康醇能抑制 B16F0 细胞的恶性增殖，并呈剂量依赖性($P<0.05$ 或 $P<0.01$)；细胞致死率也不断上升($P<0.05$ 或 $P<0.01$)；

在荧光显微镜下发现维康醇作用于 B16F0 细胞后出现明显的凋亡形态；随着维康醇浓度的增加，细胞凋亡率呈剂量依赖性增长（$P<0.05$ 或 $P<0.01$）；Caspase－3、Caspase－9 活性逐渐升高（$P<0.05$ 或 $P<0.01$）；Bax/Bcl－2 表达的比率上调（$P<0.01$）。得出了如下结论：维康醇能够通过抑制 B16F0 细胞的恶性增殖，最终诱导细胞的凋亡。其机制是通过上调 Bax/Bcl－2 表达的比率，活化 Caspase－9 并进一步激活 Caspase－3 诱导 B16F0 细胞凋亡。推测维康醇诱导 B16F0 细胞凋亡是通过线粒体调控的内源通路介导的。

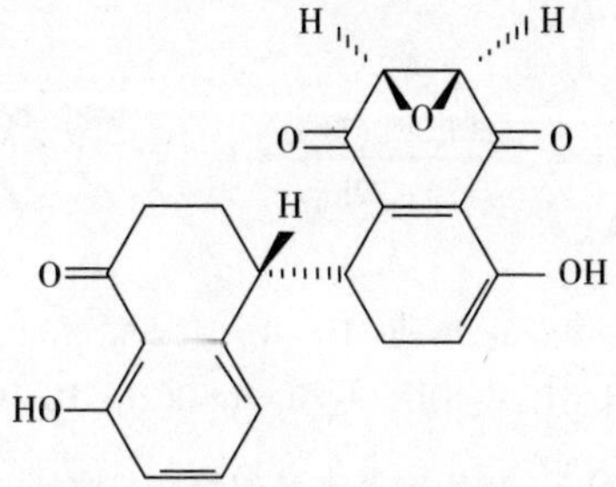

图 2－10　维康醇化学结构式

姜黄素属于植物多酚，主要是从姜黄、天南星科中的部分植物（如菖蒲）的根茎中提取出来，是姜黄发挥药理作用最重要的活性成分；姜黄素是具有对称酮的二酮类化合物（图 2－11）。研究表明，姜黄素是一种新型的酪氨酸酶抑制剂，同时具有抗炎、抗菌、抗氧化、清除自由基、抗肿瘤等广泛的药理作用。此外，对黑素瘤细胞同样有一定的抑制作用。但由于姜黄素结构不稳定，水溶性差，血药浓度低，在人体内容易被分解，导致其生物活性利用率极低，且姜黄素本身对酪氨酸酶的抑制活性不高，这些严重妨碍了姜黄素在医疗方面作为酪氨酸酶抑制剂的进一步研究应用。2013 年，韦星船等人以芳香醛为起始原料之一，在适宜的条件下与 2－茚酮进行缩合反应，合成 5 种含 2－茚酮基的对称姜黄素类似物（图 2－12）[15,16]，利用紫外分光光度法测定其对酪氨酸酶的抑制作用。结果表明：合成的对称性姜黄素类似物均对酪氨酸酶有较强的抑制作用，其中四个化合物的半数抑制浓度（IC_{50}）比广泛使用的食品添加剂曲酸（IC_{50} 为 28.59μmol/L）的抑制作用强。2014 年韦星船等人以姜黄素为先导化合物，设计并合成了一类含氮的姜黄素类似物（图 2－13），并考察了目标化合物的酪氨酸酶的抑制特性。结果表明，3－(4－羟基－3,5－二甲氧基苯基)－1－(4－吡啶基)－2－烯－1－酮对酪氨酸酶具有强抑制活性，半数抑制浓度（IC_{50}）为 45.1μmol/L，是姜黄素（IC_{50} = 97.1μmol/L）的抑制活性的 2.2 倍。抑制动力学研究表明，3－(4－羟基-3,5-二甲氧基苯基)-1-(4-吡啶基)-2-烯-1-酮对酪氨酸酶的抑制作用类型属于竞争性抑制。

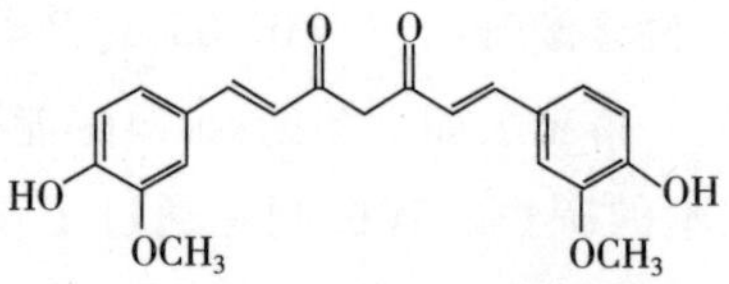

图2-11 姜黄素化学结构式

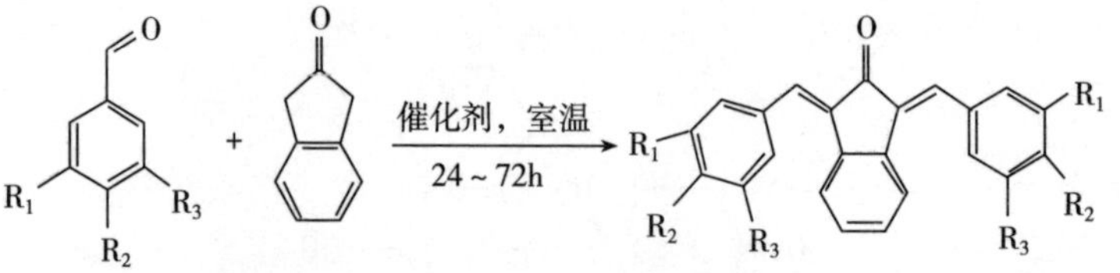

A_1. R_1=R_3=H，R_2=F；A_2. R_1=R_3=H，R_2=Cl；A_3. R_1=R_3=H，R_2=Br；A_4. R_1=R_3=H，R_2=OH；A_5. R_1=R_3=OCH_3，R_2=OH。

图2-12 姜黄素类似物合成路线

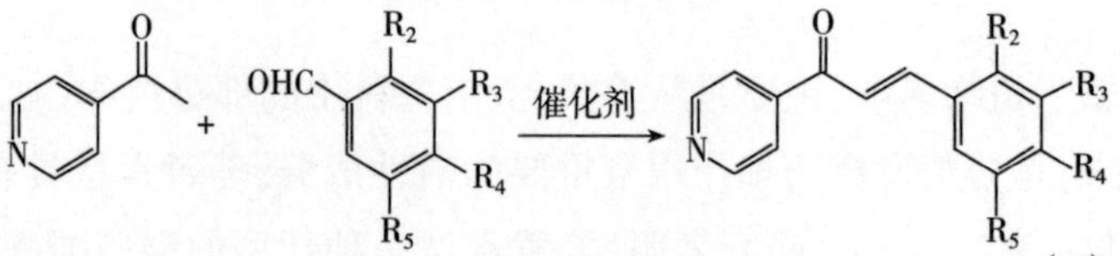

如式(1)，A_1，R_2=R_3=R_5=H，R_4=OCH_3；A_2，R_2=H，R_3=R_4=R_5=OCH_3；A_3，R_2=R_3=R_5=H，R_4=N(CH_3)$_2$；A_4，R_2=H，R_3=R_5=OCH_3，R_4=OH。

图2-13 含氮姜黄素类似物合成路线

2012年广东工业大学的莫容清等人采用活性基团拼接法将间苯二酚官能团引入姜黄素类似物中,设计合成了一个新型单羰基类似物2,5-双(2,4-二羟基苯亚甲基)环戊酮[17],并对其酪氨酸酶抑制活性进行了研究,其合成路线如图2-14所示。以2,4-二羟基苯甲醛和环戊酮为原料,无水乙醇为溶剂,40%的NaOH水溶液为催化剂,经Claisen-Schmidt反应催化合成了标题化合物,其结构通过1 HNMR、MS进行表征。化合物的初步生物活性测试结果表明,该化合物具有很强的酪氨酸酶抑制活性,半抑制浓度(IC_{50})为0.78μmol/L,比曲酸(IC_{50} = 28.59 μmol/L)的活性强36倍多。抑制动力学研究表明,该化合物对酪氨酸酶的抑制属于竞争性抑制类型。

2013年Yongfu Jiang等人合成了一类非对称姜黄素类似物(图2-15)[18],酪氨酸酶抑制活性研究表明,大部分这类化合物具有酪氨酸酶抑制活性,在这类

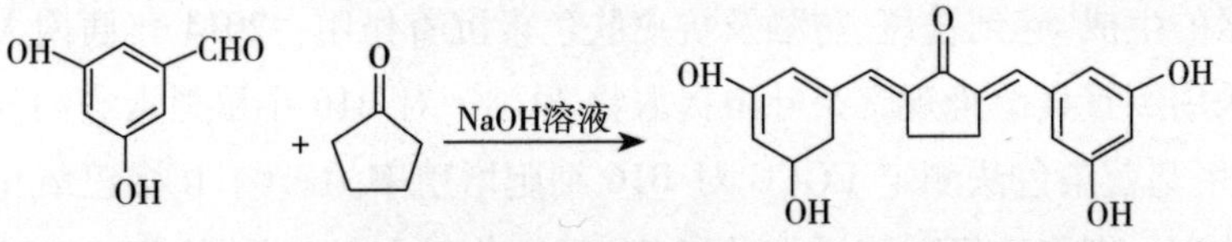

图2－14　2,5－双(2,4－二羟基苯亚甲基)环戊酮合成路线

化合物中,4－羟基取代的苯环和碳－2/碳－4－或碳－3/碳－4－二羟基取代的苯环是抑制酪氨酸酶活性最好的化合物(IC_{50} = 1.74～16.74μmol/L),比阳性对照曲酸活性好,因此可以推测羟基对于这类化合物在酪氨酸酶抑制活性中发挥了关键作用。抑制动力学分析表明,含有邻苯二酚结构的化合物是混合竞争抑制剂,而含有间苯二酚结构的化合物是竞争型抑制剂。优选了两个活性好、结构具有代表性的化合物进行了毒性试验,结果表明,该类化合物剂量达到1200 mg/kg未见毒性,这一研究说明了该类化合物具备应用于临床的酪氨酸酶抑制剂的潜力。

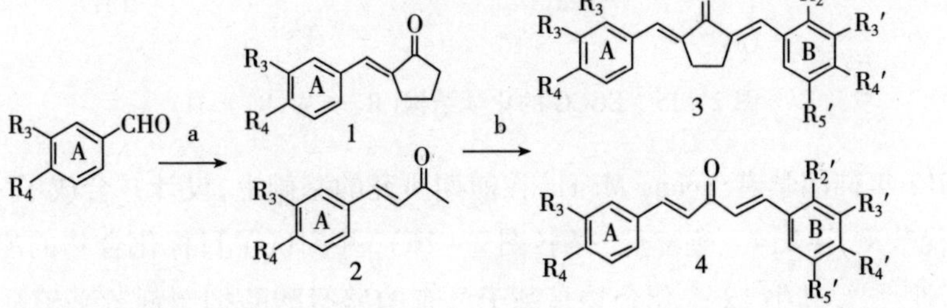

3a: $R_3=R_2'=R_3'=R_5'=H$, $R_4=OH$, $R_4'=OCH_3$
3b: $R_3=R_2'=R_5'=H_3R_4=R_4'=OH$, $R_3'=OCH_3$
3c: $R_3=R_2'=R_5'=H$, $R_4=R_3'=R_4'=OH$
3d: $R_3=R_3'=R_5'=H$, $R_4=R_2'=R_4'=OH$
3c: $R_3=R_2'=H$, $R_4=R_4'=OH$, $R_3'=R_5'=OCH_3$
3f: $R_3=R_2'=H$, $R_4=R_4'=OH$, $R_3'=R_5'=C(CH_3)_3$
3g: $R_3=R_2'=H$, $R_4=R_4'=OH$, $R_3'=Br$, $R_5'=OCH_3$
3h: $R_3=R_2'=H$, $R_4=R_4'=OH$, $R_3'=R_5'=Br$
3i: $R_2'=R_5'=H$, $R_4=R_3'=R_4'=OH$, $R_3=OCH_3$
3j: $R_3'=R_5'=H$, $R_4=R_2'=R_4'=OH$, $R_3=OCH_3$
3k: $R_2'=H$, $R_4=R_4=OH$, $R_3=R_3'=R_5'=OCH_3$

4a: $R_3=R_2'=R_3'=R_5'=H$, $R_4=OH$, $R_4'=OCH_3$
4b: $R_3=R_2'=R_5'=H$, $R_4=R_4'=OH$, $R_3'=OCH_3$
4c: $R_3=R_2'=R_5'=H$, $R_4=R_3'=R_4'=OH$
4d: $R_3=R_2'=H$, $R_4=R_4'=OH$, $R_3'=R_5'=OCH_3$
4c: $R_3=R_2'=H$, $R_4=R_4'=OH$, $R_3'=R_5'=Br$
4f: $R_2'=R_5'=H$, $R_4=R_3'=R_4'=OH$, $R_3=OCH_3$
4g: $R_2'=H$, $R_4=R_4'=OH$, $R_3=R_3'=R_5'=OCH_3$
4h: $R_3=R_2'=R_5'=H$, $R_3'=R_4'=OH$, $R_4=OCH_3$
4i: $R_2'=R_3'=R_5'=H$, $R_4=OH$, $R_3=R_4'=OCH_3$

图2－15　姜黄素类似物的合成路线

茶多酚作为茶叶中黄烷醇类、黄酮、黄酮醇类、花青素类、茶白素类和酚酸及缩酚酸类组成的混合物,主要成分是黄烷醇衍生物——儿茶素。根据结构不同,儿茶素分为表儿茶素(EC)、表没食子儿茶素(EGC)、表儿茶素没食子酸酯(ECG)、表没食子儿茶素没食子酸酯(EGCG)。体内外实验研究表明,儿茶素中的EGCG是茶叶中一个非常重要且高效的化学预防多酚化合物,结构见图2－15,属于酯型儿茶素,能清除人体内或皮肤中自由基、抑制过氧化物,减少或清除皮肤

内脂褐素的生成,起到祛斑、防晒及抗皮肤色素沉着作用。2014 年西南大学的曾亮等人采用倒置显微镜观察茶叶中提取物 EGCG 对 B16 小鼠黑素瘤细胞形态的影响,亚甲基蓝染色法测定 EGCG 对 B16 细胞增殖率的影响,L 多巴氧化法测定 EGCG 对 B16 细胞酪氨酸酶活力的影响,氢氧化钠裂解法测定 EGCG 对 B16 细胞黑色素生成量的影响。结果表明:EGCG 作用24h,高浓度组(>60μmol)细胞胞膜融合现象明显;作用 72h,高浓度组大部分细胞死亡。EGCG 对 B16 细胞增殖率、酪氨酸酶活性、黑色素生成量均有显著抑制,呈明显的剂量效应关系,且 EGCG 抑制 B16 细胞增殖,IC_{50}为 33.99μmol[19]。

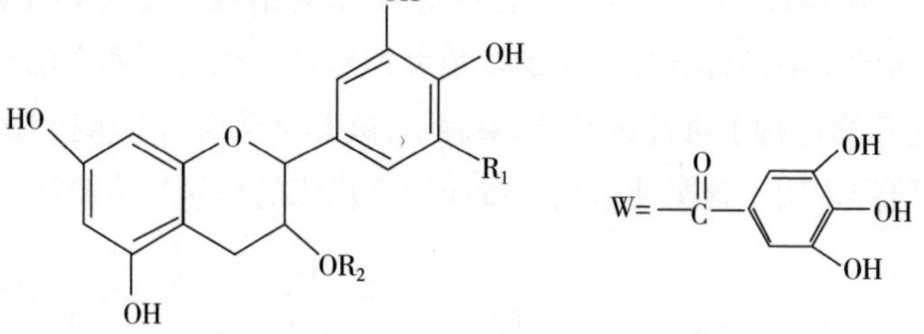

图 2-15　EGCG 的化学结构(R_1 = W,R_2 = H)

2012 年韩国学者 Young Mi Ha 在前期研究的基础上,设计并合成了一系列 5-(取代苄叉)-2,4-噻唑烷二酮(图 2-16),并测试了目标化合物的酪氨酸酶抑制活性[20]。其中 12 个化合物是基于已有的酪氨酸酶抑制剂苯基硫脲、L-多巴、酪氨酸、酪氨酸酶的天然底物的化学结构设计并合成的,其中(Z)-5-(4-羟基苄叉)-2,4-噻唑烷二酮和(Z)-5-(3-羟基-4-甲氧基苄叉)-2,4-噻唑烷二酮具有最后的酪氨酸酶抑制活性,强于阳性对曲酸(IC_{50} = 24.72μmol/L),其IC_{50}分别为 13.36μmol/L 和 9.87μmol/L。抑制动力学分析表明,这两个化合物均为竞争性抑制剂,从分子对接结果来看,这两个化合物可能与酶的活性中心形成了某种化学键。另外对这两个化合物进行了抑制黑色素瘤细胞 B16 的能力评价,结果表明,能够通过抑制酪氨酸酶的活性而达到抑制黑色素合成的目的。

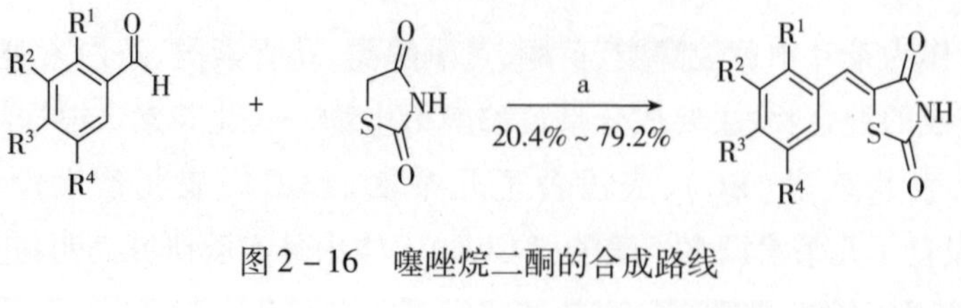

图 2-16　噻唑烷二酮的合成路线

2014 年 Hye Rim Kim 等人基于苄叉乙内酰脲、苄叉吡唑烷二酮、苄叉噻唑烷二酮具有酪氨酸酶抑制活性，为了找到能够用于食品工业、化妆品行业及临床的高效低毒的酪氨酸酶抑制剂，以芳香醛和 2－乙内酰硫脲为起始原料通过脑文格尔反应合成了一类具有苄叉结构和 2－乙内酰硫脲杂环结构化合物（图 2－17）[21]。通过对所合成的目标化合物进行酪氨酸酶抑制活性评价表明，其中三个化合物的酪氨酸酶抑制活性强于曲酸。这类化合物的抑制模式与苄叉乙内酰脲、苄叉吡唑烷二酮、苄叉噻唑烷二酮是存在区别的，可能是乙内酰硫脲的硫酮基团失去了形成氢键的能力。对于 5－（取代苄叉）乙内酰硫脲衍生物，乙内酰硫脲结构是乙内酰脲、吡唑烷二酮、噻唑烷二酮最相近的结构。（Z）－5－（2，4－二羟基苄叉）－2－乙内酰硫脲比曲酸的抑制活性强 18 倍，研究结果显示，β－苯基－α，β－不饱和羰基是活性必须基团。

图 2－17　苄叉乙内酰硫脲的合成路线

2016 年法国学者 Abbes Benmerache 等人从刺芹属植物地上部分分离出一种含有糖苷键的芳香酮类化合物（图 2－18），经过质谱、核磁共振一维谱和二维谱证实该化合物为 2－hydroxy－ 3，5－dimethyl－acetophenon－4－O－β－D－glucopyranoside。并考察了这一化合物对酪氨酸酶的抑制活性，结果表明其具有一定的酪氨酸酶抑制活性[22]。

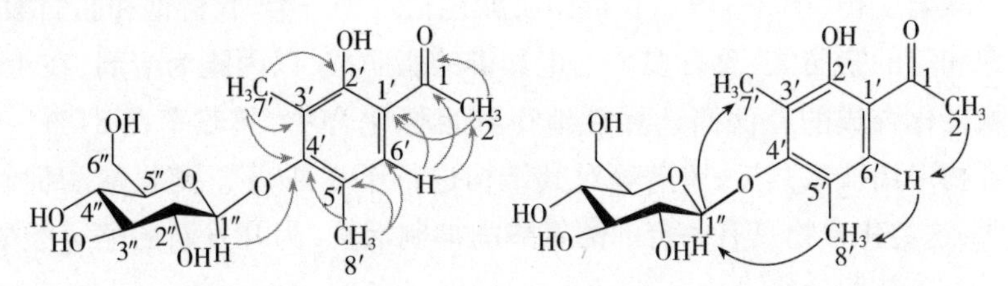

图 2－18　化合物 HMBC 和 ROESY 的关系

以氘代丙酮为溶剂，其核磁数据如下表：

1			
	δ_Hm（J in Hz）	δ_c	HMBC（H to C）
1	—	205.0	—
2	2.60s	25.3	C－1，C－1′

续表

	δ_Hm(J in Hz)	δ_c	HMBC(H to C)
1′	—	115.0	—
2′	—	160.0	—
3′	—	120.0	
4′	—	159.9	
5′	—	122.0	
6′	7.60 s	129.7	C-1,C-2′,C-8′
7′	2.30 s	8.4	C-2,C-3′,C-4
8′	2.35 s	15.8	C-4′,C-5′,C-6′
8′	2.35 s	15.8	C-4′,C-5′,C-6′
glc			
1″	4.75 d(7.7)	103.9	C-4″
2″	3.54 t(7.8)	74.3	C-1″,C-2″
3″	3.45 t(7.9)	76.5	C-2″,C-4″
4″	3.40 t(7.9)	70.1	C-3″
5″	3.17 m	76.7	C-4″
6″	3.67 dd(11.8,5.7)	61.3	C-5″
	3.79 dd(11.80,2.1)		C-5″

2013 年加拿大学者 Ziaullah, Khushwant S. Bhullar 等人合成了一种根皮苷长链脂肪酸衍生物(图2-19),长链脂肪酸包括六种三类,长链饱和脂肪酸、含有一个不饱和键的脂肪酸、含有多个不饱和键的脂肪酸,以丙酮为溶剂,在 45 ℃下采用生物催化合成的办法高选择性地得到目标化合物,其收率在 81%~98%。目标化合物采用核磁共振、质谱等对其结构进行了表征[23]。酪氨酸酶抑制活性结果表明,这类化合物具有一定的酪氨酸酶抑制活性,为开发新的酪氨酸酶抑制剂开辟了一条新的途径。

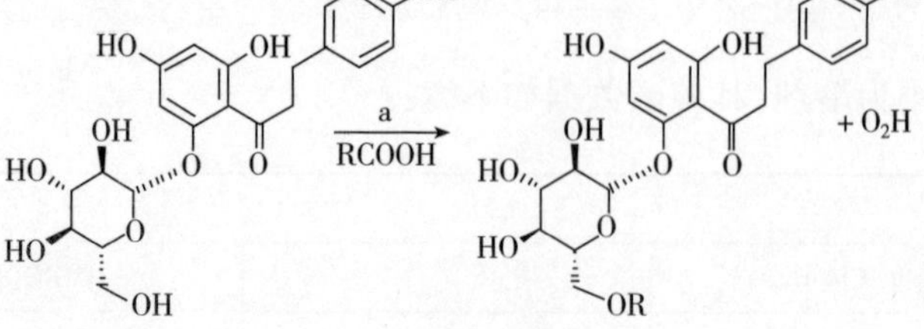

图2-19 根皮苷长链脂肪酸衍生物合成路线

2012 年 Xiaofang Wu 等人从翠叶芦荟中提取一种含有酮羰基的苦芦荟素(如图 2－20),并对其结构通过了 1－D,2－D NMR,HRMS,UV 和 IR 光谱进行表征,通过化学方法进行验证[24]。对该化合物的酪氨酸酶抑制活性进行了评价,该化合物的 IC_{50} 值为 108.62μg·mL^{-1},而且是一个非竞争性抑制剂(图 2－21)。

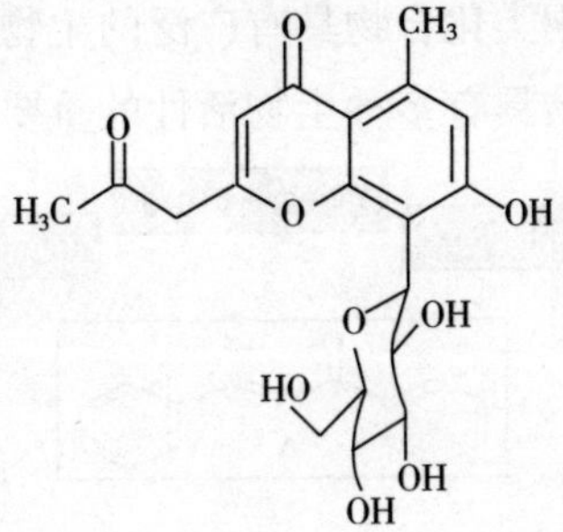

图 2－20　苦芦荟素的化学结构

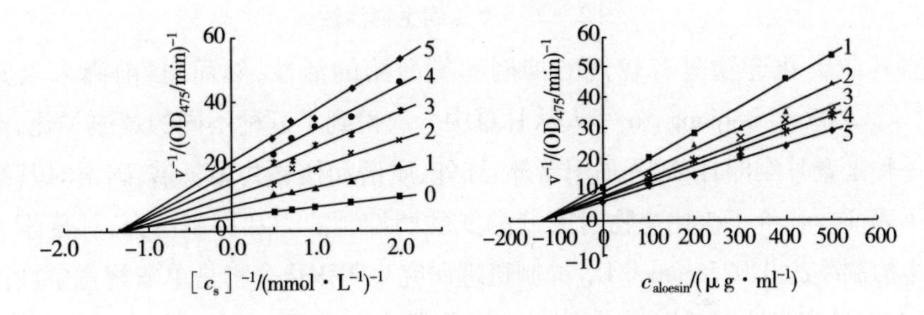

图 2－21　苦芦荟素对酪氨酸酶的抑制特性图

2016 年南非学者 Namrita Lall 等人从当地的一种植物 Greyia radlkoferi 分离得到二种二氢查耳酮类化合物(图 2－22),测试了他们抑制酪氨酸酶的活性及抑制黑色素产生的能力[25]。2',4',6'－三羟基二氢查耳酮的酪氨酸酶抑制活性最好,其半抑制浓度为 68.48μmol/L。分子对接结果表明,该类化合物对酪氨酸酶的抑制能力主要来源于其能够与活性中心铜离子发生作用,该化合物可以作为设计新的酪氨酸酶抑制剂分子的先导化合物。

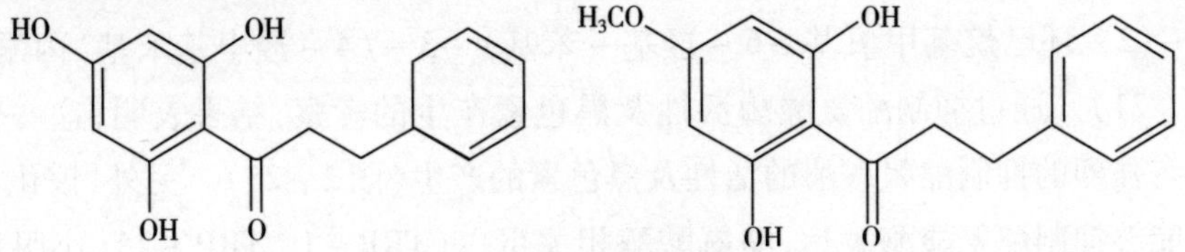

图 2－22　二氢查耳酮结构式

2.3 查耳酮类酪氨酸酶抑制剂

查尔酮是α,β-不饱和酮类化合物,不仅是重要的有机合成中间体,也是天然产物的重要成分。查尔酮类化合物具有广泛的生物活性(如图2-23),共轭体系的存在是查尔酮类化合物具有多种生物活性的重要因素[26]。

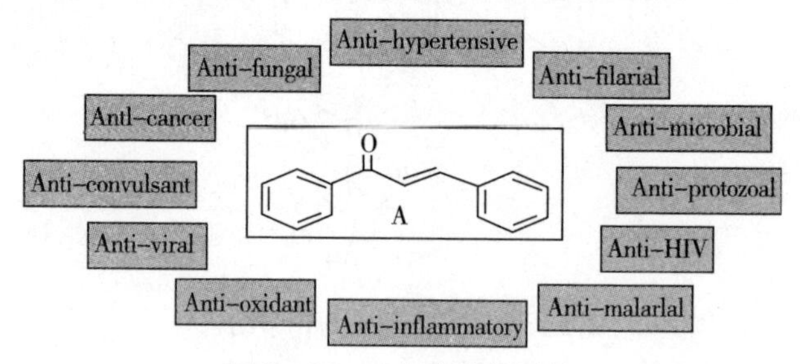

图2-23　查尔酮生物活性

查耳酮类化合物具有较好的抑制酪氨酸酶的活性,然而,目前尚未见其应用[27,28]。2005年Boqiang Fu等人从甘草中分离得到了五种不同的黄酮类化合物,其中一种是查耳酮即甘草酮,采用紫外、红外、质谱和核磁共振氢谱、碳谱对其结构进行了表征。评价了该化合物对蘑菇酪氨酸酶抑制能力和抑制潜力,甘草酮对单酚的半抑制率为0.0258mmol/L,抑制机理研究表明该化合物是单酚酶竞争性抑制剂,而对二酚酶无抑制活性[29]。该化合物具备进一步开发成祛黑增白剂的潜力。

2015年厦门大学陈清西教授考察了苯甲醛与丙酮缩合物对酪氨酸酶的抑制活性及抑制类型[30]。结果表明,该化合物具有较强的抑制酪氨酸酶活性的能力,对于单酚酶其半抑制浓度为1.5mmol/L,对于二酚酶其半抑制浓度为2.0mmol/L。动力学分析表明,该化合物对于单酚酶、二酚酶都是混合型抑制剂。

Byung-Hak Kim等人于2016年芳香醛和含有酚羟基芳香酮在碱性条件下发生羟醛缩合反应得到相应的查耳酮(图2-24),然后将所得的查耳酮进一步衍生,将酚羟基转化为环己烷甲基醚[31]。通过核磁共振、质谱分析后的确定其结构为1-(2-环己烷基甲氧基-6-羟基-苯基)-3-(4-羟甲基苯基)丙酮(查耳酮-21)。通过抑制酪氨酸酶活性及黑色素产生的考察,结果表明,这一化合物具有强烈的抑制酪氨酸酶的活性及黑色素的产生(图2-25)。另外,该化合物不仅能够抑制酪氨酸酶表达、酪氨酸酶相关蛋白(TRP-1, TRP-2)、小眼相关转录因子,而且能够抑制酪氨酸酶的转录活性,最后得出了该化合物是一种有效增白剂的结论。

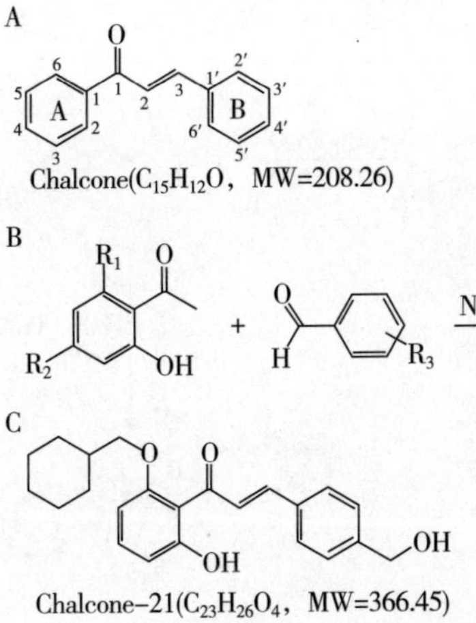

A

Chalcone($C_{15}H_{12}O$, MW=208.26)

B

NaOH/KOH

Ethanol

C

Chalcone-21($C_{23}H_{26}O_4$, MW=366.45)

图 2-24　查耳酮-21 的合成路线

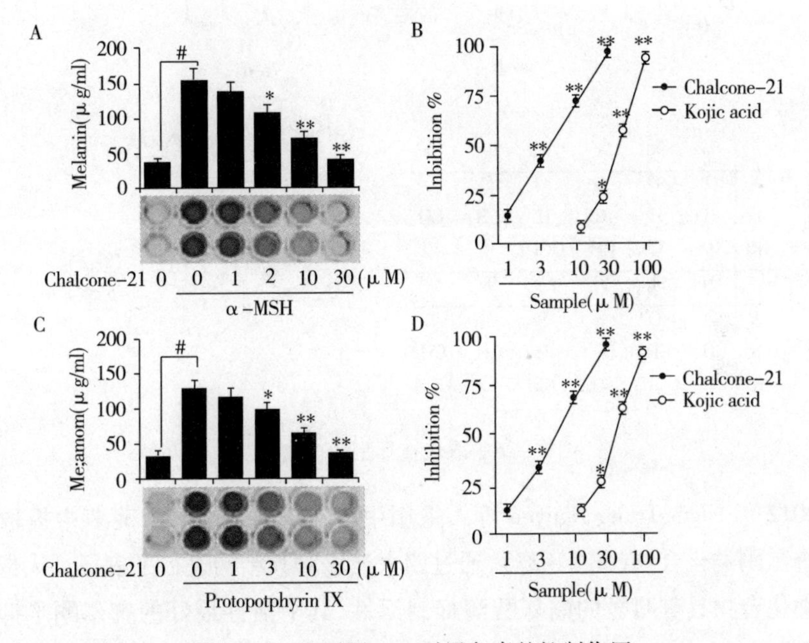

图 2-25　查耳酮-21 对黑色素的抑制作用

2011 年土耳其学者合成了一类新型结构的查耳酮衍生物(图 2-26),并对其抑制酪氨酸酶的活性进行了研究,证实了部分化合物具有较好的来源于香蕉的二酚酶的抑制活性,得出了该类化合物对酪氨酸酶的抑制活性与电子亲和能力的强弱有很好的相关性。

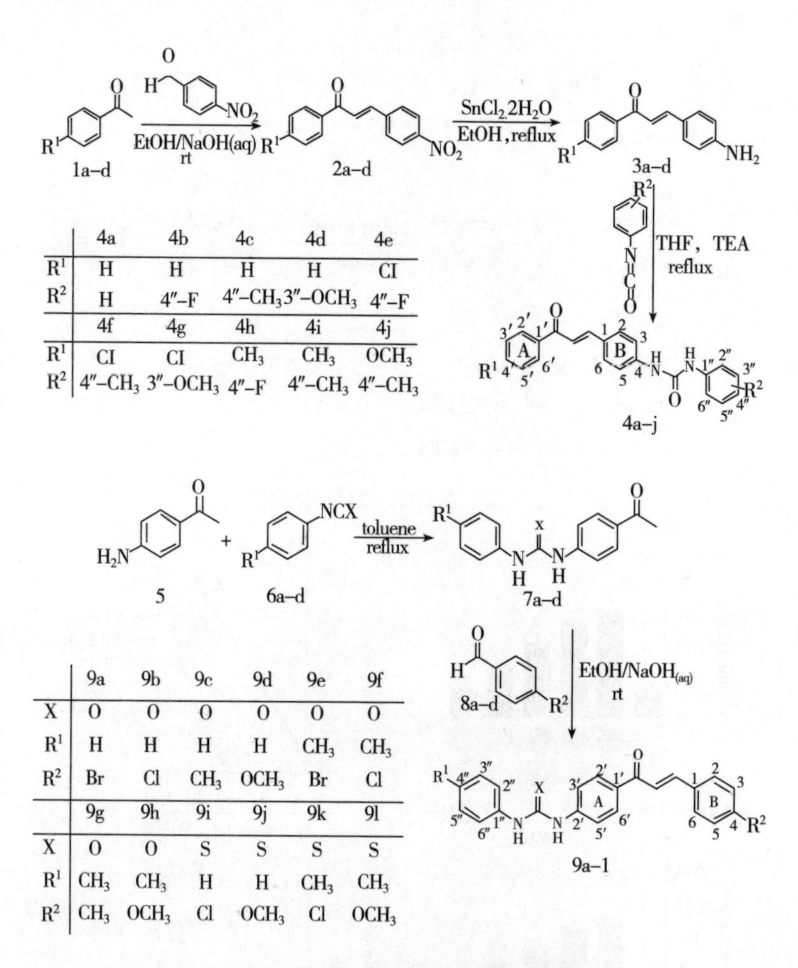

	4a	4b	4c	4d	4e
R^1	H	H	H	H	Cl
R^2	H	4″-F	4″-CH_3	3″-OCH_3	4″-F
	4f	4g	4h	4i	4j
R^1	Cl	Cl	CH_3	CH_3	OCH_3
R^2	4″-CH_3	3″-OCH_3	4″-F	4″-CH_3	4″-CH_3

	9a	9b	9c	9d	9e	9f
X	O	O	O	O	O	O
R^1	H	H	H	H	CH_3	CH_3
R^2	Br	Cl	CH_3	OCH_3	Br	Cl
	9g	9h	9i	9j	9k	9l
X	O	O	S	S	S	S
R^1	CH_3	CH_3	H	H	CH_3	CH_3
R^2	CH_3	OCH_3	Cl	OCH_3	Cl	OCH_3

图 2-26　取代查耳酮的合成路线

2012 年 Nhan Trung Nguyen 等人采用甲醇为溶剂[32]，从木菠萝中提取一类新的查尔酮类化合物(图 2-27)，通过四大光谱对其结构进行了表征。大部分查尔酮类化合物具有明显的酪氨酸酶抑制活性，其中活性最好的查尔酮半抑制浓度为 0.013μmol，比阳性对照曲酸的抑制活性强 3000 倍。

2012 年 Babasaheb P. Bandgar 等人以 3-甲酰基-9-甲基咔唑与含有酰胺基团的芳香酮发生 Claisen-Schmidt 缩合反应得到一种新型结构的查尔酮(图 2-28)[33]。对所有合成化合物进行了酪氨酸酶及黑色素产生的抑制活性考察，部分化合物具有一定的酪氨酸酶抑制活性。

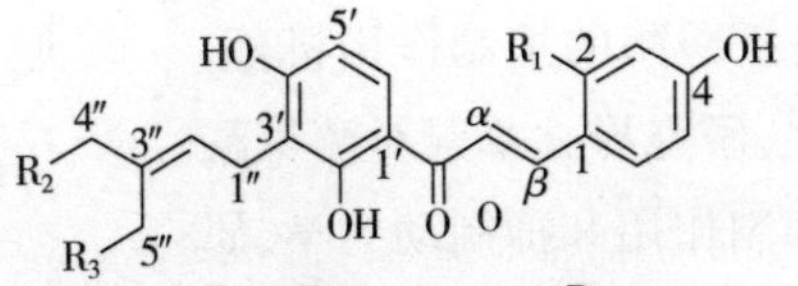

	R_1	R_2	R_3
5	H	OH	H
6	H	ferulate	H
7	H	H	sinapate
8	H	H	OH
9	H	H	ρ–coumarate
10	H	H	ferulate
11	H	ρ–coumarate	H
12	OH	H	H

图2－27　查尔酮分子结构

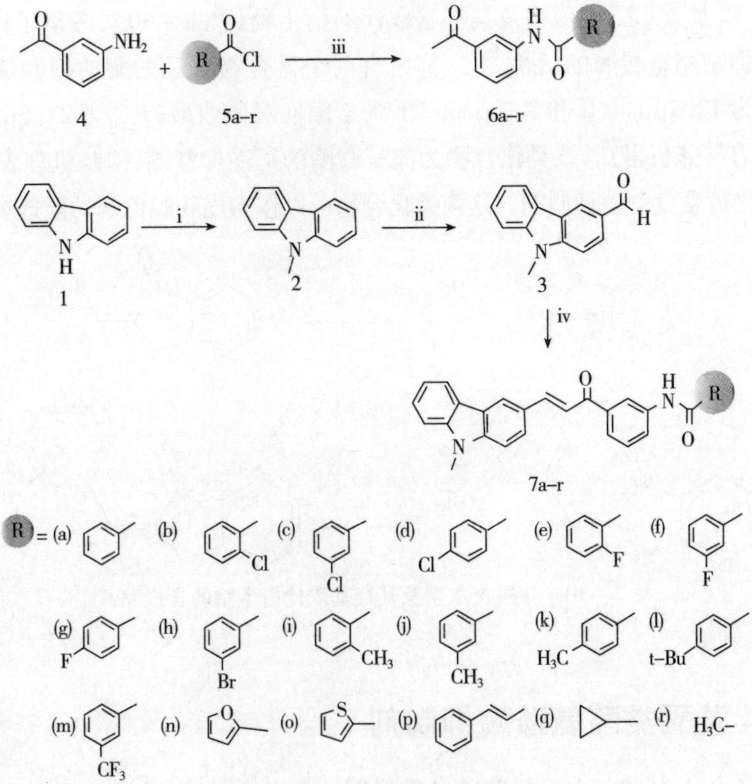

图2－28　目标化合物合成路线

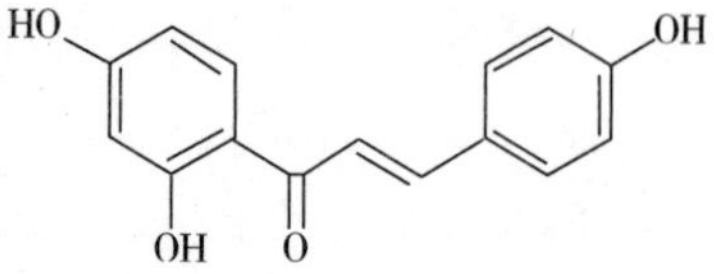

2015年庞得全等人为了探讨黑色素瘤药物异甘草素(图2-29)对酪氨酸酶的抑制作用机理,采用酶抑制动力学方法,研究异甘草素对酪氨酸酶单酚酶和二酚酶的抑制作用和抑制动力学,研究甘草素对黑色素瘤细胞A375生长的抑制作用[34]。结果表明,异甘草素对单酚酶和二酚酶都有良好的抑制作用,IC_{50}分别为(11.22±0.92)μmol/L和(48.53±4.75)μmol/L,且异甘草素是一种可逆的竞争型酪氨酸酶二酚酶抑制剂,抑制常数Ki为(12.14±0.54)μmol/L,异甘草素能显著抑制黑色素瘤细胞A375的生长,酪氨酸酶活性和黑色素含量不断降低。因此可以得到异甘草素是良好的酪氨酸酶抑制剂,能抑制黑色素瘤细胞的生长的结论。

图2-29 异甘草素化学结构

2015年澳大利亚学者Sini Radhakrishnan等人合成了一类2,3-二氢-1H-茚-1-羰基查尔酮类似物及其羟基取代的衍生物(图2-30),考察了所合成的化合物抑制酪氨酸酶的活性[35]。其中有两个化合物对二酚酶的抑制活性(IC_{50}值分别为12.3μmol/L和8.2μmol/L)强于阳性对照曲酸(IC_{50}为27.5μmol/L)。抑制动力学分析得知,该类化合物为酪氨酸酶的可逆抑制剂,抑制机理表明,这两类化合物都是竞争型抑制剂。这两类化合物可以作为皮肤脱色剂的候选物质。

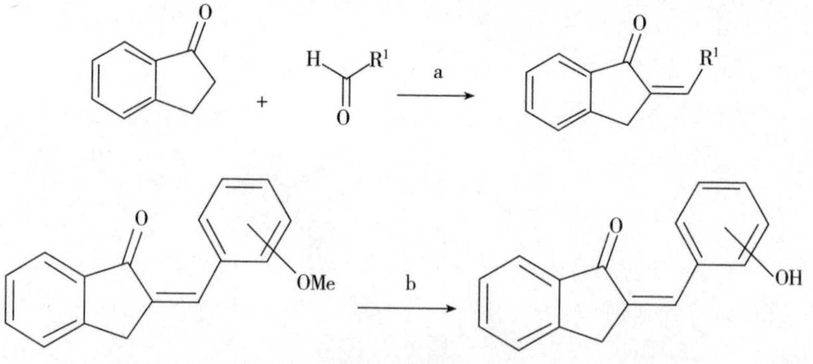

图2-30 茚酮查尔酮及其羟基取代衍生物的合成路线

2.4 黄酮类酪氨酸酶抑制剂

2015年上海理工大学杨城通过酶抑制动力学实验、荧光猝灭实验以及分子对接技术研究了木犀草素(图2-31)对酪氨酸酶活性的抑制作用[36]。酶抑制动力学实验结果表明:木犀草素是酪氨酸酶的非竞争性抑制剂,抑制常数K_I与IC_{50}分别为86mmol/L和778.2 mol/L;荧光猝灭实验结果表明:木犀草素对酪氨酸酶

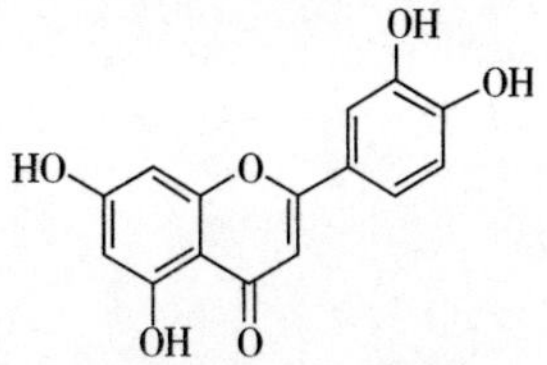

产生静态荧光猝灭作用,疏水作用与氢键作用共同稳定其复合物结构,结合位点数为1;分子对接结果表明:木犀草素在酪氨酸酶疏水口袋边缘与其相互作用,相互作用力包括疏水作用力与氢键。

图2－31　木犀草素化学结构

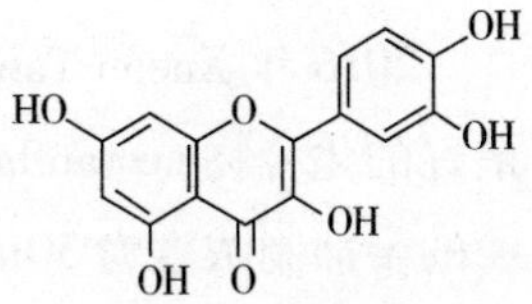

2015年管骁等人以L－多巴为底物,采用酶抑制动力学法研究了槲皮素(图2－32)对酪氨酸酶的抑制作用大小及类型,并采用荧光光谱技术分析其与酪氨酸酶的猝灭作用类型、结合位点、作用力类型。在此基础上,进一步利用柔性分子对接技术分析槲皮素对酪氨酸酶的抑制机理[37]。结果表明,槲皮素对酪氨酸酶具有抑制作用,抑制常数 K_I 为36mmol/L,以竞争性抑制剂形式抑制酪氨酸酶活性,是一种可逆性抑制剂;槲皮素以1∶1比例通过氢键和疏水作用力结合于酪氨酸酶活性中心,且对酪氨酸酶的荧光产生静态淬灭作用,具有氢键及疏水作用力;分子对接结果验证了以上实验结论:槲皮素占据了酪氨酸酶活性中心,且与活性中心部位的Asn260和Gly62残基形成了强烈的氢键作用,同时伴有疏水作用共同稳定复合物的结构。

图2－32　槲皮素的化学结构

2016年Nouha Nasr Bouzaiene等人考察了天然产物芹菜素、芫花素、柚皮素(图2－33)抑制酪氨酸酶活性及抑制黑色素合成的效果[38]。结果表明,三个化合物都具有酪氨酸酶抑制活性,同时也都具有抑制黑色素瘤细胞B16F10增生的作用。随着G0/G1阶段细胞比例的下降,芹菜素、芫花素、柚皮素可以引起亚G0/G1, S和G2/M阶段细胞比例的增加。芹菜素和柚皮素能够提高黑色素合成能力和B16F10黑色素瘤细胞酪氨酸酶的活性,而芫花素能够通过抑制酪氨酸酶的活性导致黑色素合成能力下降。芹菜素可以作为化妆品行业的增白剂,而芹菜素和柚皮素可以作为化妆品的天然鞣剂。

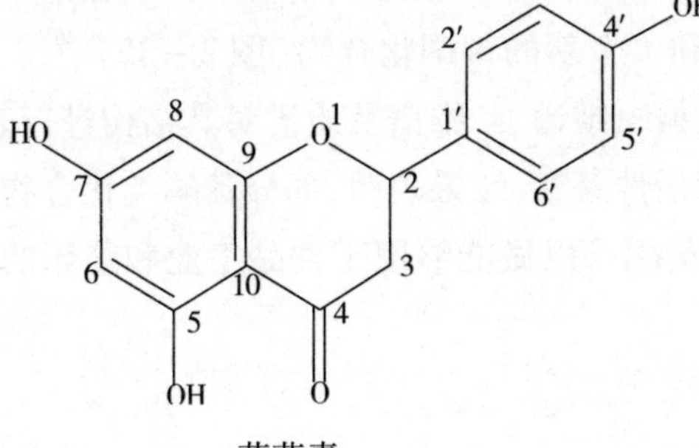

芹菜素 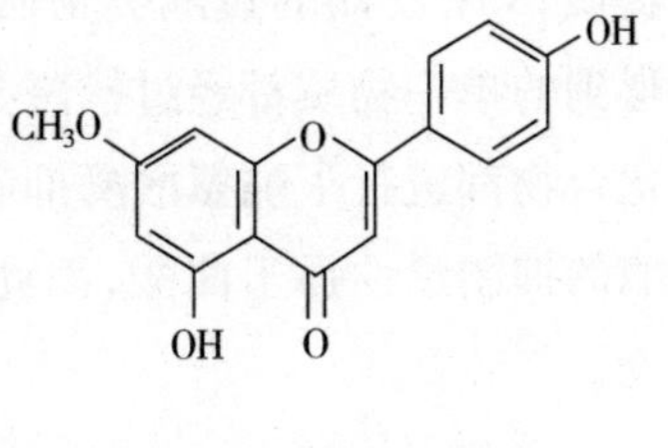　　　　芫花素

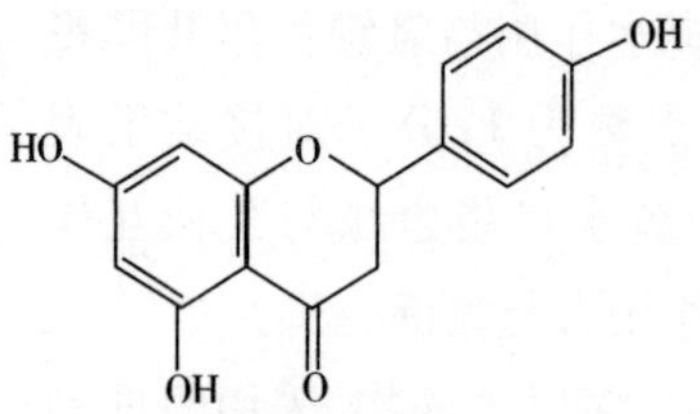

柚皮素

图 2-33　芹菜素、芫花素、柚皮素化学结构

2016 年 Xuefei Tan 等人从植物提取得到四个黄酮类化合物(图 2-34),活性最好的化合物 neorauflavane 可以作为酪氨酸酶抑制剂的先导化合物,其对于单酚酶的半抑制浓度为 30nmol/L,抑制活性比曲酸强 400 倍,对于二酚酶的半抑制浓度为 500Nm[39]。动力学研究表明,neorauflavane 对于单酚酶和二酚酶都是竞争型抑制剂,同时也证明了该化合物对于单酚酶是一种可逆抑制剂,动力学参数为 K_i^{app} = 1.48nmol/L, k_3 = 0.0033nmol/L · min 和 k_4 = 0.0049nmol/L · min。Neorauflavane 能够有效减少 B16 黑色素瘤细胞中黑色素的含量,其 IC_{50} 值为 12.95μmol/L。分子对接结果表明,B 环的间苯二酚的结构,A 环甲氧基在抑制酪氨酸酶活性过程中发挥了关键作用。

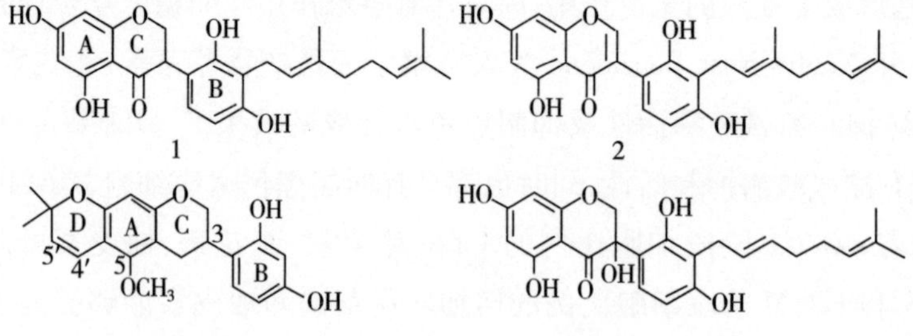

图 2-34　从 C. hirtella 中得到的四种黄酮类化合物结构式

2013 年 Ozlem Demirkiran 等人从三叶草中分离得到 12 种物质,包括 3 个羧酸类物质、8 个已知的黄酮类物质和 1 个新的黄酮化合物(图 2-35)[40]。所有分离得到的单一物质都经过核磁共振一维谱、二维谱及质谱对其结构进行表征,所有化合物都进行了酪氨酸酶抑制活性考察,结果表明,部分黄酮类化合物对酪氨酸酶的抑制活性强于曲酸,因此黄酮可以做能够用于食品工业和临床的潜在来源物。

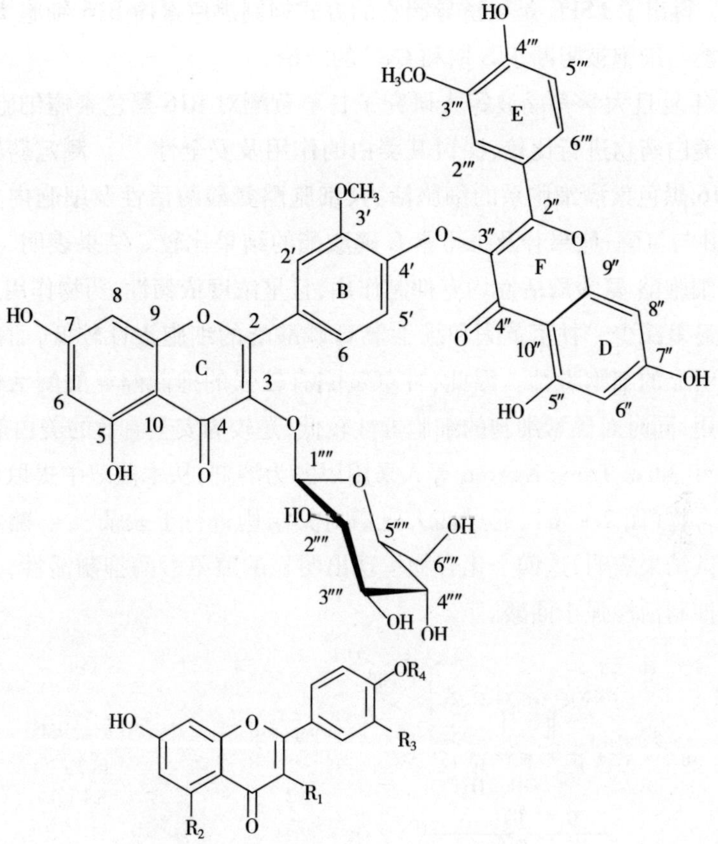

R_1=H，R_2=H，R_3=H，R_4=H(5)
R_1=OH,R_2=H,R_3=H，R_4=OCH_3(6)
R_1=O-β-glucose，R_2=OH，R_3=OH，R_4=H(7)
R_1=O-β-glucose，R_2=OH，R_3=OCH_3，R_4=H(8)
R_1=O-β-ga;actose，R_2=OH，R_3=OH，R_4=H(9)
R_1=O-β-glucouse-6″-acetyl，R_2=OH，R_3=OH，R_4=H(10)
R_1=O-β-galactose，R_2=OH，R_3=OCH_3，R_4=H(11)
R_1=O-β-glucose，R_2=OCH_3，R_3=OH，R_4=H(12)

图2-35　从三叶草中分离得到的黄酮类化合物

2009年段玉清等人研究了莲房原花青素(procyanidins from the seedpod of the lotus,LSPC)对体外培养的黑色素瘤B16细胞活性的影响[41]。以MTT法和细胞平板集落形成实验测定LSPC的抗肿瘤活性;电镜观察LSPC对瘤细胞形态学的影响;流式细胞仪检测LSPC对瘤细胞凋亡和细胞周期的影响;激光扫描共聚焦显微镜检测LSPC对瘤细胞内Ca^{2+}浓度的影响。结果LSPC对B16细胞生长的抑制作用与作用时间和浓度呈时效和量效关系,并可诱导瘤凋亡效应,最大抑制率达84.5%,使瘤细胞阻滞于S期;瘤细胞内Ca^{2+}浓度显著升高;细胞有典型的凋亡形态。得出了 LSPC 是以诱导凋亡的方式抑制黑色素瘤 B16 细胞增殖，其凋亡作用可能与细胞被阻断于 S 期和 Ca^{2+} 的结论。

2003 年复旦大学马晶波等人研究了甘草黄酮对 B16 黑色素瘤细胞的作用，并与一些美白药物进行比较，探讨其美白的作用及安全性[42]。测定药物对体外培养的 B16 黑色素瘤细胞系的细胞活力、细胞酪氨酸酶活性及细胞内黑素含量的影 响，并与氢醌、熊果苷及维生素 C 磷酸酯的结果比较。结果表明，几种化合物对 B16 细胞酪 氨酸酶活性均有抑制作用，且呈浓度依赖性；药物作用后细胞黑素生成量显著减少。甘草黄酮和维生素 C 磷酸酯的细胞毒性较低，而氢醌和熊果苷具有明显的细胞毒性。因此，甘草黄酮有较强的抑制酪氨酸酶活性和黑素生成的作用，同时对黑素细胞的细胞毒性较低，是较为安全有效的美白剂。

2012 年 Nhan Trung Nguyen 等人采用甲醇为溶剂，从木菠萝中提取四种新的黄酮类化合物（图 2－36），通过四大光谱对其结构进行了表征[43]。酪氨酸酶抑制活性测试结果表明，这四个化合物呈现出明显的酪氨酸酶抑制活性，其中三个化合物的抑制活性强于曲酸。

	R_1	R_2
1	H	OH
2	H	ρ-coumarate
3	H	ferulate
4	OH	OH

compound	IC_{50}, μM[a]
1	18.7 ± 0.1
2	8.4 ± 0.1
3	40.0 ± 0.4
4	47.3 ± 0.4

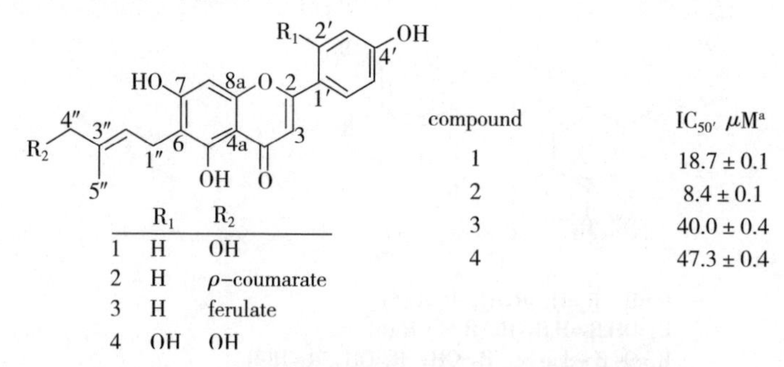

图 2－36　黄酮类化合物结构式及酪氨酸酶抑制活性

2016 年南非学者 Namrita Lall 等人从当地的一种植物 Greyia radlkoferi 分离得到二种黄酮类化合物（图 2－37），测试了他们抑制酪氨酸酶的活性及抑制黑色素产生的能力[44]。这二个化合物只有 3，5，7－三羟基黄酮具有一定的酪氨酸酶抑制活性。

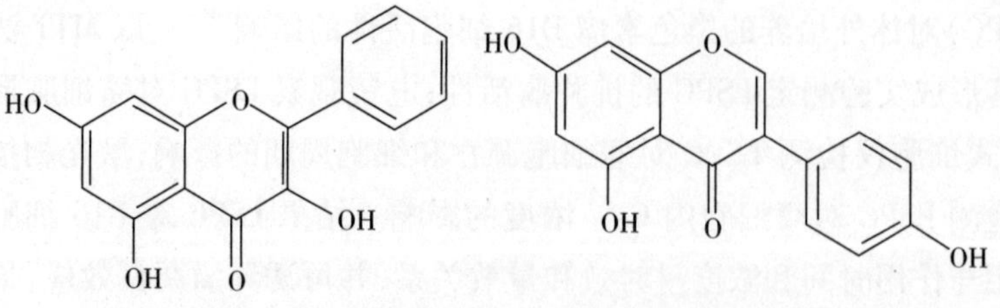

图 2－37　从 Greyia radlkoferi 分离得到的黄酮类化合物

2016年Beata G－sowska－Bajger和Hubert Wojtasek通过酪氨酸酶活性的光谱分析来探讨黄酮和喹啉的反应过程[45]。黄酮是一类具有抗氧化特性重要食品组分，大多数黄酮类化合物具有酪氨酸酶抑制活性。开展了槲皮素、山奈酚、桑色素、儿茶素、柚皮素（图2－38）通过酪氨酸酶的氧化能力以及其对于L－酪氨酸、L－多巴氧化能力的影响研究。

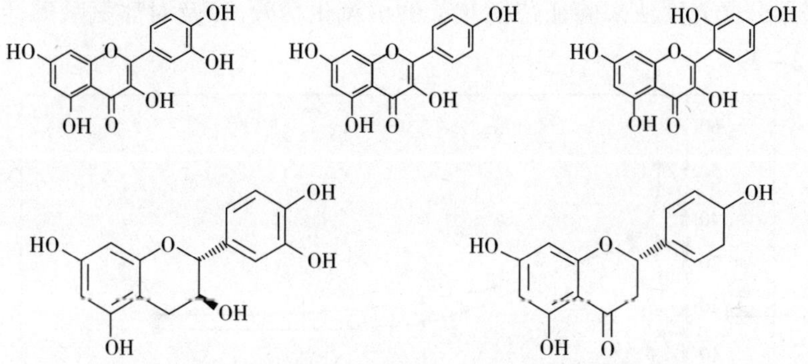

图2－38 槲皮素、山奈酚、桑色素、儿茶素、柚皮素化学结构

2007年王锡元采用化学合成的办法合成了芒柄花苷[46]，具体合成过程如下：在250mL三颈圆底烧瓶中，加入1.321 g（0.012 mol）间苯二酚，1.662 g（0.01 mol）对甲氧基苯乙酸，6.78 g催化剂，15 mL溶剂。搅拌、加热使固体溶解，在90℃下反应2 h，减压蒸馏除去溶剂，冷至室温后加入10％盐酸30 mL，搅拌30min，待油相充分固化后，抽滤，用去离子水洗涤至中性，室温下真空干燥，得黄色固体。用乙醇水溶液重结晶，得灰白色晶体。称取1中所得的2,4－二羟基－4'－甲氧基脱氧安息香于四颈烧瓶中，加入催化剂、DMF，搅拌溶解。在DMF中搅拌分批加入PCl_5，加完升温至50～55℃，搅拌20min，得浅桔红色的溶液。用氮气保护，搅拌，将3溶液滴加至2溶液中，加完升温至55℃，反应2h。反应混合物冷至室温后，倒入去离子水中，析出固体，抽滤，用去离子水洗涤至中性，室温下真空干燥，得浅黄色固体。于无水乙醇中重结晶，得浅苍黄色固体。

在100 mL三颈烧瓶中加入DMF（30 mL），丙酮（20 mL），混和均匀后加入无水碳酸钾（2.9 g，0.021 mol）。然后，边搅拌边加入芒柄花素（0.268g，1mmol），四丁基溴化氨（0.0806 g，0.25 mmol），五乙酰溴代葡萄糖（0.5g，1.2 mmol），20 ℃、搅拌反应5 h。对反应液进行薄层层析（TLC）检测。接着减压蒸馏出溶剂丙酮，将20 mL去离子水加入烧瓶中，混合物用乙酸乙酯（5×10 mL）萃取，分离，合并有机层。然后有机层用饱和食盐水(3×20 mL)洗涤,再用无水硫酸镁干燥,减压蒸馏至 3 mL 左右,放入真空干燥箱干燥后得产品。

称取 84 mg 四乙酰芒柄花苷、27 mg 无水 $Zn(OAc)_2$ 于 100 mL 三颈烧瓶中,再加入 5.5 mL 甲醇,70 ℃ 回流反应 7 h,TLC 跟踪,然后通过强酸树脂 001×7(730)进行中和,减压蒸馏后,放入真空干燥箱干燥得产品。

采用分光光度法来验证芒柄花苷的抗氧化功效,以及对酪氨酸酶活性的抑制(图 2-39)。

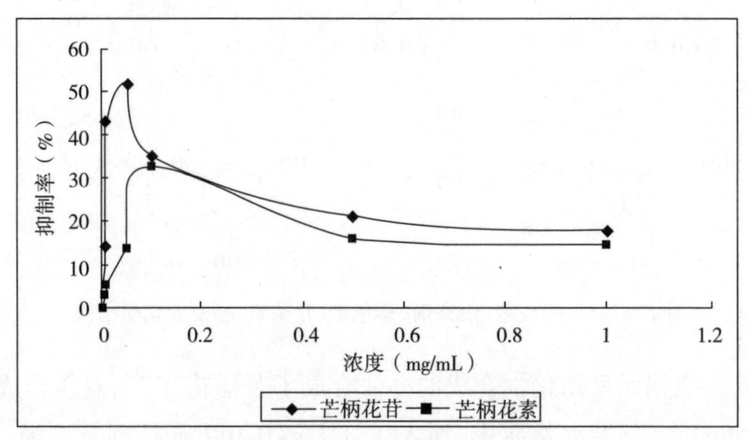

图 2-39　芒柄花苷对酪氨酸酶的抑制作用

结果表明:在低浓度时,芒柄花苷对酪氨酸酶具有较大的抑制作用,0.05 mg/mL 时的最大抑制率为 51.7 %,与芒柄花素相比,其对酪氨酸酶的抑制作用有一定的增长。

2006 年江城梅等人研究了三羟异黄酮抑制黑色素瘤 B16 细胞增殖作用的分子机制[47]。以 4,5,7 三羟异黄酮处理黑色素瘤 B16 细胞 1~4d 后,以生长曲线反映其增殖活力,以 B16 细胞形态、黑色素含量及以流式细胞仪检测细胞周期变化等观察三羟异黄酮对黑色素瘤 B16 细胞的抑制增殖作用。结果表明用 10、30、90μmol/L 的三羟异黄酮作用肿瘤细胞均有不同程度的抑瘤作用($P<0.05 \sim P<0.01$)。表现为黑色素生成能力增加,细胞生长缓慢,可使 B16 细胞阻断在 S 期。因此,三羟异黄酮不同剂量对体外黑色素瘤 B16 细胞增殖有明显的抑制作用。

2.5 席夫碱类酪氨酸酶抑制剂

Ley 等于 2001 年合成了一系列肟和肟醚类 Schiff 碱(图 2-40),考察了该类化合物对酪氨酸酶的抑制活性,结果表明:3,4-二羟基苯甲醛的乙氧基胺 Schiff 碱的 IC_{50} 为 0.3μmol/L。经过抑制机理研究,得出了这类化合物是直接和酶作用,而不是简单的还原 L-多巴的氧化产物,其对酪氨酸酶的抑制类型为非竞争型的结论。苯基硫脲、二硫苏糖醇和巯基乙醇等含硫化合物也具有酪氨酸酶抑制活性,其抑制作用主要是通过硫脲上的硫取代 Emet 活性中心两个铜离子之间的氢氧化物桥联配体,从而与酶活性中心形成很牢固的络合物[48,49]。

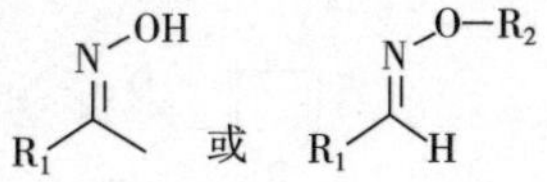

图 2-40 Schiff 碱结构

2012 年 Sung Jin Bae 等人设计并合成了一类 N-取代苄叉芳胺衍生物(图 2-41),考察了这类化合物抑制酪氨酸酶的活性及抑制小鼠黑色素瘤细胞 B16F10 的黑色素合成[50]。构效关系研究表明,含有 4-羟基苯胺和含有 4-甲氧基苯胺衍生物表现出较强的酪氨酸酶抑制活性,比 2-羟基苯胺衍生物的抑制活性好。(E)-4-((4-羟基苯亚胺)甲基)1,2-苯二酚对于酪氨酸酶的抑制活性最强,其 IC_{50} 值为(17.22±0.38)μmol/L,强于阳性对照曲酸的抑制活性,而且此化合物为非竞争型抑制剂,同时也具备减少黑色素合成的能力。Sung Jin Bae 认为(E)-4-[(4-羟基苯亚胺)甲基]1,2-苯二酚可以作为用于临床治疗的安全有效新的潜在酪氨酸酶抑制剂候选药。

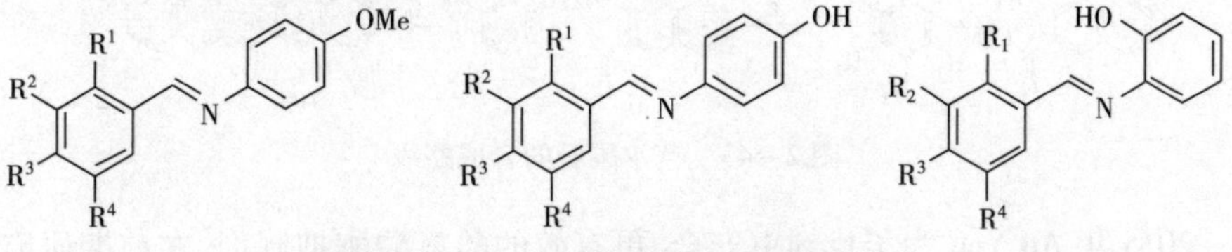

图 2-41 N-取代苄叉芳胺衍生物结构式

2016 年 Juan Xie 等人以 2-噻吩甲醛、2-呋喃甲醛、2-吡咯甲醛为起始原料之一,分别与氨基硫脲、甲基取代氨基硫脲、苯基取代氨基硫脲反应合成席夫碱(图 2-42),考察了目标化合物的酪氨酸酶抑制活性[51]。用荧光、核磁共振氢谱滴定法及分子对接研究了其抑制酪氨酸酶的特性,荧光光谱和核磁共振氢谱滴定法表明该类化合物中氨基硫脲的硫原子能够与活性中心的铜离子形成络合物而达到抑制酪氨酸酶活性的目的,而且核磁共振谱进一步证明了该类化合物能够和酶分子形成氢键，使得其更有利于与铜离子形成络合物。上述分析能够和分子对接结果有很好的一致性，2－2－噻吩甲醛氨基硫脲席夫碱、2－呋喃甲醛氨基硫脲席夫碱、2－吡咯甲醛氨基硫脲席夫碱是一类值得进一步研究的潜在酪氨酸酶抑制剂。

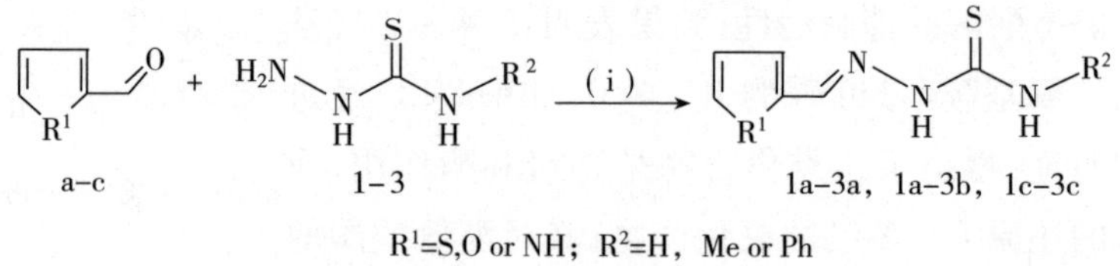

R^1=S,O or NH；R^2=H，Me or Ph

图2－42　杂环氨基硫脲席夫碱的合成路线

2016年Sini K. Radhakrishnan等人合成了一系列羟基取代查尔酮肟衍生物（图2－43），考察了这类化合物对于酪氨酸酶及小鼠黑色素瘤细胞B16F10黑色素合成的抑制活性[52]。所有合成化合物都通过核磁共振氢谱、碳谱、红外、高分辨质谱对其结构进行表征。其中有两个化合物的抑制活性强于阳性对照曲酸，抑制动力研究表明，该类化合物是竞争型酪氨酸酶抑制剂，而且能够抑制黑色素的合成及B16细胞的酪氨酸酶活性，分子对接说明了该类化合物能够与酪氨酸酶氨基酸残基很好地发生作用。

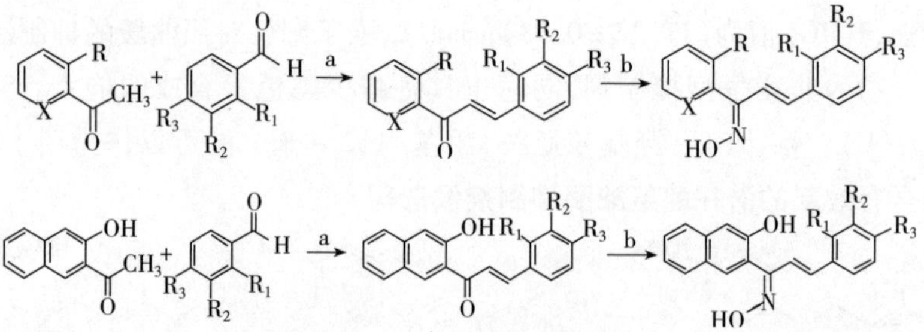

图2－43　查尔酮肟的合成路线

2015年Ao You为了找到更安全、更有效的酪氨酸酶抑制剂，在前期研究的基础上，以4－烷氧基取代、4－乙酰氧基取代的芳香酮和氨基硫脲反应，得到一系列氨基硫脲席夫碱（图2－44）[53]，考察了这类化合物的酪氨酸酶抑制活性，有趣的是大部分化合物具有较强的抑制活性，其IC_{50}值1.0μmol/L以下。讨论了构效关系（图2－45），开展了抑制机理和抑制动力学研究，证实该类化合物是可逆竞争型抑制剂。

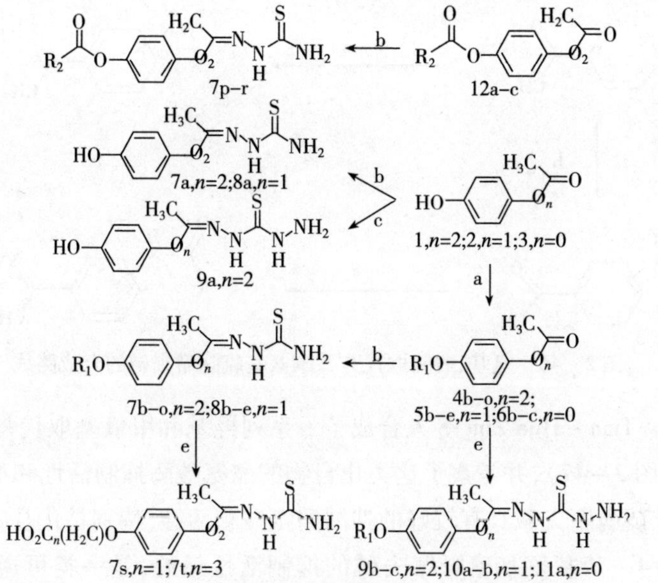

图 2－44　氨基硫脲席夫碱的合成路线

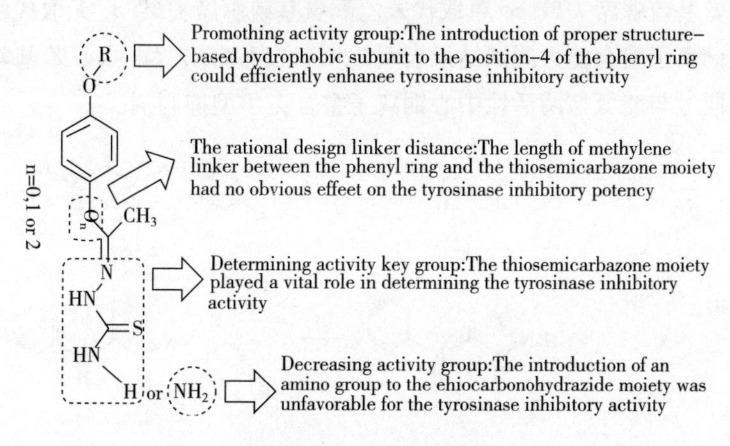

图 2－45　构效关系图

Ao You 等人于 2015 年采用氨基取代苯乙酮和酰胺取代苯乙酮为起始原料之一分别和氨基硫脲反应合成了一系列氨基硫脲席夫碱（图 2－46）[54]，并证实了这类化合物具有较好的酪氨酸酶抑制活性。通过构效关系分析发现，氨基是酪氨酸酶致活基团；氨基硫脲片段的存在是使该类化合物由酪氨酸酶致活剂转化为抑制剂的关键因素；氨基硫脲席夫碱是氨基硫脲抑制活性的决定因素；酰胺的酰基对抑制活性无明显影响；氨基在苯环上的位置对抑制活性有影响。

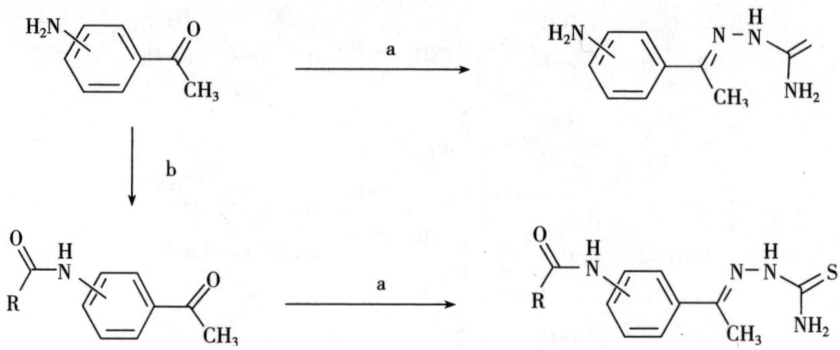

图 2－46　氨基、酰胺取代芳香酮氨基硫脲席夫碱的合成路线

2013 年 Tian－Hua Zhu 等人合成了一系列羟基和甲氧基取代丹皮酚氨基硫脲席夫碱(图 2－47)，并考察了这类化合物的酪氨酸酶抑制活性和抑制机理[55]。丹皮酚氨基硫脲席夫碱具有较好的抑制酪氨酸酶活性，特别是化合物 2,4－二羟基苯乙酮－4－苯基氨基硫脲席夫碱的抑制活性最好，是一类可逆竞争型抑制剂。邻位、对位取代苯乙酮氨基硫脲席夫碱的酪氨酸酶抑制活性顺序如下：二取代苯乙酮氨基硫脲席夫碱 ＞ 单取代苯乙酮氨基硫脲席夫碱 ＞ 无取代氨基硫脲席夫碱。铜离子螯合能力测试结果表明 2,4－二羟基苯乙酮－4－苯基氨基硫脲席夫碱是通过与酪氨酸酶活性中心铜离子螯合竞争型抑制剂。

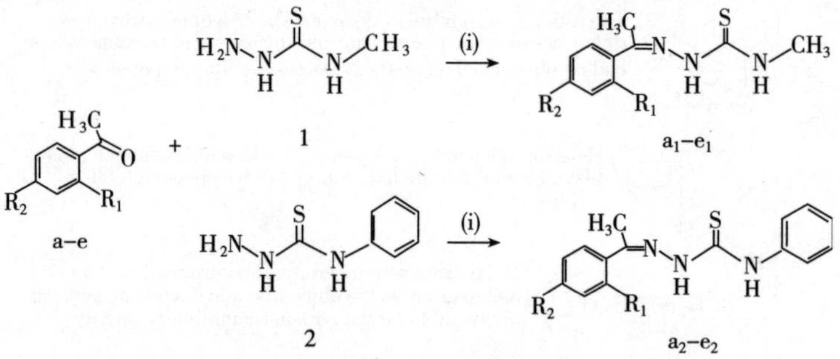

$R_1=R_2=H,OH$ or OCH_3

图 2－47　取代丹皮酚氨基硫脲席夫碱合成路线

2012 年徐焱等人在前期研究的基础上[56]，设计了一系列取代苯丙醛缩氨基硫脲类化合物，以取代苯甲醛为起始原料，经 5 步反应合成了 15 个目标化合物(图 2－48)，其结构通过红外、核磁氢谱、质谱及元素分析确认。选择托酚酮作为活性测试的对照药剂。活性测试结果表明，目标化合物对棉铃虫酪氨酸酶具有明显的抑制效果，其 IC_{50} 值均低于对照药剂托酚酮，说明所设计的化合物活性均

优于对照药剂托酚酮。保留硫脲活性结构,用碳链增加硫脲和芳香环之间的距离,增加了化合物疏水性,设计合成的目标物对棉铃虫酪氨酸酶具有明显的抑制效果,尤其是化合物3－(2－氯苯基)－丙醛缩氨基硫脲和3－(2－硝基苯基)－丙醛缩氨基硫脲,可以作为先导结构进行进一步优化。在进一步结构优化时,应考虑继续保持硫脲片段,在芳香环上增加吸电子基团,同时应考虑减小位阻,或者引入其他具有生物活性的芳香环,最终有望发现更高活性的新型昆虫生长调节剂先导化合物。

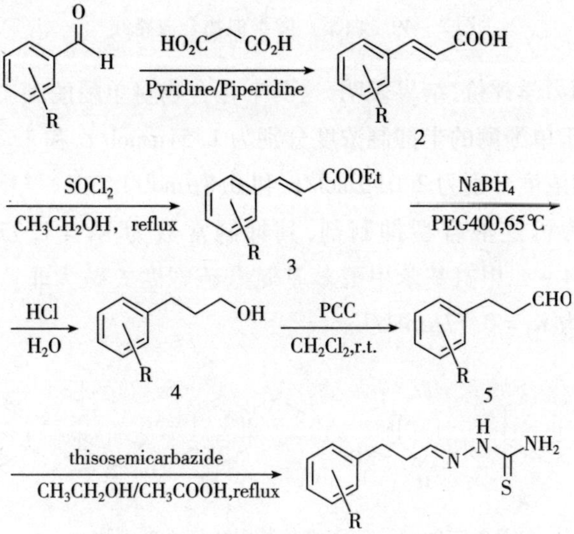

R=H, 4-F, 4-Cl, 2-Cl, 2, 4-Cl_2, 4-OCH_3, 2-OCH_3, 3-OCH_3, 4-OC_2H_5, 4-NO_2, 2-NO_2, 4-CH_3, 4-*i*-Pr, 3-Br, 2, 4-$(OCH_3)_2$

图2－48　取代苯丙醛缩氨基硫脲类化合物合成路线

2013年Larissa Lavorato Lima等人为了改善天然白藜芦醇的去黑色素的能力,以邻氨基苯酚和取代苯甲醛为起始原料合成了六个白藜芦醇的类似物(图2－49)[57]。所有合成的白藜芦醇类似物都进行了酪氨酸酶抑制活性测试,结果显示由对羟基苯甲醛所得到的席夫碱抑制活性最强,强于天然白藜芦醇。其他化合物抑制活性顺序依次为:对甲氧基苯甲醛席夫碱,苯甲醛席夫碱,2,3,4－三甲氧基苯甲醛席夫碱,间硝基苯甲醛席夫碱,3－二甲氨基苯甲醛席夫碱。

2013年Mei－Hua Yang等人合成了4－二甲氨基苯甲醛氨基硫脲席夫碱和4－二甲氨基苯甲醛苯基氨基硫脲席夫碱(图2－50),通过核磁共振氢谱、碳谱和质谱对这两个化合物进行了结构表征[58]。对两个化合物都进行酪氨酸酶抑制活

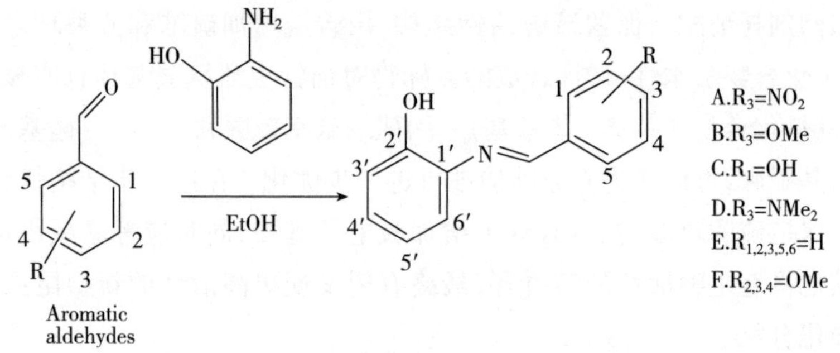

NH2 HO O 5 1 4 2 R 3 Aromatic aldehydes EtOH OH 2′ 1′ N 3′ 4′ 6′ 5′ 2 R 1 3 4 5

A.R_3=NO_2
B.R_3=OMe
C.R_1=OH
D.R_3=NMe_2
E.$R_{1,2,3,5,6}$=H
F.$R_{2,3,4}$=OMe

图 2-49　白藜芦醇类似物合成路线

性测试和抑制动力学评价，结果表明，这两个化合物对单酚酶和二酚酶都有较好的抑制活性，对于单酚酶的半抑制浓度分别为 1.54μmol/L 和 1.78μmol/L，对于二酚酶的半抑制浓度分别为 2.01μmol/L 和 0.8μmol/L。4-二甲氨基苯甲醛氨基硫脲席夫碱为可逆混合型抑制剂，其抑制常数为 K_I = 1.77μmol/L 和 K_{IS} 6.49μmol/L，而 4-二甲氨基苯甲醛苯基氨基硫脲席夫碱为非竞争型可逆抑制剂，其抑制常数为 K_I = 0.77μmol/L。

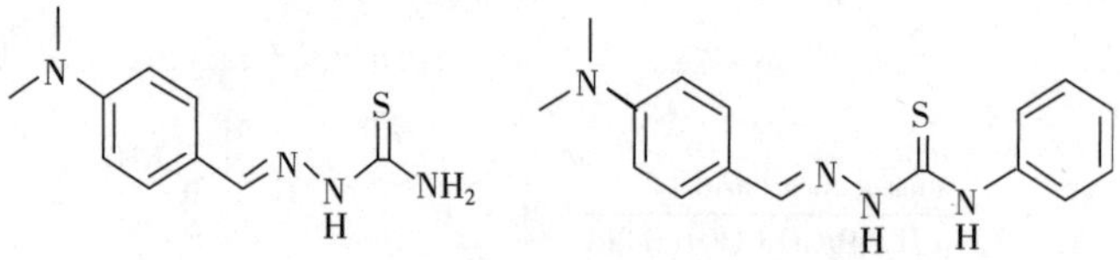

N S N N H NH2　N S N N H N H

图 2-50　4-二甲氨基苯甲醛席夫碱结构式

2017 年 Jian Xu 等人合成了一类 4-位或 5-位官能团化噻吩甲醛氨基硫脲席夫碱（图 2-51），通过核磁共振氢谱、碳谱、荧光光谱和红外光谱表征，对所合成的化合物进行了酪氨酸酶抑制活性评价[59]。核磁共振和荧光光谱结果表明，氨基硫脲片段能够酪氨酸酶的活性中心铜离子形成络合物，是这类化合物能够抑制酪氨酸酶的重要因素。5-位官能团化会使抑制活性降低，4-位甲氧乙酰基能够提高化合物的抑制活性，甲氧乙酰基能够促进氨基硫脲的硫原子和酪氨酸酶活性中心铜离子发生螯合，因此 4-位官能团化利用抑制活性的提高，4-甲氧乙酰基取代的噻吩甲醛氨基硫脲席夫碱有望成为一种新的酪氨酸酶抑制剂。

Sini K. Radhakrishnan 等人于 2016 年采用固体化学合成法合成了氮杂查尔酮及其肟（图 2-52），以 L-3,4-二羟基苯丙胺为底物测试了所合成的化合物酪氨酸酶抑制活性[60]。其中两个合成的新型肟衍生物的酪氨酸酶抑制活性比阳性对照曲酸强。化合物 1b 和 2b 抑制二酚酶的 IC_{50} 分别为 15.3μmol/L 和

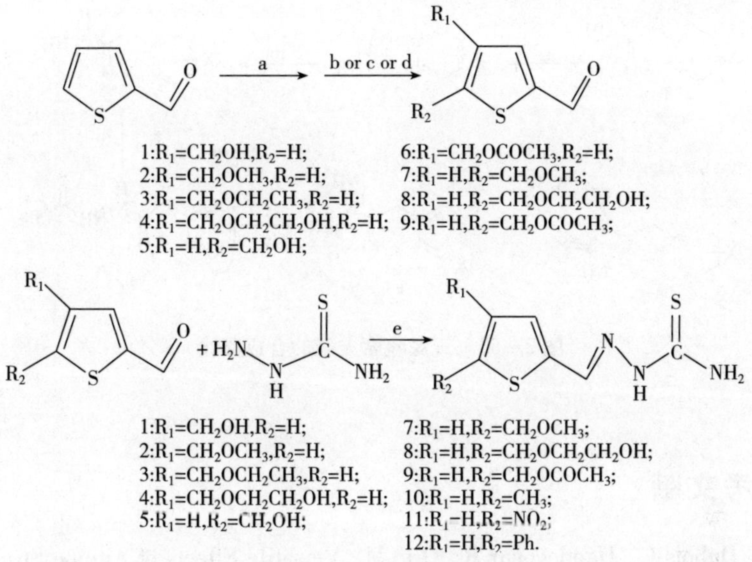
a
b or c or d
1:R₁=CH₂OH,R₂=H;
2:R₁=CH₂OCH₃,R₂=H;
3:R₁=CH₂OCH₂CH₃,R₂=H;
4:R₁=CH₂OCH₂CH₂OH,R₂=H;
5:R₁=H,R₂=CH₂OH;
6:R₁=CH₂OCOCH₃,R₂=H;
7:R₁=H,R₂=CH₂OCH₃;
8:R₁=H,R₂=CH₂OCH₂CH₂OH;
9:R₁=H,R₂=CH₂OCOCH₃;
e
1:R₁=CH₂OH,R₂=H;
2:R₁=CH₂OCH₃,R₂=H;
3:R₁=CH₂OCH₂CH₃,R₂=H;
4:R₁=CH₂OCH₂CH₂OH,R₂=H;
5:R₁=H,R₂=CH₂OH;
7:R₁=H,R₂=CH₂OCH₃;
8:R₁=H,R₂=CH₂OCH₂CH₂OH;
9:R₁=H,R₂=CH₂OCOCH₃;
10:R₁=H,R₂=CH₃;
11:R₁=H,R₂=NO₂;
12:R₁=H,R₂=Ph.

图 2－51　噻吩甲醛氨基硫脲席夫碱合成路线

12.7μmol/L。抑制动力学和抑制机理分析表明，这类新型肟衍生物是可逆竞争型抑制剂，化合物 1b 和 2b 的抑制动力学常数分别为 5.1μmol/L 和 2.5μmol/L，这类化合物可以作为皮肤增白剂或食品抗褐变的添加剂。

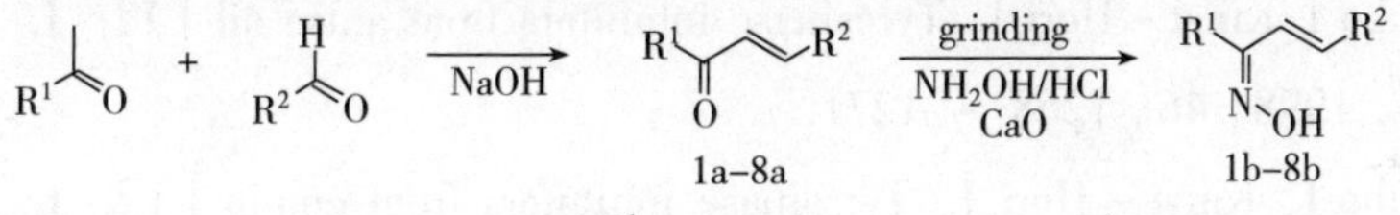
NaOH
1a-8a
grinding
NH2OH/HCl
CaO
1b-8b

	R^1	R^2
1	2-HO-C_6H_4	pyridin-2-yl
2	2-HO-C_6H_4	pyridin-3-yl
3	2-HO-C_6H_4	pyridin-4-yl
4	3-hydroxynaphthalen-2-yl	pyridin-2-yl
5	3-hydroxynaphthalen-2-yl	pyridin-3-yl
6	3-hydroxynaphthalen-2-yl	pyridin-4-yl
7	pyridin-2-yl	2,4-$(MeO)_2$-C_6H_3
8	pyridin-2-yl	4-(dimethylamino)phenyl

图 2－52　氮杂查尔酮肟的合成

2016 年 Muhammad Rafiq 等人以三氮唑胺和不同醛为起始原料，采用微波辅助合成法合成了十个三氮唑席夫碱（图 2－53）[61]，所有的化合物都进行了酪氨酸酶抑制活性考察，其中氟原子取代的化合物的抑制活性最好，强于阳性对照曲酸，对氟原子取代三氮唑席夫碱进行了抑制动力学和抑制机理分析，具有该结构特征的化合物是非竞争型抑制剂。

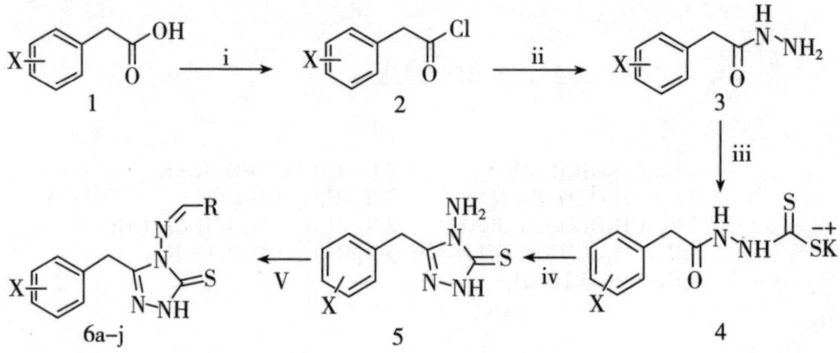

图 2 - 53　三氮唑席夫碱的合成路线

参考文献

[1] Dubois C, Haudecoeur R, Orio M. Versatile Effects of Aurone Structure on Mushroom Tyrosinase Activity [J]. *ChemBioChem*, 2012, 13, 559 - 565.

[2] Jiménez M, García - Carmona F. Competitive Inhibition of Mushroom Tyrosinase by 4 - Substituted Benzaldehydes [J]. J. Agric. Food Chem., 2001, 49, 4060 - 4063.

[3] Kubo I, Kinst - Hori I. Tyrosinase inhibitors from anise oil [J]. J. Agric. Food. Chem, 1998, 46, 1268 - 1271.

[4] Kubo I, Kinst - Hori I. Tyrosinase inhibitors from cumin [J]. J. Agric. Food. Chem, 1998, 46, 5338 - 5341.

[5] Nihei K, Yamagiwa Y, Kamikawa, T. 2 - Hydroxy - 4 - isopropylbenzaldehyde, a potent partial tyrosinase inhibitor [J]. Bioorg. Med. Chem. Lett, 2004, 14, 681 - 683.

[6]赵连梅,耿艺曼,孙士萍,等. 木鳖子单体化合物对羟基桂皮醛诱导小鼠黑素瘤 B16 细胞的分化及其机制[J]. 中国肿瘤生物治疗杂志, 2014,21,282 - 287.

[7] Cui Y, Liang G, Hu Y H, et al. Alpha - Substituted Derivatives of Cinnamaldehyde as Tyrosinase Inhibitors: Inhibitory Mechanism and Molecular Analysis[J]. J. Agric. Food Chem., 2015, 63, 716 - 722.

[8]李佳青,王旭瑞,李占海. 肉桂醛对黑色素瘤小鼠的抑瘤作用及其机制研究[J]. 解放军医药杂志,2014,26,51 - 54.

[9] Punam B, Sujeetkumar J. Effect of substitution on the inhibition of mushroom tyrosinase by pyridine carboxaldehydes [J]. Progress in Reaction Kinetics and Mechanism, 2013, 38, 75 – 85.

[10] Zheng Z P, Zhang Y N, Zhang S, et al. One – pot green synthesis of 1,3, 5 – triarylpentane – 1,5 – dione and triarylmethane derivatives as a new class of tyrosinase inhibitors [J]. Bioorganic & Medicinal Chemistry Letters, 2016, 26, 795 – 798.

[11] Dong X, Zhang Y, He J L, et al. Preparation of tyrosinase inhibitors and antibrowning agents using green technology [J]. Food Chemistry, 2016, 197, 589 – 596.

[12] Vivek K B, Sun C K. Tyrosinase and a – glucosidase inhibitory effects of an abietane type diterpenoid taxodone from metasequoia glyptostroboides [J]. Natl Acad. Sci. Lett., 2015, 38, 399 – 402.

[13] Liu X, Jia Y, Chen J, et al. Inhibition effects of benzylideneacetone, benzylacetone, and 4 – phenyl – 2 – butanol on the activity of mushroom tyrosinase [J]. Journal of Bioscience and Bioengineering, 2015, 119, 275 – 279.

[14]杨帆,孙秋艳,刘亮亮,等. 维康醇诱导小鼠 B16F0 细胞凋亡的研究[J]. 中国药理学通报, 2013,29,1269 – 1274.

[15]韦星船,缪绸雨,段彦飞,等. 2 – 茚酮基姜黄素类似物对酪氨酸酶的抑制作用[J]. 食品科学,2013,34,66 – 68.

[16]韦星船,霍梦月,郑成,等. 含氮姜黄素类似物的合成及其酪氨酸酶抑制效应[J]. 化工进展,2014,33,2155 – 2160.

[17]莫容清,杜志云,涂增清,等. 姜黄素类似物 2,5 – 双(2,4 – 二羟基苯亚甲基)环戊酮的合成及酪氨酸酶抑制活性[J]. 化学试剂, 2012,34,101 – 103;107.

[18] Jiang Y, Du Z, Xue G, et al. Synthesis and Biological Evaluation of Unsymmetrical Curcumin Analogues as Tyrosinase Inhibitors [J]. Molecules, 2013, 18, 3948 – 3961.

[19]曾亮,马梦君,官兴丽,等. 茶叶提取物 EGCG 对 B16 小鼠黑素瘤细胞的影响[J]. 西南大学学报(自然科学版),2014,36,1 – 7.

[20] Ha Y M, Park Y J, Kim J A, et al. Design and synthesis of 5 – (substituted benzylidene) thiazolidine – 2,4 – dione derivatives as novel tyrosinase

inhibitors [J]. European Journal of Medicinal Chemistry, 2012,49,245 -252.

[21]Kim H R, Lee H J, Choi Y J, et al. Benzylidene - linked thiohydantoin derivatives as inhibitors of tyrosinase and melanogenesis: importance of the b - phenyl - a, b - unsaturated carbonyl functionality[J]. Med. Chem. Commun., 2014,5, 1410 -1417.

[22] Abbes B, Abdulmagid A M, Djemaa B. Chemical composition, antibacterial, antioxidant and tyrosinase inhibitory activities of glycosides from aerial parts of Eryngium tricuspidatum L [J]. Phytochemistry Letters, 2016, 18, 23 -28.

[23] Ziaullah K S, Bhullar S N, Warnakulasuriya H P, et al. Biocatalytic synthesis, structural elucidation, antioxidant capacity and tyrosinase inhibition activity of long chain fatty acid acylated derivatives of phloridzin and isoquercitrin [J]. Bioorganic & Medicinal Chemistry, 2013,21,684 -692.

[24]Wu X, Yin S, Zhong J, et al. Mushroom tyrosinase inhibitors fromAloe barbadensisMiller[J]. Fitoterapia, 2012, 83, 1706 -1711.

[25] Namrita L, Elizabeth M, Marco N C, et al. Insights into tyrosinase inhibition by compounds isolated from Greyia radlkoferi Szyszyl using biological activity, molecular docking and gene expression analysis [J]. Bioorganic & Medicinal Chemistry, 2016, 24, 5953 -5959.

[26] Singh P, Anand A, Kumar V. Recent developments in biological activities of chalcones: A mini review [J]. Eur. J. Med. Chem, 2014, 85, 758 -777.

[27] Ohad N, Ruth B A, Tal L, et al. Prevention of Agaricus bisporus postharvest browning with tyrosinase inhibitors [J]. Postharvest. Bio. Tech., 2006, 39, 272 -277.

[28] Quintin J. Synthesis and biological evaluation of a series of tangeretin - derived chalcones[J]. Bioorg. Med. Chem. Lett., 2009, 19, 167 -169

[29] Fu B, Li H, Wang X R. Isolation and Identification of Flavonoids in Licorice and a Study of Their Inhibitory Effects on Tyrosinase [J]. J. Agric. Food. Chem., 2005, 53, 7408 - 7414.

[30] Liu X, Jia Y, Chen J, et al. Inhibition effects of benzylideneacetone, benzylacetone, and 4 - phenyl - 2 - butanol on the activity of mushroom tyrosinase [J]. J. Biosci. Bioeng. 2015, 119, 275 -279.

[31] Kim B H, Park K C, Park J H, et al. Inhibition of tyrosinase activity and

melanin production by the chalcone derivative 1 - (2 - cyclohexylmethoxy - 6 - hydroxy - phenyl) - 3 - (4 - hydroxymethyl - phenyl) - propenone [J]. Biochemical and Biophysical Research Communications, 2016,480,648 - 654.

[32] Nguyen N T, Nguyen M H K, Nguyen H X, et al. Tyrosinase Inhibitors from the Wood ofArtocarpus heterophyllus [J]. J. Nat. Prod., 2012, 75, 1951 - 1955.

[33] Badasaheb P B, Shrikant S G, Ragini G B, et al. Synthesis and biological evaluation of a novel series of pyrazole chalcones as anti - inflammatory, antioxidant and antimicrobial agents [J]. Bioorganic & Medicinal Chemistry, 2009, 17, 8168 - 8173.

[34] 庞得全,王英曼,郑维国,等. 抗黑色素瘤药物异甘草素对酪氨酸酶抑制机理的研究[J]. 中国生化药物杂志,2015,35,45 - 47.

[35] Radhakrishnan S, Shimmon R, Costa C, et al. Inhibitory kinetics of novel 2,3 - dihydro - 1H - inden - 1 - one chalcone - like derivatives on mushroom tyrosinase [J]. Bioorganic & Medicinal Chemistry Letters, 2015, 25, 5495 - 5499.

[36] 杨城,管骁,韩飞,等. 木犀草素抑制酪氨酸酶活性的分子机制[J]. 分析测试学报, 2015,34,532 - 538.

[37] 管骁,杨城,刘静,等. 槲皮素对酪氨酸酶的抑制作用与分子机理[J]. 现代食品科技,2015, 31,71 - 77.

[38] Nouha N B, Fadwa C, Aicha S, et al. Effect of apigenin - 7 - glucoside, genkwanin and naringenin on tyrosinase activity and melanin synthesis in B16F10 melanoma cells [J]. Life Sciences, 2016,144, 80 - 85.

[39] Tan X, Song Y H, Park C, et al. Highly potent tyrosinase inhibitor, neorauflavane from Campylotropis hirtella and inhibitory mechanism with molecular docking [J]. Bioorganic & Medicinal Chemistry, 2016,24, 153 - 159.

[40] Ozlem D, Temine S, Mehmet O, et al. Antioxidant and Tyrosinase Inhibitory Activities of Flavonoids from Trifolium nigrescens Subsp. Petrisavi [J]. J. Agric. Food Chem., 2013, 61, 12598 - 12603.

[41] 段玉清,周密,张海晖,等. 莲房原花青素对黑色素瘤 B16 细胞的抑制作用[J]. 中国药学杂志, 2009, 44, 103 - 106.

[42] 马晶波,冯树芳,李锋,等. 甘草黄酮对 B16 黑色素瘤细胞代谢的影响[J]. 复旦学报(医学版),2003,30,353 - 355.

[43] Nguyen N T, Nguyen M H K, Nguyen H X, et al. Tyrosinase inhibitors from the wood of artocarpus heterophyllus [J]. J. Nat. Prod., 2012, 75, 1951 – 1955.

[44] Namrita L, Elizabeth M, Marco N C, et al. Insights into tyrosinase inhibition by compounds isolated from Greyia radlkoferi Szyszyl using biological activity, molecular docking and gene expression analysis [J]. Bioorganic & Medicinal Chemistry, 2016,24, 5953 – 5959.

[45] Beata G B, Hubert W. Reactions of Flavonoids with oQuinones Interfere with the Spectrophotometric Assay of Tyrosinase Activity [J]. J. Agric. Food Chem., 2016, 64, 5417 – 5427.

[46]王锡元. 酪氨酸酶抑制剂和激活剂的研究[D].2007,江南大学硕士学位论文.

[47]江城梅,赵文红,孟 灿,等. 三羟异黄酮抑制黑色素瘤 B16 细胞增殖作用的机制研究[J]. 蚌埠医学院学报,2006, 31,121 – 123.

[48] Ley J P, Bertram H J. Hydroxy – or Methoxy – Substituted Benzaldoximes and Benzaldehyde – O – alkyloximes as Tyrosinase Inhibitors [J]. Bioorg. Med. Chem., 2001, 9, 1879 – 1885.

[49] Criton M, Mellay – Hamon V L. Analogues of N – hydroxy – N' – phenylthiourea and N – hydroxy – N' – phenylurea as inhibitors of tyrosinase and melanin formation [J]. Bioorg. Med. Chem. Lett., 2008, 18, 3607 – 3610

[50]Bae S J, Ha Y M, Park Y J, et al. Design, synthesis, and evaluation of (E) – N – substituted benzylideneeaniline derivatives as tyrosinase inhibitors [J]. European Journal of Medicinal Chemistry, 2012, 57, 383 – 390.

[51] Xie J, Dong H, Yu Y, et al. Inhibitory effect of synthetic aromatic heterocycle thiosemicarbazone derivatives on mushroom tyrosinase: Insights from fluorescence, 1H NMR titration and molecular docking studies [J]. Food Chemistry, 2016, 190, 709 – 716.

[52]Radhakrishnan, S K, Shimmon R G. Evaluation of Novel Chalcone Oximes as Inhibitors of Tyrosinase and Melanin Formation in B16 Cells [J]. Arch. Pharm. Chem. Life Sci., 2016, 349, 20 – 29.

[53]You A, Zhou J, Song S, et al. Rational design, synthesis and structure – activity relationships of 4 – alkoxy – and 4 – acyloxy –

phenylethylenethiosemicarbazone analogues as novel tyrosinase inhibitors [J]. Bioorganic & Medicinal Chemistry, 2015, 23, 924 -931.

[54] You A, Zhou J, Song S, et al. Structure - based modification of 3 - /4 - aminoacetophenones giving a profound change of activity on tyrosinase: From potent activators to highly efficient inhibitors [J]. European Journal of Medicinal Chemistry, 2015, 93, 255 -262.

[55] Zhu T H, Cao S W, Yu Y Y. Synthesis, characterization and biological evaluation of paeonol thiosemicarbazone analogues as mushroom tyrosinase inhibitors [J]. International Journal of Biological Macromolecules, 2013, 62, 589 - 595.

[56] 徐燚,王振,凌云,等. 取代苯丙醛缩氨基硫脲类化合物的合成及其对棉铃虫酪氨酸酶的生物活性研究[J]. Chin. J. Org. Chem., 2012,32, 1278 -1283.

[57] Larissa LL, Rebeca M L, Annelisa F S, et al Analogs as Tyrosinase Inhibitors: New Molecules with Depigmenting Potential [J]. The Scientific World Journal, Volume 2013, Article ID 274643, 7 pages http://dx.doi.org/10.1155/2013/274643.

[58] Yang M H, Chen C M, Hu Y H, et al. Inhibitory kinetics of DABT and DABPT as novel tyrosinase inhibitors [J]. Journal of Bioscience and Bioengineering, 2013, 115, 514 -517.

[59] Xu J, Liu J, Zhu X, et al. Novel inhibitors of tyrosinase produced by the 4 - substitution of TCT [J]. Food Chemistry, 2017, 221, 1530 - 1538.

[60] Radhakrishnan S K, Shimmon R G. Inhibitory Kinetics of Azachalcones and their Oximes on Mushroom Tyrosinase: A Facile Solid - state Synthesis [J]. Chem. Biodiversity, 2016, 13, 531 -538.

[61] Muhammad R, Muhammad S, Muhammad H, et al. Synthesis, structural elucidation and bioevaluation of 4 - amino - 1,2,4 - triazole - 3 - thione's Schiff base derivatives [J]. Arch. Pharm. Res., 2016, 39, 161 -171.

2.6 编者相关科研成果

2.6.1 博士论文相关内容

1.芳香酮氨基硫脲席夫碱酪氨酸酶抑制剂

早在1931年,Pfeifer等首次合成了席夫碱,直到20世纪60年代后特别是近年来,由于席夫碱的具有各种材料和化学性能,最重要的是由独特的抗菌[1-4]、抗癌[5]和抗病毒[6,7]等生理活性,引起了人们广泛、系统、深入的理论与应用研究。席夫碱可从含羰基的衍生物与含氨基的有机物通过Lewis酸催化脱水得到。下面就氨基脲、氨基硫脲、羟氨、甲基羟氨席夫碱的反应机理和合成方法作简要的介绍。

反应机理是:由含羰基的醛、酮类化合物与一级胺类化合物进行亲核加成反应,亲核试剂为胺类化合物,其化合物结构中带有孤电子对的氮原子进攻羰基基团上带有正电荷的碳原子,完成亲核加成反应,形成中间物α-羟基胺类化合物,然后进一步脱水形成席夫碱[8]。

$$>C=O + -NH_2 \rightleftharpoons >C(OH)-NH- \rightleftharpoons >C=N- + H_2O$$

2.氨基脲和氨基硫脲类席夫碱的合成

氨基脲[9-14]和氨基硫脲类席夫碱[15-20]的合成方法基本相同,一般都是在甲醇、乙醇、水、DMF,吡啶等极性溶剂或其混合溶剂中,在冰醋酸的催化下回流进行反应,这些反应的产率往往都很高。1976年Eugene A. Coats等人[21]以对位取代溴代苯乙酮为原料,合成了苯基α-醛酮,再和甲基取代的氨基硫脲反应得到双缩合物。

$$ArCHO + NH_2NHC(=R)NH_2 \xrightarrow{acid} ArCH=NNHC(=R)NH_2$$

R=O, S

3.羟胺和烷氧基胺类席夫碱的合成

羟胺或烷氧基胺与醛反应,一般是盐酸盐的形式参加反应,需要加入弱碱或者弱酸加速反应进程。

T. -S. Jeong 等[22]用不同的芳香醛酮与盐酸羟胺反应,加入等摩尔量三乙胺,在乙醇中室温下搅拌,产率在 60% ~95%;Jain, Nidhi 等[23]在吡啶存在下,在乙醇中回流 2h,产率可达 95%,Julia Kaffy 等[24]也进行了同样的反应,但产率只有 75%;Lei Fang 等[25]利用邻取代苯甲醛与盐酸羟胺反应,盐酸羟胺用 KOH 溶液中和,在 70°C 反应 2h,达到很好的产率; Peter Beak 等[26]在 1989 年利用苯甲醛和 NH_2OCH_3 - HCl 反应,回流过夜,产率达 71%。

$$ArCHO + RONH_2 \cdot HCl \xrightarrow{base} ArCH{=}NHOR$$

R=H, CH3

1971 年美国芝加哥的 Sou - yin chu 采用了将肟和硫酸二甲酯或硫酸二乙酯反应来合成烷氧基胺席夫碱[27]。

$$R^1C(=O)CH_2R^2 \longrightarrow R^1C(=N{-}OH)CH_2R^2 \longrightarrow R^1C(=N{-}O{-}R^3)CH_2R^2$$

2004 年美国加利福尼亚的 H. J. Peter de Lijser 采用了酮直接和烷氧基胺反应合成席夫碱[28]:将 33.3mmol 烷氧基胺盐酸盐和 16.7mmol 苯乙酮加入到反应瓶中,加入 50mL 95% 乙醇,搅拌,滴加几滴浓盐酸,回流反应 24 小时,减压蒸除溶剂,柱层析分离得到产品。

$$PhC(=O)CH_3 \xrightarrow{HCl} PhC(=N{-}O{-}R)CH_3$$

1989 年 Tetsuyoshi Hayashi 同样采用了醛酮和烷氧基胺盐酸盐反应得到所需的烷氧基胺席夫碱[29]:将 18mmol 烷氧基胺盐酸盐和 6mmol/L 盐酸 0.2mL 加入到含有 14mmol 酮的 70mL 乙醇中,于 50℃反应 12h,加入水稀释,苯萃取,水洗,无水硫酸镁干燥,收率 98%。

$$R^1C(=O)R^2 \xrightarrow[NH_2OR^3 \cdot HCl]{HCl} R^1C(=N{-}O{-}R^3)R^2$$

Chern 报道[30]了以醋酸钠为原料,水做溶剂,于 80℃取代苯乙酮可以烷氧基胺反应得到烷氧基胺席夫碱。

$$\text{PhCOCH}_3 \xrightarrow[\text{NH}_2\text{OR}]{\text{NaOAc}} \text{PhC(=N-O-R)CH}_3$$

1998 年吴晓艺以草酸二乙酯为原料[31]，合成了二乙氧基乙二肟中间体，再水解得到二羟基乙二肟。

$$NH_2OH + (COOC_2H_5)_2 \longrightarrow (NOHCOC_2H_5)_2 + H_2O$$

$$(NOHCOC_2H_5)_2 + H_2O \xrightarrow{(COOH)_2} (NOHCOH)_2$$

1974 年 Jerald K. Rasmussen 等[32]人以三甲基硅烯醚为原料，和亚硝酰氯反应合成 α－二羰基单肟化合物，所得的单肟化合物继续和羟胺反应生成了双肟产物，总收率 24.5%。

$$R^1C(OSiMe_3)=CHR^2 \xrightarrow{NOCl} R^1COC(=NOH)R^2 \xrightarrow{NH_2OH} R^1C(=NOH)C(=NOH)R^2$$

1971 年 Josephv. Burakevicha 以苯基 α－醛酮为原料[33]，合成单肟化合物，再继续和羟胺反应合成了双肟化合物。并且分离出来了正反异构体，以紫外光谱和核磁共振确证了其结构。

$$PhCOCHO \xrightarrow{NH_2OH} PhCOCH=NOH \xrightarrow{NH_2OH} PhC(=NOH)CH=NOH$$

1969 年 Davidt. Manningan 等人[34]以 3,3－二取代 2,4－戊二酮为原料和过量的羟胺反应生成双肟化合物。

$$CH_3COC(CH_3)_2COCH_3 \xrightarrow{\text{excess } NH_2OH} CH_3C(=N-OH)C(CH_3)_2C(=N-OH)CH_3$$

4.芳香酮氨基硫脲席夫碱的合成

H3-1 ~ H3-13

R =：H3－1，苯基；H3－2，4－甲基苯基；3，H3－3，4－羟基苯基；H3－4，2，4－二羟基苯基；H3－5，4－氟苯基；V3－6，4－溴苯基；　H3－7，4－异丙基苯基；H3－8，4－甲氧基苯基；H3－9，2－吡嗪基；H3－10，2－噻吩基；H3－11，3－吡啶基；H3－12，(4－甲氧基苯基)甲基；H3－13，2－(4－羟基苯基)乙基.

芳香酮氨基硫脲席夫碱的合成路线

仪器与试剂

对羟基苯乙酮(化学纯，上海达瑞化学品有限公司)；2，4－二羟基苯乙酮(化学纯，上海达瑞化学品有限公司)；2，4，6－三羟基苯乙酮(化学纯，上海达瑞化学品有限公司)，4－甲氧基苯丙酮(美国 ACROS 有机试剂公司)；4－(4－羟基苯基)－2－丁酮(化学纯，上海达瑞化学品有限公司)；2－乙酰基噻吩(化学纯，上海达瑞化学品有限公司)，2－乙酰基吡嗪(化学纯，上海达瑞化学品有限公司)，3－乙酰基吡啶(化学纯，上海达瑞化学品有限公司)，4－乙酰基苯乙酮(美国 ACROS 有机试剂公司)。

实验步骤

将 50mmol 芳香族醛酮加入到反应瓶中，加入 50mL 无水甲醇，搅拌下加入 50mmol(或 100mmol)氨基硫脲和催化量的冰乙酸，升温至回流，TLC 跟踪反应，反应完毕后，冷至室温，过滤，滤饼以 95% 乙醇重结晶，干燥，得产品。

化合物表征

苯乙酮氨基硫脲缩合物（H3－1）. Yield 86%；mp132 ~ 134℃；IR(KBr)：3379，3217，3144，2939，2858，1585，1509，1460，1431，1228，1074，836 cm^{-1}. ^{1}HNMR (300 MHz, d_6－DMSO)：δ 8.95(1H, bs, NH), 7.70(2H, d, J = 1.2Hz, phH), 7.48(1H, bs, NH_2), 7.41(3H, m, J = 1.2Hz, phH), 6.53(1H, bs, NH_2), 2.31(3H, s, CH_3). ESI－MS：m/z(100%) = 194 (M+1).

对甲苯乙酮氨基硫脲缩合物（H3－2）. Yield 78%；mp158 ~ 160℃；IR (KBr)：3380，3246，3153，2839，1890，1581，1514，1375，1251，1179，1081，

1030, 855 cm^{-1}. ^{1}HNMR (300 MHz, d_6 – DMSO): δ 8.77(1H, bs, NH), 7.61 (2H, d, J = 7.5Hz, phH), 7.36(1H, bs, NH_2), 7.21(2H, d, J = 7.5Hz, phH), 6.42(1H, bs, NH_2), 2.39(3H, s, $COCH_3$), 2.29(3H, s, ph – CH_3). ESI – MS: m/z(100%) = 208 (M + 1).

对羟基苯乙酮氨基硫脲缩合物 (H3 – 3). Yield 73%; mp208 ~ 209℃; IR (KBr): 3363, 3237, 1602, 1522, 1502, 1432, 1363, 1311, 1225, 1172, 1085, 825 cm^{-1}. ^{1}HNMR (300 MHz, d_6 – DMSO): δ 10.02(1H, bs, NH), 9.69(1H, s, OH), 8.13(1II, bs, NII_2), 7.77 (2H, d, J = 8.7Hz, phH), 7.72(1H, bs, NH_2), 6.75(2H, d, J = 8.7Hz, phH), 2.22(3H, s, CH_3). ESI – MS: m/z (100%) = 208(M – 1).

2,4 – 二羟基苯乙酮氨基硫脲缩合物 (H3 – 4). Yield 68%; mp186 ~ 187℃; IR(KBr): 3387, 3300, 1630, 1605, 1520, 1440, 1420, 1374, 1331, 1209, 1184, 1143, 839 cm^{-1}. ^{1}HNMR (300 MHz, d_6 – DMSO): δ 12.57(1H, s, OH), 10.57 (1H, s, OH), 9.72(1H, bs, NH), 7.74(1H, s, NH_2), 7.70(1H, s, NH_2), 7.36(H, d,J = 8.4Hz, phH), 6.37(1H, d,J = 8.4Hz, phH), 6.23(1H, s, phH), 2.25(3H, s, CH_3). ESI – MS: m/z(100%) = 224 (M – 1).

对氟苯乙酮氨基硫脲缩合物 (H3 – 5). Yield 82%, mp154 ~ 156℃; IR (KBr): 3416, 3210, 3162, 1598, 1503, 1461, 1431, 1286, 1212, 1082, 835 cm^{-1}. ^{1}HNMR (300 MHz, d_6 – DMSO): δ 10.17(1H, bs, NH), 8.24(1H, bs, NH_2), 7.98(2H, d, J = 6.9Hz, phH), 7.78(1H, bs, NH_2), 7.20(2H, d, J = 6.9Hz, phH), 2.28(3H, s, CH_3). ESI – MS: m/z(100%) = 212 (M + 1).

对溴苯乙酮氨基硫脲缩合物 (H3 – 6). Yield 79%, mp190 ~ 192℃; IR (KBr): 3410, 3234, 3141, 1588, 1505, 1483, 1460, 1301, 1082, 814 cm^{-1}. ^{1}HNMR (300 MHz, d_6 – DMSO): δ 10.08(1H, bs, NH), 8.28(1H, bs, NH_2), 7.97(1H, bs, NH_2), 7.89(2H, d, J = 7.8Hz, phH), 7.55(2H, d, J = 7.8Hz, phH), 2.27(3H, s, CH_3). ESI – MS: m/z(100%) = 273 (M + 1).

对异丙基氨基硫脲缩合物 (H3 – 7). Yield 68%, mp 99 ~ 100℃; IR(KBr): 3407, 3370, 3145, 2957, 1589, 1500, 1460, 1409, 1293, 1090, 832 cm^{-1}. ^{1}HNMR (300 MHz, $CDCl_3$): δ 8.77(1H, bs, NH), 7.64(2H, d, J = 8.1Hz, phH), 7.37(1H, bs, NH_2), 7.25(2H, d, J = 8.1Hz, phH), 6.47(1H, bs, NH_2), 2.94(m, 1H, CH), 2.29(3H, s, CH_3), 1.28(d, 6H, $2CH_3$). ESI –

MS: m/z(100%) = 236 (M+1).

对甲氧基苯乙酮氨基硫脲缩合物(H3-8). Yield 83%, mp 175~176℃; ^{1}HNMR (300 MHz, $CDCl_3$): δ 9.04(1H, bs, NH), 7.57(2H, d, J=8.7Hz, phH), 7.27(1H, bs, NH_2), 7.26(1H, bs, NH_2), 7.25(2H, d, J=8.7Hz, phH), 3.67(3H, s, OCH_3), 2.16(3H, s, CH_3). ESI-MS: m/z(100%) = 224 (M+1).

2-乙酰基吡嗪氨基硫脲缩合物(H3-9). Yield 67%, mp190~191℃; IR (KBr): 3369, 3249, 3170, 1618, 1501, 1468, 1406, 1363, 1287, 1243, 1174, 1111, 1013, 852 cm^{-1}. ^{1}HNMR (300 MHz, d_6-DMSO): δ 10.43 (1H, bs, NH), 9.62 (1H, s, pyrazin), 8.58(2H, d, pyrazin), 8.44(1H, bs, NH_2), 8.28(1H, bs, NH_2), 2.36(3H, s, CH_3). ESI-MS: m/z(100%) = 196 (M+1).

2-乙酰基噻吩氨基硫脲缩合物(H3-10). Yield 65%, mp146~147℃; IR (KBr): 3416, 3274, 3161, 1603, 1531, 1505, 1435, 1293, 1095, 833 cm^{-1}. ^{1}HNMR (300 MHz, d_6-DMSO): δ 8.86(1H, bs, NH), 7.35(1H, d, thiophen), 7.31(1H, d, thiophen), 7.28(1H, bs, NH_2), 7.04(1H, t, thiophen), 6.59 (1H, bs, NH_2), 2.31(3H, s, CH_3). ESI-MS: m/z(100%) = 200 (M+1).

3-乙酰基吡啶氨基硫脲缩合物(H3-11). Yield 70%, mp 212~214℃; IR (KBr): 3264, 3033, 1638, 1588, 1503, 1478, 1415, 1316, 1269, 1089, 1024, 853, 704 cm^{-1}. ^{1}HNMR (300 MHz, d_6-DMSO): δ 10.29 (1H, bs, NH), 9.08 (1H, s, pyridin), 8.54(1H, d, pyridin), 8.31(1H, d, pyridin), 8.29(1H, bs, NH_2), 8.05(1H, bs, NH_2), 7.40(1H, t, pyridin), ESI-MS: m/z(100%) = 195 (M+1).

对甲氧基苯丙酮氨基硫脲缩合物(H3-12). Yield 88%, mp: 122~123 ℃. R_f = 0.68 (ethyl acetate / petroleum ether (bp 60~90 ℃) = 2 : 1, v/v). ^{1}H NMR (300 MHz, $CDCl_3$): δ 8.50(H, bs, NH), 7.25(H, bs, NH), 7.09(2H, d, J= 8.7Hz, phH), 6.86(2H, d, J= 8.7Hz, phH), 6.34(H, bs, NH), 3.80 (3H, s, CH_3), 3.50(2H, s, CH_2), 1.81(3H, s, CH_3). ^{13}C NMR(75 MHz, $CDCl_3$): δ 180.2, 157.6, 155.7, 131.2, 130.5, 114.5, 55.7, 38.6, 14.5. IR (KBr): 3405, 3210, 3137, 3031, 2833, 1587, 1514, 859 cm^{-1}. MS(ESI): m/z (100%) = 238 (M+1). Anal. Calcd for $C_{11}H_{15}N_3OS$ (237.32): C, 55.67; H, 6.37; N, 17.71. Found: C, 55.61; H, 6.32; N, 17.76.

4-(4-羟基苯基)-2-丁酮氨基硫脲缩合物(H3-13). Yield 71%,

mp154 ~ 155℃ ;[1] R_f = 0. 69 (ethyl acetate / petroleum ether (bp 60 ~ 90 ℃) = 2 : 1, v/v). ^{1}H NMR (300 MHz, d_6 - DMSO): δ 8. 49 (1H, bs, NH), 7. 12 (1H, bs, NH), 7. 02 (2H, d, J = 6. 3Hz, phH), 6. 77 (2H, d, J = 6. 3Hz, phH), 6. 22 (1H, bs, NH), 2. 80 (2H, t, CH_2), 2. 57 (2H, t, CH_2), 1. 89 (3H, s, CH_3). ^{13}C NMR(75 MHz, d_6 - DMSO): δ 179. 1, 159. 3, 148. 9, 129. 1, 128. 8, 115. 7, 32. 5, 29. 3, 19. 8. IR (KBr): 3609, 3366, 3265, 3187, 1610, 1513, 826 cm^{-1}. MS(ESI): m/z(100%) = 236 (M - 1). Anal. Calcd for $C_{11}H_{15}N_3OS$ (237. 32): C, 55. 67; H, 6. 37; N, 17. 71. Found: C, 55. 73; H, 6. 35; N, 17. 79.

5.酪氨酸酶抑制活性测试

试验材料

实验药品

酪氨酸酶:Sigma 化学公司提供的蘑菇酪氨酸酶,比活力为5370units/mg;

L - 多巴:Sigma 化学公司提供

DMSO:分析纯,广州化学试剂厂

磷酸二氢钠:分析纯,广州化学试剂厂

磷酸氢二钠:分析纯,广州化学试剂厂

蒸馏水:双蒸水

溶液配制

(1)0. 1mol/L pH 值为 6. 8 磷酸盐缓冲溶液:

向 500mL 0. 1mol/L K_2HPO_4 中加入 0. 1M 的 KH_2PO_4水溶液混匀,用 pH 计调节溶夜 pH 值到 6. 8。

(2)酪氨酸酶溶液(现用现配):

称取 0. 5mg 的酪氨酸酶,用 0. 1mol/L pH 值为 6. 8 的磷酸盐缓冲溶液溶解,定容至 1mL。

(3)样品 DMSO 溶液:

称取一定量样品用 DMSO 溶剂溶解,定容浓度为 1. 0mmol/L。

(4)L - DOPA 溶液(现用现配):

称取 1. 5mg 的 L - DOPA,用 pH 6. 8 的磷酸盐缓冲液溶解,定容至 1mL。

主要仪器

UV - 2501PC 型紫外测试仪(Shimadzsu)

实验方法

在1.5mL的离心管中加入955μl pH=6.8的磷酸盐缓冲溶液（0.1mol/L）和一定体积的样品溶液（DMSO溶解），用DMSO补齐至975μl，然后加入5μl酪氨酸酶溶液（0.55mg/mL）于25℃孵化10分钟，将20μl L-DOPA溶液（2.58mg/mL）加入到上述混合液中，迅速混合均匀，测定波长为475 nm处的吸光度，由其随时间增长而上升的斜率计算出反应速率（消光系数 $\varepsilon = 3600M^{-1}cm^{-1}$）[51]。由于酶的活力会随着放置时间增长而下降，因此测量时可适当调整酶的浓度（底物浓度固定为0.26mmol/L），使吸光度的变化保持在每分钟改变0.1个单位。测定时温度恒定为25℃。用下式计算酪氨酸酶活性的抑制率，并依据对数浓度-酶抑制率回归方程求得半数作用浓度（IC_{50}）。

$$酪氨酸酶抑制率(\%) = [B - S/B] \times 100$$

其中B为有底物和酪氨酸酶，没加样品时的吸光度，S为有底物、酪氨酸酶和样品时的吸光度。以酪氨酸酶活力对样品浓度作图，按照作图法求出每个样品的IC_{50}值。

实验结果及作用规律

芳香族醛酮氨基硫脲席夫碱对蘑菇酪氨酸酶二酚酶的抑制活性以IC_{50}表示，其IC_{50}值结果总结于下表，以熊果苷和对甲氧基肉桂酸为对照物。

表　芳香族氨基硫脲席夫碱酪氨酸酶抑制活性

Table1　Inhibitory activity of 1-(1-arylethylidene) thiosemicarbazide against tyrosinase

Compounds	*H3-1*	*H3-2*	*H3-3*	*H3-4*	*H3-5*	*H3-6*	*H3-7*	H3-8
IC_{50} (μM)	0.34	0.27	0.31	0.58	0.17	0.52	1.0	0.11
Compounds	H3-9	H3-10	H3-11	H3-12	H3-13			
IC_{50} (μM)	0.88	0.14	0.82	0.42	0.54			
Compounds	4-甲氧基肉桂酸		熊果苷					
IC_{50} (mM)	0.41[a]		10.4[b]					

注：a. 文献报道4-甲氧基肉桂酸对酪氨酸酶的抑制活性为0.34-0.43mmol/L[23]；

b. 文献报道熊果苷对酪氨酸酶的抑制活性为30mmol/L，本文测得结果为当酪氨酸酶抑制了30%时的浓度[52]。

从表中可以发现以下规律：

（1）所有这类化合物对酪氨酸酶具有明显的抑制作用，苯环取代基对取代苯基酮氨基硫脲缩合物的酪氨酸酶抑制活性有明显的影响；

（2）衍生物对酪氨酸酶的抑制活性与酚羟基的位置有关，而与酚羟基的数目关系不大；

(3)当苯环对位为烷基取代时,直链烷基取代苯环对酪氨酸酶的抑制活性强于苯环未取代的衍生物,而支链取代取代苯环的抑制活性弱于苯环未取代的衍生物,这可能与取代基的位阻有关;

(4)苯环为卤素取代时,随着原子序数的增加,化合物的活性降低;

(5)含硫原子杂环化合物对酪氨酸酶的抑制活性明显强于其他杂环化合物。硫原子是这类衍生物的活性中心,因为硫原子可以与铜离子形成络合物,从而使酪氨酸酶失去活性;

(6)电子效应同样是影响化合物酪氨酸酶抑制活性重要因素,吸电子基团的存在,会使化合物的抑制活性明显降低;

参考文献

[1] Seshaiah Krishnan Sridhar. Synthesis and antibacterial screening of hydrazones, Schiff and Mannich bases of isatin derivatives. Eur. J. Med. Chem, 2001, 36(7): 615 -625.

[2] Alaaddin Cukurovali. Synthesis, antibacterial and antifungal activity of some new thiazolylhydrazone derivatives containing 3 - substituted cyclobutane ring. European Journal of Medicinal Chemistry, 2006, 41(2): 201 -207.

[3] L - X. Zhang. Inhibitory study of some novel Schiff base derivatives on Staphylococcus aureus by microcalorimetry. Thermochimica. Acta, 2006, 440(1): 51 - 56.

[4]谢复新,刘远芳,董华泽,等. 一个新席夫碱的合成、晶体结构及抑菌活性中国药物化学杂志. 2002, 12(5): 279 -280.

[5] Hodnett M E ,Dunn J W. Structure - antitumor activity correlation of some Schiff bases. J. Med. Chem ,1970, 13(4): 768 -770.

[6] Paola Vicini. Synthesis and Biological Evaluation of Benzoisothiazole, Benzothiazole and Thiazole Schiff Bases. Bioorganic & Medicinal Chemistry, 2003, 11(21): 4785 - 4789.

[7] Arima Das. Inhibition of herpes simplex virus type 1 and adenovirus type 5 by heterocyclic Schiff bases of aminohydroxyguanidine tosylate. Antiviral Research, 1999, 44(2): 201 - 208.

[8]南光明,刘德蓉. 浅述席夫碱及其金属配合物的由来、产生机理、合成方法及展望. 伊犁师范学院学报, 2005, 3: 58 -59.

[9] Euikyung Kim, Hyung - Ok Chun. Chemical control on the coordination

mode of benzaldehyde semicarbazone ligands synthesis, structure, and redox properties of Ruthenium complexes. Bioorganic & Medicinal Chemistry Letters, 2003, 13(14): 2355 - 2358.

[10] Tenorio, Romulo P.; Carvalho, Cristiane S. Synthesis of thiosemicarbazone and 4 - thiazolidinone derivatives and their in vitro anti - Toxoplasma gondii activity. Bioorganic&Medicinal Chemistry Letters, 2005, 15 (10): 2575 - 2578.

[11] Kostas, Ioannis D.; Andreadaki, Fotini J. Suzuki - Miyaura cross - coupling reaction of aryl bromides and chlorides with phenylboronic acid under aerobic conditions catalyzed by palladium complexes with thiosemicarbazone ligands. Tetrahedron Letters, 2005, 46(12): 1967 - 1970.

[12] Yilmaz, Ibrahim; Cukurovali, Alaaddin. Salicylaldehyde thiazolyl hydrazones as ligands. Heteroatom Chemistry, 2003, 14(7): 617 - 621.

[13]陆绍荣. 含硫席夫碱的合成研究. 合成化学, 2003, 11(4): 349 - 351.

[14] 高生华,赵晓霞. 茴香醛苯基缩氨基硫脲金属配合物的合成. 内蒙古石油化工,1998, 24(4): 78 - 79.

[15] Basuli, Falguni; Peng, Shie - Ming. Chemical Control on the Coordination Mode of Benzaldehyde Semicarbazone Ligands. Synthesis, Structure, and Redox Properties of Ruthenium Complexes. Inorganic Chemistry, 2001, 40 (6): 1126 - 1133.

[16] Hajipour, Abdol Reza; Mohammadpoor - Baltork, Iraj; Bigdeli, Mansour. A convenient and mild procedure for the synthesis of hydrazones and semicarbazones from aldehydes or ketones under solvent - free conditions. Journal of Chemical Research, Synopses, 1999, 9: 570 - 571.

[17] Yogeeswari, Perumal; Thirumurugan, athinasabapathy. 3 - Chloro - 2 - methylphenyl - substituted semicarbazones: synthesis and anticonvulsant activity. European Journal of Medicinal Chemistry, 2004, 39(8): 729 - 734.

[18] Surendra N. Pandeya, Perumal Yogeeswari. Synthesis and anticonvulsant activity of 4 - bromophenyl - substituted aryl semicarbazones. European Journal of Medicinal Chemistry, 2000, 35(10): 879 - 886.

[19] J. R Dimmock, K. K Sidhul. Some aryl semicarbazones possessing anticonvulsant activities. Eur. J. Med. Chem, 1995, 30(4): 287 - 301.

[20]李冬青,周健. Schiff 碱吡啶－2－甲醛缩氨基脲铅(Ⅱ)配合物的合成和晶体结构. 化学研究与应用, 2006, 18(7):207－210.

[21] Eugene A. Coats, Stanley R. Milstein, Gary Holbein, John Mc Donald, Ronald Reed, Harold G. Petering. Comparative analysis of the cytotoxicity of substituted [phenylglyoxal bis(4－methyl－3－thiosemicarbazone)] copper(II) helates. Journal of Medicinal Chemistry, 1976, 19(1): 131－136.

[22] T.－S. Jeong. (E)－Phenyl－ and －heteroaryl－substituted O－benzoyl－(oracyl) oximes as lipoprotein－associated phospholipase A2 inhibitors. Bioorg. Med. Chem. Lett, 2005, 15(5): 1525－1527.

[23] Jain, Nidhi; Kumar, Anil; Chauhan, Shive M. S. Metalloporphyrin and heteropoly acid catalyzed oxidation of CNOH bonds in an ionic liquid: biomimetic models of nitric oxide synthase. Tetrahedron Letters, 2005, 46(15): 2599－2602.

[24] Julia Kaffy, Claude Monneret. 1,3－Dipolar cycloaddition route to novel isoxazole－type derivatives related to combretastatin A－4. Tetrahedron Letters, 2004, 45(17):3359－3362.

[25] Fang, Lei; Chan, Wing－Hong; He, Yong－Bing. Selective complexation of metals with isoxazolidine－containing fluorophores. Tetrahedron Letters, 2004, 46(1): 173－176.

[26] Beak, Peter; Selling, Gordon W. Displacements at the nitrogen of lithioalkoxyl amides by organometallic reagents. Journal of Organic Chemistry, 1989, 54(23): 5574－80.

[27] Sou－Yie, Chu; Dominick A, Coviello. Preparation of 2－alkoxyiminoalkyl bromides by the bromination of o－alkyl oximes with N－bromosuccinimide. J. Org. Chem., 1971, 36(22): 3467－3469.

[28] H. J. Peter de Lijser.; Chao－Kuan Tsai. Photosensitized reactions of oxime Ethers: A steady－state and laser flash photolysis study. J. Org. Chem, 2004, 69(9): 3057－3067.

[29] Tetsuyoshi Hayashi, Hajime Iwamura. Development of (4－alkoxyphenoxy)－and (4－alkylphenoxy) alkanaldoxime O－ethers as potent insect juvenile hormone mimics and their structure－activity relationships. J. Agric. Food. Chem, 1989, 37(2):467－472.

[30]Jyh－Haur Chern, Chung－Chi Lee, Chih－Shiang Chang, Yen－Chun

Lee, Chia - Liang Tai, Ying - Ting Lin. Synthesis and antienteroviral activity of a series of novel, oxime ether - containing pyridyl imidazolidinones. Bioorganic & Medicinal Chemistry Letters, 2004,14(20): 5051 - 5056.

[31]吴晓艺,王亚军. 二羟基乙二肟的合成. 内蒙古石油化工. 1998, 24(1): 20 -23.

[32] Jerald K. Rasmussen ; Alfred Hassner. Synthetic methods. VI. Addition of nitrosyl chloride to trimethylsilyl enol ethers. New general method for nitrosation of carbonyl compounds. J. Org. Chem., 1974, 39(17): 2558 -2562.

[33] Joseph V. Burkevich, Anthony M. Lore, Gert P. Volpp. Phenylglyoxime. Separation, characterization, and structure of three isomers. J. Org. Chem, 1971, 36 (1): 1 -4.

[34] David T. Manning, Harold A. Coleman. Reaction of hydroxylamine with 3,3 - disubstituted 2,4 - pentanediones. Formation of novel isoxazole derivatives. J. Org. Chem, 1969, 34(11): 3248 -3253.

2.6.2　化学通报论文(2017,80(1):77 - 83)

卤代查尔酮氨基硫脲席夫碱的合成及生物活性研究

刘进兵* 唐君源

(邵阳学院,湖南 邵阳 422000)

摘要:本文合成了一系列卤代查尔酮及其氨基硫脲席夫碱,并考察了所合成化合物酪氨酸酶抑制活性及抗氧化活性。结果表明:部分化合物具有较好的酪氨酸酶抑制活性,其中 3 个化合物的抑制活性强于阳性对照曲酸,化合物 9(4 -溴代查尔酮氨基硫脲席夫碱)表现出最好的抑制活性,其 IC_{50} 值为 2. 02 μmol/L;部分化合物还具有一定的抗氧化活性。抑制机理研究表明化合物 9 属于不可逆抑制剂。优选化合物 9 进行了分子对接探讨。初步构效关系分析为进一步研究具有酪氨酸酶抑制活性的类似化合物提供了参考。

关键词:查尔酮衍生物;生物活性;酪氨酸酶抑制机理;分子对接

基金项目:湖南省教育厅重点项目(15A172)

作者简介:刘进兵(1974 -),男,副教授,博士,E - mail:syuliujb@163. com.

Bioactivities of chalcones withhalogen atoms and their thiosemicarbazide derivatives

Jinbing Liu*, Junyuan Tang

Department of Biology and Chemical Engineering, Shaoyang University, Shao Shui Xi Road, Shaoyang 422100, *PRC*

Abstract

A series ofhalogen atom substituted chalcones and their derivatives were synthesized. All of the compounds were evaluated for their antioxidant activities and inhibitory activities against mushroom tyrosinase. The results showed that some of the synthesized compounds exhibited significant tyrosinase inhibitory activities, three compounds exhibited more potent inhibitory activities than the reference standard inhibitor. Especially, compound 9 exhibited the most potent tyrosinase inhibitory activity with an IC_{50} value of 2. 02 μmol/L. In addition, some compounds showed certain antioxidant activities. The inhibition mechanism analysis of compound 9 showed that the inhibitory effect of the compound on the tyrosinase was irreversible. Preliminary structure activity relationships' (SARs) analysis suggested that further development of such compounds might be of interest. Docking studies showed sulfur atom and phenyl ring B could form π bond with residue of the tyrosinase, and the amino could form hydrogen bond with HIS178.

Keywords: chalcone derivatives; bioactivity; tyrosinase inhibition mechanism; docking studies.

Foundation item: Education Department of Hunan Province, China (15A172).

酪氨酸酶(EC 1. 14. 18. 1)是一种多酚氧化酶,在其分子结构中存在两个铜离子活性中心,广泛分布于微生物、动物、植物[1,2]。酪氨酸酶是黑色素生物合成的关键酶,影响着人和哺乳动物的毛发、皮肤、眼睛的颜色[3]。酪氨酸酶的过量表达会造成黑色素堆积,引起色素沉着性疾病,如雀斑、黄褐斑、老年斑以及恶性黑色素瘤等。最近有报道酪氨酸酶与帕金森病及其他神经衰弱症也有重要的关联,通过调节酪氨酸酶的催化活性可以促进或抑制其表达,达到治疗相关神经衰弱疾病的效果[4]。研究表明,酪氨酸酶还和果蔬褐变、昆虫蜕皮及伤口愈合有着重要的联系,酪氨酸酶抑制剂可以作为食品保鲜剂来延长食品货架期,同时也逐渐成为了研究生物杀虫剂的一个方向[5,6]。鉴于酪氨酸酶抑制剂在医药、农业、食品工业及化妆品领域有重要应用,已经引起了相关领域研究者的广泛关注,然而由于抑制活性不理想及毒副作用的原因，到目前为止，只有曲酸、熊果苷、苯基硫脲等少数几种酪氨酸酶抑制剂得到了应用，因此开发高效低毒的酪氨酸酶抑制剂已成为市场需求。

查尔酮类化合物具有 $\alpha-\beta$-不饱和酮结构，是一类重要的天然来源的化合物，具有抗菌、抗氧化、抗病毒、抗惊厥、抗肿瘤等生物活性，目前认为双键与羰基构成的共轭体系是查尔酮具有多种生物活性的主要原因[7]。最近有报道从天然产物中提取或者化学合成的查尔酮化合物具有酪氨酸酶抑制活性[8]。最近报道了一些醛酮氨基硫脲席夫碱、香豆素类氨基硫脲席夫碱具有酪氨酸酶抑制活性[9,10]。为了获得高效低毒的酪氨酸酶抑制剂，采用活性片段组合方式，在已有的研究基础上，本文合成了一类卤代查尔酮氨基硫脲席夫碱，并考察了其酪氨酸酶抑制活性和抗氧化活性，初步讨论构效关系及抑制机理，为进一步研究查尔酮席夫碱类酪氨酸酶抑制剂提供参考。

1.实验部分

1.1　试剂与仪器

醛、酮购置于上海达瑞试剂有限公司，酪氨酸酶、*L*-多巴，曲酸从 Sigma - Aldrich 公司购得，其他试剂及溶剂为市售分析纯，未进一步纯化。熔点仪为 SGW® X-4（上海精密科学仪器有限公司），未校正；核磁共振仪为 Bruker 400 or 300 型（德国 Bruker 公司）；质谱仪为 LCMS - 2010A 型（日本岛津）；红外为 VECTOR 22 型（德国 Bruker 公司）；元素分析仪为 Vario EL（德国 Elementar 公司）紫外分光光度计为 UV-2100（北京莱伯泰科仪器有限公司）。

1.2　化合物合成[11]

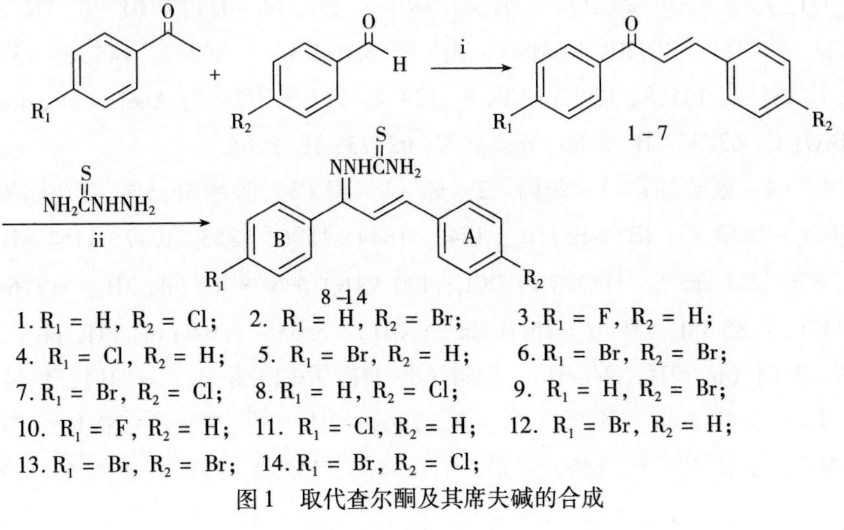

1. R_1 = H, R_2 = Cl;　2. R_1 = H, R_2 = Br;　3. R_1 = F, R_2 = H;
4. R_1 = Cl, R_2 = H;　5. R_1 = Br, R_2 = H;　6. R_1 = Br, R_2 = Br;
7. R_1 = Br, R_2 = Cl;　8. R_1 = H, R_2 = Cl;　9. R_1 = H, R_2 = Br;
10. R_1 = F, R_2 = H;　11. R_1 = Cl, R_2 = H;　12. R_1 = Br, R_2 = H;
13. R_1 = Br, R_2 = Br;　14. R_1 = Br, R_2 = Cl;

图1　取代查尔酮及其席夫碱的合成

反应条件:i. NaOH/EtOH, RT; ii. EtOH, acetic acid, reflux.

Scheme 1. Synthesis of the substituted chalcones and their thiosemicarbazide Schiff bases.

1.2.1 卤代查尔酮的合成

分别将5 mmol芳基取代甲基酮和5 mmol芳香族的醛加入到50 mL反应瓶中,加入20 mL无水乙醇,搅拌溶解,缓慢加入1 mL 30%氢氧化钠溶液。于室温下搅拌反应,TLC(乙酸乙酯/石油醚 = 1/8)跟踪反应。反应完成后,以10%的稀盐酸中和至沉淀产生,经抽滤,水洗得粗产品。粗品以95%乙醇重结晶得查尔酮。

3-(4-氯苯基)-1-苯丙-2-烯-1-酮(1),收率89.2%,黄色固体,熔点119.8~121.3 ℃; IR(KBr) ν_{max} 1653, 1602, 1488, 1443, 1403, 1331, 1219, 1090, 1014, 986, 823, 775 cm^{-1}; ^{1}HNMR ($CDCl_3$, 300 MHz) δ = 7.89 (d, 1H, J=12.9 Hz, =CH), 7.63 (d, 2H, J=8.4 Hz, ph-H), 7.60 (d, 2H, J=6.9 Hz, ph-H), 7.39 (d, 2H, J=7.2 Hz, ph-H), 7.31-7.22 (m, 3H, ph-H, =CH); ^{13}CNMR ($CDCl_3$, 75 MHz) δ = 189.1, 145.6, 138.7, 134.7, 133.8, 133.2, 129.6, 129.1, 128.7, 128.3, 121.1; MS (ESI): m/z(100%) 243 (M^+); Anal. Calcd for $C_{15}H_{11}ClO$: C, 74.23; H, 4.57; found: C, 74.21; H, 4.53.

3-(4-溴苯基)-1-苯丙-2-烯-1-酮(2),收率85.7%,黄色固体,熔点190.5~192.8 ℃; IR(KBr) ν_{max} 1656, 1606, 1482, 1446, 1395, 1328, 1222, 1070, 1031, 975, 821, 770 cm^{-1}; ^{1}H NMR (300 MHz, DMSO-d_6): δ =7.94 (d, 1H, J=8.6Hz, =CH), 7.91-7.34 (m, 9H, Ph-H), 7.01 (d, 1H, J=8.4Hz, =CH); ^{13}CNMR (DMSO-d_6, 75 MHz) δ = 189.7, 145.6, 137.8, 134.7, 134.2, 131.8, 129.3, 128.9, 128.4, 122.5, 121.5; Anal. Calcd for $C_{15}H_{11}BrO$: C, 62.74; H, 3.86; found: C, 62.73; H, 3.85.

1-(4-氟苯基)-3-苯丙-2-烯-1-酮(3),收率56.5%,黄色固体,熔点76.2~78.3 ℃; IR(KBr) ν_{max} 1742, 1644, 1536, 1333, 1217, 1151, 1035, 981, 836, 761 cm^{-1}; ^{1}HNMR ($CDCl_3$, 400 MHz) δ = 8.10 (d, 2H,J =7.6 Hz, ph-H), 7.85 (d, 1H, J=15.6 Hz, =CH), 7.65-7.62 (m, 2H, ph-H), 7.66-7.64 (m, 2H, ph-H), 7.54 (d, 1H, J=15.6 Hz, =CH), 7.44(d, 2H, J=7.6 Hz, ph-H), 7.21-7.17(m, 4H, ph-H); ^{13}CNMR ($CDCl_3$, 100 MHz) δ = 188.9, 145.1, 134.8, 134.6, 131.2, 131.0, 130.6, 129.0,

128.5, 121.6, 115.8, 115.6; MS (ESI): m/z(100%) 227 (M^+); Anal. Calcd for $C_{15}H_{11}FO$: C, 79.63; H, 4.90; found: C, 79.62; H, 4.87.

1-(4-氯苯基)-3-苯丙-2-烯-1-酮 (4),收率50.9%,黄色固体,熔点97.1~98.3 ℃; IR(KBr) ν_{max} 1736, 1653, 1549, 1333, 1217, 1151, 1084, 1030, 977, 827, 761 cm^{-1}; ^{1}HNMR (DMSO-d_6, 300 MHz) δ = 7.98 (d, 2H, J =8.4 Hz, ph-H), 7.84 (d, 1H, J = 15.6 Hz, =CH), 7.66-7.63 (m, 2H, ph-H), 7.50 (d, 1H, J = 15.6 Hz, =CH), 7.49 (d, 2H, J = 8.4 Hz, ph-H), 7.43-7.42 (m, 3H, ph-H); ^{13}CNMR (DMSO-d_6, 75 MHz) δ = 189.2, 145.3, 139.2, 136.5, 134.7, 130.7, 129.9, 129.1, 128.9, 128.5, 121.5; MS (ESI): m/z(100%) 243 (M^+); Anal. Calcd for $C_{15}H_{11}ClO$: C, 74.23; H, 4.57; found: C, 74.24; H, 4.55.

1-(4-溴苯基)-3-苯丙-2-烯-1-酮 (5),收率75.4%,黄色固体,熔点100.5~102.7 ℃; IR(KBr) ν_{max} 1736, 1648, 1557, 1391, 1337, 1213, 1155, 1072, 1030, 977, 827, 761 cm^{-1}; ^{1}HNMR (CDCl$_3$, 400 MHz) δ = 7.89 (d, 2H, J = 8.8 Hz, ph-H),7.83 (d, 1H, J = 15.6 Hz, =CH), 7.65-7.63 (m, 4H, ph-H), 7.49 (d, 1H,J =15.6 Hz, =CH), 7.43-7.41 (m, 3H, ph-H);^{13}CNMR (CDCl$_3$, 100 MHz) δ = 189.3, 145.4, 136.9, 134.6, 131.9, 130.7, 130.0, 129.0, 128.5, 127.9, 121.4; Anal. Calcd for $C_{15}H_{11}BrO$: C, 62.74; H, 3.86; found: C, 62.75; H, 3.83.

1,3-双(4-溴苯基)丙-2-烯-1-酮 (6),收率74.6%,黄色固体,熔点190.5~192.8 ℃; IR(KBr) ν_{max} 1653, 1600, 1479, 1395, 1325, 1216, 1070, 1030, 1006, 983, 812, 739 cm^{-1}; ^{1}HNMR (DMSO-d_6, 300 MHz) δ = 7.92 (d, 2H, J = 8.8 Hz, ph-H), 7.94 (d, 1H, J = 15.6 Hz, =CH), 7.69 (d, 2H, J = 9.2 Hz, ph-H), 7.60-7.46 (m, 5H, ph-H, =CH);^{13}CNMR (DMSO-d_6, 75 MHz) δ = 189.1, 144.0, 136.7, 133.6, 132.3, 132.0, 130.0, 129.9, 128.1, 125.1, 121.9; MS (ESI): m/z(100%) 367 (M^+); Anal. Calcd for $C_{15}H_{10}Br_2O$: C, 49.22; H, 2.75; found: C, 49.21; H, 2.74.

1-(4-溴苯基)-3-(4-氯苯基)丙-2-烯-1-酮 (7),收率67.8%,黄色固体,熔点89.5~90.2 ℃; IR(KBr) ν_{max} 1648, 1488, 1401, 1336, 1213, 1098, 1030, 980, 812, 739 cm^{-1}; ^{1}HNMR (DMSO-d_6, 300 MHz) δ = 7.84-7.79(m, 3H, ph-H, =CH), 7.74 (d, 2H, J = 8.7 Hz, ph-H), 7.50-7.44

(m, 4H, ph - H); ^{13}CNMR (DMSO - d_6, 75 MHz) δ = 189.5, 146.3, 136.1, 134.6, 133.8, 132.0, 131.2, 130.9, 128.8, 127.6, 122.6, 119.5; Anal. Calcd for $C_{15}H_{10}BrClO$: C, 56.02; H, 3.13; found: C, 56.01; H, 3.15.

1.2.2 卤代查尔酮氨基硫脲席夫碱的合成

将5 mmol卤代查尔酮和5 mmol氨基硫脲加入到50 mL反应瓶中,加入20 mL无水乙醇,搅拌下加入催化量醋酸。于80 °C回流反应8~24 h,TLC(乙酸乙酯/石油醚 = 1/6)跟踪反应。反应完成后将反应体系冷至室温,有沉淀产生,抽滤,滤饼以5 mL乙醇洗涤。粗产品经无水以乙醇重结晶得卤代查尔酮氨基硫脲席夫碱。

3-(4-氯苯基)-1-苯丙-2-烯-1-酮氨基硫脲席夫碱(8),收率54.3%,黄色固体,熔点131.6~132.8 ℃; IR(KBr) ν_{max} 1633, 1601, 1552, 1487, 1403, 1325, 1224, 1161, 1089, 1013, 859, 819, 736 cm^{-1}; ^{1}HNMR (DMSO - d_6, 300 MHz) δ = 11.41(s, 1H, NH), 7.84 - 7.76(m, 5H, ph - H, NH_2, =CH), 7.65 (d, 2H, J = 8.7 Hz, ph - H), 7.50 - 7.40 (m, 4H, ph - H), 6.80(d, 1H, J = 16.2 Hz, =CH); ^{13}CNMR (DMSO - d_6, 75 MHz) δ = 178.8, 147.7, 138.0, 136.7, 134.8, 133.7, 131.9, 130.6, 129.8, 128.9, 122.6, 119.4; MS (ESI): m/z(100%) 316 (M^+); Anal. Calcd for $C_{16}H_{14}ClN_3S$: C, 60.85; H, 4.47; N, 13.31; found: C, 60.83; H, 4.46; N, 13.32.

3-(4-溴苯基)-1-苯丙-2-烯-1-酮氨基硫脲席夫碱(9),收率60.8%,黄色固体,熔点137.5~140.1 ℃; IR(KBr) ν_{max} 1708, 1599, 1488, 1398, 1322, 1246, 1120, 1069, 1005, 870, 814, 713 cm^{-1}; ^{1}HNMR (DMSO - d_6, 300 MHz) δ = 11.14 (s, 1H, NH), 7.86(d, 1H, J = 16.2 Hz, =CH), 7.72 (d, 2H, J = 8.7 Hz, ph - H), 7.64 - 7.59 (m, 5H, ph - H, NH_2), 7.46 - 7.44 (m, 4H, ph - H), 6.78(d, 1H, J = 16.2 Hz, =CH); ^{13}CNMR (DMSO - d_6, 75MHz) δ = 178.8, 147.7, 138.0, 136.7, 134.8, 133.7, 131.9, 130.6, 129.8, 128.9, 122.6, 119.4; Anal. Calcd for $C_{16}H_{14}BrN_3S$: C, 53.34; H, 3.92; N, 11.66; found: C, 53.35; H, 3.95; N, 11.63.

1-(4-氟苯基)-3-苯丙-2-烯-1-酮氨基硫脲席夫碱(10),收率32.9%,黄色固体,熔点168.9~171.0 ℃; IR(KBr) ν_{max} 1603, 1487, 1333, 1279, 1221, 1159, 1122, 1075, 964, 873, 761 cm^{-1}; ^{1}HNMR (DMSO - d_6, 400 MHz) δ = 11.09 (s, 1H, NH), 8.86 (s, 1H, NH), 8.32 (s, 1H, NH),

7.82 (d, 1H, J=15.6 Hz, =CH), 7.75 (d, 2H,J =7.5 Hz, ph－H), 7.71－7.68 (m, 2H, ph－H), 7.49－7.47 (m, 2H, ph－H), 7.47 (d, 1H, J=15.6 Hz, =CH), 7.39－7.25 (m, 3H, ph－H); ^{13}CNMR (DMSO－d_6, 100 MHz) δ = 196.4, 179.4, 147.6, 140.1, 136.3, 131.9, 129.8, 129.1, 128.5, 121.6, 119.2, 115.7; MS (ESI): m/z(100%) 300 (M^+); Anal. Calcd for $C_{16}H_{14}FN_3S$: C, 64.19; H, 4.71; N, 14.04; found: C, 64.16; H, 4.73; N, 14.02.

1－(4－氯苯基)－3－苯丙－2－烯－1－酮氨基硫脲席夫碱(11),收率56.4%, 黄色固体, 熔点 128.6～130.1 ℃; IR(KBr) ν_{max} 1601, 1483, 1328, 1279, 1221, 1159, 1125, 1075, 873, 761 cm^{-1}; ^{1}HNMR (DMSO－d_6, 400 MHz) δ = 11.09 (s, 1H, NH), 7.82 (d, 1H, J=15.6 Hz, =CH), 7.75 (d, 2H,J = 7.5 Hz, ph－H), 7.71－7.68 (m, 4H, NH_2, ph－H), 7.49－7.47 (m, 2H, ph－H), 7.47 (d, 1H, J = 15.6 Hz, =CH), 7.39－7.25 (m, 3H, ph－H); ^{13}CNMR (100 MHz, DMSO－d_6) δ = 196.4, 179.4, 147.6, 140.1, 136.3, 131.9, 129.8, 129.1, 128.5, 121.6, 119.2, 115.7; MS (ESI): m/z(100%) 317 (M^+); Anal. Calcd for $C_{16}H_{14}ClN_3S$: C, 60.85; H, 4.47; N, 13.31; found: C, 60.83; H, 4.48; N, 13.30.

1－(4－溴苯基)－3－苯丙－2－烯－1－酮氨基硫脲席夫碱 (12),收率52.5%, 黄色固体, 熔点 159.2～161.0 ℃; IR(KBr) ν_{max} 1596, 1473, 1398, 1324, 1277, 1218, 1125, 1089, 1008, 867, 822, 716 cm^{-1}; ^{1}HNMR (DMSO－d_6, 300 MHz) δ = 11.11 (s, 1H, NH), 7.87－7.73(m, 5H, ph－H, NH_2, =CH), 7.67－7.60 (m, 2H, ph－H), 7.44－7.37 (m, 5H, ph－H), 6.78(d, 1H,J =16.2 Hz, =CH); ^{13}CNMR (DMSO－d_6, 75MHz) δ = 178.9, 146.4, 139.5, 135.7, 131.8, 130.2, 129.3, 128.9, 128.0, 126.9, 121.8, 118.5; Anal. Calcd for $C_{16}H_{14}BrN_3S$: C, 53.34; H, 3.92; N, 11.66; found: C, 53.36; H, 3.91; N, 11.65.

1,3－双(4－溴苯基)丙－2－烯－1－酮氨基硫脲席夫碱 (13),收率57.9%, 黄色固体, 熔点 101.2～103.7 ℃; IR(KBr) ν_{max} 1726, 1585, 1476, 1392, 1316, 1237, 1120, 1069, 1005, 870, 805, 735 cm^{-1}; ^{1}HNMR (DMSO－d_6, 300 MHz) δ = 11.18 (s, 1H, NH), 7.84(d, 1H, J=15.9 Hz, =CH), 7.30 (d, 2H, J=8.7 Hz, ph－H), 7.65－7.58 (m, 8H, ph－H, NH_2), 6.80 (d, 1H,J =15.9Hz, =CH); ^{13}CNMR (DMSO－d_6,75MHz) δ = 178.9, 146.5,

138.2, 135.9, 135.1, 132.6, 131.8, 130.7, 129.9, 128.8, 122.5, 119.1; MS (ESI): m/z(100%) 440 (M^+); Anal. Calcd for $C_{16}H_{13}Br_2N_3S$: C, 43.76; H, 2.98; N, 9.57; found: C, 43.77; H, 2.96; N, 9.55.

1-(4-溴苯基)-3-(4-氯苯基)丙-2-烯-1-酮氨基硫脲席夫碱(14),收率 55.1%,黄色固体,熔点 175.0~177.4 ℃; IR(KBr) ν_{max} 1675, 1582, 1389, 1280, 1123, 1092, 1008, 876, 814 cm^{-1}; ^{1}HNMR (DMSO-d_6, 300 MHz) δ = 11.16 (s, 1H, NH), 7.83-7.71(m, 5H, ph-H, NH_2, =CH), 7.64 (d, 2H, J=8.7 Hz, ph-H), 7.49-7.44 (m, 4H, ph-H), 6.82(d, 1H, *J* =15.9Hz, =CH); ^{13}CNMR (DMSO-d_6, 75MHz) δ = 178.9, 146.5, 138.2, 135.9, 134.8, 133.7, 131.9, 131.2, 130.7, 128.6, 127.9, 122.5, 119.1; Anal. Calcd for $C_{16}H_{13}BrClN_3S$: C, 48.69; H, 3.32; N, 10.65; found: C, 48.66; H, 3.31; N, 10.63.

1.3 活性测试

1.3.1 酪氨酸酶抑制活性

采用本课题组已经建立起的酪氨酸酶抑制活性测试方法对所合成卤代查尔酮及其氨基硫脲席夫碱进行了酪氨酸酶抑制活性测试[9]。曲酸是常用的酪氨酸酶抑制剂,所以采样曲酸为阳性对照,结果如表1。

表1 合成化合物酪氨酸酶抑制活性

Table 1Tyrosinase inhibitory activities of the synthesized compounds

Compounds	CLog*P* [d]	IC_{50} (μmol/L) [a]
1	4.337	NA[b]
2	4.487	NA[b]
3	3.843	40.20% [c]
4	4.413	40.25% [c]
5	4.563	37.22% [c]
6	5.426	NA[b]
7	5.276	NA[b]
8	4.297	14.10
9	4.447	2.02
10	3.811	7.56
11	4.381	117.18
12	4.531	140.50

续表

Compounds	CLogP [d]	IC_{50} (μmol/L) [a]
13	5.394	49.25
14	5.244	50.65
Kojic acid		23.67

a. 三次实验平均值

b. 在 200 μmol/L 无抑制活性.

c. 在 200μmol/L 的抑制率

d. CLog*P* 来源于 ChemBioDraw Ultra 12.0

1.3.2 抗氧化活性测试

为了初步考察所合成化合物的抗氧化活性,参照 Blois *et al*. 方法进行了 DPPH· 清除能力的测试[12],参照 Pellegrini R R N, *et al*. 方法进行了 ABTS · +清除能力的测试[13],参照 Apak R, *et al*. 方法对所合成化合物进行铜离子还原能力测试[14],2,6-二叔丁基-4-甲基苯酚(BHT),α-生育酚作为阳性对照,测试结果如表2。

表2 卤代查尔酮及其席夫碱抗氧化活性

Table 2 Antioxidant activity ofthe substituted chalcones and their thiosemicarbazide Schiff bases

Compound no	DPPH scavenging ability IC_{50} μmol/L ± SD	ABTS · + scavenging activity IC_{50} μmol/L ± SD	cupric reducing antioxidant capacity $A_{0.50}$ μmol/L ± SD
1	>200	>200	>200
2	>200	>200	>200
3	>200	192.35 ±1.78	>200
4	198.68 ±3.25	176.39 ±2.69	>200
5	191.55 ±2.97	194.27 ±2.36	>200
6	>200	>200	>200
7	>200	>200	>200
8	109.26 ±1.38	126.56 ±2.09	122.56 ±1.49
9	64.96 ±0.16	83.48 ±0.12	73.05 ±0.09
10	75.58 ±0.67	102.08 ±0.42	103.19 ±0.36
11	132.09 ±2.09	168.32 ±1.87	126.67 ±1.06
12	156.68 ±3.56	174.32 ±1.69	85.08 ±0.69

续表

Compound no	DPPH scavenging ability IC_{50} μmol/L ± SD	ABTS · + scavenging activity IC_{50} μmol/L ± SD	cupric reducing antioxidant capacity $A_{0.50}$ μmol/L ± SD
13	118.26 ±1.86	137.87 ±0.78	76.65 ±1.37
14	109.87 ±1.46	153.58 ±1.16	62.83 ±0.98
BHT	56.08 ±0.82	2.88 ±0.36	4.07 ±0.11
α - tocopherol	13.16 ±0.12	5.39 ±0.08	2.65 ±0.02

1.3.3 分子对接

参照李慕紫等的方法[15]，分别采用 AutoDock4.2 对化合物 9 进行了分子对接，对接结果如图 1。

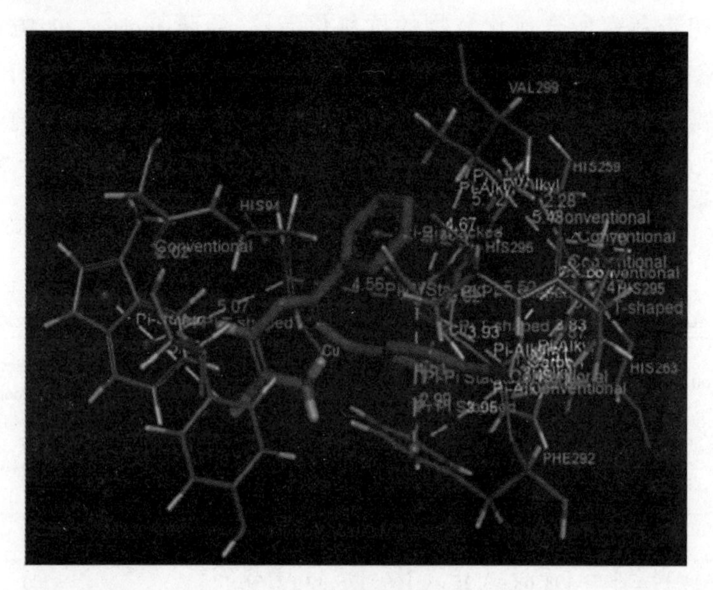

图 1　分子对接图

Figure 1. The docking simulation between tyrosinase and inhibitor compound 9.

2 结果与讨论

2.1 合成部分

以卤原子取代或未取代的芳香醛和芳香酮为起始原料，经 Claisen - Schmidt 缩合得到含有卤原子的查尔酮粗品，再用 95% 乙醇进行重结晶得到卤代查尔酮，收率中等(50.9% ~89.2%)。为了增加羰基的反应活性，以醋酸为催化剂，将相应的卤代查尔酮与氨基硫脲在乙醇中回流反应得到相应的氨基硫脲席夫碱粗品，以乙醇重结晶得产品，收率在 32.9% 到 60.8% 之间，所有化合物采用 IR，

NMR, MS 和元素分析进行表征。

2.2　酪氨酸酶抑制活性

以 L－DOPA 为底物,检测化合物对酪氨酸酶的抑制活性,结果见图 2 和表 1。图 2 表示化合物 9 对酪氨酸酶抑制作用曲线,随着抑制剂浓度的加大,酪氨酸酶的活性迅速下降,表明该化合物对酪氨酸酶具有明显的抑制活性。表 1 中所列 IC_{50}值,是通过三次试验后所得平均值。如表 1 所示,大多数卤代查尔酮及其氨基硫脲席夫碱具有一定的酪氨酸酶抑制活性,IC_{50} 值在 2. 02μmol/L ~ 140. 5μmol/L之间,从测试结果来看,卤代查尔酮氨基硫脲席夫碱(8－14)对酪氨酸酶的抑制活性比相应的查尔酮(1－7)强,这就说明氨基硫脲席夫碱片段对提高这类化合物酪氨酸酶抑制活性是有利的,很有可能是一个活性基团。化合物 8－10具有比阳性对照(曲酸)更强的抑制活性,其中化合物 9 是该类化合物中活性最强的,其 IC_{50}值为 2.02 μmol/L。化合物 8 比化合物 11 的抑制活性强,化合物 9 比化合物 12 的抑制活性强,这四个化合物在结构上的区别就是卤原子来源不一样,因此卤原子的来源对查尔酮氨基硫脲席夫碱酪氨酸酶的抑制活性存在一定的影响,卤原子来源于醛对提高酪氨酸酶抑制活性更有利。然而,对比查尔

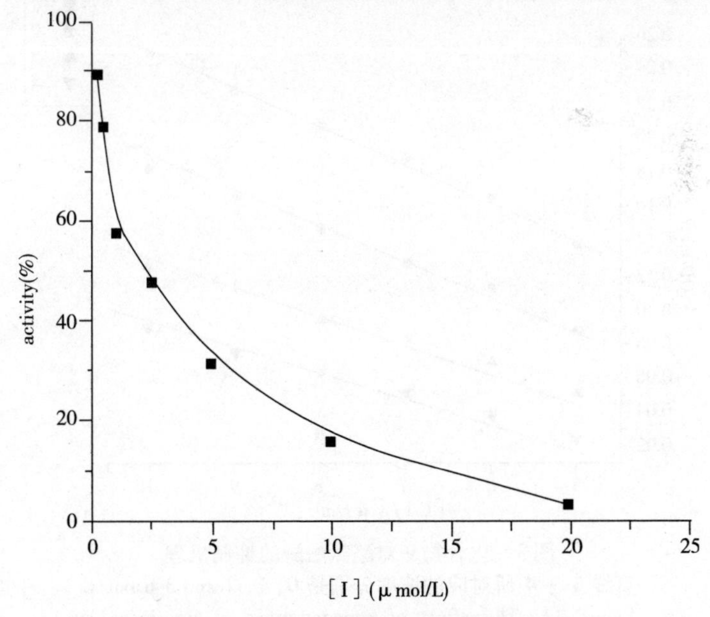

图 2　化合物 9 对酪氨酸酶的抑制作用

Figure 2　Effect of compound 9 on the diphenolase activity of mushroom tyrosinase for the catalysis of L－DOPA at 25℃.

酮化合物 1 -7 的抑制活性,当卤原子来源于醛的查尔酮没有抑制活性,而当卤原子只来源于酮时所得的查尔酮(3 -5)具有一定的酪氨酸酶抑制活性,但活性不高,同样说明卤原子的来源对酪氨酸酶的抑制活性有影响。对比化合物 10 - 12 的抑制活性可以发现,含氟原子的化合物(10)的抑制活性比其他的活性要高,导致这个结果的原因可能与氟原子体积小而电负性大的特点有关。随着化合物 12 - 14 的 CLog *P* 值的变化,其酪氨酸酶抑制活性也发生相应的变化,所以这些化合物的抑制活性与其脂溶性也存在一定的关系。

2.3 抗氧化活性

为了初步考察所合成的抗氧化活性,以 2,6 - 二叔丁基 -4 - 甲基苯酚(BHT)和 α - 生育酚为阳性对照,分别评价了 14 个化合物对 DPPH· 清除能力、ABTS · + 清除能力以及铜离子还原能力,结果如表 2。从表 2 可以发现,部分化合物具有一定的抗氧化活性,但是抗氧化活性都不是很强,没有一个化合物的活性强于阳性对照 2,6 - 二叔丁基 -4 - 甲基苯酚(BHT)和 α - 生育酚。特别是化合物 1,2,6, 7 没有任何抗氧化活性。

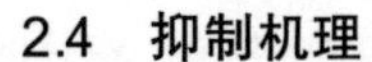

2.4 抑制机理

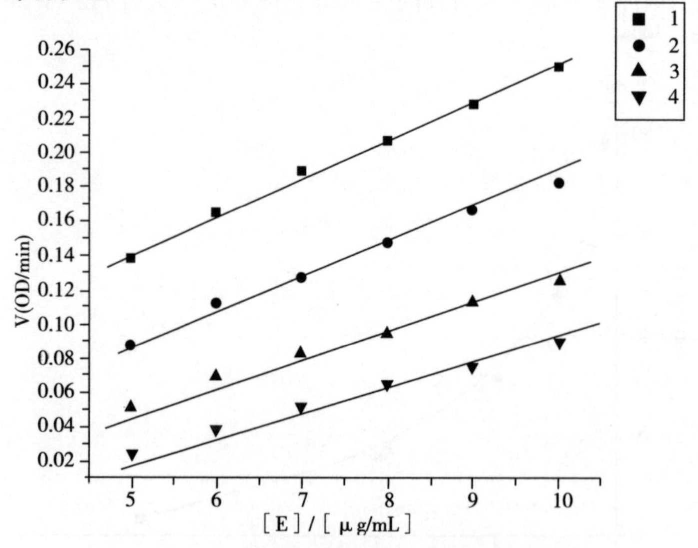

图 3 化合物 9 对酪氨酸酶的抑制机理

直线 1 - 4 所对应的浓度分别是 0, 1, 2 and 3 μmol/L

Figure 3 The effect of concentrations of tyrosinase on its activity for the catalysis of L - DOPA at different concentration of compound 9. The concentrations of compound 9 for curves 1 - 4 are 0, 1, 2 and 3 μmol/L, respectively.

化合物 9 是所有化合物中酪氨酸酶抑制活性最好的,故优选化合物 9 进行了抑制机理探讨,结果如图 3。图 3 显示的是一组平行线,因此化合物 9 属于不可逆抑制剂,也就是该化合物与酪氨酸酶活性位点结合是不可逆的。这个可能与酪氨酸酶的分子结构有关,化合物 9 能够与酪氨酸酶形成一种结构稳定的不可逆的络合物。

2.5　分子对接

采用 AutoDock4.2 对化合物 9 进行了分子对接研究。通过分子对接计算出化合物 9 与酪氨酸酶的结合能是 -30.56kJ/mol,这就说明化合物 9 能够与酪氨酸酶具有相互作用力。从图 1 可以发现,化合物 9 中的氨基可以与酪氨酸酶 HIS178 形成氢键,其键长为 2.38 Å;苯环 B 可以 LYS180 残基形成 π 键,其键长为 4.04 Å;硫原子与 HIS178 残基也存在 π 键,其键长为 5.55 Å。这就说明氨基硫脲席夫碱中的硫原子和氨基对抑制酪氨酸酶活性起着重要作用,但并不是和酪氨酸酶中铜离子活性中心形成络合物。

3　结论

本文报道了一系列卤代查尔酮及其氨基硫脲席夫碱的酪氨酸酶抑制活性,其中化合物 9 对酪氨酸酶抑制活性最好,其 IC_{50} 值为 2.02 μmol/L,优于阳性对照曲酸的抑制活性。初步构效关系分析表明:(1)卤原子的来源对卤代查尔酮及卤代查尔酮氨基硫脲席夫碱的抑制活性有重要影响;(2)氨基硫脲席夫碱片段对酪氨酸酶抑制活性起着决定性作用,这可能与酪氨酸酶的分子结构有关;(3)酪氨酸酶抑制活性与脂溶性存在一定的关系。部分化合物具有一定的抗氧化活性,但都不如阳性对照。优选化合物 9 进行了抑制机理探讨,该化合物是不可逆抑制剂,分子对接表明化合物 9 的硫原子可以与酪氨酸酶氨基酸残基形成 π 键,氨基能够与 HIS178 形成氢键,化合物 9 能够与酪氨酸酶形成稳定的络合物。

参考文献

[1] Michael R L S, Christopher A R, A R Patrick. Bioorg. Med. Chem., 2013, 21(5): 1166 - 1173.

[2] K K Song, H Huang, P Han, C L Zhang, Y Shi, Q X Chen. Biochem. Bioph. Res. Co., 2006, 342(4): 1147 - 1151.

[3] R Matsuura, H Ukeda, M Sawamura. J. Agric. Food. Chem., 2006, 54(6): 2309 - 2313.

[4] Y J Zhu, H T Zhou, Y H Hu. Food. Chem., 2011, 124(1): 298 - 302.

[5] P Thanigaimalai, T A L Hoang, K C Lee, et al. Bioorg. Med. Chem.

Lett., 2010, 20(9): 2991 -2993.

[6] A Ortiz - Urquiza, N O Keyhani. Insects, 2013, 4: 357 -374.

[7] P Singh, A Anand, V Kumar. Eur. J. Med. Chem., 2014, 85: 758 -777.

[8] M R E S Aly, H A E R Fodah, S Y Saleh, et al. Eur. J. Med. Chem., 2014, 76: 517 -530.

[9] J Liu, F Wu, L Chen, L Zhao, et al. Food. Chem., 2012, 135(4): 2872 -2878.

[10] A You, J Zhou, S Song, G Zhu, H Song, W Yi. Bioorg. Med. Chem., 2015, 23(5):924 -931

[11]刘进兵,吴凤艳,王子厚. 化学通报,2015,78(12):1096 -1101.

[12] M S Blois. Nature, 1958, 181: 1199 -1200.

[13] R R N Pellegrini, A Proteggente, A Pannala, M Yang, C Rice - Evans. Free. Radical. Biol. Med., 1999, 26(9 -10): 1231 -1237.

[14] R Apak, K Gu c lu , M Ozyu rek, S E Karademir. J. Agric. Food. Chem., 2004, 52 (26): 7970 -7981.

[15] 李慕紫,王庆华,王晓艺,何云. 食品科学,2016,37(03): 87 -90

2.6.3 European Journal of Medicinal Chemistry(2009, 44,1773 - 1778)

A class of potent tyrosinase inhibitors: Alkylidenethiosemicarbazide compounds

Jinbing Liu, Rihui Cao *, Wei Yi, Chunming Ma, Yiqian Wan, Binhua Zhou, Lin Ma, Huacan Song *

Abstract

A series of alkylidenethiosemicarbazide compounds were synthesized and their inhibitory effects on the diphenolase activity of mushroom tyrosinase were evaluated. The results showed that most of the synthesized compounds exhibited significant inhibitory activities. Especially, compound1fwas found to be the most potent inhibitor with IC_{50} value of 0. 086mmol/L, suggesting that further development of such compounds may be of interest.

1. Introduction

Tyrosinase (EC 1. 14. 18. 1) is a multifunctional coppercontaining enzyme which is widely distributed in plants and animals[1]. It is well known that tyrosinase can catalyze the hydroxylation of monophenols to o - diphenols (monophenolase activity) and the oxidation of o - diphenols to o - quinone (diphenolase activity). And the enzymatic oxidation of L - tyrosine into melanin is of considerable importance in coloring of skin, hair and eyes, and in food browning [2,3]. In addition, tyrosinase is also related to the molting process of insects [4] and adhesion of marine organisms [3,5]. Nowadays, tyrosinase inhibitors are thought to be clinically useful for the treatment of some dermatological disorders associated with melanin hyperpigmentation [6,7], and useful in cosmetic products and food industry [8]. So far, a large number of potential tyrosinase inhibitors have been discovered from natural, synthetic and even from semi - synthetic sources, such as hydroquinone [9], ascorbic acid derivates [10], kojic acid (Fig. 1)[11], azelaic acid [12], corticosteroids [12], retinoids[12], arbutin (Fig. 1)[12] and tropolone (Fig. 1)[13,14]. Among all the known tyrosinase inhibitors, tropolone

(IC_{50} = 0.4μM) was found to be one of the most potent tyrosinase inhibitors [14]. Unfortunately, only few of the reported compounds are used in medicinal and cosmetic products because of their lower activities or serious side effects. Therefore, it is still necessary to search and discover novel tyrosinase inhibitors with higher activity and lower side effect.

Previous literatures described that thiourea derivatives, such as phenylthioureas [15,16], alkylthioureas [17] and 1,3 - bis - (5 - methanesulfonylbutyl) thiourea, displayed weak or moderate tyrosinase inhibitory activity. More recently, our investigation [18] also demonstrated that 1 - (1 - arylethylidene) thiosemicarbzide derivatives exhibited potent inhibitory activities against mushroom tyrosinase. Stimulated by these results, in the present investigation, we designed and synthesized a series of novel alkylidenethiosemicarbazide compounds bearing various alkyl substituents. We report here the preparation of alkylidenethiosemicarbazide compounds and their inhibitory effects on the diphenolase activity of mushroom tyrosinase.

2. Chemistry

The alkylidenethiosemicarbazides were prepared by the condensation of alkyl ketones or aldehydes with thiosemicarbazide in the presence of acetic acid in ethanol (Scheme 1). To investigate the effects of thiosemicarbazido - group on tyrosinase inhibiting activity, pentane - 2, 4 - dione, hexane - 2, 5 - dione, 2 - acetylcyclohexanone and cyclohexane - 1, 3 - dione were selected to prepare dithiosemicarbazide compounds. Unfortunately, only mono - thiosemicarbazido - group products were obtained.

3. Biology

Taking all the newly synthesized compounds as the effectors, we investigated their inhibitory effects on the diphenolase activity of mushroom tyrosinase. The IC_{50} value of these compounds are summarized in Table 1. From the data shown in Table 1, the following conclusions were drawn.

(1) All the synthesized compounds show potent inhibiting activities against tyrosinase, especially compounds1a(IC_{50} = 0.23 μM), 1b(IC_{50} = 0.20 μM), 1f (IC_{50} = 0.086 μM) and 1g(IC_{50} = 0.28 μM) demonstrate higher inhibitory activity than arbutin(30% inhibition at 10.4mmol/L) and 3 - (4 - methoxyphenyl) acrylic acid (IC_{50} = 0.41μM), both of the later were used as reference inhibitors, even more potent than the best reported one, tropolone(IC_{50} = 0.4 μM)[14,15].

(2) With the increase of thelength of chain R (for 1 - alkylidenethiosemicarbazide compounds 1a to 1e and for 1 - (1 - arylethylidene) thiosemicarbazide compounds 1f - 1k), the inhibition activity decreased. Compound 1f bearing methyl group, the shortest substituted group, shows the most potent inhibitory activity with an IC_{50} value of 0.086μM. These observations suggested that the increase of the length of alkyl chain might cause stereo - hindrance for the approach of inhibitors to the active site of the enzyme.

(3) Compound2a was 5.5 - fold more active than compound 2b, which indicated that the smaller ring size showed the more potent activity, in orther words, the smaller size of the ring would be beneficial for the molecules of synthesized compounds to approach the active centre of enzyme as stereo - hindrance. It was further verified that the shorter chain R would be favourable to increase the inhititing activity of obtained compounds.

(4) The synthesized compounds with a saturated R would show a stronger inhibiting activity than those compounds with an unsaturated R, which suggested that the double bonds might lead to a decline in their activity.

(5) Compounds1i, 1j and 3, with a carbonyl group in their moleculars, were

also less active than those corresponding thiosemicarbazides without a carbonyl group, which indicated that electron - withdrawn effects of carbonyl group might result in the decrease of inhibiting activity of these synthesized compounds.

4. Conclusions

The present investigation reported that, for the first time, alkylidene - thiosemicarbazide compounds had potent inhibitory effects on the diphenolase activity of mushroom tyrosinase. Interestingly, compound 1f was found to be the most potent inhibitor with IC_{50} value of 0.086μM. Preliminary structure activity relationships' (SARs) analysis indicated that (1) the increase of the length of alkyl chain might cause stereohindrance for the inhibitors approaching the active site of the enzyme resulting in the decrease of inhibitory activities, and the shorter alkyl chain was more favorable; (2) the double bonds might be detrimental to their activities; and (3) the electron - withdrawing substituents might play a vital role in determining their inhibitory activities. These results suggested that further development of such compounds may be of interest.

5. Experimental protocols

5.1 Instruments and Materials

Melting points (mp) were determined withWRS - 1B melting point apparatus and the thermometer was uncorrected. NMR spectra were recorded on Mercury - Plus300 spectrometers at 25℃ in $CDCl_3$ or DMSO - d_6. All chemical shifts (δ) are quoted in ppm downfield from TMS and coupling constants (J) are given in Hz. LC - MS spectra were recorded using the LCMS - 2010A. All reactions were monitored by TLC (Merck Kieselgel 60 F_{254}) and the spots were visualized under UV light. Elemental analyses were performed on a Vario EL instrument and were within ± 0.4% of the theoretical values. Infrared (IR) spectra were recorded on VECTOR 22 spectrometer. The appropriate aldehyde, ketone, thiosemicarbazide, 4 - methoxycinnamic acid were purchased from Darui Chemical Co. (ShangHai, China), arbutin were obtained from Brillian Biochemical Co. The other commercially available reagents and solvents were used without further purification. Mushroom tyrosinase, *L* - DOPA were purchased from Sigma Chemical Co.

5.2 The general procedures for the synthesis of alkylidenethiosemicarbazide compounds:

The appropriate aldehyde or ketone (10mmol) was dissolved in anhydrous ethanol(10mL), thiosemicarbazide (10mmol) and acetic acid (0.5mL) were added into the solution. After being refluxed for 24h, the reaction mixture was cooled to room temperature and then the precipitate appeared. The corresponding pure compounds were obtained by filtraton. For some special cases, the target compounds could be purified by recrystallization from ethanol.

5.2.1 1 – ethylidenethiosemicarbazide (1a).

mp: 135 ~ 136 ℃. R_f = 0.58 (ethyl acetate / petroleum ether (bp 60 ~ 90 ℃) = 2:1, v/v). ^{1}H NMR (300 MHz, DMSO – d_6): δ 11.01 (1H, bs, NH), 7.90 (1H, bs, NH), 7.43 (1H, bs, NH), 7.40 (H, q, CH), 1.86 (3H, d, CH_3). ^{13}C NMR (75 MHz, DMSO – d_6): δ 177.9, 144.5, 19.0. IR (KBr): 3379, 3261, 3189, 3024, 1591, 1535 cm^{-1}. MS (ESI): m/z (100%) = 118 (M + 1). Anal. Calcd for $C_3H_7N_3S$ (117.17): C, 30.75; H, 6.02; N, 35.86. Found: C, 30.84; H, 5.99; N, 35.98.

5.2.2 1 – propylidenethiosemicarbazide (1b).

mp: 150 ~ 152 ℃. R_f = 0.62 (ethyl acetate / petroleum ether (bp 60 ~ 90 ℃) = 2:1, v/v). ^{1}H NMR (300 MHz, DMSO – d_6): δ 10.99 (1H, bs, NH), 7.89 (1H, bs, NH), 7.39 (1H, bs, NH), 7.38 (H, t, CH), 2.20 (2H, m, CH_2), 1.01 (3H, t, CH_3). ^{13}C NMR (75 MHz, DMSO – d_6): δ 178.1, 149.0, 26.0, 11.3. IR (KBr): 3385, 3273, 3189, 3036, 1586, 1541 cm^{-1}. MS (ESI): m/z (100%) = 132 (M + 1). Anal. Calcd for $C_4H_9N_3S$ (131.20): C, 36.62; H, 6.91; N, 32.03. Found: C, 36.84; H, 6.97; N, 32.22

5.2.3 1 – (3 – methylbutylidene)thiosemicarbazide (1c).

R_f = 0.73 (ethyl acetate / petroleum ether (bp 60 ~ 90 ℃) = 2:1, v/v). ^{1}H NMR (300 MHz, $CDCl_3$): δ 9.86 (1H, bs, NH), 7.28 (1H, bs, NH), 7.08 (1H, bs, NH), 6.50 (H, t, CH), 2.15 (2H, t, CH_2), 1.91 (1H, m, CH), 0.97 (3H, d, CH_3). ^{13}C NMR (75 MHz, $CDCl_3$): δ 177.6, 148.8, 41.4, 26.9, 22.8. IR (KBr): 3425, 3282, 3144, 2951, 1590, 1537, 1383 cm^{-1}. MS (ESI): m/z(100%) = 160 (M + 1).

5.2.4 1 – (but – 3 – enylidene)thiosemicarbazide (1d).

mp: 142 ~ 143 ℃. R_f = 0.73 (ethyl acetate / petroleum ether (bp 60 ~ 90 ℃)

= 2:1, v/v). ^1H NMR (300 MHz, $CDCl_3$): δ 9.70 (1H, bs, NH), 7.53 (1H, q, CH)7.06 (1H, bs, NH), 6.37 (1H, bs, NH), 6.16 (H, d, CH), 6.13 (H, m, CH), 1.91 (3H, d, CH_3). ^{13}C NMR (75 MHz, $CDCl_3$): δ 178.2, 145.6, 138.9, 129.1, 19.3. IR (KBr): 3384, 3275, 3163, 3021, 2961, 1585, 1527 cm^{-1}. MS (ESI): m/z (100%) = 144 (M+1). Anal. Calcd for $C_5H_9N_3S$ (143.21): C, 41.93; H, 6.33; N, 29.34. Found: C, 41.84; H, 6.18; N, 29.68.

5.2.5 1-(3-phenylallylidene)thiosemicarbazide (1e):

mp: 134~136 ℃. R_f = 0.75 (ethyl acetate / petroleum ether (bp 60~90 ℃) = 2 : 1, v/v). ^1H NMR (300 MHz, $CDCl_3$): δ 11.36 (1H, bs, NH), 8.15 (1H, bs, NH), 7.90 (H, d, CH), 7.58 (1H, bs, NH), 7.54 (2H, d, phH), 7.34 (2H, d, phH), 7.32 (H, d, CH), 7.03 (H, t, CH), 6.89 (H, t, CH). ^{13}C NMR (75 MHz, $CDCl_3$): δ 178.3, 145.4, 139.5, 136.5, 129.8, 129.5, 127.6, 125.7. IR (KBr): 3398, 3279, 3152, 3028, 1601, 757 cm^{-1}. MS (ESI): m/z (100%) = 206 (M+1).

5.2.6 1-(propan-2-ylidene)thiosemicarbazide (1f).

mp: 180~181 ℃. R_f = 0.55 (ethyl acetate / petroleum ether (bp 60~90 ℃) = 2 : 1, v/v). ^1H NMR (300 MHz, DMSO-d_6): δ 9.86 (1H, bs, NH), 7.95 (1H, bs, NH), 7.48 (1H, bs, NH), 1.91 (3H, s, CH_3), 1.89 (3H, s, CH_3); ^{13}C NMR (75 MHz, DMSO-d_6): δ 178.9, 152.3, 25.9, 18.4. IR (KBr): 3378, 3234, 3152, 2997, 1596, 1512 cm^{-1}. MS (ESI): m/z (100%) = 132 (M+1). Anal. Calcd for $C_4H_9N_3S$ (131.20): C, 36.62; H, 6.91; N, 32.03. Found: C, 36.59; H, 6.84; N, 32.25.

5.2.6 1-(butan-2-ylidene)thiosemicarbazide (1g).

mp: 98~100 ℃. R_f = 0.65 (ethyl acetate / petroleum ether (bp 60~90 ℃) = 2 : 1, v/v). ^1H NMR (300 MHz, DMSO-d_6): δ 8.58 (1H, bs, NH), 7.26 (1H, bs, NH), 6.45 (1H, bs, NH), 2.31 (2H, q, CH_2), 1.90 (3H, s, CH_3), 1.10 (3H, t, CH_3); ^{13}C NMR (75 MHz, DMSO-d_6): δ 179.1, 155.9, 32.3, 17.1, 11.4. IR (KBr): 3381, 3236, 3150, 3019, 2971, 1578 cm^{-1}. MS (ESI): m/z (100%) = 146 (M+1). Anal. Calcd for $C_5H_{11}N_3S$ (145.23): C, 41.35; H, 7.63; N, 28.93. Found: C, 41.17; H, 7.55; N, 28.71.

5.2.7 1-(4-methylpent-3-en-2-ylidene)thiosemicarbazide (1h).

mp: 119 ~ 120 ℃. R_f = 0.70 (ethyl acetate / petroleum ether (bp 60 ~ 90℃) = 2 : 1, v/v). ^{1}H NMR (300 MHz, DMSO - d_6): δ 9.76 (1H, bs, NH), 8.09 (1H, bs, NH), 7.55 (1H, bs, NH), 4.74 (1H, s, CH), 1.94 (3H, s, CH_3), 0.99 (6H, s, CH_3). ^{13}C NMR (75 MHz, DMSO - d_6): δ 179.1, 153.5, 152.9, 136.7, 25.9, 20.1. IR (KBr): 3385, 3263, 3158, 3025, 2961, 1385 cm^{-1}. MS (ESI): m/z (100%) = 172 (M + 1).

5.2.8 1 - (1 - (4 - methoxyphenyl) propan - 2 - ylidene) thiosemicarbazide (1i).

mp: 122 ~ 123 ℃. R_f = 0.68 (ethyl acetate / petroleum ether (bp 60 ~ 90 ℃) = 2 : 1, v/v). ^{1}H NMR (300 MHz, $CDCl_3$): δ 8.50 (H, bs, NH), 7.25 (H, bs, NH), 7.09 (2H, d, J = 8.7Hz, phH), 6.86 (2H, d, J = 8.7Hz, phH), 6.34 (H, bs, NH), 3.80 (3H, s, CH_3), 3.50 (2H, s, CH_2), 1.81 (3H, s, CH_3). ^{13}C NMR (75 MHz, $CDCl_3$): δ 180.2, 157.6, 155.7, 131.2, 130.5, 114.5, 55.7, 38.6, 14.5. IR (KBr): 3405, 3210, 3137, 3031, 2833, 1587, 1514, 859 cm^{-1}. MS (ESI): m/z (100%) = 238 (M + 1). Anal. Calcd for $C_{11}H_{15}N_3OS$ (237.32): C, 55.67; H, 6.37; N, 17.71. Found: C, 55.61; H, 6.32; N, 17.76.

5.2.9 1 - (4 - (4 - hydroxyphenyl) butan - 2 - ylidene) thiosemicarbazide (1j).

mp: 154 ~ 155 ℃. R_f = 0.69 (ethyl acetate / petroleum ether (bp 60 ~ 90 ℃) = 2 : 1, v/v). ^{1}H NMR (300 MHz, DMSO - d_6): δ 8.49 (1H, bs, NH), 7.12 (1H, bs, NH), 7.02 (2H, d, J = 6.3Hz, phH), 6.77 (2H, d, J = 6.3Hz, phH), 6.22 (1H, bs, NH), 2.80 (2H, t, CH_2), 2.57 (2H, t, CH_2), 1.89 (3H, s, CH_3). ^{13}C NMR (75 MHz, DMSO - d_6): δ 179.1, 159.3, 148.9, 129.1, 128.8, 115.7, 32.5, 29.3, 19.8. IR (KBr): 3609, 3366, 3265, 3187, 1610, 1513, 826 cm^{-1}. MS (ESI): m/z (100%) = 236 (M - 1). Anal. Calcd for $C_{11}H_{15}N_3OS$ (237.32): C, 55.67; H, 6.37; N, 17.71. Found: C, 55.73; H, 6.35; N, 17.79.

5.2.10 1 - cyclopentylidenethiosemicarbazide (2a).

mp: 152 ~ 154 ℃. R_f = 0.65 (ethyl acetate / petroleum ether (bp 60 ~ 90 ℃) = 2 : 1, v/v). ^{1}H NMR (300 MHz, DMSO - d_6): δ 8.45 (1H, bs, NH), 7.16

(1H, bs, NH), 6.44 (1H, bs, NH), 2.39 (2H, t, CH_2), 2.29 (2H, t, CH_2), 1.89 (2H, m, CH_2), 1.78 (2H, m, CH_2). ^{13}C NMR (75 MHz, DMSO − d_6): δ 178.5, 164.0, 33.7, 28.5, 25.3. IR(KBr): 3383, 3261, 3138, 2957, 1661 cm^{-1}. MS (ESI): m/z (100%) = 158 (M+1). Anal. Calcd for $C_6H_{11}N_3S$ (157.24): C, 45.83; H, 7.05; N, 26.72. Found: C, 45.95; H, 7.09; N, 26.85.

5.2.11 1 − cyclohexylidenethiosemicarbazide (2b).

mp: 154 ~ 155 ℃. R_f = 0.65 (ethyl acetate / petroleum ether (bp 60 ~ 90 ℃) = 2 : 1, v/v). ^1H NMR (300 MHz, DMSO − d_6): δ 8.82 (1H, bs, NH), 7.26 (1H, bs, NH), 6.46 (1H, bs, NH), 2.31 (4H, m, $2CH_2$), 1.71 (6H, m, $3CH_2$). ^{13}C NMR(75 MHz, DMSO − d_6): δ 178.8, 157.5, 35.7, 27.4, 26.3, 25.8. IR (KBr): 3379, 3217, 3144, 2939, 2858, 1585 cm^{-1}. MS (ESI): m/z (100%) = 172 (M+1).

Anal. Calcd for $C_7H_{13}N_3S$ (171.26): C, 49.09; H, 7.65; N, 24.54. Found: C, 49.06; H, 7.63; N, 24.75.

5.2.12 1 − (3 − oxocyclohexylidene)thiosemicarbazide (3).

mp: 175 ~ 176 ℃. R_f = 0.66 (ethyl acetate / petroleum ether (bp 60 ~ 90 ℃) = 2 : 1, v/v). ^1H NMR (300 MHz, DMSO − d_6): δ 11.85 (1H, bs, NH), 9.20 (1H, bs, NH), 7.38 (1H, bs, NH), 2.41 (2H, t, CH_2), 2.16 (2H, t, CH_2), 1.90 (2H, s, CH_2), 1.72 (2H, m, CH_2). ^{13}C NMR(75 MHz, DMSO − d_6): δ 208.7, 178.6, 159.2, 46.8, 44.7, 22.5, 19.2. IR (KBr): 3416, 3275, 3161, 1716, 1603 cm^{-1}. MS (ESI): m/z (100%) = 186 (M+1).

5.2.13 1 − (2,5 − dimethyl − 1H − pyrrol − 1 − yl)thiourea (4).

mp: 217 ~ 218 ℃. R_f = 0.75 (ethyl acetate / petroleum ether (bp 60 ~ 90 ℃) = 2 : 1, v/v). ^1H NMR (300 MHz, DMSO − d_6): δ 10.19 (1H, bs, NH), 8.18 (1H, bs, NH), 6.54 (1H, bs, NH), 5.68 (2H, s, 2CH), 1.99 (6H, s, $2CH_3$). ^{13}CNMR (75 MHz, DMSO − d_6): δ 183.15, 127.64, 104.71, 11.73. IR (KBr): 3385, 3231, 3146, 1956, 1616 cm^{-1}. MS (ESI): m/z (100%) = 170 (M+1). Anal. Calcd for $C_7H_{11}N_3S$ (169.25): C, 49.68; H, 6.55; N, 24.83. Found: C, 49.59; H, 6.51; N, 25.21.

6. Tyrosinase assay

Tyrosinase inhibition assays were performed according to the method developed by Hearing with slight modification [19]. Briefly, all the synthesized compounds were screened for the *o* – diphenolase inhibitory activity of tyrosinase using *L* – DOPA as substrate. All the active inhibitors from the preliminary screening were subjected to IC_{50} studies. All the synthesized compounds were dissolved in DMSO to a concentration of 2.5%. Phosphate buffer pH 6.8 was used to dilute the DMSO stock solution of test compound. Thirty units of mushroom tyrosinase (28nM) was first pre – incubated with the compounds, in 50 nM phosphate buffer (pH 6.8), for 10min at 25 ℃. Then the *L* – DOPA (0.5mmol/L) was added to the reaction mixture and the enzyme reaction was monitored by measuring the change in absorbance at 475 nm of formation of the DOPAchrome for 10 min. Doseresponse curves were obtained by performing assays in the presence of increasing concentrations of inhibitors (0, 1.6, 3.2, 6.3, 12.5, 25, 50, 100, 200 μM). IC_{50} value, a concentration giving 50% inhibition of tyrosinase activity, was determined by interpolation of the doseresponse cureves. The percent of inhibition of tyrosinase reaction was calculated as following:

$$\text{percent inhibition}(\%) = [(B - S)/B] \times 100.$$

Here, the *B* and *S* are the absorbances for the blank and samples, respectively. 4 – methoxycinnamic acid and arbutin were used as standard inhibitors for the tyrosinase. All the studies have been carried out at least in triplicate and the results here represent means ± SEM (standard error of the mean).

Acknowledgment

This work was supported by the Natural Science Foundation of Guangdong Province, China (2004B30101007).

References

1. K.K. Song, H. Huang, P. Han, Biochem. Biophys. Res. Commun. 342 (2006) 1147.

2. Perez – Gilabert, M.; Garcia – Carmona, F. Biochem. Biophys. Res. Commun. 285 (2001) 257.

3. S. Okombi, D. Rival. Bioorg. Med. Chem. Lett.. 16 (2006) 2252.

4. M. Shiino, Y. Watanabe, K. Umezawa, Bioorg. Med. Chem. 9 (2001) 1233.

5. A. Palumbo, G. Misuraco, Biochim. Biophysica. Acta. 1073 (1991) 85.

6. M. K. Khalid, M. M. Ghulam, Bioorg. Med. Chem. 14 (2006) 344.

7. M. Sugumarau, FEBS Lett. 293 (1991) 4.

8. K. Marumo, J. H. Waite, Biochim. Biophysica. Acta. 872 (1986) 98.

9. M. Shiino,; Y. Watanabe,; K. Umezawa, Bioorg. Med. Chem. 9 (2001) 1233.

10. A. Garcia, J. E. Fulrton, Dermatol. Surg. 22 (1996) 443.

11. S. Kojima, H. Yamaguch, K. Morita, Biol. Pharm. Bull. 18 (1995) 1076.

12. J. Cabanes, S. Chazaarra, J. Pharm. Pharmacol. 46 (1994) 982.

13. C. M. Gerardo M, T. H. K. Mahmud, M. P. Yovani, A. Arjumand, Bioorg. Med. Chem. Lett. 16 (2006) 324

14. S. M. Son, K. D. Moon, C. Y. Lee, J. Agric. Food Chem. 48 (2000) 2071.

15. K. Iida, K. Hase, K. Shimomura, S. Sudo, S. Kadota, Planta. Med. 61 (1995) 425.

16. T. Klabunde, C. Eicken, J. C. Sacchettini, Nature Structural Biology. 5 (1998) 1084

17. M. Criton, FR 2880022, Jun. 26, 2006

18. J. Daniel, US 2006135618, Jun. 22, 2006

19. Hearing, V. J. In Methods in Enzymology; Academic Press: New York, 1987; Vol. 142, pp 154.

20. Y. Shi, Q. X. Chen, Q. Wang, K. K. Song, L. Qiu, Food. Chem. 92 (2005) 707.

21. H. S. Lee, J. Agric. Food. Chem. 50 (2002) 1400.

22. S. Kazuhisa, N. Koji, J. Biosci. Bioeng. 99 (2005) 272.

$$R(H)C{=}O \xrightarrow[\text{Methanol, reflux}]{NH_2NHCSNH_2} R(H)C{=}NNHCSNH_2$$

1a–1e

1a, R = CH_3; 1b, R = C_2H_5; 1c, R = *Iso* – C_3H_7; 1d, R = $CH_3CH=CH$; 1e, R = $PhCH=CH$

$$R(H_3C)C{=}O \xrightarrow[\text{Methanol, reflux}]{NH_2NHCSNH_2} R(H_3C)C{=}NNHCSNH_2$$

1f–1j

1f, R = CH_3; 1g, R = C_2H_5; 1h, R = $CH=C(CH_3)_2$; 1i, R = 4 – $CH_3OC_6H_4CH_2$; 1j, R^2 = 4 – $HOC_6H_4CH_2CH_2$.

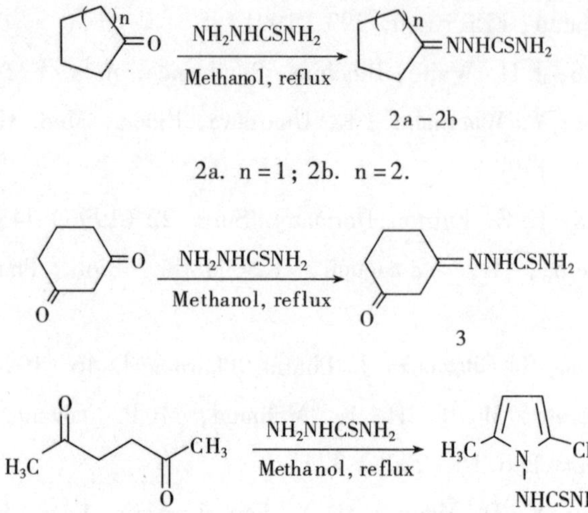

NH₂NHCSNH₂ / Methanol, reflux

2a－2b

2a. n＝1; 2b. n＝2.

3

4

Scheme 1. Synthesis of alkylidenethiosemicarbazide compounds

Table 1 Tyrosinase inhibitory activities of alkylidenethiosemicarbazide Compounds, as compared with the reference inhibitors

Samples	1a	1b	1c	1d	1e	1f	1g	1h
Yield(%)	89.2	83.6	85.1	81.5	76.2	91.5	88.3	53.6
IC_{50}(μM)	0.23	0.20	0.62	1.00	2.70	0.086	0.28	11.50
Samples	1i	1j	2a	2b	3	4	4－methoxy cinnamic acid[a]	Arbutin[a]
Yield(%)	68.1	56.8	78.3	74.6	51.0	46.8		
IC_{50}(μM)	0.42	0.54	0.17	0.95	15.10	0.85	0.41(mmol/L)[a]	10.40(mmol/L)[b]

[a] IC_{50} values in the literature is 0.34－0.43 mmol/L. [20,20]

[b] The concentration of 10.40 (mmol/L) corresponding to inhibition percentage, determined in this work, is 30%. The reported IC_{50}(μM) value of arbutin was more than 30 mmol/L [22].

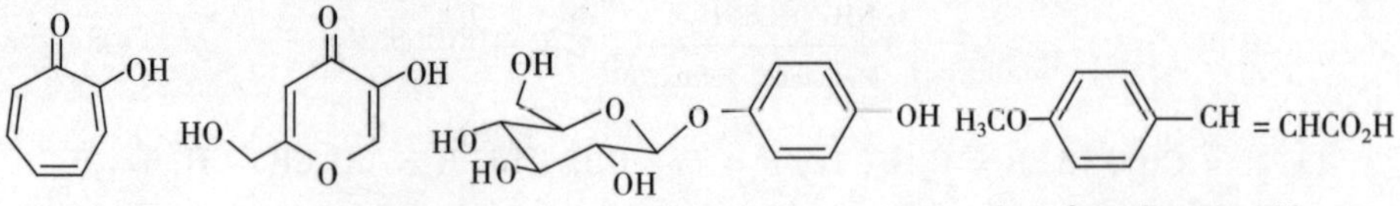

tropolone　kojic acid　arbutin　4-methoxycinnamic acid

Fig. 1 Structures of some reported tyrosinase inhibitors

2.6.4 Bioorganic & Medicinal Chemistry(2008, 16, 1096 – 1102)

1 – (1 – Arylethylidene) thiosemicarbazide derivatives: A new class of tyrosinase

inhibitors Jinbing Liu, Wei Yi, Yiqian Wan, Lin Ma and Huacan Song

Abstract—A series of 1 – (1 – arylethylidene) thiosemicarbazide compounds and their analogues were synthesized and characterized by ^{1}H NMR, MS. Their tyrosinase inhibitory activities were investigated by an assay based on the catalyzing ability of tyrosinase for the oxidation of L – DOPA, comparing with 4 – methoxycinnamic acid and arbutin. The results showed that (1) all the synthesized compounds could perform a significant inhibitory activity for tyrosinase; (2) for these compounds, the main active moiety interacting with the center of tyrosinase would be thiosemicarbazo group; (3) the inhibitory activity was close related with thiosemicarbazide moieties and the groups attached on the aromatic ring.

1. Introduction

Tyrosinase(monophenol oro – diphenol, oxygen oxidoreductase, EC 1. 14. 18. 1), also known as polyphenol oxidase (PPO), is a copper – containing monooxygenase that is widely distributed in microorganisms, animals, and plants.[1-6] Tyrosinase could catalyze two distinct reactions involving molecular oxygen[7-13] in the hydroxylation of monophenols to o – diphenols (monophenolase) and in the oxidation of o – diphenols to o – quinones (diphenolase). Due to the high reactivity, quinones could polymerize spontaneously to form high molecular weight brown – pigments (melanins) or react with amino acids and proteins to enhance brown color of the pigment produced. In addition, tyrosinase is known to be involved in the molting process of insect, and adhesion of marine organisms.[14-17]

Therefore, there is an increasing significance to discover effective tyrosinase inhibitors for clinical medicines, cosmetic products, food industry as well as agricultural purposes.[18-24] Up to now, although a large number of tyrosinase inhibitors were reported, most of them could not be used now[25-35] because of their lower individual activities or side effect. For example, 2 – (4 – hydroxyphenoxy) tetrahydro – 6 – (hydroxylmethyl) – 2H – pyran – 3,4,5 – triol(arbutin or arbutoside)[36] (IC_{50} = 30mmol/L) and kojic acid[37] (IC_{50} = 23μmol/L) (Fig. 1) were not demonstrated as their clinically efficient.[38] So far tropolone is one of the most strong tyrosinase –

inhibitors reported (Fig. 1) (IC_{50} = 0.4μmol/L),[39] but the serious side effect limited its use as medicine.

On the other hand, it was reported that phenyl thioureas[40] and alkyl thioureas[41] could exhibit weak to moderate depigmenting activity and Ley and Bertram[28] reported that benzaldoximes and benzaldehyde - o - alkyloximes possess higher tyrosinase inhibitory ability. Based on these reports, we hoped that condensation products of acetophenone homologues and thiosemicarbazide could show a high inhibition activity, as sulfur atom and nitrogen atom are able to complex the two copper atoms in the active site of tyrosinase.

In order to look for highly potent tyrosinase inhibitors, a series of 1 - (1 - arylethylidene) thiosemicarbazide derivatives were synthesized and their inhibitory activities against mushroom tyrosinase were evaluated using arbutin and 4 - methoxycinnamic acid as comparing substances. Meanwhile, the structure - activity relationships of these compounds were also primarily discussed.

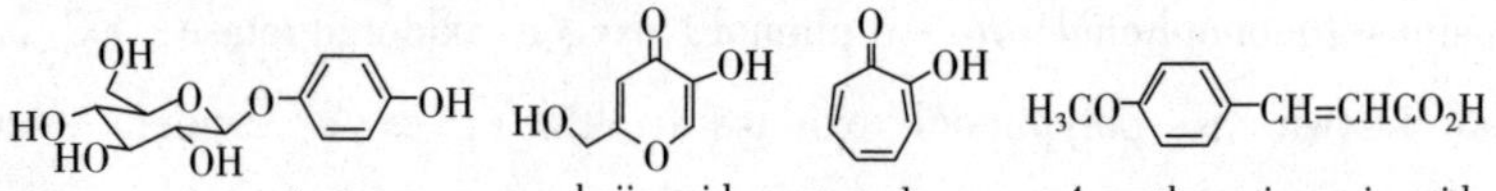

arbutin kojic acid tropolone 4-methoxycinnamic acid

Figure 1. Structures of some known tyrosinase inhibitors.

2. Results and discussion

2.1 Chemistry

According to the general procedure shown in Scheme 1, the synthesis of a series of 1 - (1 - arylethylidene) thiosemicarbazide compounds and their analogues could be carried out easily in anhydrous alcohol, using acetic acid as catalyst, just by the condensation of methyl ketone or aldehyde with thiosemicarbazide in the molecular ratio of 1 : 1 or 1 : 2. The reaction mixture was refluxed for 24 h and then cooled to room temperature, the formed precipitate was separated by filtration. If necessary, the target compounds could be purified by recrystallization from 95% alcohol. The yields for these reactions are from fair to good (65 - 86%), and all the compounds were characterized by ^{1}H NMR, MS.

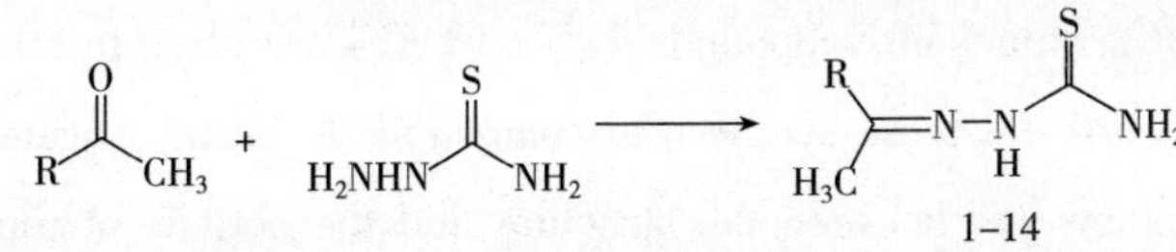

R=:1,phenyl;2,4–methylphenyl;3,4–hydroxyphenyl;4,2,4–dihydroxyphenyl;
5,2,4,6–trihydroxyphenyl;6,4–flourophenyl;7,4–bromophenyl;
8,4–isopropylphenyl;9,4–methoxyphenyl;10,2–pyrazinyl;11,2–thiophenyl;
12,3–pyridinyl;13,(4–methoxypheny)methyl;14,2–(4–hydroxyphenyl)ethyl.

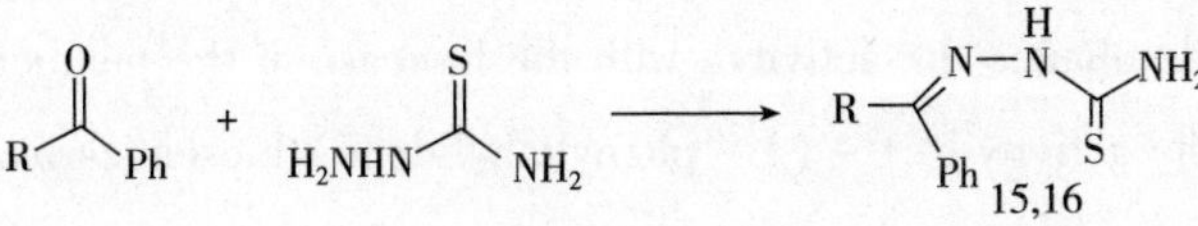

R=:15,hydroxy(pheny)methyl;16,benzoyl.

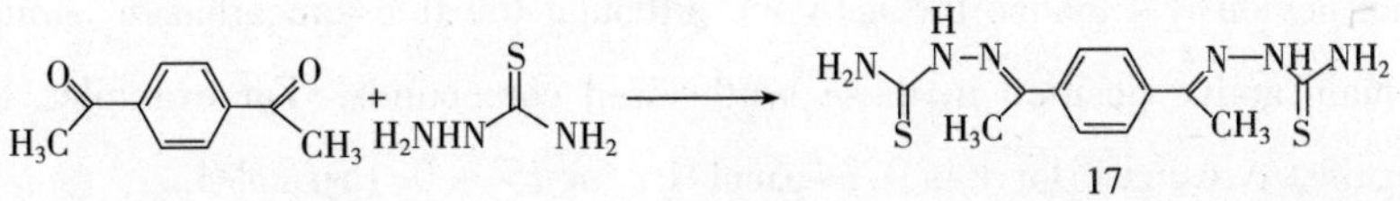

Scheme 1 Synthesis of 1 – (1 – arylethylidene) thiosemicarbazide compounds and their analogues.

2.2 Inhibiting activity

For evaluating the tyrosinase inhibitory activity, all the synthesized compounds were subjected to tyrosinase inhibition assay with L – DOPA as substrate, according to the typical assay protocol developed by Hearing.[42] The tyrosinase inhibitory activity of arbutin and 4 – methoxycinnamic acid was ever reported,[6,36] therefore, they were selected as comparing substances. The IC_{50} values of 1 – (1 – arylethylidene) thiosemicarbazide compounds and their analogues against tyrosinase were summarized in Table 1.

The IC_{50} value of 4 – methoxycinnamic acid, determined in this work, is 0. 41 mmol/L, the reported IC_{50} value of 4 – methoxycinnamic acid is 0. 34 mmol/L and 0. 43 mmol/L;[6,43] therefore, both the values reported and determined in this work are close to each other, which means that the protocol employed in this work could be reasonable. Tropolone (in Fig. 1), as tyrosinase inhibitor, could show an IC_{50} value of 0. 4μmol/L.[39] It could be found from Table 1that compounds 1, 2, 3, 6, 9, 11, and 17 showed a little higher inhibiting ability against tyrosinase than tropolone.

Meantime, the IC_{50} values of compounds 4, 7, 13, 14 are close to that of tropolone. Therefore, it is worth for these synthesized compounds to be investigated further.

For the relationships between the structure and the activity of these synthesized compounds, some results could be obtained from the data listed in Table 1:

(1) For the 1 - (1 - phenylethylidene) thiosemicarbazide compounds, the substituents attached on the benzene ring could change the inhibition activity. Methyl, hydroxyl, fluoro, and methoxy on the 4 - positon of benzene ring could enhance the activity, but bromo and isopropyl could decrease the activity. Although hydroxyl on 4 - position could enhance the activity, with the increase of the number of hydroxyl on benzene ring, the activity of 1 - (1 - phenylethylidene) thiosemicarbazide compounds would decrease.

(2) The increase in the number of thiosemicarbazido group in one molecule could not obviously enhance the activity, although the thiosemicarbazido group could be the main active position in these synthesized compounds. For example, the IC_{50} value against tyrosinase for 1 is 0.34μmol/L, for 17 is 0.15μmol/L.

(3) When the benzene ring of phenylethylidenethiosemicarbazide molecule was changed by heterocyclic ring, the obtained ethylidenethiosemicarbazide compounds could still show a stronger activity against tyrosinase. For example, the IC_{50} value for 11 is 0.14μmol/L.

(4) When thiosemicarbazido group and benzene ring are separated by one or two methene groups, the change on the inhibitory activity of the obtained ethylidenethiosemicarbazide compounds is not significant.

(5) The inhibitory activity of corresponding compounds would be decreased obviously when the methyl of 1 - (1 - phenylethylidene) thiosemicarbazide compounds was changed into hydroxy (phenyl) methyl or benzoyl.

Table 1 Inhibitory activity of 1 - (1 - arylethylidene) thiosemicarbazide against ryrosinase

Compound	1	2	3	4	5	6	7	8
IC_{50} (μmol/L)	0.34	0.27	0.31	0.58	22.0	0.17	0.52	1.0
Compound	9	10	11	12	13	14	15	17
IC_{50} (μmol/L)	0.11	0.88	0.14	0.82	0.42	0.54	55.5	0.15

续表

Compound	16[a]	Thiosemicarbazide[b]	1 – (Thiophen – 2 – yl) – ethanone	Acetophenone
IC_{50} (mmol/L)	0.10[a]	2.00[b]	0.15	0.85
Compound	1 – (4 – Fluoro phenyl) – ethanone	1 – (4 – Methoxy phenyl) – ethanone[c]	4 – Methoxycin namicacid[d]	Arbutin[e]
IC_{50} (mmol/L)	0.99	2.00[c]	0.41[d]	10.4[e]

[a] The concentration of 0.10 (mmol/L) corresponding to inhibition percentage, determined in this work, is 47%.

[b] The concentration of 2.00 (mmol/L) corresponding to inhibition percentage, determined in this work, is 39.9%.

[c] The concentration of 2.00 (mmol/L) corresponding to inhibition percentage, determined in this work, is 47.2%.

[d] The reported IC_{50} values of 4 – methoxycinnamic acid is 0.34 – 0.43 mmol/L.

[e] For arbutin, the concentration of 10.40 (mmol/L) corresponding to inhibition percentange, determined in this work, is 30%. The reported IC_{50} value of arbutin was more than 30 mmol/L.[36]

2.3 Inhibiting mechanism

It was reported that the structure of tyrosinase was determined[44] (shown in Fig. 2), there are two copper ions in the active center of tyrosinase and it was deduced that there is a lipophilic long – narrow gorge near to the active center.[45]

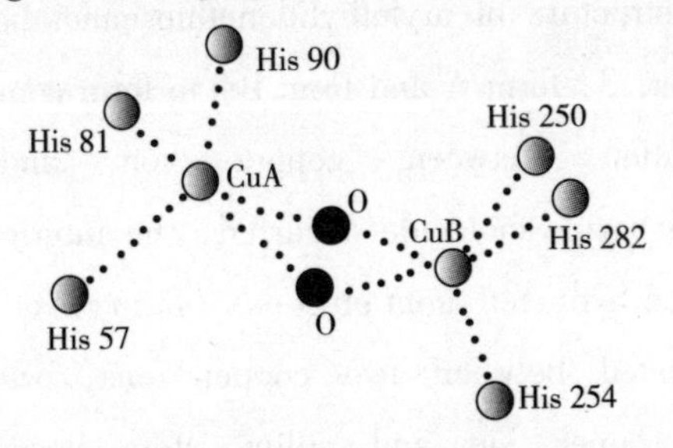

Figure 2 The active center structure of tyrosinase.

There are sulfur atom and nitrogen atom in the molecule of arylethylidenethiosemicarbazide compounds and their analogues, and sulfur atom and nitrogen atom could exhibit strong affinity for copper ion. Therefore, it could be supposed that complexes would be formed between arylethylidenethiosemicarbazide compounds and tyrosinase when both substrates were mixed together in solution. In addition, sulfur atom and nitrogen atom, especially sulfur atom, of arylethylidenethiosemicarbazide compounds and copper ion of tyrosinase could be the center of complexation. The structures of the complexes are suggested as shown in

Figure 3, based on the active center structure of tyrosinase.

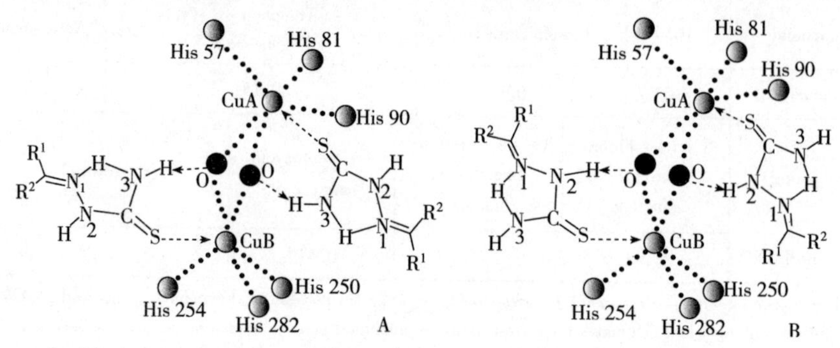

Figure 3 The proposed structure of the formed complexes between tyrosinase and synthesized compounds.

In the complex, the active center of tyrosinase could coordinate with two arylethylid enethiosemicarbazide molecules at the same time in two opposite directions. It is suggested that the intramolecule hydrogen bond was formed between hydrogen atom on 3 – N and nitrogen atom (2 – N) in arylethylidenethiosemicarbazide molecule, which are beneficial to decreasing molecular energy state.

According to the structure of arylethylidenethiosemicarbazide molecules, there would be two forms (Fig. 3, form A and form B) to form complexes with tyrosinase. When the coordination between copper ion and sulfur atom of arylethylidenethiosemicarbazide molecule occurred, the intermolecule hydrogen bond would be formed between hydrogen atom on 3 – N (form A) or on 2 – N (formB) and two oxygen atoms located between two copper ions, which would make the coordination between copper ion and sulfur atom become more tight, and consequently, which could make the free oxygen molecule decrease its reaction ability and even unable to take part in the hydroxylation with monophenols and in the oxidation witho – diphenols, as the free oxygen molecule was surrounded closely by two copper ions from tyrosinase and two hydrogen atoms from two arylethylidenethiosemicarbazide molecules. Therefore, tyrosinase would lose its catalyzing ability.

As for the complex form A and form B (Fig. 3), which one is the best beneficial form? Based on the molecular energy state, form A should be the best one. For the form A, comparing with form B, larger substituted – benzene ring (R^1) and methyl (or H) (R^2) could depart from tyrosinase for a longer distance, and consequently, the bulk inhibition from R^1 and R^2 would decrease, which would decrease the energy of the complex and which is just beneficial to the formation of the complexes between arylethylidenethiosemicarbazide and tyrsoinase.

In order to confirm this conclusion, four compounds, acetophenone, 1 – (4 – fluorophenyl) ethanone, 1 – (4 – methoxyphenyl) ethanone, and 1 – (thiophen – 2 – yl) ethanone, which are corresponding to the starting materials of compounds1, 6, 9, and 11were selected and their inhibiting abilities against tyrosinase determined at the same condition, their IC_{50} value are all more than 0. 15 mmol/L. At the same time, when the concentration of thiosemicarbazide is 2. 00 mmol/L, the corresponding inhibition percentage, determined in this work, is only 39. 9%.

On the other hand, there is a sulfur atom in molecule of 1 – (thiophen – 2 – yl) – ethanone, and this sulfur atom should exhibit strong affinity for copper atom of tyrosinase. However, 1 – (thiophen – 2 – yl) ethanone just expressed a weaker inhibiting ability against tyrosinase at the same condition, which means that the sulfur atom of 1 – (thiophen – 2 – yl) ethanone possesses a weaker affinity or a weaker coordination ability than that of thiosemicarbazido group of compounds 11, 1 – (1 – (thiophen – 2 – yl) ethylidene) thiosemicarbazide, although both sulfur atoms of compounds 11 could possess chance to take part in the complexation at the same time. The reason could be that there is a larger bulk inhibition for the coordination between copper ion and sulfur atom of thiophene; meantime, there is a strong competition from the coordination between copper ion and sulfur atom of thiosemicarbazido group. Therefore, the thiosemicarbazido group could be the important active position in the molecule of arylethylidenethiosemicarbazide compounds as tyrosinase inhibitor, and the major form of the complexation of tyrosinase with compound 11 is supposed as B, instead of A (Fig. 4).

As to the lower IC_{50} value of thiosemicarbazide, comparing with that of arylethylidenethiosemicarbazide compounds and their homologues, the following two main reasons could be suggested here: (1) Because thiosemicarbazide was hydrophilic and tyrosinase was lipophilic and all the tyrosinase inhibition assays were determined almost in phosphate buffer, there was a less chance for thiosemicarbazide and tyrosinase to interact to form complex; (2) The presence of substituted – aromatic ring in arylethylidenethiosemicarbazide compounds would be the main beneficial reason, because it was deduced that there is a lipophilic longnarrow gorge near to the active center[45] and the lipophilic aromatic ring of arylethylidenethiosemicarbazide compounds could easily get close to the gorge, which would make the thiosemicarbazido group of arylethylidenethiosemicarbazide compounds get close to the active center of tyrosinase much more easily. In other words, lipophilic moiety of arylethylidenethiosemicarbazide compounds could bring the thiosemicarbazido group to the site close to the active center of tyrosinase, which was beneficial for the coordination between the active center of tyrosinase and the thiosemicarbazido group of thiosemicarbazide compounds. It is the stronger coordination between the active center of tyrosinase and arylethylidenethiosemicarbazide compounds than that between the active center of tyrosinase and thiosemicarbazide, arylethylidenethiosemicarbazide compounds would show a stronger inhibiting ability than thiosemicarbazide against tyrosinase.

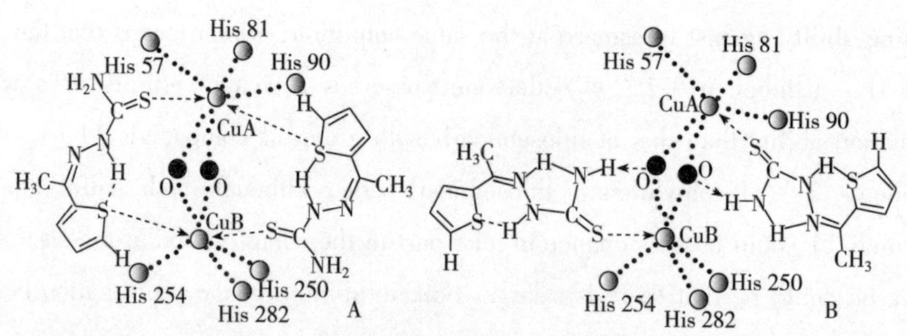

Figure 4 The proposed structures of the complexation of tyrosinase with arylethylidenethiosemicarbazide compounds and 1 - (1 - (thiophen - 2 - yl) ethylidene) thiosemicarbazide.

3. Conclusions

In this study, a series of 1 - (1 - arylethylidene) thiosemicarbazide compounds and their analogues were synthesized and characterized by ^{1}H NMR, MS. Their inhibitory activities against tyrosinase were studied based on the catalyzing ability of tyrosinase for the oxidation ofL - DOPA, using arbutin and 4 - methoxycinnamic acid as comparing substrate. The following conclusions would be obtained: (1) 1 - (1 - arylethylidene) thiosemicarbazide compounds could perform a significant inhibitory activity for tyrosinase; seven compounds could show stronger inhibiting ability than reported tropolone, four compounds showed nearly same IC_{50} values with tropolone. (2) The form of the interaction between these compounds and tyrosinase could be supposed as in Figure 3 (form A) in which the formation of hydrogen bond between hydrogen atom on 3 – N (form A) and free oxygen molecule located between two copper ions of tyrosinase would make the free oxygen molecule decrease its reaction ability and even unable to take part in the hydroxylation and in the oxidation. (3) For these investigated compounds, it was suggested that the main active moiety interacting with the center of tyrosinase would be thiosemicarbazido group. (4) The liposoluble group of arylethylidenethiosemicarbazide compounds would make the interaction between the active center of tyrosinase and thiosemicarbazido moiety become much easier, which would make tyrosinase lose its catalyzing ability. (5) Although thiosemicarbazido group could be the main active part, the increase in the number of thiosemicarbazido group could not obviously increase their activity. (6) The changes of the substituent on benzene ring would exert an influence on the inhibitory activity. Meanwhile, when benzene was replaced by heterocyclic ring, the corresponding activity would have little change.

4. Experimental

4.1 Equipments and reagents

Melting points (mp) were determined with WRS – 1B melting point apparatus and are uncorrected. NMR spectra were recorded on Mercury – Plus300 spectrometers at 25℃ using $CDCl_3$ or DMSO – d_6 as a solvent. All chemical shifts (d) are quoted in ppm downfield from TMS and coupling constants (J) are given in Hz. LC – MS spectra were recorded using the LCMS – 2010A. All reactions were monitored by TLC (Merck Kieselgel 60 F254) and the spots were visualized under UV light. The appropriate aldehyde or ketone, thiosemicarbazide, and 4 – methoxycinnamic acid were purchased from Darui Chemical Co. (ShangHai, China), arbutin was obtained from Brillian Biochemical Co. (Beijing, China). The other commercially available reagents and solvents were used without further purification. Mushroom tyrosinase and L – DOPA were purchased from Sigma Chemical Co. (Sigma – Aldrich China, Beijing, China).

4.2 Synthesis of thiosemicarbazides

4.2.1 The general procedures for the synthesis of 1 – (1 – arylethylidene) – thiosemicarbazide compounds. The appropriate aldehyde or ketone (10 mmol) was dissolved in anhydrous ethanol (10 mL), thiosemicarbazide (10 mmol) and acetic acid (0.5 mL) were added to the above solution. The reaction mixture was refluxed for 24 h and then was cooled to room temperature. The appearing precipitate was filtered and recrystallized from 95% alcohol to obtain the corresponding 1 – (1 – arylethylidene) thiosemicarbazide compounds.

4.2.2 1 – (1 – Phenylethylidene) thiosemicarbazide (1). Yield 86%; mp 132 ~134℃; ^1H NMR (300 MHz, DMSO – d_6): δ 8.95 (1H, br s, NH), 7.70 (2H, d, J = 1.2 Hz, phH), 7.48 (1H, br s, NH2), 7.41 (3H, m, J = 1.2 Hz, phH), 6.53 (1H, br s, NH2), 2.31 (3H, s, CH3). MS (ESI): m/z (100%) = 194 (M + 1).

4.2.3 1 – (1 – p – Tolylethylidene) thiosemicarbazide (2). Yield78%; mp 158 ~160℃; ^1H NMR (300 MHz, DMSO – d_6): δ 8.77 (1H, br s, NH), 7.61 (2H, d, J = 7.5 Hz, phH), 7.36 (1H, br s, NH2), 7.21 (2H, d, J = 7.5 Hz, phH), 6.42 (1H, br s, NH2), 2.39 (3H, s, $COCH_3$), 2.29 (3H, s, ph – CH3). MS (ESI): m/z (100%) = 208 (M + 1).

4.2.4 1 – (1 – (4 – Hydroxyphenyl) ethylidene) thiosemicarbazide (3). Yield 73%; mp 208 ~209℃; ^1H NMR (300 MHz, DMSO – d_6): δ 10.02 (1H, br s, NH), 9.69 (1H, s, OH), 8.13 (1H, br s, NH_2), 7.77 (2H, d, J = 8.7 Hz, phH), 7.72 (1H, br s, NH_2), 6.75 (2H, d, J = 8.7 Hz, phH), 2.22 (3H, s, CH_3). MS (ESI): m/z (100%) = 208 (M – 1).

4.2.5 1 – (1 – (2,4 – Dihydroxyphenyl) ethylidene) thiosemicarbazide (4). Yield 68%; mp 186 ~187℃; ^1H NMR (300 MHz, DMSO – d_6): δ 12.57 (1H, s, OH), 10.57 (1H, s, OH), 9.72 (1H, br s, NH), 7.74 (1H, s, NH_2), 7.70 (1H, s, NH_2), 7.36 (H, d, J = 8.4 Hz, phH), 6.37 (1H, d, J = 8.4 Hz, phH), 6.23 (1H, s, phH), 2.25 (3H, s, CH_3). MS (ESI): m/z (100%) = 224 (M – 1).

4.2.6 1 – (1 – (2,4,6 – Trihydroxyphenyl) ethylidene) thiosemicarbazidc (5). Yield 65%; mp 211 ~213℃; ^1H NMR (300 MHz, DMSO – d_6): δ 12.18 (2H, s, OH), 10.32 (1H, s, OH), 9.72 (1H, br s, NH), 7.73 (1H, br s, NH_2), 7.36 (1H, br s, NH_2), 5.78 (2H, s, phH), 2.54 (3H, s, CH_3). MS (ESI): m/z (100%) = 240 (M – 1).

4.2.7 1 – (1 – (4 – Fluorophenyl) ethylidene) thiosemicarbazide (6). Yield 82%, mp 154 ~156℃; ^1H NMR (300 MHz, DMSO – d_6): δ 10.17 (1H, br s,

NH), 8.24 (1H, br s, NH_2), 7.98 (2H, d, J = 6.9 Hz, phH), 7.78 (1H, br s, NH_2), 7.20 (2H, d, J = 6.9 Hz, phH), 2.28 (3H, s, CH_3). MS (ESI): m/z (100%) = 212 (M+1).

4.2.8 1-(1-(4-Bromophenyl) ethylidene) thiosemicarbazide (7). Yield 79%, mp 190 ~ 192℃; ^{1}H NMR (300 MHz, DMSO-d_6): δ 10.08 (1H, br s, NH), 8.28 (1H, br s, NH_2), 7.97 (1H, br s, NH_2), 7.89 (2H, d, J = 7.8 Hz, phH), 7.55 (2H, d, J = 7.8 Hz, phH), 2.27 (3H, s, CH_3). MS (ESI): m/z (100%) = 273 (M+1).

4.2.9. 1-(1-(4-Isopropylphenyl) ethylidene) thiosemicarbazide (8). Yield 68%, mp 99 ~ 100℃; ^{1}H NMR (300 MHz, $CDCl_3$): δ 8.77 (1H, br s, NH), 7.64 (2H, d, J = 8.1 Hz, phH), 7.37 (1H, br s, NH_2), 7.25 (2H, d, J = 8.1 Hz, phH), 6.47 (1H, br s, NH_2), 2.94 (m, 1H, CH), 2.29 (3H, s, CH_3), 1.28 (d, 6H, $2CH_3$). MS (ESI): m/z (100%) = 236 (M+1).

4.2.10 1-(1-(4-Methoxyphenyl) ethylidene) thiosemicarbazide (9). Yield 83%, mp 175 ~ 176℃; ^{1}H NMR (300 MHz, $CDCl_3$): δ 9.04 (1H, br s, NH), 7.57 (2H, d, J = 8.7 Hz, phH), 7.27 (1H, br s, NH_2), 7.26 (1H, br s, NH_2), 7.25 (2H, d, J = 8.7 Hz, phH), 3.67 (3H, s, OCH_3), 2.16 (3H, s, CH_3). MS (ESI): m/z (100%) = 224 (M+1).

4.2.11 1-(1-(Pyrazin-2-yl) ethylidene) thiosemicarbazide (10). Yield 67%, mp 190 ~ 191℃; ^{1}H NMR (300 MHz, DMSO-d_6): δ 10.43 (1H, br s, NH), 9.62 (1H, s, pyrazine proton), 8.58 (2H, d, pyrazine protons), 8.44 (1H, br s, NH_2), 8.28 (1H, br s, NH_2), 2.36 (3H, s, CH_3). MS (ESI): m/z (100%) = 196 (M+1).

4.2.12 1-(1-(Thiophen-2-yl) ethylidene) thiosemicarbazide (11). Yield 65%, mp 146 ~ 147℃; ^{1}H NMR (300 MHz, DMSO-d_6): δ 8.86 (1H, br s, NH), 7.35 (1H, d, thiophen), 7.31 (1H, d, thiophene proton), 7.28 (1H, br s, NH_2), 7.04 (1H, t, thiophene proton), 6.59 (1H, br s, NH_2), 2.31 (3H, s, CH_3). MS (ESI): m/z (100%) = 200 (M+1).

4.2.13 1-(1-(Pyridin-3-yl) ethylidene) thiosemicarbazide (12). Yield 70%, mp 212 ~ 214℃; ^{1}H NMR (300 MHz, DMSO-d_6): δ 10.29 (1H, br s, NH), 9.08 (1H, s, pyridine proton), 8.54 (1H, d, pyridine proton), 8.31 (1H, d, pyridine proton), 8.29 (1H, br s, NH_2), 8.05 (1H, br s, NH_2), 7.40 (1H, t, pyridine proton), MS (ESI): m/z (100%) = 195 (M+1).

4.2.14 1-(1-(4-Methoxyphenyl) propan-2-ylidene) thiosemicarbazide (13). Yield 88%, mp 122~123℃; ^{1}H NMR (300 MHz, DMSO-d_6): δ 8.50 (1H, br s, NH), 7.26(1H, br s, NH_2), 7.10 (2H, d, J = 8.7 Hz, phH), 6.86 (2H, d, J = 8.7 Hz, phH), 6.34 (1H, br s, NH_2), 3.80(3H, s, OCH_3), 3.50 (2H, s, CH_2), 1.84 (3H, s, CH_3). MS (ESI): m/z (100%) = 238 (M+1).

4.2.15 1-(4-(4-Hydroxyphenyl) butan-2-ylidene) thiosemicarbazide (14). Yield 71%, mp154~155℃; ^{1}H NMR (300 MHz, DMSO-d_6): δ 9.87 (1H, br s, NH), 8.50 (1H, s, OH), 7.12 (1H, br s, NH2), 7.02 (2H, d, J = 6.3 Hz, phH), 6.77 (2H, d, J = 6.3 Hz, phH), 6.22 (1H, br s, NH_2), 2.80 (2H, t, CH_2), 2.60 (2H, t, CH_2), 1.89 (3H, s, CH_3). MS (ESI): m/z (100%) = 236 (M-1).

4.2.16 1-(2-Hydroxy-1,2-diphenylethylidene) thiosemicarbazide (15). Yield 58%, mp 178~179℃; ^{1}H NMR (300 MHz, DMSO-d_6): δ 11.44 (1H, br s, NH), 8.40 (1H, br s, NH_2), 7.97 (1H, br s, NH_2), 7.90 (t, 2H, phH), 7.36 (m, 8H, phH), 7.29 (s, 1H, OH), 6.27 (d, 1H, CH). MS (ESI): m/z (100%) = 286 (M+1).

4.2.17 1-(2-Oxo-1,2-diphenylethylidene) thiosemicarbazide (16). Yield 62%, mp 203~204℃; ^{1}H NMR (300 MHz, DMSO-d_6): δ 11.87 (1H, br s, NH), 8.20 (1H, br s, NH_2), 7.39 (m, 10H, phH), 6.75 (1H, br s, NH). MS (ESI): m/z (100%) = 284 (M+1).

4.2.18 1-(1,4-Diacetylphenyl) dithiosemicarbazide (17). Yield 82%, mp > 280℃; ^{1}H NMR (300 MHz, $CDCl_3$): 810.22 (1H, br s, NH), 8.28 (2H, br s, NH_2), 7.94 (2H, br s, NH_2), 2.30 (6H, s, 2 · CH_3). MS (ESI): m/z (100%) = 309 (M+1).

4.3 Tyrosinase assay

Tyrosinase inhibition assays were performed according to the method developed by Hearing with slight modification.[42] Briefly, all the synthesized compounds were screened for the o-diphenolase inhibitory activity of tyrosinase using L-DOPA as substrate. All the active inhibitors from the preliminary screening were subjected to IC_{50} studies. All the synthesized compounds were dissolved in DMSO to a concentration of 2. 5%. Phosphate buffer, pH 6. 8, was used to dilute the DMSO stock solution of test compound. Thirty units of mushroom tyrosinase (28 nM) was first pre-incubated with the compounds, in 50 nM phosphate buffer (pH 6. 8), for 10 min at 25℃. Then the L-DOPA (0. 5 mmol/L) was added to the reaction mixture and the enzyme reaction was monitored by measuring the change in absorbance at 475 nm of formation of the

DOPAchrome for 10 min. Dose - response curves were obtained by performing assays in the presence of increasing concentrations of inhibitors (0, 1.6, 3.2, 6.3, 12.5, 25, 50, 100, 200μmol/L). IC_{50} value, a concentration giving 50% inhibition of tyrosinase activity, was determined by interpolation of the dose - response curves. The percent of inhibition of tyrosinase reaction was calculated as follows:

percent inhibition (%) = $[(B-S)/B] \times 100$.

Here, theBandSare the absorbances for the blank and samples, respectively. 4-Methoxycinnamic acid and arbutin were used as standard inhibitors for the tyrosinase.

References and notes

1. Seo, S. Y.; Sharma, V. K.; Sharma, N. J. Agric. Food. Chem. 2003, 51, 2837.

2. Van Gelder, C. W. G.; Flurkey, W. H.; Wichers, H. Phytochemistry,, 1997, 45, 1309.

3. Olivares, C.; Jime'nez-Cervantes, C.; Lozano, J. A. F. Biochemistry, 2001, 354, 131.

4. Friedman, M. J. Agric. Food. Chem. 1996, 44, 631.

5. Whitaker, J. R. Polyphenol Oxidase. InFood Enzymes, Structure and Mechanism; Wong, D. W. S., Ed.; Chapman & Hall: New York, 1995; p 271.

6. Shi, Y.; Chen, Q. X.; Wang, Q.; Song, K. K. Food. Chem. 2005, 92, 707.

7. Li, B.; Huang, Y.; Paskewitz, S. M. FEBS Lett. 2006, 580, 1877.

8. Seo, S. Y.; Sharma, V. K.; Sharma, N. J. Agric. Food. Chem. 2003, 51, 2837.

9. Hearing, V. J.; Tsukamoto, K. FASEB. J. 1991, 5, 2902.

10. Nerya, O.; Musa, R. Postharvest. Biol. Tech. 2006, 39, 272.

11. Khan, K. M.; Maharvi, G. M.; Khan, M. T. H. Bioorg. Med. Chem.

2006, 14, 344.

12. Nerya, O. ; Musa, R. ; Khatib, S. Phytochemistry, 2004, 65, 1389.

13. Kubo, I. J. Agric. Food Chem. 1999, 47, 4121.

14. Kramer, K. J. ; Kanost, M. R. ; Hopkins, T. L. Tetrahedron, 2001, 57, 385.

15. (a) Barrett, F. M. Can. J. Zool. 1984, 62, 834; (b) Sugumaran, M. Adv. Insect Physiol. 1988,21, 179; (c) Lee, S. E. ; Kim, M. K. ; Lee, S. -G. ; Ahn, Y. J. ; Lee, H. S. Food. Sci. Biotechnol. 2000, 9, 330. 16. Sugumarau, M. FEBS Lett. 1991, 293,4.

17. Marumo, K. ; Waite, J. H. Biochim. Biophys. Acta, 1986, 872, 98.

18. Gruber, P. ; Vieths, S. ; Wangorsch, A. ; Nerkamp, J. ; Hofmann, T. J. Agric. Food. Chem. 2004, 52, 4002.

19. Huang, X. H. ; Chen, Q. X. ; Wang, Q. Food. Chem. 2006, 94,1.

20. Jime′nez, M. J. Agric. Food. Chem. 1997, 45, 2061.

21. Hollman, P. C. H. ; Gaag, M. V. D. ; Mengelers, M. J. B. Free Radic. Biol. Med. 1996, 21, 703.

22. Mosher, D. B. ; Pathak, M. A. ; Zpatrick, T. B. In Update: Dermatology in General Medicine; Fitzpatrick, A. Z. , Eisen, T. B. , Wolff, K. , Freedberg, I. M. , Austen, K. F. ,

Eds. ; McGraw - Hill: New York, 1983; Vol. 1, p 205.

23. Maeda, K. ; Fukuda, M. J. Soc. Cosmet. Chem. 1991, 42, 361.

24. Friedman, M. J. Agric. Food. Chem. 1996, 44, 631.

25. Nihei, K. ; Yamagiwa, Y. Bioorg. Med. Chem. Lett. 2004, 14, 681.

26. Okombi, S. ; Rival, D. Bioorg. Med. Chem. Lett. 2006, 16, 2252.

27. Isao, K. ; Chen, Q. X. ; Nihei, K. Food. Chem. 2003, 81, 241.

28. Ley, J. P. ; Bertram, H. J. Bioorg. Med. Chem. 2001, 9, 1879.

29. Khatib, S. ; Nerya, O. Bioorg. Med. Chem. 2005, 13, 433.

30. Cho, S. J. ; Roh, J. S. Bioorg. Med. Chem. Lett. 2006, 16, 2682.

31. Shiino, M. ; Watanabe, Y. Bioorg. Med. Chem. 2001, 9, 1233.

32. Khan, M. T. H. ; Choudhary, M. I. Bioorg. Med. Chem. 2005, 13, 3385.

33. Shiino, M. ; Watanabe, Y. Bioorg. Chem. 2003, 31, 129.

34. Um, S. J. ; Park, M. S. Bioorg. Med. Chem. 2003, 11, 5345.

35. Khan, K. M. ; Maharvi, G. M. Bioorg. Med. Chem. 2006, 14, 344.

36. Sugimoto, K. ; Nomura, K. J. Biosci. Bioeng. 2005, 99, 272.

37. Iida, K. ; Hase, K. ; Shimomura, K. ; Sudo, S. ; Kadota, S. ; Namba, T. Planta Medica, 1995, 61, 425.

38. Briganti, S. ; Camera, E. ; Picardo, M. Pigm. Cell. Res. 2003, 16, 101.

39. Kahn, V. ; Andrawis, A. Phytochemistry1985, 24, 905.

40. Klabunde, T. ; Eicken, C. Nature Struct. Biol. 1998, 5, 1084.

41. Daniel, J. US Patent 2006135618, 2006.

42. Hearing, V. J. In Methods in Enzymology; Academic Press: New York, 1987; Vol. 142, p 154.

43. Lee, H. S. J. Agric. Food. Chem. 2002, 50, 1400.

44. Longa, S. D. ; Ascone, I. ; Bianoconi, A. J. Biol. Chem. 1996, 271, 21025.

45. Chen, Q. X. ; Lin, G. F. ; Song, K. K. J. Xiamen Univ. (Natural Sci.) 2007, 46, 274

2.6.5 Bioorganic & Medicinal Chemistry Letters(2015, 25, 5142 - 5146)

Design and synthesis of aloe - emodin derivatives as potent anti - tyrosinase, antibacterial and anti - inflammatory agents

Jinbing Liu*, Fengyan Wu, Changhong Chen

Department of Biology and Chemical Engineering, Shaoyang University, Shao Shui Xi Road, Shaoyang 422100, *PRC*

* Corresponding author: Department of Biology and Chemical Engineering, Shaoyang University, Shao Shui Xi Road, Shaoyang 422100, PRC

Phone: +86 - 739 - 5431768; Fax: +86 - 739 - 5431768

E - mail address: syuliujb@163. com

Abstract

Twentyaloe - emodin derivatives were designed, synthesized, and their biological activities were evaluated. Some compounds displayed potent tyrosinase inhibitory activities, especially, compounds with thiosemicarbazide moiety showed more potent inhibitory effects than the other compounds. The structure - activity relationships (SARs) were preliminarily discussed. The inhibition mechanism of selected compounds 1 and 13 were investigated. The results showed compound 1 was reversible inhibitor, however, compound 13 was irreversible. Kinetic analysis indicated that compound 1 was competitive tyrosinase inhibitor. Furthermore, the antibacterial activities and anti - inflammatory activities of some selected compounds were also screened. The results showed that compound 3 exhibited more potent antibacterial activity than the aloe - emodin, compounds 5 and 6 possessed more potent anti - inflammatory activities than the diacerein.

Keywords: Aloe - emodin derivatives; tyrosinase inhibitory activities; antibacterial activity; anti - inflammatory activity.

Tyrosinase (EC 1. 14. 18. 1) is a kind of oxidoreductase with dinuclear copper ions, belong to the type - 3 copper protein family, namely the two copper ions of tyrosinase active center linked together with three specific histidine residues form a tetrahedron structure.[1,2] Tyrosinase is widely distributed in microorganisms, animals, and plants.[3] Tyrosinase is a key enzyme in the biosynthesis of melanin pigments, it could catalyze two distinct reactions involving molecular oxygen in the hydroxylation of monophenols to o - diphenols (monophenolase) and in the oxidation of o - diphenols to o - quinones (diphenolase).[4] Due to the high reactivity, quinines could polymerize spontaneously to form high molecular weight brown - pigments (melanins), which widely spreads in skin, hair and eyes of mammals, or react with amino acids and proteins to enhance brown color of the pigment produced.[5] Melanin plays an essential role in protection against UV injury under normal physiological conditions.[6,7] Previous reports confirmed that tyrosinase not only was involved in melanizing in animals, but also was one of the main causes of most fruits and vegetables quality loss during post - harvest handling and processing, leading to faster degradation and shorter shelf life.[8,9] This can cause a decrease in nutritional values and economic loss in food industry. Thus, preventing this unfavorable browning reaction has always been a challenge in food science.[10] Recently, investigation also demonstrated that various dermatological disorders, such as age spots and freckle, were caused by the accumulation of an excessive level of epidermal pigmentation.[11,12] Skin health and appearance is a major concerned issue for people globally. As early as last century, pharmaceuticals and traditional remedies have become incrementally popular for this purpose.[13] Tyrosinase has also been linked to Parkinson's and other neurodegenerative diseases.[14,15] In insects, tyrosinase is uniquely associated with three different biochemical processes, including sclerotization of cuticle, defensive encapsulation and melanization of foreign organism, and wound healing.[16,17] These processes provide potential targets for developing safer and effective tyrosinase inhibitors as insecticides and ultimately for insect control. Thus, the development of safe and effective tyrosinase inhibitors is of great concern in the medical, agricultural, food and cosmetic industries in relation to hyperpigmentation.[18] So far, some natural and synthetic compounds as tyrosinase inhibitors were reported, such as hydroquinone, ascorbic acid, arbutin, kojic acid, aromatic aldehydes, aromatic acids, aromatic alcohol, tropolone, and polyphenols.[19] However, only a few such as kojic acid, arbutin, tropolone, and 1-phenyl-2-thiourea (PTU) are used as cosmetic products.[20] Furthermore, it has been reported that arbutin is decomposed by temperature (10% decomposition at 20℃ for 15 days) and it has shown very low activity. Arbutin and tropolone have not been demonstrated to be clinically efficacious when critically analyzed in carefully controlled studies.[21] The use of kojic acid is controversial due to its toxicity. The use of the most popular drug hydroquinone in cosmetic products was banned in the European Union and is under scrutiny by the United States Food and Drug Administration (FDA) because it can cause DNA damage and carcinogenic effect.[13] Therefore, it is still necessary to search and discover novel tyrosinase inhibitors with higher activities and lower side effects.

On the other hand, it was reported that phenyl thioureas and alkyl thioureas could exhibit weak to moderate depigmenting activities.[22] Because of tyrosinase belongs to catechol oxidase with the type-3 copper protein group, the sulfur atom of phenyl thioureas or alkyl thioureas can bind to both copper ions in the active site of the enzyme.[23] Ley and Bertram also reported that benzaldoximes and benzaldehyde-o

– alkyloximes possessed higher tyrosinase inhibitory ability.[24] The above information was beneficial to design new tyrosinase inhibitor.

Natural products (NPs), including semi – synthetic NPs and NP derived compounds, with great characteristics of high chemical diversity and biochemical specificity, are all through major entities among the FDA – approved drugs as anticancer, anti – inflammation, anti – infective agents and so on.[25] Rheum officinale Baill, is one of the most popular traditional medicinal herbs that is officially listed in the Chinese Pharmacopoeia.[26] There has been great clinical interest in using Rheum officinale Baill in treatment of many kinds of diseases. Aloe – emodin (AE, 1,8 – dihydroxy – 3 – (hydroxymethyl) anthracene – 9, 10 – dione), is one of major bioactive anthraquinone derivatives from the roots of Rheum officinale Baill, which has been reported to possess antibacterial, antiviral, anticancer, hepatoprotective, antiproliferative activity, anti – angiogenic effects, laxative and anti – inflammation effects.[27-29] In addition, aloe – emodin, as an antioxidant, can reduce oxidative damage caused by free radicals.

Taking advantage of above information, in order to find some new bioactivities of aloe – emodin derivatives, such as inhibitory activities of tyrosinase, antibacterial activities and anti – inflammatory activities. Based on the structural feature of aloe – emodin, a series of aloe – emodin derivatives were designed and synthesized. Evaluations of inhibitory activities of tyrosinase, antibacterial activities and anti – inflammatory activities of these compounds were carried out. Meanwhile, the structure – activity relationships between these compounds and inhibitory activities of tyrosinase were also primarily discussed.

We hope that these findings can lead to the discovery of potential pharmacological agents for treating the tyrosinase – related disorders and also offer key and useful information for future design of highly potent tyrosinase inhibitors, antibacterial and anti – inflammatory reagents.

The general procedure for the synthesis ofaloe – emodin derivatives are described in Schemes 1. Aloe – emodin could be alkylated smoothly by using a simple procedure to give 1, 8 – alkoxy substituted aloe – emodin (1 – 4), but the yields were low to moderate. Aloe – emodin and the 1, 8 – alkoxy substituted aloe – emodin (1 – 4) were oxidized with silica supported PCC to provide the corresponding 4, 5 –

disubstituted - 9,10 - dioxo - 9,10 - dihydroanthracene - 2 - carbaldehyde (5 - 8). These aldehyde compounds reacted with thiosemicarbazide or the other substituted amines in absolute ethanol to provide the corresponding Schiff base compounds in poor to moderate yields. All the target compounds were characterized by chemical and spectral methods.

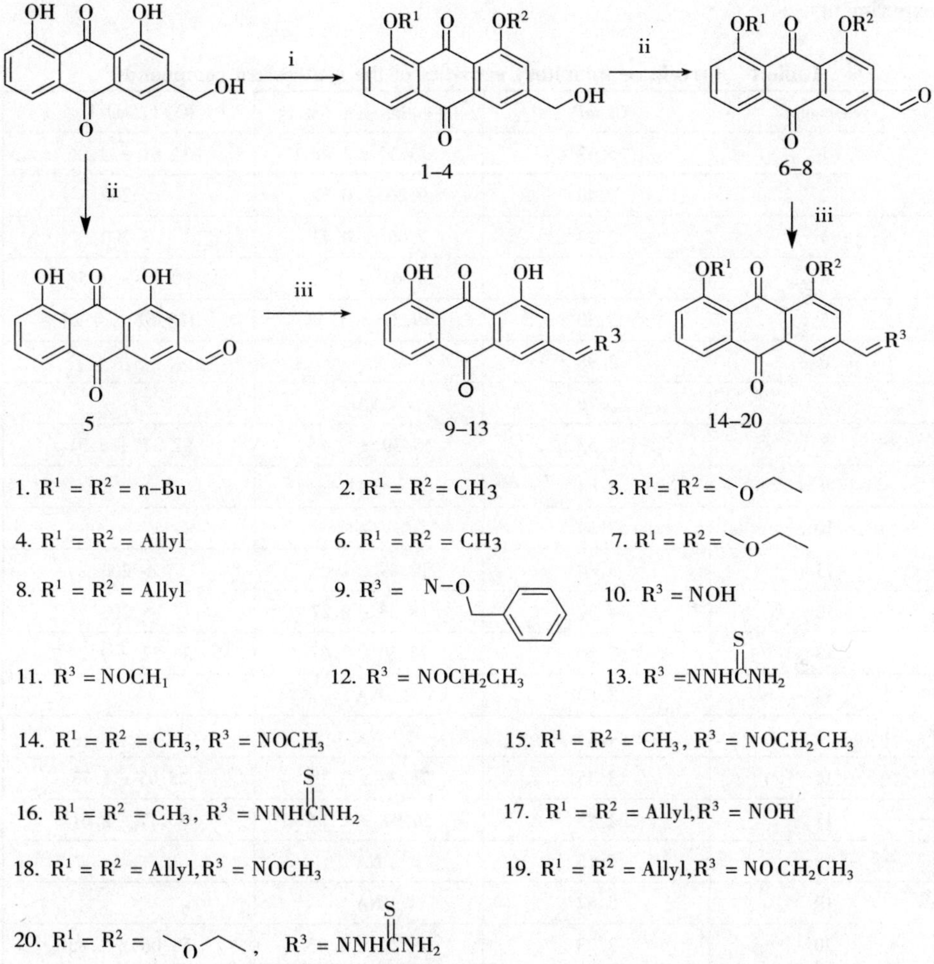

1. $R^1 = R^2 = n\text{-}Bu$ 2. $R^1 = R^2 = CH_3$ 3. $R^1 = R^2 = CH_2CH_2OCH_3$

4. $R^1 = R^2 = Allyl$ 6. $R^1 = R^2 = CH_3$ 7. $R^1 = R^2 = CH_2CH_2OCH_3$

8. $R^1 = R^2 = Allyl$ 9. $R^3 = N{-}O{-}CH_2C_6H_5$ 10. $R^3 = NOH$

11. $R^3 = NOCH_1$ 12. $R^3 = NOCH_2CH_3$ 13. $R^3 = NNHC(=S)NH_2$

14. $R^1 = R^2 = CH_3$, $R^3 = NOCH_3$ 15. $R^1 = R^2 = CH_3$, $R^3 = NOCH_2CH_3$

16. $R^1 = R^2 = CH_3$, $R^3 = NNHC(=S)NH_2$ 17. $R^1 = R^2 = Allyl$, $R^3 = NOH$

18. $R^1 = R^2 = Allyl$, $R^3 = NOCH_3$ 19. $R^1 = R^2 = Allyl$, $R^3 = NOCH_2CH_3$

20. $R^1 = R^2 = CH_2CH_2OCH_3$, $R^3 = NNHC(=S)NH_2$

Scheme 1 The synthesis of aloe - emodin derivatives. Reagents and conditions: i. RX / K_2CO_3/ DMF / 55℃, or 2 - methoxyethanol / tosyl chloride / THF / 0℃ / K_2CO_3/ 80℃; ii. Silica supported PCC / acetone / reflux; iii. Thiosemicarbazide / ethanol / reflux, or substituted amine / ethanol / room temperature.

For evaluating the tyrosinase inhibitory activity, all the synthesized compounds were subjected to tyrosinase inhibition assay with L - DOPA as substrate, according tothe method reported by our groups with some slight modifications.[30] The tyrosinase inhibitory activity of kojic acid was ever reported, therefore, it was selected as comparing substance. The IC_{50} values of aloe – emodin derivatives against tyrosinase were summarized in Table 1, and IC_{50} values of all these compounds were determined from logarithmic concentration - inhibition curves and given as means of three experiments.

Table 1 Tyrosinase inhibitory activities of the synthesized compounds

Compounds	CLog*P*	Inhibition ratio (%)[a]	IC_{50} (μmol/L)[b]
1	5.63	73.57 ± 1.48	32.81 ± 1.06
2	2.46	9.86 ± 0.37	> 200
3	2.24	7.45 ± 0.43	> 200
4	4.01	58.63 ± 2.12	90.99 ± 4.58
5	3.30	49.37 ± 1.38	108.62 ± 4.26
6	2.98	NA[c]	
7	2.76	NA	
8	4.53	55.39 ± 2.85	92.37 ± 3.91
9	4.84	NA	
10	3.39	NA	
11	3.84	12.89 ± 1.05	> 200
12	4.37	18.33 ± 0.87	> 200
13	3.50	78.59 ± 3.67	24.52 ± 1.24
14	3.60	NA	
15	4.13	NA	
16	3.45	76.38 ± 3.35	28.05 ± 1.73
17	4.83	56.97 ± 2.55	83.17 ± 2.61
18	5.15	NA	
19	5.67	NA	
20	3.23	65.27 ± 2.83	58.06 ± 1.85
Kojic acid			23.6 ± 0.52

[a] Percent inhibition at a concentration of 100μmol/L. Mean values ± SD.

[b] Inhibitor concentration (mean ± SD of three independent experiments) required for 50% inactivation of tyrosinase.

[c] NA = not active.

Our results showed that some compounds exhibited certain inhibition activities on mushroom tyrosinase with IC_{50} values ranged from 24.52μmol/L to 108.62μmol/L.

Especially, compounds 13, 16, 20 bearing thiosemicarbazide showed more potent inhibitory activities than most of the other compounds. In addition, compounds 13, 16 demonstrated similar inhibitory activities to the reference standard inhibitor kojic acid, 2 – ((4,5 – dihydroxy – 9,10 – dioxo – 9,10 – dihydroanthracen – 2 – yl) methylene) hydrazine – 1 – carbothioamide (13) exhibited the most potent tyrosinase inhibitory activity with IC_{50} value of 24. 52μmol/L. The results showed that the group of thiosemicarbazide played important role in determining activities against tyrosinase. These may be related to the structure of tyrosinase contained a type – 3 copper center with a coupled dinuclear copper active site in the catalytic core. Tyrosinase inhibition ability of compounds 13, 16, 20 depended on the competency of the sulfur atom to chelate with the dicopper nucleus in the active site, and tyrosinase would lose its catalyzing ability after forming complex. These results are consistent with our previous research.[30] From the result of Table 1, the other Schiff bases showed weak tyrosinase inhibitory activities except for the compound 17 (4,5 – bis(allyloxy) – 9,10 – dioxo – 9,10 – dihydroanthracene – 2 – carbal – dehyde oxime) with IC_{50} value of 83. 17μmol/L. However, it still showed lower inhibitory activity than the Schiff bases with thiosemicarbazide moiety. Comparing to the tyrosinase inhibitory activities of compounds 1 – 4, with the increase of lipophilicity of compound, the inhibitory activities increased gradually, 1,8 – dibutoxy – 3 – (hydroxymethyl) anthracene – 9, 10 – dione (1) exhibited the most potent tyrosinase inhibitory activity with IC_{50} value of 32. 81μmol/L. These results showed that the inhibitory activities of this kind of compounds with alkoxyl group at the 1 – position and 8 – position of anthracene ring might be relate to the lipophilicity. Because lipophilic groups can affect both the inhibition potency, as well as the ability of the tyrosine to compete with the inhibitor. Lipophilic group may interact with the enzyme hydrophobic pocket and augment binding affinity. According to the inhibitory activities of compounds 5 – 8, with the increase of lipophilicity of compounds, the inhibitory activities also increased gradually.

The inhibition mechanism of compounds 1 and 13 on mushroom tyrosinase for the oxidation of L – DOPA was determined. Figure 1 showed the relationship between enzyme activity and concentration in the presence of different concentrations of compound 1, respectively. The results demonstrated that the plots gave a family of straight lines, which all passed through the origin. Increasing the inhibitor concentration resulted in a decrease in the slope of the line. This result indicated that the inhibition of compound 1 on mushroom tyrosinase was reversible. Figure 2 showed the relationship between enzyme activity and concentration in the presence of different concentration of compound 13. The results displayed that the plots of V versus $[E]$ gave a family of parallel straight lines with the same slopes. It demonstrated that the inhibitory effect of compound 13 on the tyrosinase was irreversible.

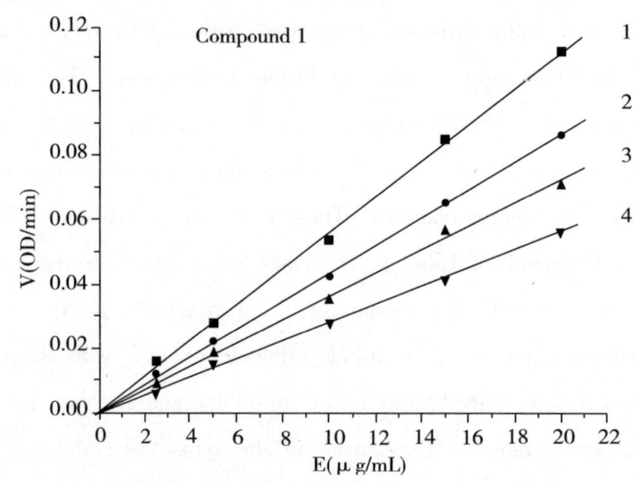

Figure1 Determination of the inhibitory effect of compound 1 on mushroom tyrosinase for the oxidation of L－DOPA. The concentrations of compound 1 for curves 1 － 4 were 0, 10, 30 and 50μmol/L, respectively.

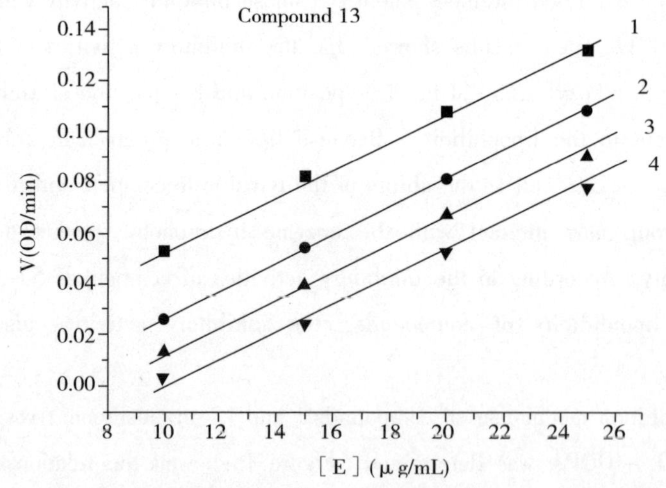

Figure2 Determination of the inhibitory effect of compound 13 on mushroom tyrosinase for the oxidation of L－DOPA. The concentrations of compound 13 for curves 1 － 4 were 0, 10, 30 and 50 μmol/L, respectively.

To further insight into the inhibition mechanism, finally the inhibitory type of the selected compound 1 on mushroom tyrosinase for the oxidation of L – DOPA was determined by the Line Weaver – Burk double reciprocal plots. Figure 3 showed the double – reciprocal plots of the enzyme inhibited by compound 1. The result displayed that the plots of 1/V versus 1/[S] gave three straight lines with different slopes, but they intersected at the same point on the ordinate. The values of V_{max} remained the same and the values of Km increased with increasing concentrations of the inhibitor, which indicated that compound 1 was the competitive inhibitor of tyrosinase. The result showed that compound 1 could only bind with the free enzyme.

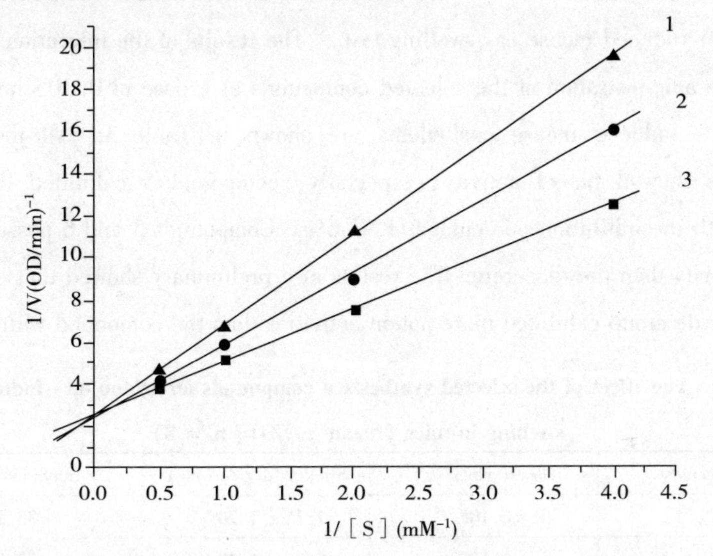

Figure3 Determination of the inhibitory types of selected compound 1 on mushroom tyrosinase for the oxidation of L – DOPA. The concentrations of compound 1 for curves 1 – 3 were 45, 30 and 15μmol/L, respectively.

Three compounds were selected to test the antibacterial activities,[31] and the results were summarized in Table 2. The MIC data in Table 2 indicated that the selected compounds possessed moderate antibacterial activities, especially compound 3 exhibited more potent antibacterial activity than the aloe – emodin using as positive control drug at the same conditions, MIC value of the compound was 1μg/mL. This result showed the antibacterial activities of aloe – emodin derivatives were worth further studying.

Table 2　Antibacterial activities of the selected synthesized compounds

Compounds	CLog*P*	MIC (μg/mL)
2	2.46	64
3	2.24	1
4	4.01	32
Aloe – emodin	2.70	2

Due to rhein and diacetylrhein, low molecular weight anthraquinone derivatives, have been shown to be useful anti – inflammatory agents in the treatment of rheumatoid arthritis, compounds2, 5 and 6 were selected for further investigation in vivo in the croton oil – induced mouse ear swelling test.[32] The results of the inhibition caused by intragastric administration of the selected compounds at a dose of 0.105 mmol/kg on croton oil – induced mouse ear edema are shown in Table 3. All the selected compounds showed potent activity, especially, compound 5 exhibited the highest activity with the inhibition percentage of 50.6%. Compounds 5 and 6 possessed more potent activity than the diacerein. The results also preliminary showed that compounds with aldehyde group exhibited more potent activities than the compound with hydroxyl.

Table 3　The effect of the selected synthesized compounds on croton oil – induced ear swelling in mice (mean ± SD), n = 8)

Compounds	Doses (mmol/kg)	Swelling degree (mg)	Inhibition (%)
2	0.105	2.75 ± 1.20**	35.2
5	0.105	2.10 ± 1.12**	50.6
6	0.105	2.44 ± 0.91**	42.6
Diacerein	0.105	2.58 ± 0.84**	39.1
Control		4.25 ± 0.91	

In conclusion, a series of new aloe – emodin derivatives were designed and synthesized. The results showed that some synthesized compounds exhibited potent tyrosinase inhibition activities. SARspreliminary analysis suggested that: (1) the thiosemicarbazone moiety played a very key role in determining the tyrosinase inhibitory activity; (2) tyrosinase inhibitory activity might be relate to the lipophilicity; (3) the substituent type of anthraquinone had effect on the inhibitory activity. Moreover, the inhibition mechanism and inhibition kinetics study revealed that compound 1 exhibited inhibitory effects on tyrosinase by acting as the reversible and competitive inhibitor, and the compound 13 was irreversible inhibitor. These results suggested that tyrosinase inhibitory activities of aloe - emodin derivatives were worth further studying. Compound 3 exhibited more potent antibacterial activity than the aloe - emodin, the MIC value of the compound was 1μg/mL. This compound might be used as one lead compound with antibacterial activity. Compounds 5 and 6 possessed more potent anti - inflammatory activities than the diacerein. Therefore, these compounds may be considered as the agents with potential anti - inflammatory activities and good candidates for more advanced screening.

Acknowledgments

We thank the innovation team of Shaoyang University (2012) for funding this project. This work was also financially supported by the Foundation of Education Department of Hunan Province, China (15A172).

References

1. Xie, J. ; Dong, H. ; Yu, Y. ; Cao, S. *Food Chem*. 2016, *190* , 709.

2. Li, Y. C. ; Wang, Y. ; Jiang, H. B. ; Deng, J. P.*P. Natl. Acad. Sci . USA*. 2009, *106* , 17002.

3. Song, K. K. ; Huang, H. ; Han, P. ; Zhang, C. L. ; Shi, Y. ; Chen, Q. X.*Biochem. Biophys. Res. Commun*. 2006, *342* , 1147.

4. Chen, Q. X. ; Liu, X. D. ; Huang, H. *Biochemistry (Mosc)*. 2003, *68* , 644.

5. Matsuura, R. ; Ukeda, H. ; Sawamura, M.*J. Agric. Food Chem* . 2006, *54* , 2309.

6. Kumar, C. M. ; Sathisha, U. V. ; Dharmesh, S. ; Rao, A. G. ; Singh, S. A. *Biochimie*. 2011, *93* , 562.

7. Pinon, A. ; Limami, Y. ; Micallef, L. ; Cook - Moreau, J. ; Liagre, B. ; Delage, C. ; Duval, R. E. ; Simon, A. *Exp. Cell Res.* 2011, *317* , 1669.

8. Yi, W. ; Cao, R. ; Peng, W. ; Wen, H. ; Song, H. *Eur. J. Med. Chem.* 2010, *45* , 639.

9. Chai, W. M. ; Shi, Y. ; Feng, H. L. ; Xu, L. ; Xiang, Z. H. ; Gao, Y. S. ; Chen. Q. X. *J. Agric. Food Chem*. 2014, *62* , 6382.

10. Arung, E. T. ; Kusuma, I. W. ; Iskandar, Y. M. ; Yasutake, S. ; Shimizu, K. ; Kondo, R. *J. Wood Sci.* 2005, *51* , 520.

11. Gupta, A. K. ; Gover, M. D. ; Nouri, K. N. ; Taylor, S. *J. Am. Acad. Dermatol* . 2006, *55* , 1048.

12. Thanigaimalai, P. ; Hoang, T. A. L. ; Lee, K. C. *Bioorg. Med. Chem. Lett.* 2010, *20* , 2991.

13. Akhtara, M. N. ; Sakeh, N. M. ; Zareen, S. ; Gul, S. ; Lo, K. M. ; Ul - Haq, Z. *J. Mol. Struct.* 2015, *1085* , 97.

14. Pan, T. ; Li, X. ; Jankovic, J. *Int. J. Cancer* 2011, *128* , 2251.

15. Zhu, Y. J. ; Zhou, H. T. ; Hu, Y. H. *Food Chem* . 2011, *124* , 298.

16. Ashida, M. ; Brey, P. *P. Natl. Acad. Sci . USA*. 1995, 92, 10698.

17. Ortiz - Urquiza, A. ; Keyhani, N. O. *Insects* 2013, 4, 357.

18. Ortiz - Ruiz, C. V. ; Berna, J. ; Garcia - Molina, M. M. ; Tudela, J. ; Tomas, V. ; Garcia - Canovas, F. *Bioorg. Med. Chem.* 2015, *23* , 3738.

19]. You, A. ; Zhou, J. ; Song, S. ; Zhu, G. ; Song, H. ; Yi, W. *Bioorg. Med. Chem.* 2015, *23* , 924.

20. Battaini, G. ; Monzani, E. ; Casella, L. ; Santagostini, L. ; Pagliarin, R. *J. Biol. Inorg. Chem.* 2000, *5* , 262.

21. Kim, H. R. ; Lee, H. J. ; Choi, Y. J. ; Park, Y. J. *Med. Chem. Commun.* 2014, *5* , 1410.

22. Klabunde, T. ; Eicken, C. ;Sacchettini, J. C. ; Krebs, B. *Nature Struct. Biol.* 1998, *5* , 1084.

23. Matsuura, R. ; Ukeda, H. ; Sawamura, M. *J. Agric. Food Chem.* 2006, 54, 2309.

24. Ley, J. P. ; Bertram, H. J.*Bioorg. Med. Chem.* 2001, *9* , 1879.

25. Liang, Y. K. ; Yue, Z. Z. ; Li, J. X. ; Tan, C. ; Miao, Z. H. ; Tan, W. F. ; Yang, C. H. *Eur. J. Med. Chem.* 2014, *84* , 505

26. Tao, L. ; Xie, J. ; Wang, Y. ; Wang, S. ; Wu, S. ; Wang, Q. ; Ding, H. *Bioorg. Med. Chem. Lett.* 2014, *24* , 5385.

27. Guo, J. M. ; Xiao, B. X. ; Liu, Q. ; Zhang, S. ; Liu, D. H. ; Gong, Z. H. *Acta Pharmacol. Sin.* 2007, *28* , 1991.

28. Park, M. Y. ; Kwon, H. J. ; Sung, M. K.*Biosci. Biotechnol. Biochem.*

2009, 73, 828.

29. Zembower, D. E. ; Kam, C. M. ; Powers, J. C. ; Zalkow, L. H. *J. Med. Chem*. 1992, *35* , 1597.

30. Liu, J. B. ; Yi, W. ; Wan, Y. Q. ; Ma, L. ; Song, H. C. *Bioorg. Med. Chem*. 2008, *16* , 1096.

31. National Committee for Clinical Laboratory Standards, Methods for Dilution Antimicrobial Susceptibility Tests for Bacteria that Grow Aerobically Approved Standard; NCCLS Document M7 – A3, third ed. NCCLS, 771 Lancaster Avenue, Villanova, PA, 1993, ISBN 1 – 56238 – 209 – 8.

32. Segura, L. ; Vila, R. ; Gupta, M. P. ; Espósito – Avella, M. ; Adzet, T. ; Cañigueral, S. *J. Ethnopharmacol*. 1998, *61* , 243.

Supporting information

Methods and Materials

Chemical reagents and instruments

Melting points (m. p.) were determined with WRS – 1B melting point apparatus and the thermometer was uncorrected. NMR spectra were recorded on Bruker 400 spectrometers at25℃ in DMSO – d_6. All chemical shifts (δ) are quoted in parts per million downfield from TMS and coupling constants (J) are given in hertz. Abbreviations used in the splitting pattern were as follows: s = singlet, d = doublet, t = triplet, q = quintet, m = multiplet, LC – MS spectra were recorded using the LCMS – 2010A. All reactions were monitored by TLC (Merck Kieselgel 60 F254) and the spots were visualized under UV light. Infrared (IR) spectra were recorded as potassium bromide pellets on VECTOR 22 spectrometer. Tyrosinase, L – 3, 4 – dhydroxyphenylalanine (L – DOPA) and kojic acid were purchased from Sigma – Aldrich Chemical Co. Other chemicals were purchased from commercial suppliers and were dried and purified when necessary.

The general procedures for the synthesis of 1 , 8 – alkoxy substitutedaloe – emodin (1 – 4)

A Procedure : 1. 08g (4mmol) of aloe – emodin was dissolved in50mL of dry DMF and added to a 100 mL dry flask. 9. 6mmol of halohydrocarbon was added to the mixture. Under intensive stirring conditions, anhydrous potassium (16mmol) carbonate was added to the above reaction mixture and warm up to 55℃. The reaction mixture was stirred at the temperature for further about 20h, and the reaction process was monitored by TLC (EtOAc/PE (60 - 90℃) = 1 : 2). When no starting material could be detected, the reaction mixture was cooled to room temperature. The solution was poured into water (40 mL). The water layer was extracted by ethyl acetate (3 × 30mL), the organic layer was merged and washed with water (3 × 30mL). The organic layer was dried with anhydrous sodium sulfate and evaporated under reduced pressure to get the crude product, which was purified by column chromatography with ethyl acetate in petroleum ether.

B Procedure: 6g (0.15mol) of sodium hydroxide was dissolved in 20mL of water and cooled to room temperature. 2 - Methoxyethanol (15.22g, 0.2mol) was added to the solution and stirred for 30 min at 0 ℃. Tosyl chloride (19.1g, 0.1mol) in 25mL of dry THF was added to the reaction mixture by dropwise, then the mixture was stirred for 5h at 0 ℃. The appearing precipitate was filtered and washed with ether, the filtrate was extracted by ether (2 × 20mL). The combined organic phases were washed with brine and dried by anhydrous sodium sulfate, the solvent was evaporated under reduced pressure to get 2 - methoxyethyl 4 - methylbenzenesulfonate

2 - Methoxyethyl 4 - methylbenzenesulfonate (2.3g, 0.01mol) in 30mL of dry DMF was added to a 50mL dry flask. Aloe - emodin (2.7g, 0.01mol) and anhydrous potassium carbonate (6.9g, 0.05mol) were added to the solution and stirred for 16h at 80℃. The reaction mixture was cooled to room temperature and poured into 80mL of water. The mixture was extracted with DCM (3 × 30mL). The combined organic phases were washed with brine and dried by anhydrous sodium sulfate, the solvent was evaporated under reduced pressure to get crude product, which was purified by column chromatography with ethyl acetate in petroleum ether.

1,8 - dibutoxy - 3 - (hydroxymethyl)anthracene - 9,10 - dione (1)

Yellow solid, M. p.: 123 ~ 125℃, yield: 9.3%; IR (cm^{-1}, KBr): 2985, 2879, 2821, 1617, 1454, 1047; ^{1}HNMR: (DMSO - d_6, 400 MHz) δ (ppm): 8.15 - 7.42 (m, 5H, Ar H), 5.57 (s, 1H, OH), 4.24 (t, 4H, J = 6.4 Hz, $2CH_2$), 3.32 (s, 2H, CH_2), 1.82 - 1.73 (m, 4H, 2CH2), 1.59 - 1.52 (m, 4H, $2CH_2$), 0.99 (t, 6H, J = 7.4 Hz, CH_3); ^{13}CNMR (DMSO - d_6, 100 MHz) δ (ppm): 183.8, 179.6, 159.3, 157.4, 153.8, 135.1, 133.0, 132.5, 124.5, 121.3, 118.6, 117.5, 116.8, 114.9, 69.5, 63.4, 32.8, 19.6, 14.5. ESI - MS m/z: 383 $(M+H)^+$.

3 - (hydroxymethyl) - 1,8 - dimethoxyanthracene - 9,10 - dione (2)

Yellow solid, M. p.: 223 ~ 225℃, yield: 53.1%; IR (cm^{-1}, KBr): 3464, 2933, 1657, 1584, 1459, 1325, 1277, 1174, 746; ^{1}HNMR: (DMSO - d_6, 400 MHz) δ (ppm): 7.75 - 7.65 (m, 3H, Ar - H), 7.52 (d, 1H, J = 8.1Hz, ArC7 - H), 7.43 (s, 1H, ArC4 - H), 5.57 (t, 1H, J = 5.7Hz, OH), 4.63 (d, 2H, J = 5.7Hz, CH_2), 3.92 (s, 6H, 2CH_3); ^{13}CNMR (DMSO - d_6, 100 MHz) δ (ppm): 183.6, 181.7, 162.2, 158.9, 152.1, 135.6, 133.4, 131.0, 122.8, 120.5, 119.6, 118.3, 118.1, 117.4, 115.9, 66.4. ESI - MS m/z: 299 $(M+H)^+$.

3 - (hydroxymethyl) - 1,8 - bis (2 - methoxyethoxy) anthracene - 9,10 - dione (3)

Yellow solid, M. p.: 95 ~ 97℃, yield: 19.7%; IR (cm^{-1}, KBr): 2978, 2943, 2876, 1660, 1467, 1447, 1127, 752; ^{1}HNMR: (DMSO - d_6, 400 MHz) δ (ppm): 7.71 - 7.57 (m, 3H, Ar - H), 7.54 (d, 1H, J = 6.9Hz, ArC7 - H), 7.44 (s, 1H, ArC4 - H), 5.56 (t, 1H, J = 5.7Hz, OH), 4.62 (d, 2H, J = 5.7Hz, CH_2), 4.26 - 4.23 (m, 4H, 2CH_2), 3.75 - 3.72 (m, 4H, 2CH_2), 3.38 (s, 3H, OCH_3), 3.37 (s, 3H, OCH_3); ^{13}CNMR (DMSO - d_6, 100 MHz) δ (ppm): 182.5, 182.0, 160.4, 158.7, 152.2, 134.9, 133.1, 132.5, 124.9, 120.5, 119.4, 118.8, 118.3, 116.3, 73.0, 68.5, 66.3, 58.9. ESI - MS m/z: 387 $(M+H)^+$.

1,8 - bis (allyloxy) - 3 - (hydroxymethyl) anthracene - 9,10 - dione (4)

Yellow solid, M. p.: 119 ~ 120℃, yield: 19.7%; IR (cm^{-1}, KBr): 2982, 2885, 1662, 1598, 1281, 1082, 1057, 989, 925, 887, 749; ^{1}HNMR: (DMSO - d_6, 400 MHz) δ (ppm): 7.75 - 7.69 (m, 3H, Ar - H), 7.52 (t, 1H, J = 6.4 Hz, ArC6 - H), 7.47 (s, 1H, ArC4 - H), 6.13 - 6.05 (m, 2H, C = C - H), 5.66 - 5.34 ((m, 5H, OH, 2C = CH_2), 4.74 (d, 4H, J = 6.8 Hz, 2CH_2), 4.62 (d, 2H, J = 5.6 Hz, CH_2); ^{13}CNMR (DMSO - d_6, 100 MHz) δ (ppm): 184.1, 182.2, 162.1, 158.6, 152.8, 136.5, 134.3, 133.6, 133.0, 121.5, 120.8, 120.2, 119.4, 118.7, 117.6, 116.1, 73.4, 66.5. ESI - MS m/z: 351 $(M+H)^+$.

The general procedures for the synthesis of4,5 - disubstituted - 9,10 - dioxo - 9, 10 - dihydroanthracene - 2 - carbaldehyde (5 - 8)

Aloe - emodin or 4, 5 - alkoxy substituted aloe - emodin in acetone was added to a 250 mL flask and stirred. Silica supported PCC (20 g) was added to the above solution. The mixture was stirred under reflux for about 24h, the reaction process was monitored by TLC (EtOAc/PE (60 ~ 90℃) = 1 : 3). When no starting material could be detected, the reaction mixture was filtered and washed with acetone. Solvent of the combined organic phases was removed under reduced pressure to get crude product, which was purified by column chromatography with ethyl acetate in petroleum ether.

4,5 – dihydroxy – 9,10 – dioxo – 9,10 – dihydroanthracene – 2 – carbaldehyde (5)

Yellow solid, M. p.: 207 ~ 209℃, yield: 60.7%; IR(cm^{-1}, KBr): 3132, 1705, 1673, 1604, 1447, 1274, 762; ^{1}HNMR: ($DMSO-d_6$, 400 MHz) δ (ppm): 11.95(s, 1H, OH), 11.84(s, 1H, OH), 10.09(s, 1H, –CHO), 8.11 (d, 1H,J =8.4Hz, ArC8 – H), 7.85 –7.73(m, 3H, Ar – H), 7.42(d, 1H,J = 8.4Hz, ArC6 – H); ^{13}CNMR ($DMSO-d_6$, 100 MHz) δ (ppm): 192.1, 187.6, 182.5, 162.7, 161.3, 142.8, 137.2, 134.5, 132.1, 125.8, 122.6, 121.4, 119.5, 118.0, 116.5. ESI – MS m/z: 269 $(M+H)^+$.

4,5 – dimethoxy – 9,10 – dioxo – 9,10 – dihydroanthracene – 2 – carbaldehyde (6)

Yellow solid, M. p.: 198 ~ 200℃, yield: 83.7%; IR(cm^{-1}, KBr): 2920, 2846, 1697, 1662, 1442, 1381, 1234, 1074; ^{1}HNMR: ($DMSO-d_6$, 400 MHz) δ (ppm): 10.14(s, 1H, CHO), 8.17(dd, 1H, J =8.2Hz, J =7.6Hz, ArC7 – H), 7.91(d, 1H, J =7.6Hz, ArC8 – H), 7.78(s, 1H, ArC3 – H), 7.71(d, 1H,J =8.2Hz, ArC6 – H), 7.56(s, 1H,ArC1 – H), 3.99(s, 3H, OCH_3), 3.91 (s, 3H, OCH_3); ^{13}CNMR ($DMSO-d_6$, 100 MHz) δ (ppm): 191.8, 183.2, 182.5, 161.2, 158.9, 142.6, 137.4, 134.8, 132.3, 125.9, 122.8, 121.2, 119.5, 117.8, 116.7, 56.1. ESI – MS m/z: 297 $(M+H)^+$.

4,5 – bis (2 – methoxyethoxy) – 9,10 – dioxo – 9,10 – dihydroanthracene – 2 – carbaldehyde (7)

Yellow solid, M. p.: 167 ~ 168℃, yield: 18%; IR(cm^{-1}, KBr): 2975, 2943, 2876, 1624, 1467, 1447, 1127, 1079, 752; ^{1}HNMR: ($DMSO-d_6$, 400 MHz) δ (ppm): 10.07(s, 1H,CHO), 8.04(dd, 1H, J =8.0Hz, J =7.6Hz, ArC7 – H), 7.86(d, 1H, J =7.6Hz, ArC8 – H), 7.81 –7.75 (m, 2H, Ar – H), 7.65

(d, 1H,J =8.0Hz, ArC6 – H), 4.26 – 4.23(m, 4H, 2CH_2), 3.75 – 3.72(m, 4H, 2CH_2), 3.43(s, 6H, 2OCH_3); ^{13}CNMR (DMSO – d_6, 100 MHz) δ (ppm): 191.6, 183.2, 181.5, 160.5, 158.8, 142.5, 135.7, 133.2, 132.4, 130.2, 125.3, 120.8, 119.1, 118.7, 118.2, 73.5, 69.2, 58.6. ESI – MS m/z: 385 (M + H)$^+$.

4,5 – bis (allyloxy) – 9,10 – dioxo – 9,10 – dihydroanthracene – 2 – carbaldehyde (8)

Yellow solid, M. p.: 152 ~ 153℃, yield: 62.6%; IR (cm^{-1}, KBr): 3065, 3014, 2863, 1694, 1582, 1230, 1130, 1002, 929, 746; ^{1}HNMR: (DMSO – d_6, 400 MHz) δ (ppm): 10.14(s, 1H, CHO), 8.20(dd, 1H, J = 8.0Hz, J = 7.6Hz, ArC7 – H), 7.92(s, 1H, ArC3 – H), 7.59(d, 1H, J =7.6Hz, ArC8 – H), 7.75(d, 1H,J=8.0 Hz, ArC6 – H), 7.57(s, 1H, ArC1 – H), 6.15 –6.04 (m, 2H, C = C – H), 5.67 – 5.35((m, 4H,2C = CH_2), 4.86(d, 2H, J = 4.4 Hz, CH_2), 4.77(d, 2H, J = 4.4 Hz, CH_2); ^{13}CNMR (DMSO – d_6, 100 MHz) δ (ppm): 191.5, 182.8, 182.1, 162.5, 158.1, 143.0, 136.7, 134.3, 133.4, 130.9, 121.2, 120.9, 120.5, 119.3, 118.5, 117.9, 116.8, 72.5. ESI – MS m/z: 349 (M + H)$^+$.

The general procedures for the synthesis of4,5 – disubstituted – 9,10 – dioxo – 9, 10 – dihydroanthracene – 2 – carbaldehyde derivatives

A Procedure: The corresponding 4,5 – disubstituted – 9,10 – dioxo – 9,10 – dihydroanthracene – 2 – carbaldehyde derivatives (5mmol) was dissolved in anhydrous ethanol (10mL) and dioxane (10mL). Thiosemicarbazide (5mmol) was added to the solution. The reaction was hastened by adding 2 drops of dry acetic acid and refluxing continued (5 ~ 8 h) until the disappearance of the starting material. The reaction mixture was cooled to room temperature, the precipitate was separated by filtration and recrystallized from ethanol to give the product.

B Procedure: The corresponding 4,5 – disubstituted – 9,10 – dioxo – 9,10 – dihydroanthracene – 2 – carbaldehyde derivatives (5mmol) was dissolved in anhydrous ethanol (10mL) and dioxane (10mL). The substituted amine (5mmol) was added to the above solution. The reaction was carried out at room temperature for 6 – 9 h, and the reaction process was monitored by TLC. After completion of the reaction, the reaction mixture was concentrated in vacuo. Water (30mL) was added to the reaction mixture, and extracted with dichloromethane (3 ×50mL). This extract was washed with water and brine, dried over anhydrous sodium sulfate, and the solvent was removed by rotary evaporation under reduced pressure to give oil, which was purified by column chromatography with ethyl acetate in petroleum ether.

4,5 – dihydroxy – 9,10 – dioxo – 9,10 – dihydroanthracene – 2 – carbaldehyde O – benzyl oxime (9)

Yellow solid, M. p.: 203 ~ 205℃, yield: 30.4%; IR(cm^{-1}, KBr):) 2889, 2834, 1665, 1617, 1268, 759; ^{1}HNMR: (DMSO – d_6, 400 MHz) δ (ppm): 11.92 (s, 2H, – OH), 8.46(s, 1H, HC = N), 7.97(s, 1H, ArC3 – H), 7.84(t, 1H, J = 7.6 Hz, ArC7 – H), 7.74(s, 1H, ArC1 – H) 7.54(d, 1H, J = 7.6 Hz, ArC6 – H), 7.44 – 7.34 (m, 6H, Ar – H), 5.31(s, 2H, CH_2); ^{13}CNMR (DMSO – d_6, 100 MHz) δ (ppm): 188.1, 182.8, 163.5, 160.3, 155.2, 140.7, 138.2, 137.8, 134.5, 132.9, 130.2, 128.5, 126.8, 125.4, 122.6, 120.9, 119.5, 118.4, 116.8, 75.2. ESI – MS m/z: 374 $(M + H)^+$.

4,5 – dihydroxy – 9,10 – dioxo – 9,10 – dihydroanthracene – 2 – carbaldehyde oxime (10)

Yellow solid, M. p.: 267 ~ 268℃, yield: 67%; IR(cm^{-1}, KBr):) 3353, 1620, 1598, 1450, 756; ^{1}HNMR: (DMSO – d_6, 400 MHz) δ (ppm): 11.96(s, 2H, 2OH), 11.79(s, 1H, OH), 8.30(s, 1H, ArC3 – H), 7.99(s, 1H, ArC1 – H), 7.84(t, 1H, J = 7.6 Hz, ArC7 – H), 7.75(d, 1H, J = 7.6 Hz, Ar – C6H), 7.52(s, 1H, HC = N), 7.41(d, 1H, J = 7.6 Hz, ArC8 – H); ^{13}CNMR (DMSO – d_6, 100 MHz) δ (ppm): 188.2, 181.7, 162.9, 160.7, 151.2, 139.6, 137.8, 134.5, 132.9, 125.7, 122.2, 120.9, 119.3, 118.6, 116.8. ESI – MS m/z: 284 $(M + H)^+$.

4,5 – dihydroxy – 9,10 – dioxo – 9,10 – dihydroanthracene – 2 – carbaldehyde O – methyl oxime (11)

Yellow solid, M. p.: 215 ~ 217℃, yield: 63%; IR(cm^{-1}, KBr):) 2943, 2885, 1665, 1281, 1236, 1047, 756; ^{1}HNMR: (DMSO – d_6, 400 MHz) δ (ppm): 11.91(s, 1H, OH), 11.69(s, 1H, OH), 8.39(s, 1H, CH), 8.06(s, 1H, ArC3 – H), 7.92(s, 1H, ArC1 – H), 7.83(t, 1H, J = 8.4 Hz, ArC7 – H), 7.72(d, 1H, J = 8.4 Hz, ArC6 – H), 7.39(d, 1H, J = 8.4 Hz, ArC8 – H), 3.55(s, 3H, OCH_3); ^{13}CNMR (DMSO – d_6, 100 MHz) δ (ppm): 187.9, 182.8, 163.2,

160.9, 155.6, 141.3, 137.5, 134.7, 132.6, 125.5, 122.3, 121.2, 119.8, 118.7, 116.6, 55.3. ESI－MS m/z: 298 (M＋H)$^+$.

4,5－dihydroxy－9,10－dioxo－9,10－dihydroanthracene－2－carbaldehyde O－ethyl oxime (12)

Yellow solid, M. p.: 189 ~ 191℃, yield: 34%; IR (cm^{-1}, KBr):) 2966, 2870, 1666, 1605, 1580, 1413, 1387, 1325, 1047, 822; ^{1}HNMR: (DMSO－d_6, 400 MHz) δ (ppm): 11.91(s, 2H, OH), 8.39(s, 1H, CH), 7.95(s, 1H, ArC3－H), 7.83(t, 1H, J = 8.0 Hz, ArC7－H), 7.72(d, 1H, J = 8.0 Hz, ArC6－H), 7.51(s, 1H, ArC1－H), 7.40(d, 1H, J = 8.0 Hz, ArC8－H), 4.28(q, 2H,J = 7.2 Hz, CH2), 1.31(t, 3H,J = 7.2 Hz, CH3); ^{13}CNMR (DMSO－d_6, 100 MHz) δ (ppm): 188.1, 182.5, 163.0, 160.7, 154.8, 140.6, 137.8, 135.2, 132.3, 126.4, 122.6, 120.7, 119.5, 118.4, 116.9, 69.8, 14.2. ESI－MS m/z: 312 (M＋H)$^+$.

2－((4,5－dihydroxy－9,10－dioxo－9,10－dihydroanthracen－2－yl) methylene)hydrazine－1－carbothioamide (13)

Yellow solid, M. p.: 218 ~ 220℃, yield: 62%; IR (cm^{-1}, KBr):) 2940, 2885, 1630, 1505, 1470, 1454, 746; ^{1}HNMR: (DMSO－d_6, 400 MHz) δ (ppm): 11.94(s, 2H, OH), 11.70(s, H, NH), 8.55(s, 2H, NH2), 8.38(s, 1H, CH), 7.93(s, 1H, ArC3－H), 7.85(s, 1H, ArC1－H), 7.76(t, 1H,J = 8.0Hz, ArC7－H), 7.68(d, 1H,J = 8.0Hz, ArC6－H), 7.53(d, 1H,J = 8.0Hz, ArC8－H); ^{13}CNMR (DMSO－d_6, 100 MHz) δ (ppm): 188.5, 183.1, 182.6, 162.9, 160.5, 151.8, 139.7, 137.2, 134.5, 132.3, 124.9, 121.8, 121.5, 119.3, 118.5, 116.9. ESI－MS m/z: 342 (M＋H)$^+$.

4,5－dimethoxy－9,10－dioxo－9,10－dihydroanthracene－2－carbaldehyde O－methyl oxime (14)

Yellow solid, M. p.: 213 ~ 215℃, yield: 74%; IR (cm^{-1}, KBr):) 2943, 2885, 1665, 1281, 1236, 1047, 756; ^{1}HNMR: (DMSO－d_6, 400 MHz) δ (ppm): 8.38(s, 1H, ArC3－H), 7.92(s, 1H, ArC1－H), 7.80－7.74(m, 1H, ArC7－H), 7.72(d, 1H,J = 6.8Hz, ArC6－H), 7.68(s, 1H, CH), 7.57(d, 1H,J = 6.8Hz, ArC8－H), 4.01(s, 3H, OCH_3), 3.93(s, 3H, OCH_3), 3.91(s, 3H, OCH_3); ^{13}CNMR (DMSO－d_6, 100 MHz) δ (ppm): 183.5, 182.3, 161.7, 160.6, 154.8, 140.6, 137.3, 134.5, 132.3, 123.8, 122.7, 120.9, 119.5,

118.4, 116.9, 60.7, 54.5. ESI - MS m/z: 326 (M + H)$^+$.

4,5 - dimethoxy - 9,10 - dioxo - 9,10 - dihydroanthracene - 2 - carbaldehyde O - ethyl oxime (15)

Yellow solid, M. p.: 163 ~ 164℃, yield: 45%; IR (cm^{-1}, KBr):) 2977, 2928, 1662, 1614, 1515, 1450, 1382, 1046, 755; ^{1}HNMR: (DMSO - d_6, 400 MHz) δ (ppm): 8.36 (s, 1H, ArC3 - H), 7.98 (s, 1H, ArC1 - H), 7.75 - 7.73 (m, 1H, ArC7 - H), 7.69 (d, 1H, J = 8.0Hz, ArC6 - H), 7.66 (s, 1H, CH), 7.53 (d, 1H, J = 8.0Hz, ArC8 - H), 4.27 - 4.21 (m, 2H, OCH_2), 3.92 (s, 3H, OCH_3), 3.89 (s, 3H, OCH_3), 1.32 (t, 3H, J = 6.4Hz, CH_3); ^{13}CNMR (DMSO - d_6, 100 MHz) δ (ppm): 183.3, 182.4, 161.1, 160.8, 154.6, 140.9, 136.8, 133.9, 132.5, 123.4, 121.6, 120.5, 119.7, 118.6, 116.5, 69.7, 55.6, 13.2. ESI - MS m/z: 340 (M + H)$^+$.

2 - ((4,5 - dimethoxy - 9,10 - dioxo - 9,10 - dihydroanthracen - 2 - yl) methylene)hydrazine - 1 - carbothioamide (16)

Yellow solid, M. p.: 260 ~ 262℃, yield: 47%; IR (cm^{-1}, KBr):) 2946, 1666, 1580, 1511, 1411, 1364, 1109, 1066, 763; ^{1}HNMR: (DMSO - d_6, 400 MHz) δ (ppm): 11.63 (s, H, NH), 8.40 (s, 2H, NH2), 8.30 (s, 1H, CH), 8.12 (s, 1H, ArC3 - H), 7.93 (s, 1H, ArC1 - H), 7.83 (t, 1H, J = 8.0Hz, ArC7 - H), 7.68 (d, 1H, J = 8.0Hz, ArC6 - H), 7.55 (d, 1H, J = 8.0Hz, ArC8 - H), 3.97 (s, 3H, OCH_3), 3.55 (s, 3H, OCH_3); ^{13}CNMR (DMSO - d_6, 100 MHz) δ (ppm): 183.5, 183.1, 182.6, 161.5, 159.8, 148.2, 139.3, 136.7, 134.4, 132.0, 123.5, 122.7, 121.6, 120.8, 118.0, 116.5, 55.9. ESI - MS m/z: 370 (M + H)$^+$.

4,5 - bis (allyloxy) - 9,10 - dioxo - 9,10 - dihydroanthracene - 2 - carbaldehyde oxime (17)

Yellow solid, M. p.: 171 ~ 173℃, yield: 50%; IR (cm^{-1}, KBr):) 3340, 2889, 1665, 1233, 977, 932, 890, 746; ^{1}HNMR: (DMSO - d_6, 400 MHz) δ (ppm): 11.76 (s, 1H, - OH), 8.29 (s, 1H, ArC3 - H), 7.93 (s, 1H, ArC1 - H), 7.79 (d, 1H, J = 8.4 Hz, ArC6 - H), 7.76 - 7.63 (m, 2H, ArC7 - H, CH), 7.54 (d, 1H, J = 8.4 Hz, ArC8 - H), 6.13 - 6.05 (m, 2H, C = C - H), 5.65 (d, 2H, J = 5.6 Hz, C = CH_2), 5.33 (d, 2H, J = 5.6 Hz, C = CH_2), 4.77 - 4.74 (m, 4H, 2CH_2); ^{13}CNMR (DMSO - d_6, 100 MHz) δ (ppm): 183.3,

181.7, 162.5, 160.6, 151.7, 139.3, 136.7, 134.8, 133.9, 132.6, 124.5, 122.8, 120.5, 119.8, 119.1, 118.2, 116.5, 72.1. ESI－MS m/z: 364 $(M+H)^+$.

4,5 － bis (allyloxy) － 9,10 － dioxo － 9,10 － dihydroanthracene － 2 － carbaldehyde O－methyl oxime (18)

Yellow solid, M. p.: 133 ~ 135℃, yield: 52%; IR (cm^{-1}, KBr):) 2892, 1668, 1585, 1233, 999, 932, 900, 752; ^{1}HNMR: ($DMSO-d_6$, 400 MHz) δ (ppm): 8.39(s, 1H, ArC3－H), 7.97(s, 1H, ArC1－H), 7.75－7.69(m, 3H, ArC7－H, ArC6－H, CH), 7.54(d, 1H, J = 8.4 Hz, ArC8－H), 6.12－6.05 (m, 2H, C=C－H), 5.67(d, 2H, J = 8.0 Hz, $C=CH_2$), 5.34(d, 2H, J = 8.0 Hz, $C=CH_2$), 4.77－4.75(m, 4H, $2CH_2$), 4.01(s, 3H, CH_3); ^{13}CNMR ($DMSO-d_6$, 100 MHz) δ (ppm): 183.5, 181.6, 162.8, 160.6, 153.4, 140.1, 136.5, 135.2, 134.4, 132.8, 124.3, 122.9, 120.7, 120.1, 119.3, 118.8, 116.6, 71.8, 60.8. ESI－MS m/z: 378 $(M+H)^+$.

4,5 － bis (allyloxy) － 9,10 － dioxo － 9,10 － dihydroanthracene － 2 － carbaldehyde O－ethyl oxime (19)

Yellow solid, M. p.: 107 ~ 108℃, yield: 71%; IR (cm^{-1}, KBr):) 2892, 1668, 1585, 1233, 999, 932, 900, 752; ^{1}HNMR: ($DMSO-d_6$, 400 MHz) δ (ppm): 8.38(s, 1H, ArC3－H), 8.02(s, 1H, ArC1－H), 7.75－7.69(m, 3H, ArC7－H, ArC6－H, CH), 7.54(d, 1H, J = 8.0 Hz, ArC8－H), 6.13－6.05 (m, 2H, C=C－H), 5.67(d, 2H, J = 8.0 Hz, $C=CH_2$), 5.34(d, 2H, J = 8.0 Hz, $C=CH_2$), 4.77－4.75(m, 4H, $2CH_2$), 4.27－4.21(m, 2H, CH_2), 1.33－1.26(m, 3H, CH_3); ^{13}CNMR ($DMSO-d_6$, 100 MHz) δ (ppm): 184.1, 182.6, 163.5, 161.5, 155.1, 141.6, 137.2, 136.7, 135.2, 133.6, 125.5, 123.6, 121.6, 121.5, 120.1, 119.6, 117.2, 72.6, 69.5, 14.3. ESI－MS m/z: 392 $(M+H)^+$.

2－((4,5－bis (2－methoxyethoxy)－9,10－dioxo－9,10－dihydroanthracen－2－yl)methylene)－hydrazine－1－carbothioamide (20)

Yellow solid, M. p.: 218 ~ 219℃, yield: 30%; IR (cm^{-1}, KBr):) 3164, 2978, 2879, 1665, 1600, 1508, 1281, 1233, 1069, 749; ^{1}HNMR: ($DMSO-d_6$, 400 MHz) δ (ppm): 11.57(s, H, NH), 8.43(s, 2H, NH2), 8.14(s, 1H, CH), 7.99(s, 1H, ArC3－H), 7.89(s, 1H, ArC1－H), 7.75(t, 1H, J = 8.0Hz, ArC7－H), 7.73(d, 1H, J = 8.0Hz, ArC6－H), 7.57(d, 1H, J =

8.0Hz, ArC8 – H), 4.39(t, 4H, J = 4.8Hz, $2CH_2$), 4.28(t, 4H, J = 4.8Hz, $2CH_2$), 3.77(s, 6H, $2CH_3$); ^{13}CNMR (DMSO – d_6, 100 MHz) δ (ppm): 183.9, 182.4, 178.5, 161.7, 159.2, 148.3, 139.8, 137.0, 134.3, 132.6, 125.5, 122.8, 121.2, 119.7, 118.3, 116.5, 74.1, 69.5, 59.8. ESI – MS m/z: 458 $(M+H)^+$.

Tyrosinase assay

The spectrophotometric assay for tyrosinase was performed according to the method reported by our groups with some slight modifications[30]. Briefly, all the synthesized compounds were screened for the diphenolase inhibitory activity of tyrosinase using L – DOPA as substrate. All the compounds were dissolved in DMSO. The final concentration of DMSO in the test solution was 2.0%. Phosphate buffer, pH 6.8, was used to dilute the DMSO stock solution of test compounds. Thirty units of mushroom tyrosinase (0.5 mg/mL) were first pre – incubated with the compounds, in 50 mmol/L phosphate buffer (pH 6.8), for 10 min at 25°C. Then the L – DOPA (0.5 mmol/L) was added to the reaction mixture and the enzyme reaction was monitored by measuring the change in absorbance at 475 nm of formation of the DOPA chrome for 1 min. The measurement was performed in triplicate for each concentration and averaged before further calculation. IC_{50} value, a concentration giving 50% inhibition of tyrosinase activity, was determined by interpolation of the dose – response curves. The percent of inhibition of tyrosinase reaction was calculated as following:

$$\text{Inhibition rate}(\%) = [(B - S)/B] \times 100$$

Here, the B and S are the absorbances for the blank and samples. Kojic acid was used as reference standard inhibitors for comparison.

In vitro antibacterial activity assays

In vitro antimicrobial activity was determined by the agar dilution methodaccording to the Clinical and Laboratory Standard Institute (CLSI) (formerly National Committee for Clinical Laboratory Standards) (NCCLS 1993)[31]. The overnight cultures (After 16 – 18h of incubation at 37°C) of all the bacteria were used for the assay and adjusted to the turbidity of a 0.5 McFarland Atandard. The stock solution (1mg/mL) of all the test chemicals was prepared by dissolving 1mg of the test chemical in 1 ml of dimethyl sulfoxide (DMSO), DMSO was used as control for all the test compounds. Twenty milliliter of MHA and 500μl of each test bacterial culture of 16 – 18h incubation adjusted at 0.5 McFarland were mixed and poured into sterilized and labeled plates. The wells of 6 mm diameter were punched in the solidified agar plates. Test chemicals of 100μl were added to individual wells. The loaded plats were incubated at 37℃ for 24 h. The diameter of zone of growth inhibition around each well was measured after incubation using a Vernier Caliper. The minimum inhibitory concentrations (MIC) is the lowest concentration of the antimicrobial agent that prevents the development of viable growth after16 – 18h incubation. Minimum inhibitory concentrations (MIC's, mg/ml) were determined on MuellereHinton (MH) agar with medium containing dilutions of antibacterial agents by the literature method. Stock solution of 4 mg/mL was prepared in DMSO and was appropriately diluted to get final concentrations of 128, 64, 32, 16, 8, 4, 2, 1, 0.5, 0.25 and .012μg/mL. Standard antibiotics were also diluted in same manner to make comparison。

Croton oil – induced ear edema

The experiments applied with animals were approved by Research Ethic Committee of Anhui province, China. Male Kunming mice weighing 18 – 22 g were obtained from Nanjing Qinglongshan Laboratory Animal Co. Ltd., (Nanjing, China), and fed with rat food and water ad libitum. All animals were fasted for 10 h before the experiments. The temperature (25℃) and humidity (60%) in the animal room were well controlled.

Croton oil – induced ear swelling was carried out according to the procedures described by Segura[32]. Forty male mice were divided randomly into five groups. Three groups were tested by intragastric administration with compound 2, 5, 6 at the dose of 0.105mmol/kg, respectively. The mice in the remaining groups were used as either vehicle controls or positive controls and were given 0.5% CMC – Na or Diacerein (0.105mmol/kg). One hour after the third day of intragastric administration, each animal received 0.05mL of a 2% solution of croton oil (croton oil : ethanol : ether : water = 2 : 20 : 73 : 5), on the inner surface of the right ear. The left ear was considered as control. Four hours after croton oil application, mice were killed and both ears were removed. Circular sections were taken using a cork borer with a diameter of 9 mm, weighed and measured. The degree of ear swelling was calculated based on the weight and thickness of left ear without croton oil.

The degree of ear swelling was calculated according to the following formula:

Percentage swelling = [(B – A)/ A] ×100%

Where A represented the weight of the mice´s left ear and B represented the weight of the right ear. The measurement data were expressed as the mean ± SD. Data were subjected to one – way analysis of variance (ANOVA), followed by multiple comparison with least significant differences (LSD) test as appropriate. Statistical significance was considered with $P < 0.01$. The analysis of data was performed by software SPSS 13.0.

2.6.6 Chem. Biol. Drug. Des(2013,82: 39 – 47)

Microwave – assisted synthesis and tyrosinase inhibitory activity of chalcone derivatives

Jinbing Liu*, Changhong Chen, Fengyan Wu, Liangzhong Zhao

[a]*Department of Biology and Chemical Engineering*, *Shaoyang University*, *Shao Shui Xi Road*, *Shaoyang* 422100, *PRC*

* Corresponding author: Department of Biology and Chemical Engineering, Shaoyang University, Shao Shui Xi Road, Shaoyang 422100, PRC

Phone: +86 – 739 – 5431768; Fax: +86 – 739 – 5431768

E – mail address: xtliujb@ yahoo. com. cn; jliu1238@ umd. edu

Abstract

A series of chalcones and their derivatives were synthesized and their inhibitory effects on the diphenolase activity of mushroom tyrosinase were evaluated. The results showed that some of the synthesized compounds exhibited significant inhibitory activity, four compounds exhibited more potent tyrosinase inhibitory activity than the reference standard inhibitor kojic acid (5 – hydroxy – 2 – (hydroxymethyl) – 4H – pyran – 4 – one). Especially, 1 – (– 1 – (4 – methoxyphen – yl) – 3 – phenylallylidene) thiosemicarbazide (18) exhibited the most potent tyrosinase inhibitory activity with IC_{50} value of 0.274μmol/L. The inhibition mechanism analysis of 1 – (– 1 – (2,4 – dihydroxyphenyl) – 3 – phenylallylidene) thiosemicarbazide (16) and 1 – (– 1 – (4 – methoxyphenyl) – 3 – phenylallylidene) thiosemicarbazide (18) demonstrated that the inhibitory effects of the two compounds on the tyrosinase were irreversible. Preliminary structure activity relationships' (SARs) analysis suggested that further development of such compounds might be of interest.

Keywords: microwave – assisted synthesis, chalcone derivatives, tyrosinase inhibitors, inhibition mechanism.

Introduction

Tyrosinase (monophenol oro – diphenol, oxygen oxidoreductase, EC 1.14.18.1, syn. polyphenol oxidase), also known as polyphenol oxidase (PPO), is a copper – containing monooxygenase that is widely distributed in microorganisms, animals, and plants (1). Tyrosinase could catalyze two distinct reactions involving molecular oxygen in the hydroxylation of monophenols to o – diphenols (monophenolase) and in the oxidation of o – diphenols to o – quinones (diphenolase) (2). Due to the high reactivity, quinones could polymerize spontaneously to form high molecular weight brown – pigments (melanins) or react with amino acids and proteins to enhance brown color of the pigment produced (3, 4). Hyperpigmentations, such as senile lentigo, melasma, freckles, and pigmented acne scars, are of particular concern to women. Therefore, the regulation of melanin synthesis by inhibiting the tyrosinase enzyme is a current research topic in the context of preventing hyperpigmentation. The treatment usually involves the use of medicines or medicinal cosmetics containing depigmenting agents or skin whitening agents (5, 6). In clinical usage, tyrosinase inhibitors are used for treatments of dermatological disorders related to melanin hyperaccumulation and are essential in cosmetics for depigmentation (7 – 11), such as age spots and freckle, were caused by the accumulation of an excessive level of epidermal pigmentation (12, 13).

Previous reports confirmed that tyrosinase not only was involved in mealism animals, but alsowas one of the main causes of most fruits and vegetables quality loss during post harvest handling and processing, leading to faster degradation and shorter shelf life (14). Tyrosinase has also been linked to Parkinson's and other neurodegenerative diseases (15). In insects, tyrosinase is uniquely associated with three different biochemical processes, including sclerotization of cuticle, defensive encapsulation and melanization of foreign organism, and wound healing (16). These processes provide potential targets for developing safer and effective tyrosinase inhibitors as insecticides and ultimately for insect control. Thus, the development of safe and effective tyrosinase inhibitors is of great concern in the medical, agricultural, and cosmetic industries. However, only a few such as kojic acid (5 – hydroxy – 2 – (hydroxymethyl) – 4H – pyran – 4 – one), arbutin, tropolone, ascorbic acid derivatives, hydroxylstilbene derivatives and 1 – phenyl – 2 – thiourea (PTU) are used as therapeutic agents and cosmetic products (12, 17).

Chalcones (1,3 – diaryl – 2 – propen – 1 – ones) constitute an important class of natural products belonging to the flavonoids familyof widespread occurrence in plants. They have been reported to exhibit a wide range of pharmaceutical effects including antioncogenetic, anti – inflammatory, anti – ulcerative, antimalarial, antiviral, antifungal, and antibacterial activities (18 – 20). Recently, chalcones were highlighted as a new class of tyrosinase inhibitor in several publications (21, 22). The literature survey revealed that compounds with thiourea moieties have been reported to demonstrate a wide range of pharmacological activities, which include antibacterial, antifungal, anticonvulsant and tyrosinase inhibitory activity, such as phenylthioureas, alkylthioureas and 1,3 – bis – (5 – methanesulfonylbutyl) thiourea, displayed weak or moderate tyrosinase inhibitory activity (23, 24). More recently, our investigations also demonstrated that thiosemicarbazide derivatives exhibited potent inhibitory activities against mushroom tyrosinase (25, 26). Stimulated by these results, in continuation of our research work on safe and effective tyrosinase inhibitors, in the present investigation, we synthesized a series of chalcones and their Schiff bases, their inhibitory activities against mushroom tyrosinase were evaluated using kojic acid (5 – hydroxy – 2 – (hydroxymethyl) – 4H – pyran – 4 – one) as comparing substance. Meanwhile, the structure – activity relationships of these compounds were also primarily discussed. Microwave assisted synthesis is an efficient and eco – friendly synthetic strategy and has now become a powerful tool for green chemistry. Microwave irradiation has been applied to organic reactions in the absence of solvent or in the presence of a solid support, such as clays, alumina and silica, resulting in shorter reaction times and better product yields than those obtained by using conventional heating (27 – 30). In this paper, some compounds were also synthesized by microwave irradiation.

Methods and Materials

Chemical reagents and instruments

Melting points (m. p.) were determined with WRS – 1B melting point apparatus and the thermometer was uncorrected. NMR spectra were recorded on Bruker 400 spectrometersat25℃ in $CDCl_3$ or DMSO – d_6. All chemical shifts (δ) are quoted in parts per million downfield from TMS and coupling constants (J) are given in hertz. Abbreviations used in the splitting pattern were as follows: s = singlet , d = doublet, t = triplet, q = quintet, m = multiplet, LC – MS spectra were recorded using the LCMS – 2010A. All reactions were monitored by TLC (Merck Kieselgel 60 F254) and the spots were visualized under UV light. Infrared (IR) spectra were recorded as potassium bromide pellets on VECTOR 22 spectrometer.

Tyrosinase, L – 3, 4 – dhydroxyphenylalanine (L – DOPA) andkojic acid were purchased from Sigma – Aldrich Chemical Co. Other chemicals were purchased from commercial suppliers and were dried and purified when necessary.

The general procedures for the synthesis of substituted chalcone compounds (1 – 12).

Method 1

To the mixture of substitutedacetophenone (5.0 mmol) and benzaldehyde (5.0 mmol), was added ethanol (20 mL) and 30% NaOH in H_2O (1 mL). The reaction mixture was taken in round – bottomed flask placed in a microwave reactor and irradiated at 800W for 3 min. Then this reaction mixture was poured over crushed ice and acidified with 10% HCl solution (10 mL) to form precipitate. The precipitate was filtered and washed with appropriate amount of water. The crude product was recrystallized from 95% ethanol to afford substituted chalcone 1 – 12.

Method 2

Substitutedacetophenone (5. 0 mmol) and benzaldehyde (5. 0 mmol) were dissolved in ethanol (20 mL), and 30% NaOH (1 mL) was added to the mixture. The reaction mixture was stirred at room temperature. The completion of the reaction was checked by TLC. The mixture was neutralized with 10% HCl solution (10 mL) to form precipitate. The precipitate was filtered and washed with appropriate amount of water. The crude product was recrystallized from 95% ethanol to afford substituted chalcone 1 – 12.

1 – (2 – hydroxyphenyl) – 3 – phenylprop – 2 – en – 1 – one (1).

Yield 63.2%. Yellow solid, mp 78. 3 ~ 80. 3 ℃; IR (KBr): 1736, 1653, 1549, 1333, 1217, 1151, 1084, 1030, 977, 827, 761 cm^{-1}. ^{1}HNMR (400 MHz,

$CDCl_3$) δ 7.98(d, 2H, J=8.4 Hz, ph－H), 7.84(d, 1H, J=15.6 Hz, ＝CH), 7.66－7.63 (m, 2H, ph－H), 7.50(d, 1H, J=15.6 Hz, ＝CH), 7.49(d, 2H, J=8.4 Hz, ph－H), 7.43－7.42(m, 3H, ph－H); ^{13}CNMR (100 MHz, $CDCl_3$) δ 189.2, 145.3, 139.2, 136.5, 134.7, 130.7, 129.9, 129.1, 128.9, 128.5, 121.5.

1－(3－hydroxyphenyl)－3－phenylprop－2－en－1－one (2).

Yield 81.2%. Yellow solid, mp 128.6～130.3℃; IR(KBr): 3208, 1648, 1570, 1445, 1346, 1259, 1188, 1030, 972, 856, 757cm⁻. ^{11}HNMR (300 MHz, DMSO－d_6) δ 7.90－7.84 (m, 3H, ph－H), 7.75 (d, 1H, J=15.6 Hz, ＝CH), 7.88 (d, 1H, J=8.1 Hz, ph－H), 7.48－7.46 (m, 4H, ph－H), 7.41 (d, 1H, J=15.6 Hz, ＝CH), 7.09 (d, 1H, J=8.1 Hz, ph－H).

1－(4－hydroxyphenyl)－3－phenylprop－2－en－1－one (3).

Yield 67.2%. Yellow solid, mp 179.0～181.2 ℃; IR(KBr): 1736, 1653, 1549, 1333, 1217, 1151, 1084, 1030, 977, 827, 761cm⁻. ^{11}HNMR (400 MHz, $CDCl_3$) δ 7.98 (d, 2H, J=8.4 Hz, ph－H), 7.84 (d, 1H, J=15.6 Hz, ＝CH), 7.66－7.63 (m, 2H, ph－H), 7.50 (d, 1H, J=15.6 Hz, ＝CH), 7.49 (d, 2H, J=8.4 Hz, ph－H), 7.43－7.42 (m, 3H, ph－H); ^{13}CNMR (100 MHz, $CDCl_3$) δ 189.2, 145.3, 139.2, 136.5, 134.7, 130.7, 129.9, 129.1, 128.9, 128.5, 121.5.

1－(2,4－dihydroxyphenyl)－3－phenylprop－2－en－1－one (4)

Yield 51.3%. Yellow solid, mp 139.0～141.4 ℃; IR(KBr): 3237, 1752, 1628, 1549, 1491, 1441, 1358, 1225, 1142, 968, 860, 769 cm⁻. ^{11}HNMR (300 MHz, DMSO－d_6) δ 8.25 (d, 1H, J=9.0 Hz, ph－H), 8.03 (d, 1H, J=15.3 Hz, ＝CH), 7.93－7.90 (m, 2H, ph－H), 7.85 (d, 1H, J=15.3 Hz, ＝CH), 7.49－7.47 (m, 3H, ph－H), 6.48 (d, 1H, J=9.0 Hz, ph－H), 6.34 (s, 1H, ph－H).

1－(2－hydroxy－4－methoxyphenyl)－3－phenylprop－2－en－1－one (5)

Yield 40.4%. Yellow solid, mp 87.7～99.0 ℃; IR(KBr): 3370, 1595, 1483, 1242, 1163, 1105, 1076, 1022, 964, 860, 786 cm⁻. ^{11}HNMR (300 MHz, DMSO－d_6) δ 8.32 (d, 1H, J=9.0 Hz, ph－H), 8.06 (d, 1H, J=15.3 Hz, ＝CH), 7.94－7.91 (m, 2H, ph－H), 7.86 (d, 1H, J=15.3Hz, ＝CH), 7.49－7.48 (m, 3H, ph－H), 6.60(d, 1H, J=9.0 Hz, ph－H), 6.54 (s, 1H, ph－

H), 3.86 (s, 3H, OCH_3).

1 – (4 – methoxyphenyl) – 3 – phenylprop – 2 – en – 1 – one (6)

Yield 77.4%. White solid, mp 104.6 ~ 108.6 ℃; IR(KBr): 2894, 1736, 1653, 1503, 1342, 1259, 1225, 1184, 1101, 1035, 968, 827, 761 cm^-. [11]HNMR (400 MHz, $CDCl_3$) δ 8.06 (d, 2H, J = 8.8 Hz, ph – H), 7.83 (d, 1H, J = 15.6 Hz, = CH), 7.66 – 7.63 (m, 2H, ph – H), 7.56 (d, 1H, J = 15.6 Hz, = CH), 7.42 – 7.41 (m, 3H, ph – H), 7.00 (d, 2H, J = 8.8 Hz, ph – H), 3.89 (s, 3H, OCH3); ^{13}CNMR (100 MHz, $CDCl_3$) δ 188.7, 163.4, 144.0, 135.1, 131.1, 130.8, 130.3, 128.9, 128.3, 121.9, 113.9, 55.5.

1 – (4 – (methoxymethoxy)phenyl) – 3 – phenylprop – 2 – en – 1 – one (7)

Yield 57.2%. Yellow solid; IR(KBr): 2897, 2843, 1731, 1636, 1512, 1371, 1342, 1221, 1134, 1068, 1010, 956, 840, 790 cm^-. [11]HNMR (400 MHz, $CDCl_3$) δ 8.04 (d, 2H, J = 8.8 Hz, ph – H), 7.83 (d, 1H, J = 15.6 Hz, = CH), 7.66 – 7.63 (m, 2H, ph – H), 7.56 (d, 1H, J = 15.6 Hz, = CH), 7.42 – 7.41 (m, 3H, ph – H), 7.12 (d, 2H, J = 8.8 Hz, ph – H), 5.23 (s, 2H, OCH2), 3.50 (s, 3H, OCH3); ^{13}CNMR (100 MHz, $CDCl_3$) δ 188.9, 161.0, 144.2, 135.1, 132.0, 130.7, 130.4, 128.9, 128.4, 121.9, 115.9, 94.1, 56.3.

1 – (4 – fluorophenyl) – 3 – phenylprop – 2 – en – 1 – one (8)

Yield 56.5%. Yellow solid, mp 76.2 ~ 78.3 ℃; IR(KBr): 1742, 1644, 1536, 1333, 1217, 1151, 1035, 981, 836, 761 cm^-. [11]HNMR (400 MHz, $CDCl_3$) δ 8.10 (d, 2H, J = 7.6 Hz, ph – H), 7.85 (d, 1H, J = 15.6 Hz, = CH), 7.65 – 7.62 (m, 2H, ph – H), 7.66 – 7.64 (m, 2H, ph – H), 7.54 (d, 1H, J = 15.6 Hz, = CH), 7.44(d, 2H, J = 7.6 Hz, ph – H), 7.21 – 7.17(m, 4H, ph – H); ^{13}CNMR (100 MHz, $CDCl_3$) δ 188.9, 145.1, 134.8, 134.6, 131.2, 131.0, 130.6, 129.0, 128.5, 121.6, 115.8, 115.6.

1 – (4 – chlorophenyl) – 3 – phenylprop – 2 – en – 1 – one (9)

Yield 50.9%. Yellow solid, mp 97.1 ~ 98.3 ℃; IR(KBr): 1736, 1653, 1549, 1333, 1217, 1151, 1084, 1030, 977, 827, 761 cm^-. [11]HNMR (400 MHz, $CDCl_3$) δ 7.98 (d, 2H, J = 8.4 Hz, ph – H), 7.84 (d, 1H, J = 15.6 Hz, = CH), 7.66 – 7.63 (m, 2H, ph – H), 7.50 (d, 1H, J = 15.6 Hz, = CH), 7.49 (d, 2H, J = 8.4 Hz, ph – H), 7.43 – 7.42 (m, 3H, ph – H); ^{13}CNMR (100 MHz, $CDCl_3$) δ 189.2, 145.3, 139.2, 136.5, 134.7, 130.7, 129.9, 129.1,

128.9, 128.5, 121.5.

1 - (4 - bromophenyl) - 3 - phenylprop - 2 - en - 1 - one (10)

Yield 75.4%. Yellow solid, mp 100.5 ~ 102.7 ℃; IR(KBr): 1736, 1648, 1557, 1391, 1337, 1213, 1155, 1072, 1030, 977, 827, 761 cm^{-}. ^{1}HNMR (400 MHz, $CDCl_3$) δ 7.89 (d, 2H, J=8.8 Hz, ph - H), 7.83 (d, 1H, J=15.6 Hz, =CH), 7.65 - 7.63 (m, 4H, ph - H), 7.49 (d, 1H, J=15.6 Hz, =CH), 7.43 - 7.41 (m, 3H, ph - H); ^{13}CNMR (100 MHz, $CDCl_3$) δ 189.3, 145.4, 136.9, 134.6, 131.9, 130.7, 130.0, 129.0, 128.5, 127.9, 121.4.

1 - (3 - chlorophenyl) - 3 - phenylprop - 2 - en - 1 - one (11)

Yield 56.6%. Yellow solid, mp 95.9 ~ 97.0 ℃; IR(KBr): 1771, 1648, 1599, 1325, 1283, 1205, 1047, 972, 856, 757 cm^{-}. ^{1}HNMR (300 MHz, DMSO - d_6) δ 8.22 (s, 1H, ph - H), 8.15 (d, 1H, J=8.7 Hz, ph - H), 8.01 (d, 1H, J=15.6 Hz, =CH), 7.95 - 7.93 (m, 2H, ph - H), 7.82 (d, 1H, J=15.6 Hz, =CH), 7.75 - 7.73 (m, 2H, ph - H), 7.49 - 7.47 (m, 3H, ph - H).

1 - (3 - bromophenyl) - 3 - phenylprop - 2 - en - 1 - one (12)

Yield 67.2%. Yellow solid, mp 87.0 ~ 91.3 ℃; IR(KBr): 1736, 1653, 1541, 1337, 1205, 1151, 1068, 981, 856, 757 cm^{-}. ^{1}HNMR (400 MHz, $CDCl_3$) δ 8.14 (s, 1H, ph - H), 7.95 (d, 1H, J=8.8 Hz, ph - H), 7.84 (d, 1H, J=15.6 Hz, =CH), 7.72 (d, 1H, J=8.8 Hz, ph - H), 7.66 - 7.64 (m, 2H, ph - H), 7.48 (d, 1H, J = 15.6 Hz, = CH), 7.43 - 7.36 (m, 4H, ph - H); ^{13}CNMR (100 MHz, $CDCl_3$) δ 189.0, 145.7, 140.0, 135.6, 134.6, 131.5, 130.8, 130.2, 129.0, 128.6, 127.0, 123.0, 121.4.

The general procedures for the synthesis of substituted chalconethiosemicarbazide Schiff base compounds (13 - 21).

Method 1

To the mixture of substituted chalcone (5.0 mmol) and thiosemicarbazide (5.0 mmol), was added ethanol (20 mL) andcatalytic amount of acetic acid. The reaction mixture was taken in round - bottomed flask placed in a microwave reactor and irradiated at 800W for 8 min. The reaction mixture was allowed to cool to room temperature and precipitate was formed. The precipitate was filtered and washed with appropriate amount of ethanol. The crude product was recrystallized from 95% ethanol to afford substituted chalcone Schiff base 13 - 21.

Method 2

To the mixture of substituted chalcone (5.0 mmol) and thiosemicarbazide (5.0 mmol), was added ethanol (20 mL) andcatalytic amount of acetic acid. The reaction mixture was refluxed at 80 °C over a silicone oil bath for 6 – 24 hrs. The completion of the reaction was checked by TLC. The reaction mixture was allowed to cool to room temperature and precipitate was formed. The precipitate was filtered and washed with appropriate amount of ethanol. The crude product was recrystallized from 95% ethanol to afford substituted chalcone Schiff base13 – 21.

1 – (– 1 – (2 – hydroxyphenyl) – 3 – phenylallylidene)thiosemicarbazide (13)

Yield 42.9%. Yellow solid, mp 201.3 ~ 207.6 ℃; IR(KBr): 3395, 1648, 1557, 1499, 1283, 1217, 1113, 1076, 985, 881, 761 cm^{-1}. ^{1}HNMR (300 MHz, DMSO – d_6) δ 10.44 (s, 1H, NH), 8.64 (s, 1H, NH), 8.16 (s, 1H, NH), 8.13 (d, 1H, J = 9.0 Hz, ph – H), 7.96 (d, 1H, J = 15.6 Hz, = CH), 7.70 (d, 2H, J = 8.7Hz, ph – H), 7.43 – 7.41 (m, 4H, ph – H), 7.28 (d, 1H, J = 15.6 Hz, = CH), 6.46 (d, 1H, J = 9.0 Hz, ph – H), 6.34 (s, 1H, ph – H).

1 – (– 1 – (3 – hydroxyphenyl) – 3 – phenylallylidene)thiosemicarbazide (14)

Yield 63.1%. Yellow solid, mp 178.7 ~ 180.1 ℃; IR(KBr): 3378, 1615, 1570, 1487, 1433, 1234, 1159, 1076, 968, 827, 757 cm^{-1}. ^{1}HNMR (300 MHz, DMSO – d_6) δ 11.07 (s, 1H, NH), 9.54 (s, 1H, NH), 8.46 (s, 1H, NH), 7.79 (d, 1H, J = 15.6 Hz, = CH), 7.75 – 7.72 (m, 3H, ph – H), 7.47 (d, 1H, J = 8.1 Hz, ph – H), 7.41 – 7.37 (m, 4H, ph – H), 7.24 (d, 1H, J = 15.6 Hz, = CH), 7.01 (d, 1H, J = 8.1 Hz, ph – H).

1 – (– 1 – (4 – hydroxyphenyl) – 3 – phenylallylidene)thiosemicarbazide (15)

Yield 37.9%. Yellow solid, mp 200.1 ~ 204.5 ℃; IR(KBr): 3341, 1603, 1487, 1333, 1279, 1221, 1159, 1122, 964, 873, 836, 761 cm^{-1}. ^{1}HNMR (300 MHz, DMSO – d_6) δ 10.90 (s, 1H, NH), 9.76 (s, 1H, NH), 8.58 (s, 1H, NH), 8.02 (d, 2H, J = 8.4 Hz, ph – H), 7.75 (d, 1H, J = 15.0 Hz, = CH), 7.74 – 7.20 (m, 2H, ph – H), 7.49 (d, 1H, J = 15.0 Hz, = CH), 7.49 – 7.47 (m, 3H, ph – H), 7.00 (d, 2H, J = 8.4 Hz, ph – H).

1 – (– 1 – (2 ,4 – dihydroxyphenyl) – 3 – phenylallylidene)thiosemicarbazide (16)

Yield 32.1%. Yellow solid, mp 185.1 ~ 188.9 ℃; IR(KBr): 3361, 1599,

1516, 1495, 1445, 1288, 1230, 1163, 1101, 997, 831, 699 cm^{-1}. ^{1}HNMR (300 MHz, DMSO - d_6) δ 10.28 (s, 1H, NH), 8.34(s, 1H, NH), 8.33 (d, 1H, J = 8.4 Hz, ph - H), 8.11(s, 1H, NH), 7.57 (d, 2H, J = 8.4 Hz, ph - H), 7.47 (d, 1H, J = 15.6 Hz, = CH), 7.42 - 7.40 (m, 2H, ph - H), 7.34 - 7.30 (m, 2H, ph - H), 7.01 (d, 1H, J = 15.6 Hz, = CH), 7.01 - 6.96 (m, 2H, ph - H). MS (ESI): m/z(100%) 312 (M - 1).

1 - (- 1 - (2 - hydroxy - 4 - methoxyphenyl) - 3 - phenylallylidene) thiosemicarbazide (17)

Yield 17.5%. Yellow solid, mp 116.5 ~ 122.6℃; IR(KBr): 3407, 1599, 1487, 1433, 1259, 1196, 1155, 1097, 831, 695 cm^{-1}. ^{1}HNMR (300 MHz, DMSO - d_6) δ 10.34 (s, 1H, NH), 8.24 (s, 1H, NH), 8.21 (d, 1H, J = 8.7 Hz, ph - H), 8.03 (s, 1H, NH), 7.56 (d, 2H, J = 6.9 Hz, ph - H), 7.47 (d, 1H, J = 15.6 Hz, = CH), 7.46 - 7.37 (m, 3H, ph - H), 6.60 (d, 1H, J = 15.6 Hz, = CH), 6.59 - 6.54 (m, 2H, ph - H), 3.77 (s, 3H, OCH3). MS (ESI): m/z (100%) 326 (M - 1).

1 - (- 1 - (4 - methoxyphenyl) - 3 - phenylallylidene)thiosemicarbazide (18)

Yield 28.7%. Yellow solid, mp 158.6 ~ 165.7 ℃; IR(KBr): 1595, 1483, 1242, 1163, 1105, 1076, 1022, 964, 860, 786 cm^{-1}. ^{1}HNMR (300 MHz, DMSO - d_6) δ 10.96 (s, 1H, NH), 9.92 (s, 1H, NH), 8.53 (s, 1H, NH), 7.77 (d, 1H, J = 15.6 Hz, = CH), 7.74 - 7.72 (m, 2H, ph - H), 7.61 (d, 2H, J = 8.7 Hz, ph - H),7.41 (d, 1H, J = 15.6 Hz, = CH), 7.39 - 7.36 (m, 3H, ph - H), 7.01 (d, 2H, J = 8.7 Hz, ph - H), 3.81 (s, 3H, OCH3). MS (ESI): m/z (100%) 312 (M + 1).

1 - (- 1 - (4 - fluorophenyl) - 3 - phenylallylidene)thiosemicarbazide (19)

Yield 32.9%. Yellow solid, mp 168.9 ~ 171.0 ℃; IR(KBr): 1603, 1487, 1333, 1279, 1221, 1159, 1122, 1075, 964, 873, 761 cm^{-1}. ^{1}HNMR (400 MHz, DMSO - d_6) δ11.09 (s, 1H, NH), 8.86 (s, 1H, NH), 8.32 (s, 1H, NH), 7.82 (d, 1H, J = 15.6 Hz, = CH), 7.75 (d, 2HJ = 7.5 Hz, ph - H), 7.71 - 7.68 (m, 2H, ph - H), 7.49 - 7.47 (m, 2H, ph - H), 7.47 (d, 1H, J = 15.6 Hz, = CH), 7.39 - 7.25 (m, 3H, ph - H); ^{13}CNMR (100 MHz, DMSO - d_6) δ 196.4, 179.4, 147.6, 140.1, 136.3, 131.9, 129.8, 129.1, 128.5, 121.6, 119.2, 115.7. MS (ESI): m/z(100%) 300 (M + 1).

1 – (–1 – (3 – chlorophenyl) – 3 – phenylallylidene)thiosemicarbazide (20)

Yield 45.8%. Yellow solid, mp 175.2 ~ 178.3℃; IR(KBr): 1595, 1565, 1466, 1321, 1279, 1134, 1072, 968, 865, 836, 753 cm^{-1}. ^{1}HNMR (300 MHz, DMSO – d_6) δ 11.13 (s, 1H, NH), 9.12 (s, 1H, NH), 8.57 (s, 1H, NH), 8.37 (s, 1H, ph – H), 8.12 (d, 1H, J = 8.7 Hz, ph – H), 8.09 (d, 1H, J = 15.6 Hz, = CH), 7.78 – 7.74 (m, 2H, ph – H), 7.56 (d, 1H, J = 16.0 Hz, = CH), 7.51 – 7.49 (m, 2H, ph – H), 7.46 – 7.42 (m, 3H, ph – H).

1 – (–1 – (3 – bromophenyl) – 3 – phenylallylidene)thiosemicarbazide (21)

Yield 39.4%. Yellow solid, mp 159.7 ~ 162.2 ℃; IR(KBr): 1595, 1491, 1445, 1317, 1292, 1205, 1122, 1064, 968, 873, 757 cm^{-1}. ^{1}HNMR (300 MHz, DMSO – d_6) δ 11.16 (s, 1H, NH), 8.40 (s, 1H, NH), 8.14 (s, 1H, NH), 7.99 (s, 1H, ph – H), 7.88 (d, 1H, J = 8.7 Hz, ph – H), 7.80 (d, 1H, J = 15.9 Hz, = CH), 7.60 – 7.30 (m, 2H, ph – H), 7.64 (d, 1H, J = 8.7 Hz, ph – H), 7.44 – 7.40 (m, 4H, ph – H), 6.81 (d, 1H, J = 15.9 Hz, = CH).

Assay of the diphenolase activity

The spectrophotometric assay for tyrosinase was performed according to the method reported by our groups with some slight modifications (14, 25). Briefly, all the synthesized compounds were screened for the diphenolase inhibitory activity of tyrosinase using L – DOPA as substrate. All the compounds were dissolved in DMSO. The final concentration of DMSO in the test solution was 2.0%. Phosphate buffer, pH 6.8, was used to dilute the DMSO stock solution of test compounds. Thirty units of mushroom tyrosinase (0.5 mg/mL) were first pre – incubated with the compounds, in 50 mmol/L phosphate buffer (pH 6.8), for 10 min at 25° C. Then the L – DOPA (0.5 mmol/L) was added to the reaction mixture and the enzyme reaction was monitored by measuring the change in absorbance at 475 nm of formation of the DOPA chrome for 1 min. The measurement was performed in triplicate for each concentration and averaged before further calculation. IC_{50} value, a concentration giving 50% inhibition of tyrosinase activity, was determined by interpolation of the dose – response curves. The percent of inhibition of tyrosinase reaction was calculated as following:

$$\text{Inhibition rate}(\%) = [(B - S)/B] \times 100$$

Here, the B and S are the absorbances for the blank and samples. Kojic acid (5 – hydroxy – 2 – (hydroxymethyl) – 4H – pyran – 4 – one) was used as reference standard inhibitor for comparison.

Results and Discussion

Synthesis

In the present investigation substituted chalcones have been prepared by the Claisen - Schmidt condensation of substituted acetophenone and benzaldehyde (Scheme 1). The crude product was recrystallized from 95% ethanol to afford substituted chalcone. Chalcone Schiff bases were synthesized from corresponding chalcone and thiosemicarbazide in present of acetic acid as catalyst. The rapid, highly yielding and ecofriendly microwave assisted organic syntheses (MAOS) were also used to synthesize substituted chalcones and their Schiff bases. The majority of microwave assisted reactions was conducted in polar, highly microwave absorbing solvent. The yields of these compounds are from moderate to good, and all the target compounds were characterized by IR, NMR. Some target compounds were also characterized by MS.

Tyrosinase inhibitory activity

For evaluating the tyrosinase inhibitory activity, all the synthesized compounds were subjected to tyrosinase inhibition assay with L - DOPA as substrate, according tothe method reported by our groups with some slight modifications. The tyrosinase inhibitory activity of kojic acid (5 - hydroxy - 2 - (hydroxymethyl) - 4H - pyran - 4 - one) was ever reported, therefore, it was selected as comparing substance. Fig. 1 showed that the remaining enzyme activity rapidly decreased with the increasing concentrations of compound 18. The IC_{50} values of substituted chalcones and their Schiff bases against tyrosinase were summarized in Table 1, and IC_{50} values of all these compounds were determined by interpolation of the dose - response curves and given as means of three experiments.

Our results showed that most of compounds exhibited potent inhibition on mushroom tyrosinase with IC_{50} values ranged from 0. 274 to 148. 41 μmol/L. Especially, compounds 14 - 21 bearing thiosemicarbazide showed more potent inhibitory activities than the other compounds. In addition, compounds 16 - 19 demonstrated more potent inhibitory activity than the reference standard inhibitor kojic acid, 1 - (- 1 - (4 - methoxyphenyl) - 3 - phenylallylidene) thiosemicarbazide (18) exhibited the most potent tyrosinase inhibitory activity with IC_{50} value of 0.274

μmol/L. 1 – (– 1 – (2 – hydroxyphenyl) – 3 – phenylallylidene) thiosemicarbazide (13) bearing thiosemicarbazide exhibited low tyrosinase inhibitory activity, and the percent of inhibitory is 47.97% at the 200μmol/L, but it still showed higher activity against tyrosinase than 1 – (2 – hydroxyphenyl) – 3 – phenylprop – 2 – en – 1 – one (1). The results showed substituted chalcone Schiff bases exhibited more potent inhibitory activities than corresponding substituted chalcones. Most of the chalcone analogues showed low activities against tyrosinase. The results showed that the group of thiosemicarbazide played important role in increasing activities against tyrosinase. These may be related to the structure of tyrosinase contained a type – 3 copper center with a coupled dinuclear copper active site in the catalytic core. Tyrosinase inhibition of compounds 14 – 21 depended on the competency of the sulfur atom to chelate with the dicopper nucleus in the active site, and tyrosinase would lose its catalyzing ability after forming complex (31).

From the tyrosinase inhibitory activities of compounds 1 – 3, although these compounds show low activities, we can find 1 – (2 – hydroxyphenyl) – 3 – phenylprop – 2 – en – 1 – one (3) exhibited more potent inhibitory activity than the other compounds. This suggested that hydroxyl group at the position of benzene ring might play an important role in determining this kind of compounds inhibitory activities. This may be related to the steric hindrance. Comparing to the tyrosinase inhibitory activities of compounds 1 – 4, compound 4 exhibited the most potent tyrosinase inhibitory activity with IC_{50} value of 110.22μmol/L. This result showed that the number of hydroxyl group may be an important effecting factor of inhibitory activities. However, from the table 1, we can find that compound 6 (CLog$P$ = 3.84) exhibited higher activity than the compound 3 (CLog$P$ = 3.50). This suggested that methoxyl group at the 4 – position of benzene ring might be benefical to improve this kind of compounds inhibitory activities. This may be related to the lipophilicity.

1 – (3 – bromophenyl) – 3 – phenylprop – 2 – en – 1 – one (12) showed the most potent inhibitory activity in all the halogen substituted chalcone. Comparing the inhibitory activities of compounds 8 – 10 with halogen atom at the 4 – position of benzene ring and compounds 11 – 12 with halogen atom at the 3 – position of benzene ring, with the increase of the radius of halogen atom, the inhibitory activities decreased gradually. Comparing to the inhibitory activities of all the thiosemicarbazide

Schiff bases, 1 - (- 1 - (4 - methoxyphenyl) - 3 - phenylallylidene) thiosemicarbazide (18) exhibited the most potent tyrosinase inhibitory activity with IC_{50} value of 0.274μmol/L. From the tyrosinase inhibitory activities of compounds 13 - 18, the same result can be also obtained, hydroxyl group at the position of benzene ring and the number hydroxyl group might play an important role in determining this kind of Schiff base compounds inhibitory activities, methoxyl group at the 4 - position of benzene ring might be benefical to improve this kind of Schiff base compounds inhibitory activities. Comparing to the tyrosinase inhibitory activities of compounds 19 - 21, the result showed that 1 - (- 1 - (4 - fluorophenyl) - 3 - phenylallylidene) thiosemicarbazide (19) have better activity than the other two compounds, with the increase of the radius of halogen atom, the inhibitory activities also decreased gradually. According to the ClogP values of compounds 19 - 21, the inhibitory activities of these compounds may be also related to the lipophilicity.

Inhibitory mechanism

The inhibitory mechanisms of the selected1 - (- 1 - (2,4 - dihydroxyphenyl) - 3 - phenylallylidene) thiosemicarbazide (16) and 1 - (- 1 - (4 - methoxyphenyl) - 3 - phenylallylidene) thiosemicarbazide (18) on mushroom tyrosinase for the oxidation of L - DOPA were determined (32). Figure 2, 3 showed the relationship between enzyme activity and concentration in the presence of different concentrations of the above - mentioned compounds. The results displayed that the plots of V versus $[E]$ gave a family of parallel straight lines with the same slopes. These results demonstrated that the inhibitory effects of compound 16, 18 on the tyrosinase were irreversible, suggesting that thiosemicarbazide Schiff bases of substituted chalcone compounds effectively inhibited the enzyme by binding to its binuclear active site irreversibly. The result may be related to the structure of tyrosinase, within the structure, there are two copper ions in the active centre of tyrosinase and a lipophilic long - narrow gorge near to the active centre. Compound 16 or 18 could exhibit strong affinity for copper ions in the active centre and form a reversible non - covalent complex with the tyrosinase, and this then reacts to produce the covalently modified "dead - end complex" (13, 33 - 35).

Conclusions

In conclusion, we have design, synthesized and tested a series of chalcones and their thiosemicarbazones as inhibitors of tyrosinase for defining their structure - activity relationship (SAR) studies. As a result1 - (- 1 - (4 - methoxyphenyl) - 3 - phenylallylidene) thiosemicarbazide (18) proved the best tyrosinase inhibitor among synthesized compounds. Preliminary structure activity relationships (SARs) analysis indicated that (1) thiosemicarbzide moiety might play an important role in determining their inhibitory activities, because of the sulfur atom could chelate with the dicopper nucleus in the active site, and tyrosinase would lose its catalyzing ability after forming complex; (2) methoxyl group at the 4 - position of benzene ring might be benefical to improve the inhibitory activities of tyrosinase; (3) hydroxyl group at the position of benzene ring and the number hydroxyl group might affect inhibitory activity. The inhibition mechanism analysis of 1 - (- 1 - (2,4 - dihydroxyphenyl) - 3 - phenylallylidene) thiosemicarbazide (16) and 1 - (- 1 - (4 - methoxyphenyl) - 3 - phenylallylidene) thiosemicarbazide (18) demonstrated that the inhibitory effects of the two compounds on the tyrosinase were irreversible.

Acknowledgments

This work wasfinancially supportcd by the Foundation of Hunan Provincial Science & Technology Department, China (2012FJ4350), the Key Subject Foundation of Shaoyang University (2008XK201) and the Hunan Province University Students Research Fund (Xiang Jiao Tong [2010]244 - 320).

References

1. Song, K. K., Huang, H., Han, P., Zhang, C. L., Shi, Y., Chen, Q. X. (2006) Inhibitory effects of *cis* - and *trans* - isomers of 3,5 - dihydroxystilbene on the activity of mushroom tyrosinase. Biochem Biophys Res Commun; 342: 1147 - 1151.

2. Chen, Q. X., Liu, X. D., Huang, H. (2003) Inactivation kinetics of mushroom tyrosinase in the dimethyl sulfoxide solution. Biochemistry (Mosc); 68: 644 - 649.

3. Kubo, I., Nihei, K. I., Shimizu, K. (2004) Oxidation products of quercetin by mushroom tyrosinase. Bioorg Med Chem Lett; 12: 5343 - 5347.

4. Matsuura, R., Ukeda, H., Sawamura, M. (2006) Tyrosinase inhibitory activity of citrus essential oils. J Agric Food Chem; 54: 2309 - 2313.

5. Wang, H - M., Chen, C - Y., Chen, C - Y. (2010) (-) - *N* -

Formylanonaine from *Michelia alba* as human tyrosinase inhibitor and antioxidant. Bioorg Med Chem;18：5241 –5247.

6. Tripathi, R. K., Hearing, V. J., Urabe, K., Aroca, P., Spritz, R. A. (1992) Mutational mapping of the catalytic activities of human tyrosinase. J Biol Chem;267:23707 – 23712.

7. Shiino, M., Watanabe, Y., Umezawa, K. (2001) Synthesis of N – substituted N – nitrosohydroxylamines as inhibitors of mushroom tyrosinase. Bioorg Med Chem;9：1233 –1240.

8. Gillbro, J. M., Marles, L. K., Hibberts, N. A., Schallreuter, K. U. (2004) Autocrine catecholamine biosynthesis and the beta – adrenoceptor signal promote pigmentation in human epidermal melanocytes. J Invest Dermatol;123:346 – 353.

9. Spencer, J. D., Chavan, B., Marles, L. K., Kauser, S., Rokos, H., Schallreuter, K. U. (2005) A novel mechanism in control of human pigmentation by {beta} – melanocyte – stimulating hormone and 7 – tetrahydrobiopterin. J Endocrinol;187:293 –302.

10. Schallreuter, K. U., Hasse, S., Rokos, H., Chavan, B., Shalbaf, M., Spencer,J. D.,

Wood, J. M. (2009) Cholesterol regulates melanogenesis in human epidermal melanocytes and melanoma cells. Exp Dermatol;18:680 –688.

11. Wood, J. M., Decker, H., Hartmann, H., Chavan, B., Rokos, H. (2009) Senile hair graying：H_2O_2 –mediated oxidative stress affects human hair color by blunting methionine sulfoxide repair. FASEB J;23:2065 –2075.

12. Gupta, A. K., Gover, M. D., Nouri, K. N., Taylor, S. (2006) The treatment of melasma：A review of clinical trials. J Am Acad Dermatol;55：1048 –1065.

13 Thanigaimalai, P., Hoang, T. A. L. (2010) Structural requirement(s) of *N* – phenylthioureas and benzaldehyde thiosemicarbazones as inhibitors of melanogenesis in melanoma B 16 cells. Bioorg Med Chem Lett;20:2991 –2993.

14. Yi, W., Cao, R. (2010) Synthesis and biological evaluation of novel 4 – hydroxybenzaldehyde derivatives as tyrosinase inhibitors. Eur J Med Chem; 45:639 – 646.

15. Zhu, Y. J.; Zhou, H. T.; Hu, Y. H. (2011) Antityrosinase and antimicrobial activities of 2 - phenylethanol, 2 - phenylacetaldehyde and 2 - phenylacetic acid. Food Chemistry; 124: 298 - 302.

16. Ashida, M., Brey, P. (1995) Role of the integument in insect defense: Pro - phenol oxidase cascade in the cuticular matrix. Proc Natl Acad Sci U. S. A; 92: 10698 - 10702.

17. Battaini, G., Monzani, E., Casella, L., Santagostini, L., Pagliarin, R. (2000) Inhibition of the catecholase activity of biomimetic dinuclear copper complexes by kojic acid. J Biol Inorg Chem; 5: 262 - 268.

18. Kim, B. T., O, K. J., Chun, J. C., Hwang, K. J. (2008) Synthesis of dihydroxylated chalcone derivatives with diverse substitution patterns and their radical scavenging ability toward DPPH free radicals. Bull Korean Chem Soc; 29: 1125 - 1130.

19. Rao, Y. K., Fang, S. H., Tzeng, Y. M. (2004) Differential effects of synthesized 2′ - oxygenated chalcone derivatives: modulation of human cell cycle phase distribution. Bioorg Med Chem; 12: 2679 - 2686.

20. Bandgaretal, B. P., Gawande, S. S., Bodade, R. G., Totre, J. V. (2010) Synthesis and biological evaluation of simple methoxylated chalcones as anticancer, anti - inflammatory and antioxidant agents. Bioorg Med Chem; 18: 1364 - 1370.

21. Nerya, O., Musa, R., Khatib, S., Tamir, S., Vaya, J. (2004) Chalcones as potent tyrosinase inhibitors: the effect of hydroxyl positions and numbers. Phytochemisty; 65: 1389 - 1395.

22. Seo, W. D., Ryu, Y. B., Crutis - Long, M. J., Lee, C. W., Park, K. H. (2010) Evaluation of anti - pigmentary effect of synthetic sulfonylamino chalcone. Eur J Med Chem; 45: 2010 - 2017.

23. Klabunde, T., Eicken, C. (1998) Crystal structure of a plant catechol oxidase containing a dicopper center. Nature Struct Biol; 5: 1084 - 1090.

24. Ley, J. P., Bertram, H. (2001) Hydroxy - or methoxy - substituted benzaldoximes and benzaldehyde - *O* - alkyloximes as tyrosinase inhibitors. J Bioorg Med Chem; 9: 1879 - 1885.

25. Liu, J. B., Yi, W., Wan, Y. Q., Ma, L., Song, H. C. (2008) 1 - (1

– Arylethylidene) thiosemicarbazide derivatives: A new class of tyrosinase inhibitors. Bioorg Med Chem;14:1096 – 1102.

26. Liu, J. B., Cao, R. H., Yi, W., Ma, C. M., Wan, Y. Q., Zhou, B. H., Ma, L., Song, H. C. (2009) A class of potent tyrosinase inhibitors: Alkylidenethiosemicarbazide compounds. Eur J Med Chem;44:1773 – 1778.

27. Paweczyk, A., Zaprutko. L. (2009) Microwave assisted synthesis of unsaturated jasmone heterocyclic analogues as new fragrant substances. Eur J Med Chem; 44:3032 – 3039.

28. Kappe, C. O. (2008) Microwave dielectric heating in synthetic organic chemistry.

Chem Soc Rev;37:1127 – 1139.

29. Kappe, C. O. (2004) Controlled microwave heating in modern organic synthesis.

Angew Chem Int Ed;43:6250 – 6284

30. Shi, F., Dai, A. X., Zhang, S., Zhang, X. H., Tu, S. J. (2011) Efficient microwave – assisted synthesis of novel 3 – aminohexahydrocoumarin derivatives and evaluation on their cytotoxicity. Eur J Med Chem;46:953 – 960.

31. Gerdemann, C., Eicken, C., Krebs, B. (2002) The crystal structure of catechol oxidase: New insight into the function of type – 3 copper proteins. Accounts of Chemical Research;35:183 – 191.

32. Yin, S. J., Si, Y. X., Chen, Y. F., Qian, G. Y. (2011) Mixed – Type Inhibition of Tyrosinase from Agaricus bisporus by Terephthalic Acid: Computational Simulations and Kinetics. Protein J; 30:273 – 280.

33. Tsou, C. L. (1988) Kinetics of substrate reaction during irreversible modification of enzyme activity. Advances in Enzymology and Related Areas of Molecular Biology;61:381 – 436.

34. Garcia – Canovas, F., Tudela, J., Varon, R., Vazquez, A. M. (1989) Experimental methods for kinetic study of suicide substrates. J. Enzyme Inhib;3:81 – 90.

35. Espín, J. C. Wichers, H. J. (2001) Effect of captopril on mushroom tyrosinase activity in vitro. Biochimica et Biophysica Acta;1544:289 – 300

O

O

H

R^3 R^1 R^2 + → R^3 R^1 R^2 7,9,10

O

R^3 R^1 R^2 1–6, 8,11,12 $\xrightarrow{NH_2NHCSNH_2}$

$NNHCSNH_2$

R^3 R^1 R^2 13–21

1. R^1 = OH, R^2 = H, R^3 = H;
2. R^1 = H, R^2 = OH, R^3 = H;
3. R^1 = H, R^2 = H, R^3 = OH;
4. R^1 = OH, R^2 = H, R^3 = OH;
5. R^1 = OH, R^2 = H, R^3 = OMe;
6. R^1 = H, R^2 = H, R^3 = OMe;
7. R^1 = H, R^2 = H, R^3 = OCH_2OCH_3;
8. R^1 = H, R^2 = H, R^3 = F;
9. R^1 = H, R^2 = H, R^3 = Cl;
10. R^1 = H, R^2 = H, R^3 = Br;
11. R^1 = H, R^2 = Cl, R^3 = H;
12. R^1 = H, R^2 = Br, R^3 = H;
13. R^1 = OH, R^2 = H, R^3 = H;
14. R^1 = H, R^2 = OH, R^3 = H;
15. R^1 = H, R^2 = H, R^3 = OH;
16. R^1 = OH, R^2 = H, R^3 = OH;
17. R^1 = OH, R^2 = H, R^3 = OMe;
18. R^1 = H, R^2 = H, R^3 = OMe;
19. R^1 = H, R^2 = H, R^3 = F;
20. R^2 = H, R^2 = Cl, R^3 = H;
21. R^1 = H, R^2 = Br, R^3 = H

Scheme 1 Synthesis of the substituted chalcones and their thiosemicarbazide Schiff bases.

Table 1 Tyrosinase inhibitory activities of the synthesized compounds

Compounds	CLog*P* [d]	IC_{50} (μmol/L) [a]	percent of inhibition [c]
1	3.96		11.2%
2	3.50		18.35%
3	3.50		29.06%
4	3.45	110.22	75.2%
5	4.03		31.5%
6	3.84	141.80	54.11%
7	3.35		13.26%
8	3.84		40.2%
9	4.41		40.25%

续表

Compounds	$CLogP$ [d]	IC_{50} (μmol/L) [a]	percent of inhibition[c]
10	4.56		37.22%
11	4.41	117.18	65.91%
12	4.56		NA[b]
13	4.92		47.97%
14	4.92	148.41	59.14%
15	4.92	77.87	71.61%
16	4.42	19.79	95.49%
17	5.00	18.81	95.58%
18	5.28	0.274	98.64%
19	5.29	7.56	96.59%
20	5.86	23.91	85.68%
21	6.01	52.68	76.78%
Kojic acid		23	11.2%

a. Values were determined from logarithmic concentration - inhibition curves (at least seven points) and are given as means of three experiments

b. not active at 200 μmol/L concentration.

c. percent of inhibition of tyrosinase reaction at the 200 μmol/L

d. Value of $CLogP$ was obtained by ChemBioDraw Ultra 12.0

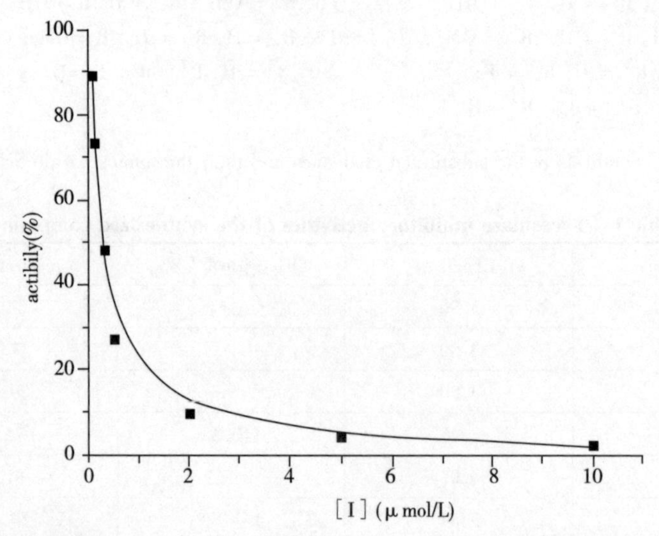

Figure 1　Effect of compound 18 on the diphenolase activity of mushroom tyrosinase for the catalysis of L - DOPA at 25℃.

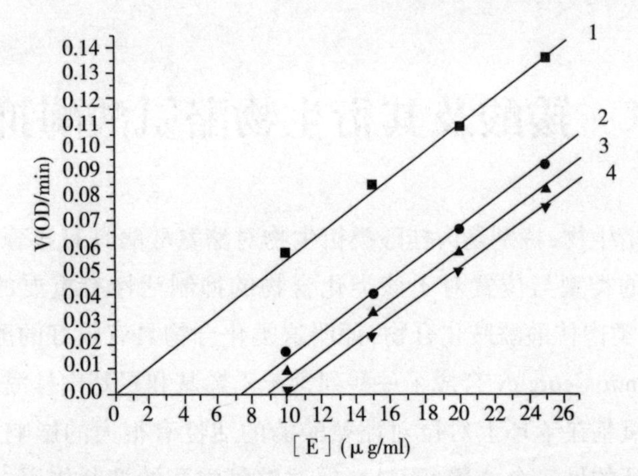

Figure 2 The effect of concentrations of tyrosinase on its activity for the catalysis of L - DOPA at different concentration of compound 16. The concentrations of compound 16 for curves 1 - 4 are 0, 5, 10 and 50μmol/L, respectively.

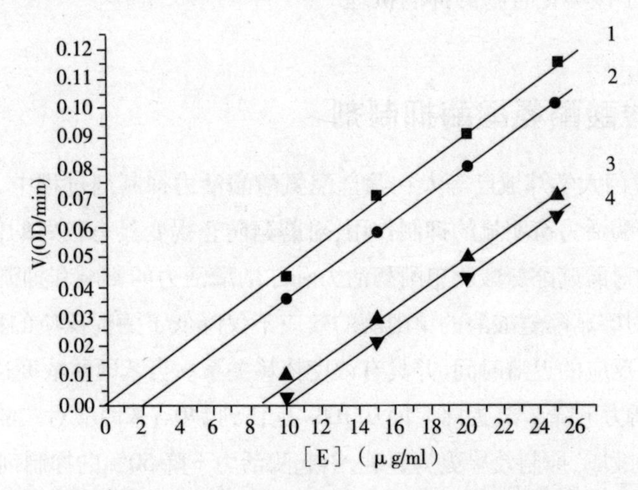

Figure 3 The effect of concentrations of tyrosinase on its activity for the catalysis of L - DOPA at different concentration of compound 18. The concentrations of compound 18 for curves 1 - 4 are 0, 0.2, 0.3 and 0.5μmol/L, respectively.

第3章　羧酸及其衍生物酪氨酸酶抑制剂

羧酸及其衍生物,特别是肉桂酸类衍生物对酪氨酸酶具有显著的抑制作用,苯环取代基团的类型与位置对于该类化合物的抑制活性有重要的影响。最近Maya合成了一类肉桂酸酰胺化合物,证明这类化合物具有较好的酪氨酸酶抑制活性[1]。Lyubomir Georgiev合成了一些列含苯乙烯基和羟基肉桂酰胺的化合物,发现羟基或甲氧基在苯环上对位对酪氨酸酶的活性有很大的影响,增加一个取代在同一个苯环的另一个位置,明显减弱对酪氨酸酶的抑制作用[2]。厦门大学陈清西教授课题组对卤代肉桂酸的酪氨酸酶抑制活性及抑制动力学进行了研究,初步探讨了该类化合物的抑制机理[3]。

3.1　羧酸酪氨酸酶抑制剂

2012年厦门大学郑成已等人在筛选酪氨酸酶活力抑制剂过程中,发现苯丙酸对蘑菇酪氨酸酶活力有明显的抑制作用,对蘑菇防止褐变具有较好的保鲜效果[4]。报道了苯丙酸对蘑菇酪氨酸酶单酚酶活力和二酚酶活力的影响和抑制作用。结果表明,苯丙酸对蘑菇酪氨酸酶的单酚酶的效应不仅降低了酶促反应的稳定态活性,也延长了酶促反应的迟滞时间,并具有浓度依赖关系。当苯丙酸浓度达10.0mmol/L时,稳定态活力下降2/3,迟滞时间从100s延长到140s;苯丙酸对二酚酶活力的抑制也显示浓度效应,抑制效果更为显著,引起酶活力下降50%的抑制剂浓度(IC_{50})为2.05mol/L,苯丙酸对二酚酶活力的抑制显示可逆的效应,抑制类型为混合型,抑制常数(K_I和K_{IS})分别为1.570mmol/L和3.070mmol/L。利用苯丙酸抑制酪氨酸酶活力及抑菌特性,将可开发应用于粮食、蔬菜、水果的保鲜与防腐。

2015年广东药学院李慕紫等人以AutoDock 4.2和iGEMDOCK 2.1分子对接软件对13种苯甲酸类似物(图3-1)与酪氨酸酶进行模拟对接研究,探讨了苯甲酸类似物对接结合自由能与实验测得的酶抑制活性的关系,并对分子对接结果进行分析[5]。结果表明,Autodock程序中铜电荷的设置对结合自由能有显著影响,铜电荷为2.0时,苯甲

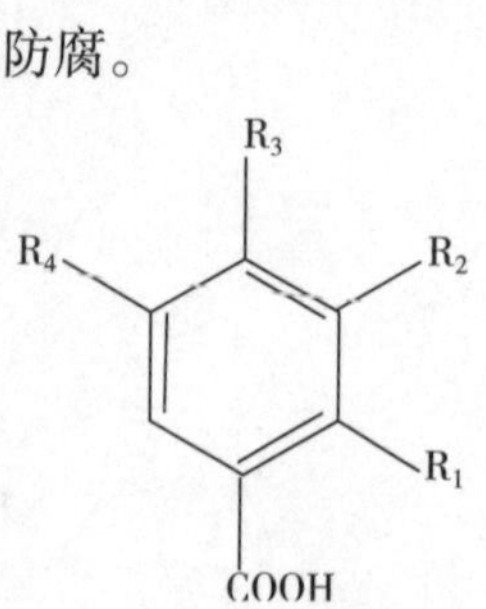

R_3 R_4 R_2 R_1 COOH

图3-1　苯甲酸类似物

酸类似物结合自由能和 pIC_{50} 线性相关系数可达 0.8036。iGEMDOCK 2.1 预测苯甲酸类似物的结果线性较差，显示 AutoDock4.2 较 iGEMDOCK 2.1 预测酪氨酸酶抑制剂的可靠性更高。

2012 年胡溪育合成了 Dawson 型多金属氧酸盐 $\alpha/\beta - K_6P_2W_{18}O_{62} \cdot 10H_2O$（$P_2W_{18}$），研究其对蘑菇酪氨酸酶的抑制作用及其抑制机理[6]。结果表明：P_2W_{18} 对酪氨酸酶单酚酶和二酚酶均有较强的抑制作用，0.067 mmol/L 的 P_2W_{18} 能使单酚酶活力下降 74.5%，0.67 mmol/L 的 P_2W_{18} 能使二酚酶活力下降 52%。导致单酚酶活力和二酚酶活力下降 50% 的抑制剂浓度（IC_{50}）分别为 0.05 和 0.64 mmol/L。P_2W_{18} 对酪氨酸酶二酚酶的抑制效应表现为不可逆抑制作用。以 Keggin 型多金属氧酸盐 $H_3PW_{12}O_{40}$ 和 $H_4SiW_{12}O_{40}$ 为效应物，研究它们对蘑菇酪氨酸酶的抑制作用及其抑制机理，结果表明：这两种化合物作为可逆性抑制剂对酪氨酸酶二酚酶抑制效果显著，导致二酚酶活力下降 50% 的抑制剂浓度（IC_{50}）分别为 1.57mmol/L和 2.31 mmol/L。$H_3PW_{12}O_{40}$ 对酶的抑制类型表现为混合型，而 $H_4SiW_{12}O_{40}$ 为竞争性。通过比较两种化合物对酪氨酸酶 IC_{50} 和抑制常数大小来衡量这两种化合物对酪氨酸酶抑制效果的强弱。合成了甘氨酸的 Keggin 结构多金属氧酸盐 $(HGly)_3PW_{12}O_{40}$ 和 $(HGly)_4SiW_{12}O_{40}$，以这两种化合物为效应物，研究其对蘑菇酪氨酸酶的抑制作用及抑制机理。结果表明：$(HGly)_3PW_{12}O_{40}$ 和 $(HGly)_4SiW_{12}O_{40}$ 对酪氨酸酶二酚酶的抑制作用均属于可逆过程，引起二酚酶活力下降 50% 的抑制剂浓度（IC_{50}）分别为 1.55 mmol/L 和 1.39 mmol/L。$(HGly)_3PW_{12}O_{40}$ 表现为反竞争性抑制，而 $(HGly)_4SiW_{12}O_{40}$ 为非竞争性抑制。

多金属氧酸盐（POMs）因其丰富多样的结构、功能特性和广阔的应用领域而备受关注。研究发现，POMs 很可能成为一种潜在有效的酶抑制剂。2016 年邢蕊研究了两类 Keggin 型磷钼酸盐，合成了两类 Keggin 型磷钼酸盐，即不同过渡金属取代的磷钼酸（$PMo_{11}Cu$，$PMo_{11}Zn$，$PMo_{11}V$，$PMo_{11}Fe$）和甘氨酸基的不同过渡金属取代的磷钼酸盐（$Gly-PMo_{11}Cu$，$Gly-PMo_{11}Zn$，$Gly-PMo_{11}V$，$Gly-PMo_{11}Fe$），并以其为效应物，通过聚丙烯凝胶电泳和分光光度法研究其对蘑菇酪氨酸酶活力的抑制影响。结果表明：效应物均明显抑制酪氨酸酶活性，其抑制效果虽不及曲酸，但均为典型酪氨酸酶抑制剂熊果苷（$IC_{50} = 5.3$ mmol/L）的 10 倍多。其中甘氨酸基的过渡金属取代的磷钼酸盐对酶的抑制效果均优于对应的过渡金属取代的磷钼酸盐。此外，4 种过渡金属取代的磷钼酸盐对酶均表现为可逆的抑制，而甘氨酸基的过渡金属取代磷钼酸盐对酶的抑制则均属不可逆过程。结合电泳和动力学结果，得出 4 种过渡金属取代的磷钼酸盐中 $PMo_{11}Fe$ 对酪氨酸酶的抑制效果最好，次之是 $PMo_{11}Cu$，而 $PMo_{11}Zn$ 和 $PMo_{11}V$ 对酶的抑制效果相差不是很大。这可能是 $PMo_{11}Fe$ 对酪氨酸酶二酚酶的抑制类型为非竞争性不同于其它三种竞争型的化合物所引起的。通过直线的斜率对效应物的浓度二次作图，得到 $PMo_{11}Cu$，$PMo_{11}Zn$，$PMo_{11}V$ 对游离酶的抑制常数 K_I 值分别为 0.104 mmol/L，0.410 mmol/L，0.172 mmol/L，通过直线的截距对 $PMo_{11}Fe$ 浓度二次作图，得到 $PMo_{11}Fe$ 对游离酶及酶－底物复合物的抑制常数 $K_I = K_{IS} = 0.467$ mmol/L。

2014 年集美大学郑阿萍等人以 $H_3PW_{12}O_{40}$ 和 $H_4SiW_{12}O_{40}$（简写为 PW_{12} 和 SiW_{12}）为效应物，测定其对酪氨酸酶活力的抑制作用[7]。通过非变性聚丙烯凝胶（Native－PAGE）电泳确定酪氨酸酶是多家族基因编码，其分子量为 3×10^4～3.4×10^4，4.2×10^4～4.6×10^4，6.4×10^4～6.8×10^4，且均具有活性，测定 PW_{12} 和 SiW_{12} 对酪氨酸酶的抑制效果，并结合酶动力学法研究其抑制机理。结果表明，当 PW_{12} 和 SiW_{12} 浓度分别达到 13 mmol/L 和 25 mmol /L 时，酪氨酸酶的活力完全被抑制，即 PW_{12} 和 SiW_{12} 对酪氨酸酶二酚酶具有不同程度的抑制效果。当所加酶量为 0.0173 mg/mL 时，PW_{12} 和 SiW_{12} 对酪氨酸酶二酚酶活力的半抑制率（IC_{50}）分别为 1.57 mmol/L 和 2.31 mmol/L，它们对酪氨酸酶二酚酶的抑制均为可逆过程。其中，PW_{12} 对二酚酶的抑制类型为混合型，其 K_I 及 K_{IS} 分别为 0.34 mmol/L和 0.43 mmol/L，SiW_{12} 对二酚酶的抑制类型表现为竞争型，其 K_I 为 0.59 mmol/L。综合考虑 IC_{50} 值和抑制常数等参数，PW_{12} 对酪氨酸酶二酚酶的抑制能力优于 SiW_{12}。

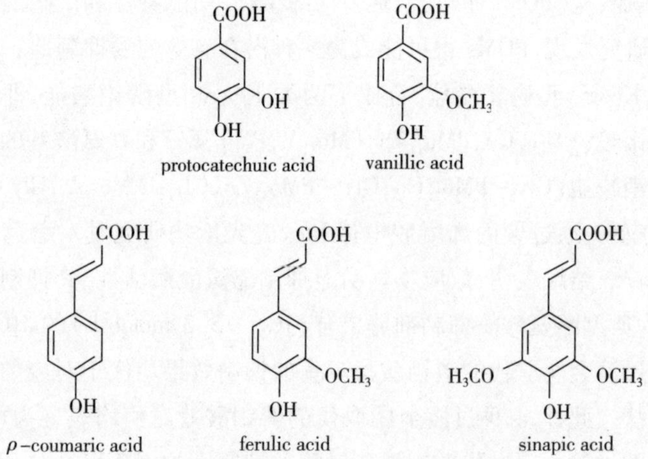

图 3－2　从米糠中分离得到的羧酸化合物

2016 年 Sukhontha Sukhonthara 等人从米糠提取了一类羧酸类物质（原儿茶酚酸、香草酸、对羟基肉桂酸、阿魏酸、3,5 - 二甲氧基 - 4 - 羟基肉桂酸，图 3 - 2），考察了其对苹果、土豆中多酚氧化酶的抑制活性[8]。结果表明，阿魏酸和对羟基肉桂酸对多酚氧化酶具有一定抑制活性，特别是对羟基肉桂酸对多酚氧化酶的抑制活性最好。

2013 年厦门大学杨美花等人以小鼠黑色素瘤 B16 细胞为模型，研究了天然药物有效成分原儿茶酸对黑色素生成和酪氨酸酶活力的影响[9]。实验结果显示：原儿茶酸（图 3 - 3）可显著降低细胞的黑色素含量，浓度为 10μmol/L 时使黑色素含量降低 60%。进一步研究发现，原儿茶酸对小鼠黑色素瘤 B16 细胞粗提取和共培养体系中的酪氨酸酶均有较强的抑制作用。对粗提酪氨酸酶的活力半抑制浓度（IC_{50}）约为 160μmol/L，与共培养体系中酪氨酸酶活力的抑制、抑制剂浓度和作用时间有关，10μmol/L 作用 54h 可使酪氨酸酶活力下降 88%。该研究为原儿茶酸作为酪氨酸酶抑制剂的开发应用奠定了基础，研究结果说明，原儿茶酸作为酶抑制剂具有一定的可行性，但是其抑制黑色素合成的分子作用机制还未阐明，值得深入研究。

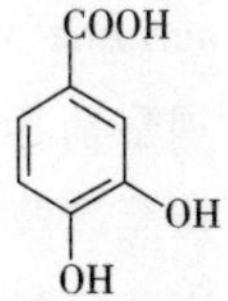

COOH

OH

OH

图 3 - 3　儿茶酚酸结构式

2004 年天津医科大学籍晓明通过应用逆转录聚合酶链反应（RT — PCR）技术在基因水平检测维甲酸、氢醌等药物对 B16F10 小鼠黑素瘤细胞酪氨酸酶 mRNA 表达的调控作用，并探讨其可能的作用机制[10]。首先根据实验设计要求将 B16F10 细胞在加有不同浓度药物的新鲜培养基中培养一定时间后，消化收集细胞；用 TRIozl 试剂盒抽提细胞总 RNA，然后将细胞总 RNA 逆转录为 DNA；再以所得 DNA 为模板，采用 PCR 技术进行核酸扩增；扩增产物进行凝胶电泳，通过凝胶扫描仪将电泳所得扩增条带扫入计算机并分析实验结果。实验结果显示：维甲酸对 Bl6F10 细胞酪氨酸酶 mRNA 表达确有影响，它能够下调 B16F10 细胞酪氨酸酶 mRNA 的表达；而且因维甲酸浓度、作用时间的不同，这种下调作用亦有所不同。随着药物浓度的加大、作用时间的延长，B16F10 细胞酪氨酸酶 mRNA 的表达具有逐渐减弱的趋势。维甲酸浓度为 100μmol/L、作用时间为 72h，酪氨酸酶 mRNA 的表达下调 30. 13%。而氢醌无论是单独作用，还是与维甲酸联合作用对酪氨酸酶 mRNA 的表达均无显著影响。通过实验我们发现维甲酸类药物抑制黑素细胞生成黑素作用的主要机制是它能够在基因转录水平抑制酪氨酸酶 mRNA 的表达。在实验中也发现维甲酸对霍乱毒素诱导的酪氨酸酶 mRNA 表达上调具有抑制作用。这说明维甲酸也可以通过 cAMP/PAK 途径调节酪氨酸酶基因表达,从而对黑素的合成产生影响。可见,维甲酸除直接作用于酪氨酸酶基因转录外,还能够通过其它途径间接影响酪氨酸酶基因转录。

2013 年张璇探讨新藤黄酸对小鼠黑色素瘤细胞株 B16 增殖抑制的影响及其抗肿瘤的可能机制,为将新藤黄酸开发为抗肿瘤药物提供了相应的理论依据[11,12]。采用 MTT 实验检测新藤黄酸对 B16 细胞生长和增殖的作用;采用 acridine orange/ethidium bromide(AO/EB) 荧光染色和 Hochest33258 荧光染色观察新藤黄酸对 B16 细胞的影响。采用流式细胞仪 Annexin V－FITC/ PI 双染检测新藤黄酸诱导 B16 细胞的凋亡发生率。采用流式细胞仪检测新藤黄酸处理 B16 细胞后细胞内活性氧(ROS)的产生;Western blotting 法检测凋亡相关蛋白 Caspase－3 的变化,并检测 PI3K、p－PI3K、Akt、p－Akt、p－mTOR、PTEN 蛋白的变化以探讨新藤黄酸诱导 B16 细胞凋亡的分子机制。MTT 法实验结果表明新藤黄酸在一定浓度下对黑色素瘤 B16 细胞的增殖有明显抑制作用,且呈一定的时效和量效依赖关系。荧光显微镜观察新藤黄酸处理黑色素瘤 B16 细胞后形态学变化;AO/EB 染色观察发现正常细胞染色正常,细胞密度较大;凋亡细胞或损伤细胞呈现橘黄色荧光,细胞密度降低。新藤黄酸处理的细胞呈现明显的凋亡状态。荧光显微镜观察新藤黄酸处理黑色素瘤 B16 细胞后,Hochest33258 染色结果显示,正常细胞呈现均一暗淡的蓝色荧光;新藤黄酸处理的细胞中则呈现蓝白色、碎块状致密浓染,具有明显的凋亡特征。流式细胞仪检测结果表明,新藤黄酸作用于黑色素瘤 B16 细胞后,细胞凋亡率随药物浓度的增大而升高。得出了新藤黄酸在一定的时间和浓度范围内能够抑制黑色素瘤 B16 细胞的增殖,诱导细胞发生凋亡,其诱导细胞凋亡的机制可能与调控 PI3K/Akt/mTOR 信号通路有关的结论。

2014 年 Yong－Hua Hu 等人合成了 2－氯肉桂酸和 2,4－二氯肉桂酸(图 3－4),考察了这两个化合物对于酪氨酸酶的抑制活性和抑制动力学[13]。结果表明,这两个化合物都具有酪氨酸酶的抑制活性,通过对抑制动力学分析,两个化合物对于单酚酶能够延长滞后期但是降低了稳态期的活性,2－氯肉桂酸延长滞后期时间刚好为 5%,而 2,4－二氯肉桂酸延长滞后期超过 30.4%。2－氯肉桂酸和 2,4－二氯肉桂酸对于二酚酶的 IC_{50} 值分别为 0.765mmol/L 和 0.295mmol/L,抑制动力学分析表明这两个化合物是可逆非竞争型抑制剂,抑制常数分别为 0.348mmol/L 和 0.159mmol/L。

2012 年陈冉等人研究了十种苯丙氨酸(Phe)的对位衍生物 L－4－溴苯丙氨酸(Bp)、L－4－碘苯丙氨酸(Ip)、L－4－磷酸基－苯丙氨酸(Pp)、L－4－氨基－

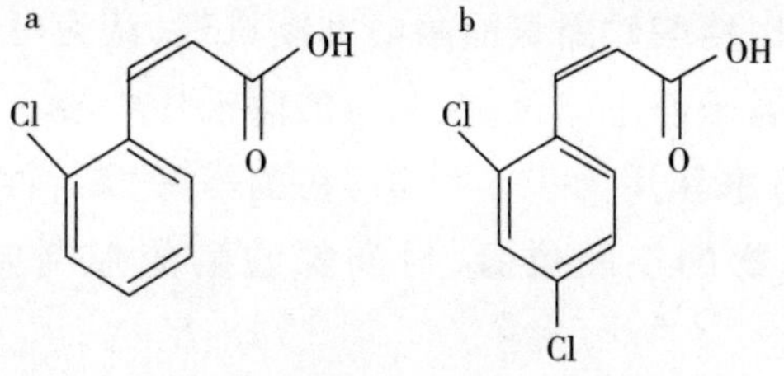

a　Cl OH O　　b　Cl OH O Cl

图3－4 氯代肉桂酸结构式

苯丙氨酸(Ap)、L－4－硝基－苯丙氨酸(Np)、L－4－磺酸基－苯丙氨酸(Sp)、L－4－氟苯丙氨酸(Fp)、L－4－氯苯丙氨酸(Clp)、L－4－甲氧基－苯丙氨酸(Mp)、L－4－氰基－苯丙氨酸(Cp)对小鼠黑色素瘤细胞系B16的细胞毒性、诱导细胞凋亡作用和抑制细胞成集落作用[14]。结果表明,它们的细胞毒性大小依次为Pp、Fp、Mp、Ip、Sp、Ap、Bp。其中Mp和Ip的毒性相近,Sp和Ap的毒性相近。它们诱导细胞凋亡的作用强弱依次为Fp、Mp、Ip。其中Mp和Ip的作用相近。它们抑制细胞成集落的作用大小依次为Pp、Fp、Ip、Mp、Sp、Bp、Np、Ap。其中Pp、Fp、Ip、Mp的作用相近, Sp、Bp、Np的作用相近。初步的细胞毒理分析表明,Fp、Mp、Ip能够诱导B16细胞凋亡和抑制B16细胞形成集落。

2008年向敏等人为了考察熊果酸(UA) 诱导B16黑色素瘤细胞的分化作用。采用MTT法测定UA对B16细胞增殖抑制作用;并以形态学、黑色素含量、酪氨酸酶活性和体内致瘤能力变化为指标,观测UA对B16细胞的诱导分化作用。结果表明,用4－64μmol/L的UA作用于肿瘤细胞,可见有不同程度的抑制作用[15]。MTT法测得UA作用B16细胞12h、24h和48h的IC_{50}分别为58.05μmol/L、35.13μmol/L和12.17μmol/L。UA对B16细胞有较强的分化诱导作用,表现为UA作用后,细胞形态发生明显变化,出现细胞核变小,规则,核浆比变小,线粒体、粗面内质网等细胞器丰富等变化;B16细胞黑色素颗粒生成能力明显增多($P<0.01$ or $P<0.05$) 并伴有酪氨酸酶活性的升高($P<0.01$),而B16细胞体内成瘤能力显著下降($P<0.05$) 。得出了UA对B16细胞有分化诱导作用的结论,其机制有待进一步研究。

2012年华南理工大学穆燕等人以L－酪氨酸和L－多巴为底物研究了对甲氧基肉桂酸(图3－5)对酪氨酸酶的抑制能力[16],结果表明,对甲氧基肉桂酸对酪氨酸酶催化单酚底物L－酪氨酸有较强的抑制作用,但对酪氨酸酶催化二酚底物L－多巴没有抑制作用。采用荧光光谱以及分子对接手

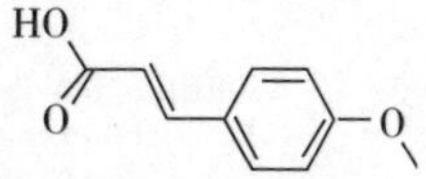

HO O O

图3－5 对甲氧基肉桂酸结构式

段探讨了对甲氧基肉桂酸对酪氨酸酶的抑制机理，认为对甲氧基肉桂酸对酪氨酸酶的荧光光谱属于静态猝灭，在酪氨酸酶活性中心与酪氨酸酶结合形成复合物，以氢键和疏水作用力稳定结构；在酪氨酸酶活性中心，对甲氧基肉桂酸占据了单酚底物的空间位置，对酪氨酸酶的抑制应该是一种竞争性抑制。

2013 年 Yue - Xiu Si 等人对胡椒酸（图 3 - 6）的酪氨酸酶抑制活性及抑制动力学进行了研究[17]。胡椒酸是一种具有抗氧化能力的苯甲酸结构天然产物，基于苯甲酸结构和抗氧化能力，对胡椒酸开展了抑制动力学和计算机模拟研究，结果表明，该化合物是可逆混合型抑制剂，胡椒酸能够快速与酪氨酸酶形成键合力，持续将底物和酪氨酸酶反应，在孵化过程中，胡椒酸能够与底物形成牢固键，荧光猝灭分析表明，并没有检测到胡椒酸破坏酪氨酸酶的三维结构。分子对接和计算机模拟结果表明，胡椒酸能够与酪氨酸酶氨基酸残基发生作用。

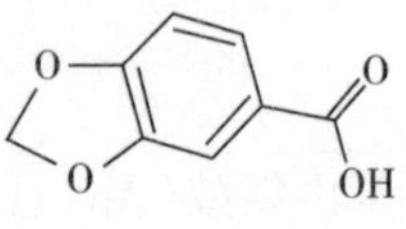

图 3 - 6　胡椒酸结构式

2013 年 Mitko Miliovsky 等人以芳香酸酐和芳香醛为起始原料通过普尔金反应一釜法合成了一系列具有二苯乙烯结构的羧酸（图 3 - 7）[18]。该方法简单、选择性好、收率高、反应时间短。所有化合物都通过了核磁共振氢谱、碳谱、红外、高分辨质谱进行表征，双键构型进一步通过核磁共振二维图谱和 NOESY 谱进行确定。对所合成的化合物进行了抗菌、抗氧化、消炎及酪氨酸酶抑制活性评价，结果表明该类化合物具有一定的酪氨酸酶抑制活性，其中（E） - 2 - （1 - 羧基 - 2 - 苯乙烯） - 4，5 - 二羟基苯甲酸的抑制活性最好，抑制活性强于氢化喹啉。

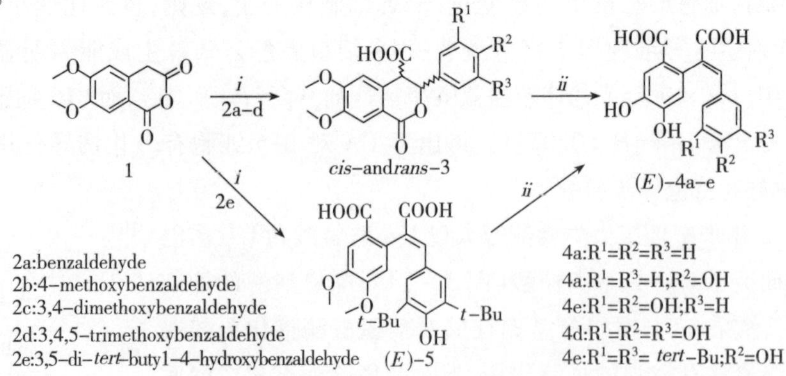

图 3 - 7　一釜法合成二苯乙烯羧酸类化合物

2013年Alessandro Venditti等人芦笋中富含的香料组份芦笋酸为原料进行还原反应得到二氢芦笋酸(DHAA)(图3-8),合成方法采用了文献报道的方法并进行了改进[19]。对所合成的化合物进行了酪氨酸酶抑制活性测试,结果表明这类化合物具有一定的抑制活性,抑制动力学是通过双倒数曲线确定的,这类化合物是竞争型抑制剂。综合实验结果,该类化合物被认为是一种潜在的可以用于医药、食品及化妆品行业的活性分子。

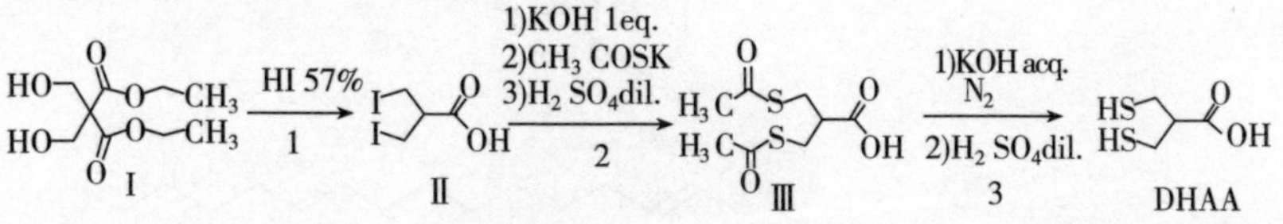

图3-8 DHAA的合成路线

2013年Wei-Ming Chai等人考察了呋喃甲醇、呋喃甲醛、呋喃甲酸(图3-9)的酪氨酸酶抑制活性及抑制动力学,结果表明,这三个化合物是酪氨酸酶的可逆抑制剂[20]。呋喃甲醇和呋喃甲醛是混合型抑制剂,而呋喃甲酸是非竞争型抑制剂,确定了其抑制常数,抑制活性顺序是呋喃甲醛 > 呋喃甲酸 > 呋喃甲醇,呋喃环在酶抑制能力中发挥关键作用。

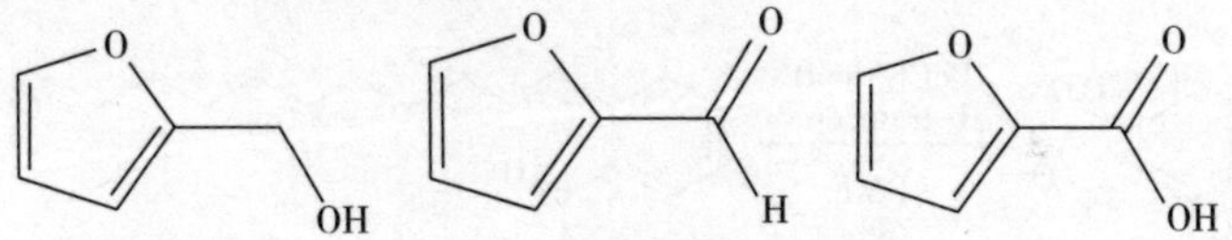

图3-9 呋喃衍生物结构式

2013年Yuanqing Fu等人采用了一种新的高效的方法,以2,6-二羟基苯甲酸为起始原料,合成了银杏酸(图3-10),该方法通过钯催化交叉偶联、催化氢化等5步反应,总收率达到34%[21]。所得的银杏酸通过核磁共振氢谱、碳谱、红外、高分辨质谱进行结构确证。该化合物具有较好的抑制酪氨酸酶的活性,其IC_{50}值为2.8mg/mL。此合成方法的开发有利于开展银杏酸的生物活性研究,而银杏酸在医药和食品行业有着很大的潜在用途。

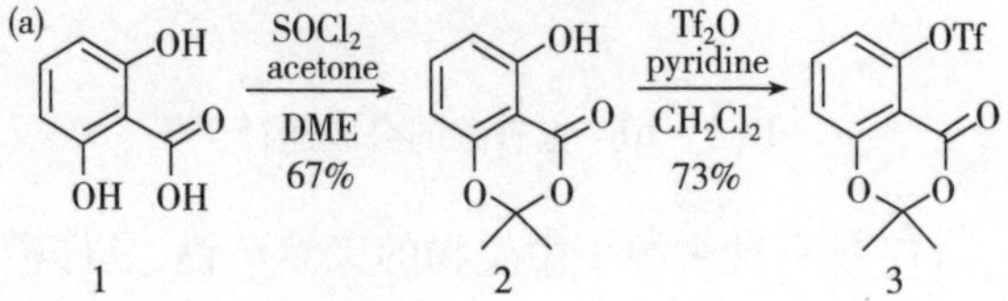

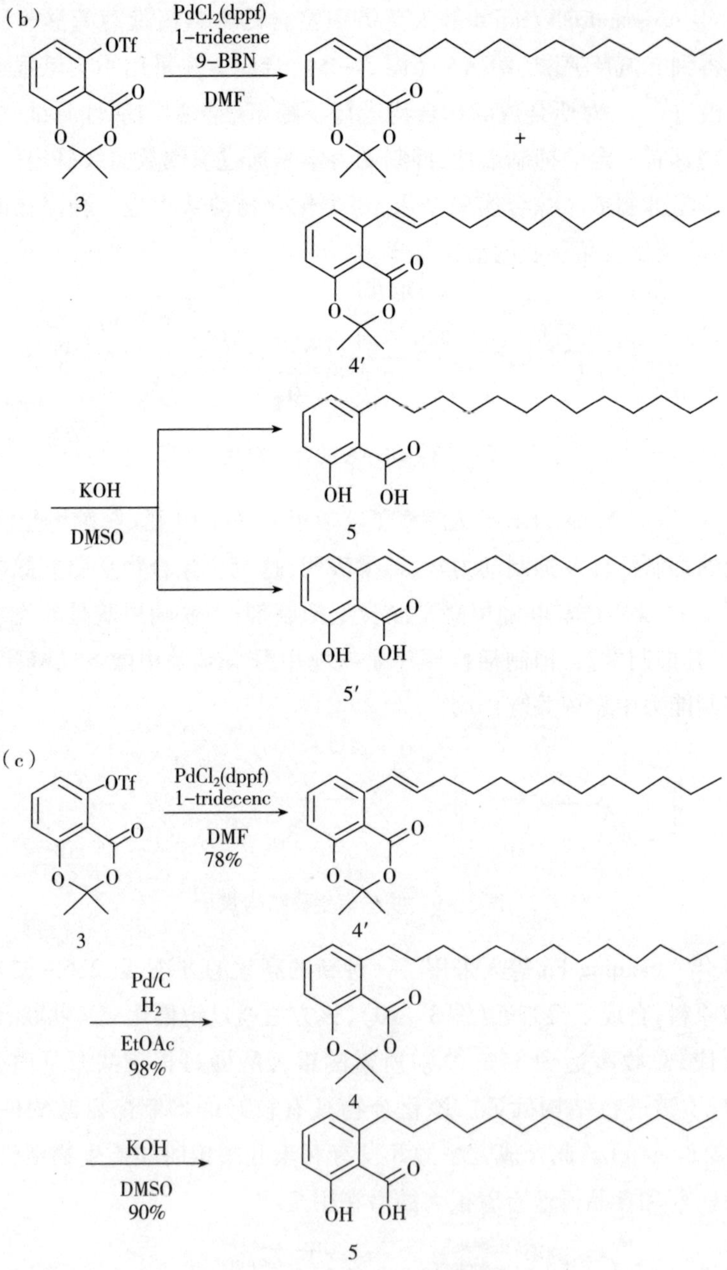

(b) 3 —[$PdCl_2(dppf)$, 1-tridecene, 9-BBN, DMF]→ 4 + 4′ —[KOH, DMSO]→ 5 + 5′

(c) 3 —[$PdCl_2(dppf)$, 1-tridecenc, DMF, 78%]→ 4′ —[Pd/C, H_2, EtOAc, 98%]→ 4 —[KOH, DMSO, 90%]→ 5

图 3-10　银杏酸的合成途径

富马酸(图 3-11)是一种来源于微生物的天然产物,是柠檬酸环一个已知的重要中间体。2016 年 Lin Gou 等人测试了其对酪氨酸酶的抑制活性,通过抑制动力学分析和计算机模拟考察了该类化合物对酪氨酸酶的抑制特性,结果表明该化合物是可逆非竞争型抑制剂，其半抑制浓度为(13.7 ± 0.25)mmol/L，抑制常数Ki slope = (12.64 ± 0.75)mmol/L[22]。动力学评价和荧光光谱表明富马酸会导致酪氨酸酶活性中心的改变，键的相互作用力在无L－DOPA的条件下得以证实，分子对接结果进一步证实了富马酸能够和活性中心残基HIS263和HIS85发生作用，而且存在四个重要的氢键。

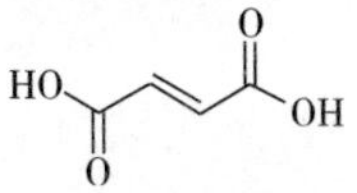

图3－11 富马酸结构式

最近Yi Cui等人考察了对位取代肉桂酸对酪氨酸酶的抑制作用，包括4－乙氧基肉桂酸(4－ECA)、4－氯肉桂酸(4－CCA)、4－硝基肉桂酸(4－NCA)，通过紫外光谱、荧光光谱、分子对接、与铜离子相互作用来考察肉桂酸衍生物对酪氨酸酶的抑制机理[23]。结果表明，这三个化合物对酪氨酸酶具有较好的抑制活性，光谱分析用于考察这些化合物抑制酪氨酸酶催化L－酪氨酸和L－多巴的能力，得出了其半抑制浓度和抑制常数。测试化合物对酪氨酸酶猝灭机制属于静态类型，而且当4－NCA加入时会发生荧光发射峰红移，铜离子相互作用及分子对接结果证实4－CCA不能直接和铜离子相互作用，但能够和活性中心的氨基酸残基相互作用，4－ECA和4－NCA能够与酪氨酸酶活性中心铜离子形成螯合物，对位取代肉桂酸衍生物的酪氨酸酶抑制活性为设计新的酪氨酸酶抑制剂奠定基础。

2016年Takahiro Oyama等人为了找到新型酪氨酸酶抑制剂，发现苯基苯甲酸是一种特殊的结构(图3－12)[24]。在三个苯基苯甲酸的异构体中，3－苯基苯甲酸的酪氨酸酶的抑制活性是最强的，其单酚酶的半抑制浓度为6.97μmol，二酚酶的半抑制浓度为36.3μmol。抑制动力学研究表明，对于单酚酶和二酚酶的表观抑制模式分别是非竞争型抑制剂和混合型抑制剂。分子对接结果表明，3－苯基苯甲酸的羧基能够有效的和酪氨酸酶的活性中心铜离子作用，证实了对羧基进行修饰会明显影响抑制酪氨酸酶的活性，与预期的一样，羧基酯化后抑制活性消失。这些研究结果证实了3－苯基苯甲酸是一个开发新型酪氨酸酶抑制剂有用的先导化合物，提供了一种从分子水平研究酪氨酸酶抑制机制的方法。

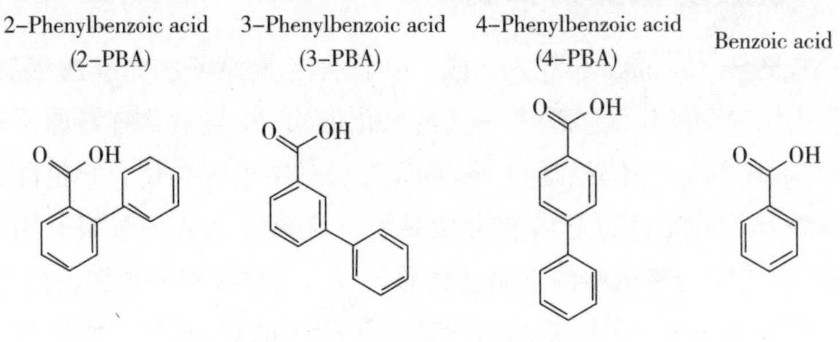

图3－12 苯基苯甲酸的结构式

2016 年 Yong - Hua Hu 等人研究了 4 - 羟基肉桂酸对于酪氨酸酶的抑制活性、抑制动力学及其在蘑菇保鲜中的应用[25]。研究结果表明,4 - 羟基肉桂酸抑制酪氨酸酶是一种缓慢可逆的分步反应,确定了 4 - 羟基肉桂酸与酪氨酸酶的微观反应速率常数,分子对接模拟 4 - 羟基肉桂酸和酪氨酸酶的相互作用。4 - 羟基肉桂酸主要是酚羟基与酪氨酸酶活性中心发生相互作用而达到抑制酪氨酸酶活性的目的,细胞毒性测试结果表明 4 - 羟基肉桂酸对正常细胞无毒性,在蘑菇保存实验中发现该化合物可以延缓蘑菇褐变时间。

2007 年张春乐发现了肉桂酸、4 - 羟基肉桂酸和 4 - 甲氧基肉桂酸(图 3 - 13)对酪氨酸酶具有单酚酶和二酚酶的抑制活性[26],实验结果表明,它们使单酚酶活力下降 50% 的抑制剂浓度(IC_{50})分别为 0. 58 mmol/L、0. 27 mmol/L 和 0. 50mmol/L。使二酚酶活力下降 50% 的抑制剂浓度(IC_{50})分别为 2. 1mmol/L、0. 5mmol/L 和 0. 42 mmol/L。研究它们对二酚酶的抑制类型,肉桂酸和 4 - 甲氧基肉桂酸属于非竞争性抑制,抑制常数分别为 1. 994 mmol/L 和 0. 458 mmol/L,4 - 羟基肉桂酸为竞争性抑制剂,抑制常数为 0. 244 mmol/ L。

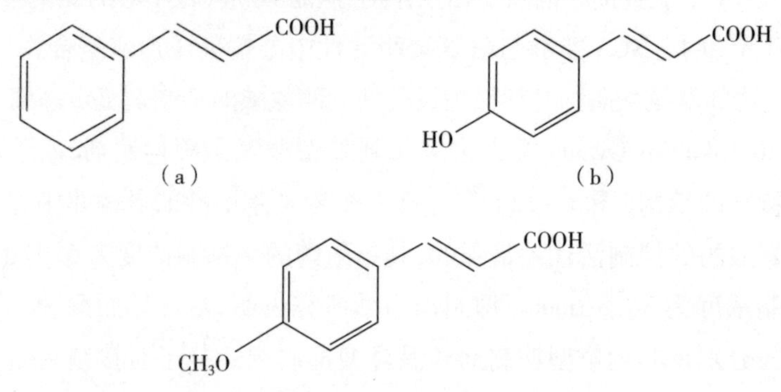

图 3 - 13　肉桂酸及其衍生物结构式

3.2　酰胺酪氨酸酶抑制剂

2013 年 Maya Chochkova 等人合成了一类含有氟原子的肉桂酸酰胺(图 3 - 14)[27],采用快速 DPPH 测试法考察其抗氧化能力,并与不含有氟原子的类似物及标准的抗氧化剂进行比较,通过 Nenadis 方法考察了该类化合物清除自由基的能力,含氟酰胺的清除自由基的能力相对于不含氟原子的酰胺略有增加。含氟原子的肉桂酸酰胺的酪氨酸酶抑制活性通过以 L - 酪氨酸为底物进行测试,结果表明,SA - Trp(6 - F) - OMe 是酪氨酸酶抑制活性最好的,两个含氟原子的酰胺

SA－Trp(6－F)－OMe 和 SA－Tyr(3－F)－OMe 与阳性对照对羟基肉桂酸相比没有明显的不同，但比另一个阳性对照氢化喹啉活性低。

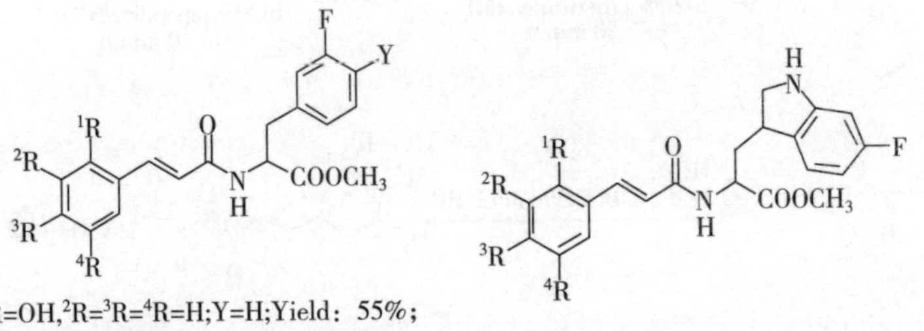

1) ^{1}R=OH,^{2}R=^{3}R=^{4}R=H;Y=H;Yield：55%;
2) ^{1}R=^{2}R=^{4}R=H,^{3}R=OH;Y=H;Yield:48%;
5) ^{1}R=H,^{2}R=^{4}R=OCH_3,^{3}R=OH;Y=OH;Yield:45%;
6) ^{1}R=^{2}R=^{3}R=^{4}R=H;Y=H;Yield:77%;
7) ^{1}R=H,^{2}R=^{4}R=OCH_3,^{3}R=OH;Y=H;Yield:29%;
8) ^{1}R=^{4}R=H,^{2}R=OCH_3,^{3}R=OH;Y=H;Yield:41%;

3) ^{1}R=^{4}R=H,^{2}R=OCH_3,^{3}R=OH;Yield:41%;
4) ^{1}R=H,^{2}R=^{4}R=OCH_3,^{3}R=OH;Yield:62%;

图3－14 肉桂酸酰胺结构式

2013年Lyubomir Georgiev等人合成了一类羟基肉桂酸酰胺（图3－15）[28]，并测试了体外其抗氧化、抑制酪氨酸酶活性。为了探明氨基在抑制酪氨酸酶中的作用，氨基酸酯的肉桂酸酰胺和天然生物胺的酰胺化合物得到了合成，并对这类化合物进行了酪氨酸酶抑制活性测试，结果表明，对位有羟基取代的化合物具有明显的酪氨酸酶抑制活性，这可能与其结构易于和L－酪氨酸酶结合有关，进一步证实了对羟基苯基是酪氨酸酶的重要活性基团，对羟基肉桂酰胺对酪氨酸酶的抑制活性强于氢化喹啉，同时也比对羟基肉桂酸对酪氨酸酶的抑制活性更强。

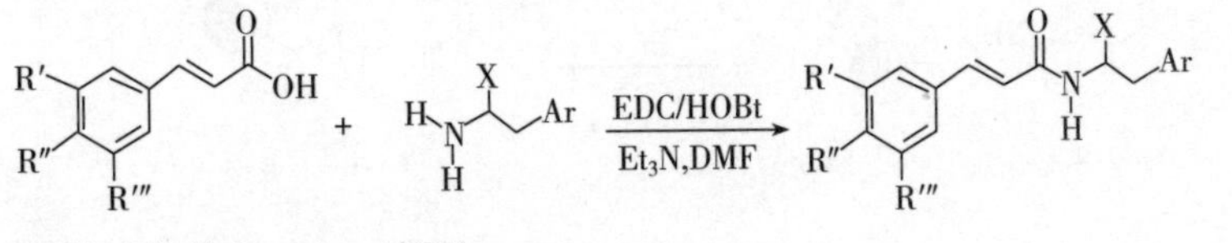

EDC/HOBt, Et_3N,DMF

R′=R″=R‴=H,cinnamic acid(CA);
R′=R‴=H,R″=OH, ρ－coumaric acid(ρ－CoumA)
R′=H,R″=OH,R‴=OCH_3,ferulic acid(FA);
R′=R‴=OCH_3,R″=OH,sinapic acid(SA);
R′=H,R″=R‴=OH,caffeic acid(CafA)

图3－15 肉桂酸酰胺的合成路线

2013年Seon－Yeong Kwak等人报道了一类咖啡酰胺酸基异羟肟酸（图3－16）具有较好的抗氧化和抑制酪氨酸酶的活性[29]，特别是当咖啡酸和脯氨酸或者是含有苯环的氨基酸，如苯丙氨酸。各种羟基肉桂酸衍生物进一步和苯丙酰异羟肟酸、脯氨酸异羟肟酸进行连接，并讨论了其酪氨酸酶抑制活性构效关系。咖啡酸和芥子酸能有效的抑制酪氨酸酶活性并能减少黑色素的合成。

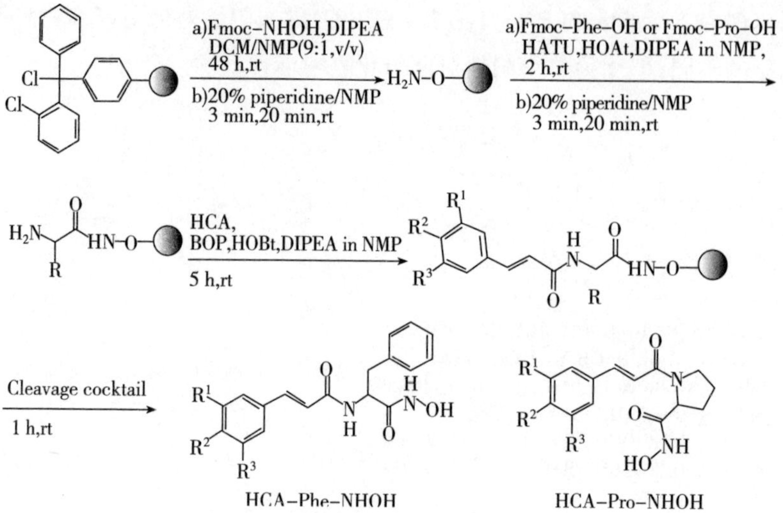

a)Fmoc-NHOH,DIPEA DCM/NMP(9:1,v/v) 48 h,rt b)20% piperidine/NMP 3 min,20 min,rt
a)Fmoc-Phe-OH or Fmoc-Pro-OH HATU,HOAt,DIPEA in NMP, 2 h,rt b)20% piperidine/NMP 3 min,20 min,rt
HCA, BOP,HOBt,DIPEA in NMP 5 h,rt
Cleavage cocktail 1 h,rt
HCA-Phe-NHOH
HCA-Pro-NHOH

图 3-16 咖啡酰胺酸基异羟肟酸的合成路线

2012 年 Babasaheb P. Bandgar 等人通过 Claisen - Schmidt 缩合反应由 3-甲酰基-5-甲基咔唑和各种 3-酰胺基苯乙酮合成了含有酰胺基团的查尔酮类化合物(图 3-17),所有化合物都进行了体外黄嘌呤氧化酶、酪氨酸酶和黑色素产生的抑制作用[30]。大部分目标化合物具有体外抑制黄嘌呤氧化酶的作用,部分化合物对酪氨酸酶具有抑制作用,这些研究成果对于未来设计新型酪氨酸酶抑制剂具有重要指导作用。

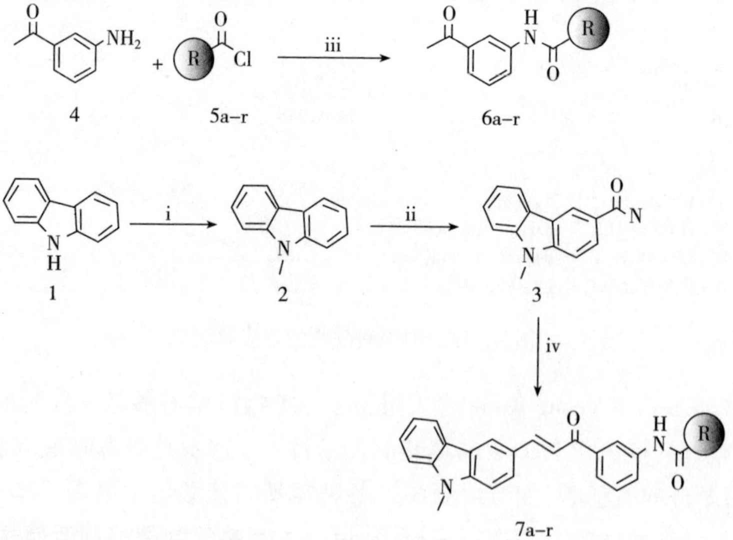

4 5a-r iii 6a-r
1 i 2 ii 3 iv 7a-r

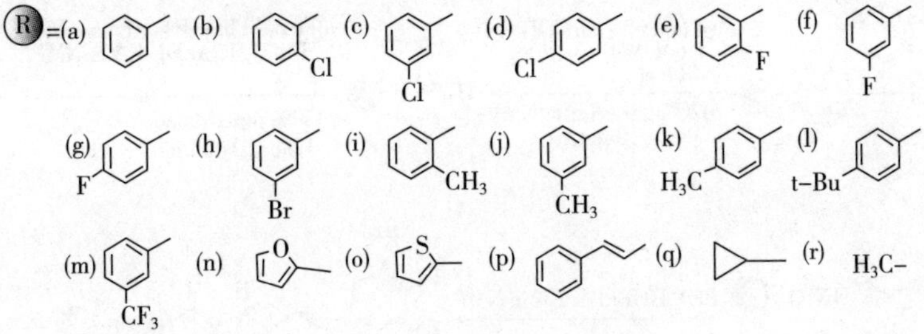

R=(a) (b) Cl (c) Cl (d) Cl (e) F (f) F (g) F (h) Br (i) CH_3 (j) CH_3 (k) H_3C (l) t-Bu (m) CF_3 (n) (o) (p) (q) (r) H_3C-

图 3 – 17　酰胺查尔酮的合成路线

2012 年 Heung Soo Baek 等人合成了一系列多羟基 N – 苄基 – 苯甲酰胺衍生物(图 3 – 18),评价了目标化合物酪氨酸酶抑制活性及去黑色素[31]。结果表明,亲脂性是提高该类酰胺化合物去黑色素能力的重要因素,分子对接法讨论了酪氨酸酶与底物相互作用的方式,阐述了抑制酪氨酸酶活性的原因。

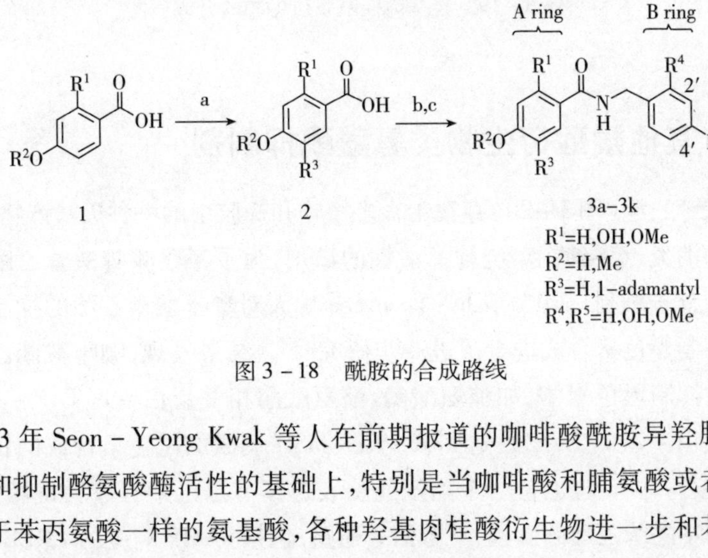

A ring　B ring

1 —a→ 2 —b,c→ 3a–3k

R^1=H,OH,OMe
R^2=H,Me
R^3=H,1-adamantyl
R^4,R^5=H,OH,OMe

图 3 – 18　酰胺的合成路线

2013 年 Seon – Yeong Kwak 等人在前期报道的咖啡酸酰胺异羟肟酸具有抗氧化性和抑制酪氨酸酶活性的基础上,特别是当咖啡酸和脯氨酸或者含有芳环的类似于苯丙氨酸一样的氨基酸,各种羟基肉桂酸衍生物进一步和苯丙氨酰异羟肟酸或脯氨酰异羟肟酸（HCAPhe – NHOH，HCA – Pro – NHOH）(图 3 – 19)进行连接,同时研究了抗氧化及酪氨酸酶抑制活性的构效关系[32]。HCAPhe – NHOH，HCA – Pro – NHOH 表现出比 HCA 更好的抗氧化活性,而且咖啡酸和芥酸衍生物能够有效的抑制酪氨酸酶活性,减少 Mel – Ab 黑色素细胞中黑色含量。

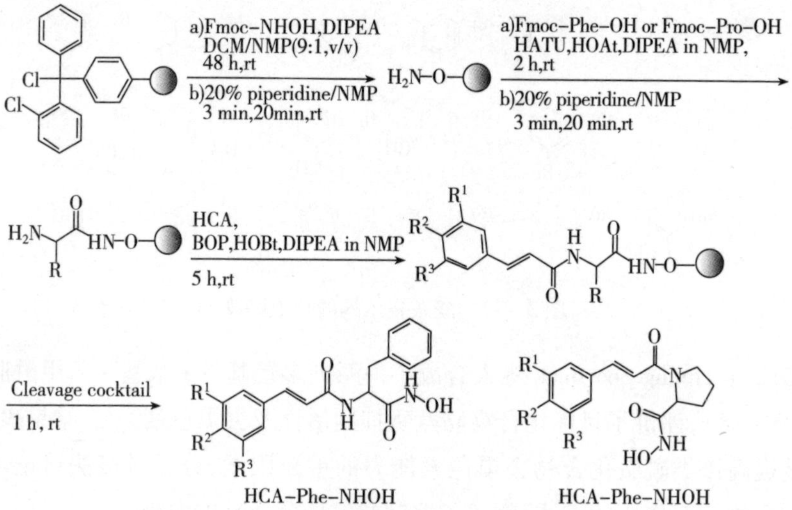

图 3－19　羟基肉桂酰胺衍生物的合成

3.3　其他羧酸衍生物类氨酸酶抑制剂

咖啡酸苯乙酯(图 3－20)存在于很多植物和蜂胶中的一种天然产物,是一种具有很好的消炎、免疫调节、抗肿瘤活性的试剂,为了考察咖啡酸苯乙酯抑制黑色素合成的分子机制, 2013 年 Ji－Yeon Lee 等人对咖啡酸苯乙酯的抑制酪氨酸酶活性及抑制黑色素合成进行了进一步研究[33]。结果发现,咖啡酸苯乙酯能够有效地通过抑制黑色素酶,如酪氨酸酶、酪氨酸酶相关蛋白－1(TRP－1)、酪氨酸酶相关蛋白－2(TRP－2),达到减少 α－MSH 刺激的黑色素合成的目的。另外,MITF 表达和核转位也是一个酪氨酸酶表达调节黑色素合成关键因素,但是并未受到咖啡酸苯乙酯的影响,下游信号通路,包括 cAMP 响应的 CREB、糖苷合成激酶 3β(GSK－3β)也为受到咖啡酸苯乙酯的影响。有趣的是咖啡酸苯乙酯能够抑制 MITF 蛋白和酪氨酸酶启动子相互作用而达到抑制酪氨酸酶启动子转录活性的目的。通过研究发现,MITF 在黑色素合成中起着重要作用,咖啡酸苯乙酯能够用于治疗黑色素合成失调的疾病。

2007 年张春乐研究肉桂酸甲酯(图 3－21)对蘑菇酪氨酸酶的抑制活性,结果表明,对单酚酶和二酚酶,它的 IC_{50}分别为 0.92 mmol/L 和 1.65 mmol/L,是非竞争性抑制剂[34]。

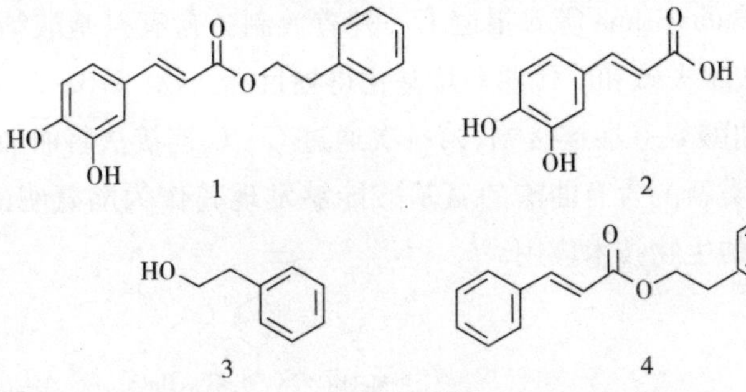

O O HO OH 1

O OH HO OH 2

HO 3

O O 4

图3-20 肉桂酸酯的结构式

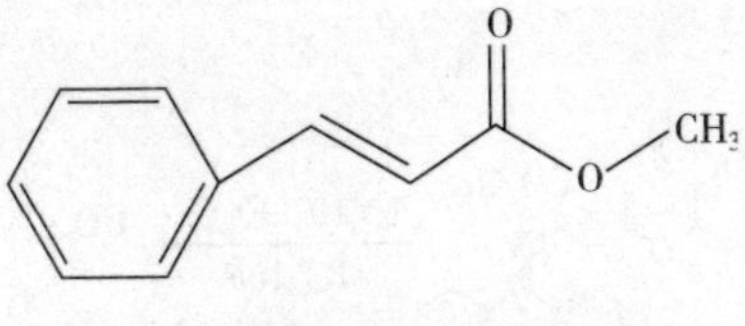

O O CH_3

图3-21 肉桂酸甲酯结构式

2013年Ana Sofia Monteiro等人合成了一系列作为黑色素细胞相关酶治疗前药(MDEPT)的三氮烯前药(图3-22)[35]。这些化合物来源于酰基酪氨酸,而该化合物是一个很好的酪氨酸酶的底物,在黑色素瘤细胞中能够过量表达[35]。我们分析了三氮烯前药的化学稳定性和血浆酶水解,同时考察了在酪氨酸酶存在下抗肿瘤药物的释放特性,接着评价随着酪氨酸酶活性强调的差异这些前药在黑色素瘤细胞的毒性,其中3-[2-(Acetylamino)-3-(4-hydroxyphenyl)propanoyl]-1-(4-cyanophenyl)-3-methyltriazene(5c)对黑色素瘤细胞具有最高的细胞毒性,而且与酪氨酸酶活性具有很好的相关性,这就说明了该类前药的细胞毒性是酪氨酸酶依赖性的。

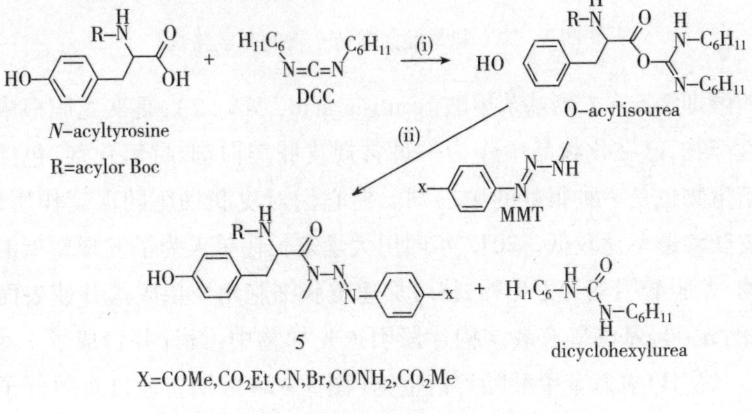

R-NH O OH HO *N*-acyltyrosine R=acylor Boc + $H_{11}C_6$-N=C=N-C_6H_{11} DCC (i) → HO R-NH O O H N-C_6H_{11} N-C_6H_{11} O-acylisourea

(ii) x-C6H4-N=N-NH MMT

HO R-NH O N-N=N x + $H_{11}C_6$-NH-C(O)-NH-C_6H_{11}

5 dicyclohexylurea

X=COMe,CO_2Et,CN,Br,$CONH_2$,CO_2Me

图3-22 三氮烯前药的合成路线

2015 年 C. Balakrishna 等人报道了一种有效制备含有氨基酸结构的曲酸衍生物，通过甘氨酸席夫碱和溴代曲酸烷基化得到目标产物（图 3－23）。采用该方法制备了单、双曲酸氨基酸连接物，第一次通过 C－C 连接法将曲酸和氨基酸直接连接起来。这类新的含有曲酸的氨基酸能够发现其作为酪氨酶的广泛应用，同时也能有其他的生物技术应用[36]。

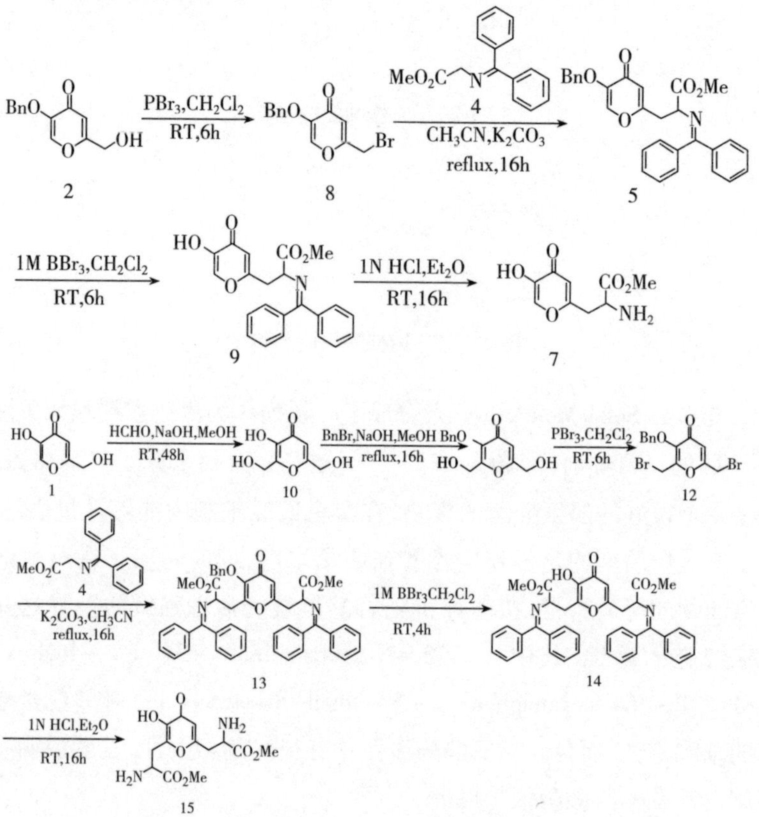

图 3－23　氨基酸曲酸衍生物的合成路线

龙胆酸即 2，5－二羟基苯甲酸（gentisic acid，MA，2），是从龙胆根中提取的天然活性产物，已在化妆品中作为一种常规皮肤美白剂、抗氧化剂，包括它的烷基酯类衍生物也是一种很好的美白剂。但它们对皮肤细胞的毒副作用均较强，尽管其皮肤渗透率比较低。2012 年四川大学宋长伟等人为了发现新型的美白活性化合物，龙胆酸甲酯为先导物，将可加速皮肤细胞角质层脱落并使表面肌肤光滑、柔细的 α－羟基酸等有效结构片段引入先导物中，设计并合成了一系列 2－羟基－5－烷（H）氧基苯甲酸酯类衍生物（图 3－24），期望通过有效分子片段的引入以增强其抑制酪氨酸酶的活性，同时得到毒性更低、活性更强的美白类化合物[37]。以龙胆酸甲酯、卤代烃和 α－羟基酸乙酯为原料，合成了10个未见文献报道的2－羟基－5－烷（H）氧基苯甲酸酯类衍生物（3a，3b 和 6a－6h），其结构经^{1}H NMR，IR，MS 和 HRMS 确认。初步生物活性测试结果表明6a，6b，6e 和6f 具有较强的抑制酪氨酸酶活性，进一步药理实验表明经取代改造后的6a－6h，其毒性和光学毒性都相对龙胆酸甲酯和氢醌更低。

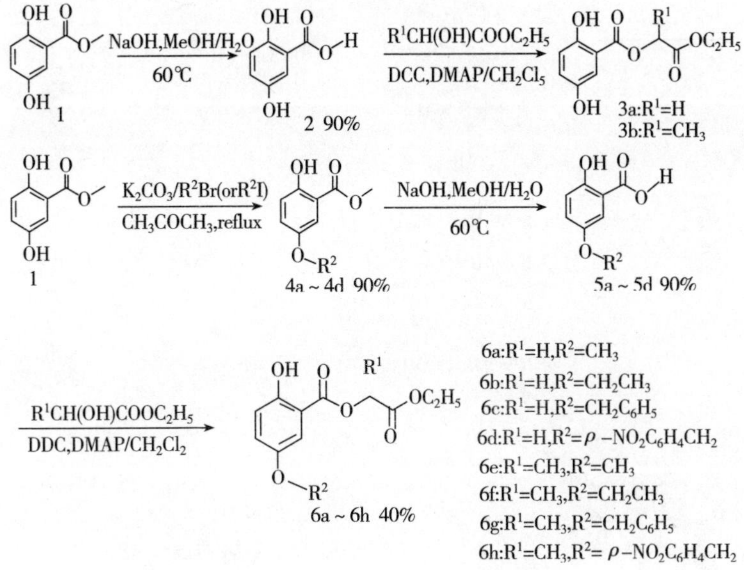

图3－24　龙胆酸衍生物的合成

2013年日本学者 Yusei Kashima 等人合成了一系列三－氧－甲基去甲基岩白菜素（图3－25），并对所合成的化合物的抗氧化及抑制酪氨酸酶的活性进行了评价[38]。在所有的被测试的化合物中，具有邻苯二酚结构的化合物抗氧化能力和酪氨酸酶抑制活性最强，部分化合物对酪氨酸酶的抑制活性在微摩尔级，其中对酪氨酸酶的半抑制浓度为9.1μmol/L，强于阳性对照，构效关系揭示了与苯甲酸相连的酯基与三－氧－甲基去甲基岩白菜素一样都是产生酪氨酸酶抑制活性有效基团，抑制机理分析证实该化合物是混合型抑制剂。研究结果表明，该化合物可以作为设计和开发潜在酪氨酸酶抑制剂的有效参考。

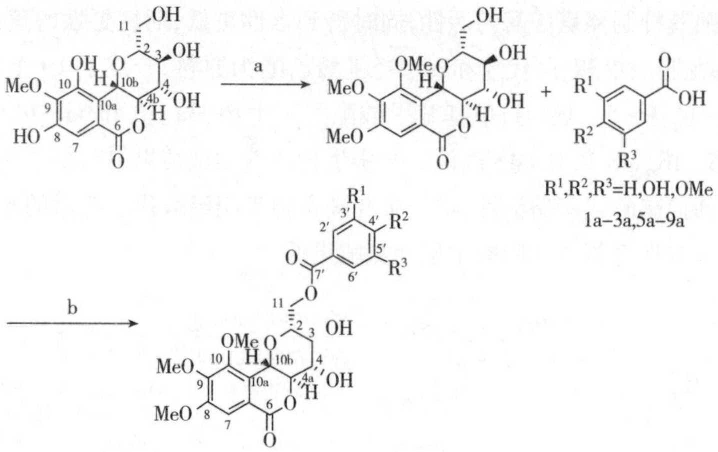

R^1,R^2,R^3=H,OH,OMe

1a-3a,5a-9a

1: R^1=H, R^2=H, R^3=H
2: R^1=H, R^2=OH, R^3=H
3: R^1=H, R^2=OMe, R^3=H
5: R^1=OMe, R^2=OMe, R^3=H
6: R^1=OMe, R^2=OH, R^3=H
7: R^1=OH, R^2=OMe, R^3=OMe
8: R^1=OMe, R^2=H, R^3=OMe
9: R^1=OMe, R^2=OH, R^3=OMe

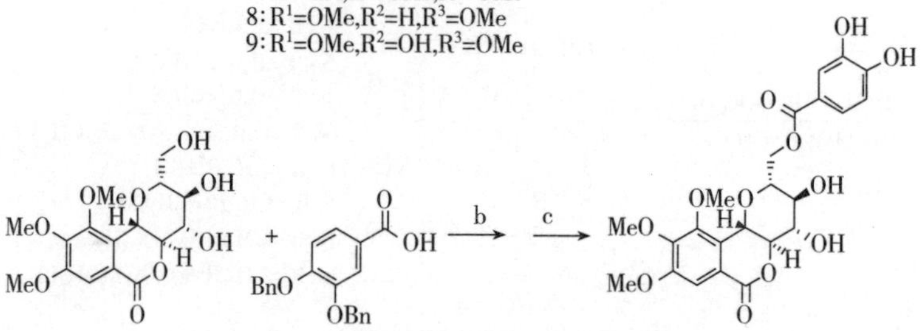

图 3－25　去甲基岩白菜素衍生物的合成

2013 年 Haroon Khan 等人从轮叶黄精的根茎中分离得到两种长链羧酸酯(图 3－26),通过红外、核磁、紫外、高分辨对所得化合物进行了结构表征,并测试了其对酪氨酸酶的抑制活性[39]。结果表明这两个化合物都具有很强的酪氨酸酶抑制活性,其 IC_{50}分别为 22.34μmol/L 和 9.45μmol/L。基于初步研究,这两个化合物可能是一种有用的临床治疗药剂,有必要开展进一步详细研究。

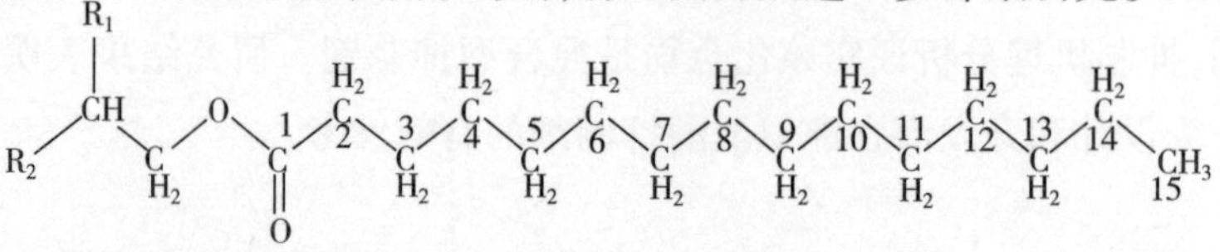

1. R_1=H, R_2=CH_3

2. R_1=OH, R_2=CH_2OH

图 3－26　长链羧酸酯的化学结构式

酶促褐变和细菌的腐败是导致中国橄榄在收获期品质下降影响货架期的重要因素。为了筛选中国橄榄保鲜的重要抗褐变物质,2016 年厦门大学 Yu – Long Jia 等人合成了咖啡酸正壬酯和咖啡酸正己酯(图 2 – 27),并测试了其对酪氨酸酶的抑制活性,其中咖啡酸正壬酯对酪氨酸酶抑制活性最强,其 IC_{50} 值为 37.5μmol/L,而咖啡酸正己酯并未体现出酪氨酸酶抑制活性[40]。抑制动力学分析表明,咖啡酸正壬酯在浓度低于 50μmol/L 时为可逆型抑制剂,在浓度高于 50μmol/L 时为不可逆抑制剂,对于可逆抑制机理,其抑制动力学常数 K_I 和 K_{IS} 分别为 24.6μmol/L 和 37.4μmol/L。对于应用于中国橄榄果研究结果表明,咖啡酸正壬酯具有有效抗褐变和防腐败效果,因此,该化合物可能是一个有效的酪氨酸酶抑制剂。

A HO HO O O B HO HO O O

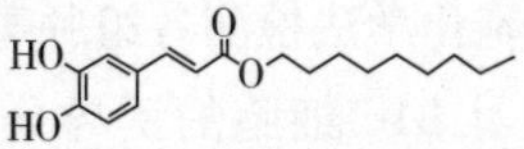

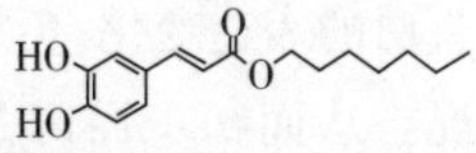

图 3 – 27 咖啡酸正壬酯和咖啡酸正己酯结构式

2009 年易伟等人合成了两个维生素 C 酯(图 3 – 28),并考察了其酪氨酸酶抑制活性。抑制结果表明化合物 1 和化合物 2 对酪氨酸酶的半抑制浓度分别为 0.58mmol/L 和 0.16mmol/L。量 – 效曲线表明这两个化合物不仅能够延长滞后时间,而且能够降低稳态速度[41]。动力学分析表明,化合物 2 是可逆抑制剂,抑制机理是混合型的,而化合物 1 是不可逆抑制剂。同时也研究了这两个化合物的抗氧化能力,包括清除双氧水的能力、清除过氧离子自由基的能力、清除 DPPH 自由基的能力,结果表明化合物 2 具有潜在的抗氧化能力。

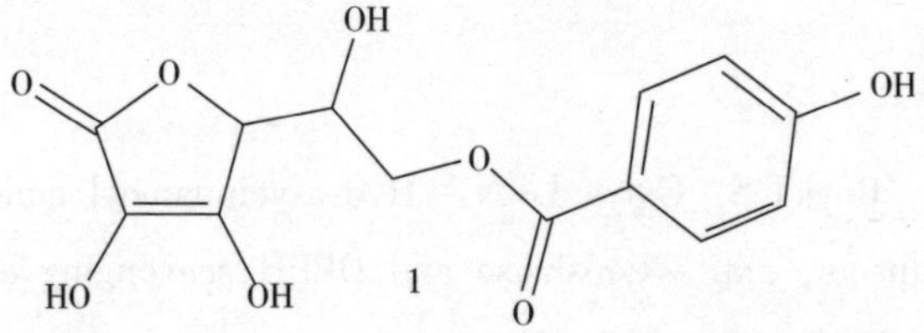

OH O O OH O HO OH 1 O

OH OH O O OH O OH HO OH 2 O

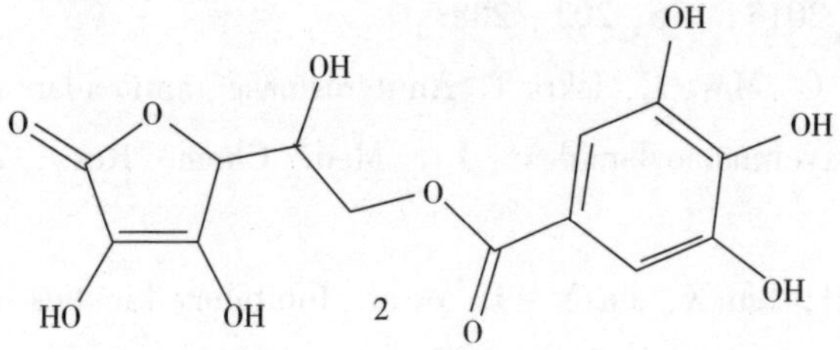

图 3 – 28 维生素 C 酯的结构式

2010 年曹春晓为了通过体外实验研究维生素 E 琥珀酸酯(Vitamin E succinate,VES;α - tocopheryl succinate,α - TOS)对黑色素瘤细胞增殖、分化、细胞周期和相关蛋白表达以及黑色素小体改变的影响并进一步探讨 VES 抑制黑色素瘤细胞生长的作用机制,从而为黑色素瘤治疗提供新的方法及相应的理论依据[42]。通过体外培养小鼠黑色素瘤 B16 细胞,采用四甲基偶氮唑蓝(MTT)比色法检测不同浓度 VES 作用于 B16 细胞 24h、48h、72h 后对其增殖抑制情况及经瑞 - 吉染色后光学显微镜观察细胞形态变化并筛选出适合的药物浓度进行后续试验。采用流式细胞术(FCM)测定 VES 对小鼠黑色素瘤 B16 细胞作用 48h 后的细胞周期分布和凋亡率;应用透射电镜观察黑色素小体;NaOH 裂解法测定黑色素含量变化;采用流式细胞术检测 VES 处理 B16 细胞 48h 后细胞周期蛋白 cyclinD1 和 P21 蛋白的表达变化;免疫细胞化学法检测各组肿瘤细胞 cyclinD1 和 P21 蛋白表达情况。从而初步探讨 VES 对 B16 细胞的诱导分化作用及相应机制。得出了如下结论:在不表现明显的细胞凋亡现象的剂量范围(5 - 20)μg/mL 内,VES 可明显抑制黑色素瘤 B16 细胞增殖,并呈剂量 - 时间依赖性;VES 可通过对黑色素瘤细胞周期阻滞诱导分化,将细胞阻滞于 G0/G1 期,该作用呈剂量依赖性,并以 48h 作用最明显;VES 对 B16 细胞有较强的分化诱导能力,形态学表现为生长缓慢,细胞连成网状结构,电镜可见 VES 处理过的 B16 细胞内含大量典型黑色素小体。功能上表现:VES 作用于 B16 细胞后,黑色素含量明显增加,尤其是 20μg/mL VES。VES 具有诱导黑色素瘤细胞分化作用,其机制可能与其下调 cyclinD1 蛋白,上调 P21 蛋白的表达有关。

参考文献

[1] Maya C, Boyka S, Galya I. N - Hydroxycinnamoyl amides of fluorinated amino acids: Synthesis, anti - tyrosinase and DPPH scavenging activities [J]. J. Fluorine. Chem., 2013, 156, 203 - 208.

[2] Lyubomir G, Maya C, Iskra T. Antityrosinase, antioxidant and antimicrobial activities of hydroxycinnamoylamides [J]. Med. Chem. Res., 2013, 22, 4173 - 4182.

[3] Hu Y - H, Liu X, Jia Y - L, et al. Inhibitory kinetics of chlorocinnamic acids on mushroom tyrosinase [J]. Journal of Bioscience and Bioengineering. 2014, 117, 142 - 146.

[4] 郑成已,郭云集,梁戈,等. 苯丙酸对蘑菇酪氨酸酶活力的抑制作用[J]. 厦门大学学报(自然科学版),2012, 51,117 - 120.

[5] 李慕紫,王庆华,王晓艺,等. 苯甲酸及其类似物与酪氨酸酶分子对接研究 [J]. 2016,37,87 - 90.

[6] 邢蕊. 多酸型酪氨酸酶抑制剂的功能性研究[D], 2016, 集美大学硕士论文.

[7] 郑阿萍,王 芳,邢 蕊,等. 两种多酸型酪氨酸酶抑制剂的性能研究[J]. 高等学校化学学报,2014,35,476 - 481.

[8] Sukhontha S, Kunwadee K, Chockchai T. Inhibitory effect of rice bran extracts and its phenolic compounds on polyphenol oxidase activity and browning in potato and apple puree [J]. Food Chemistry, 2016, 190, 922 - 927.

[9] 杨美花,石艳,李智聪,等. 原儿茶酸对小鼠黑色素瘤 B16 细胞酪氨酸酶活力及黑色素生成的抑制效应[J]. 厦门大学学报(自然科学版),2013, 52, 842 - 845.

[10]籍晓明. 维甲酸对小鼠 B16F10 黑色素瘤细胞酪氨酸酶 mRNA 表达的调控[D]. 天津医科大学硕士论文,2004.

[11]张璇. 新藤黄酸诱导黑色素瘤 B16 细胞凋亡及其分子机制研究[D]. 安徽中医药大学硕士论文,2013.

[12]张璇,王梅,程卉,等. 新藤黄酸对黑色素瘤 B16 细胞凋亡的作用[J]. 安徽中医学院学报,2013,32,3 - 56.

[13] Hua Y - H, Chen Q - X, Cui Y, et al. 4 - Hydroxy cinnamic acid as mushroom preservation: Anti - tyrosinase activity kinetics and application [J]. International Journal of Biological Macromolecules, 2016, 86, 489 - 495.

[14] 陈冉,闫春玲,王彬彬,等. 十种苯丙氨酸衍生物对小鼠黑色素瘤细胞系 B16 的作用研究[J]. 中国细胞生物学学报,2012, 34, 573 - 579.

[15] 向敏,王建梅,顾振纶. 熊果酸诱导 B16 黑色素瘤细胞分化作用的研究[J]. 中国现代医学杂志,2008,18,2315 - 2318.

[16] 穆燕,李琳,董方圆, 等. 对甲氧基肉桂酸对酪氨酸酶的抑制机理[J]. 华南理工大学学报(自然科学版),2012, 40, 134 - 138.

[17] Si Y - X, Ji S, Fang N - Y, et al. Effects of piperonylic acid on tyrosinase: Mixed - type inhibition kinetics and computational simulations [J]. Process Biochemistry, 2013, 48, 1706 - 1714.

[18] Miliovsky M, Svinyarov I, Mitrev Y, et al. A novel one – pot synthesis and preliminary biological activity evaluation ofcis – restricted polyhydroxy stilbenes incorporating protocatechuic acid and cinnamic acid fragments [J]. European Journal of Medicinal Chemistry, 2013, 66, 185 – 192.

[19] Alessandro V, Manuela M, Anna M S, et al. Dihydroasparagusic Acid: Antioxidant and Tyrosinase Inhibitory Activities and Improved Synthesis [J]. J. Agric. Food Chem., 2013, 61, 6848 – 6855.

[20] Chai W – M, Liu X, Hu Y – H, et al. Antityrosinase and antimicrobial activities of furfuryl alcohol, furfural and furoic acid [J]. International Journal of Biological Macromolecules, 2013, 57, 151 – 155.

[21] Fu Y, Hong S, Li D, et al. Novel Chemical Synthesis of Ginkgolic Acid (13:0) and Evaluation of Its Tyrosinase Inhibitory Activity [J]. J. Agric. Food Chem., 2013, 61, 5347 – 5352.

[22] Lin G, Lee J, Yang J – M, et al. Inhibition of tyrosinase by fumaric acid: Integration of inhibition kinetics with computational docking simulations [J]. International Journal of Biological Macromolecules, *in press*.

[23] Cui Y, Hua Y – H, Yu F, et al. Inhibition kinetics and molecular simulation of p – substituted cinnamic acid derivatives on tyrosinase [J]. International Journal of Biological Macromolecules, *in press*.

[24] Oyama T, Takahashi S, Yoshimori A, et al. Discovery of a new type of scaffold for the creation of novel tyrosinase inhibitors[J]. 2016, 24, 4509 – 4515.

[25] Hua Y – H, Chen Q – X, Cui Y, et al. 4 – Hydroxy cinnamic acid as mushroom preservation: Anti – tyrosinase activity kinetics and application [J]. International Journal of Biological Macromolecules, 2016, 86, 489 – 495.

[26] 张春乐. 抑制剂对蘑菇酪氨酸酶的抑制效应及抗菌活性 [D]. 厦门大学硕士论文,2007.

[27] Chochkova M, Stoykova B, Ivanova G, et al. N – Hydroxycinnamoyl amides of fluorinated amino acids: Synthesis, anti – tyrosinase and DPPH scavenging activities [J]. Journal of Fluorine Chemistry, 2013, 156, 203 – 208.

[28] Georgiev L, Chochkova M, Totseva I, et al. Anti – tyrosinase, antioxidant and antimicrobial activities of hydroxycinnamoylamides [J]. Med. Chem. Res., 2013, 22, 4173 – 4182.

[29] Kwak S – Y, Yang J – K, Choi H – R, et al. Synthesis and dual biological effects of hydroxyl – cinnamoyl phenylalanyl/prolyl hydroxamic acid derivatives as tyrosinase inhibitor and antioxidant [J]. Bioorganic & Medicinal Chemistry Letters, 2013, 23, 1136 – 1142.

[30] Bandgar B P, Adsul L K, Chavan H V, et al. Synthesis, biological evaluation, and molecular docking of N – {3 – [3 – (9 – methyl – 9H – carbazol – 3 – yl) – acryloyl] – phenyl} – benzamide/amide derivatives as xanthine oxidase and tyrosinase inhibitors [J]. Bioorganic & Medicinal Chemistry, 2012 20, 5649 – 5657.

[31] Baek H S, Hong Y D, Lee C S, et al. Adamantyl N – benzylbenzamide: New series of depigmentation agents with tyrosinase inhibitory activity [J]. Bioorganic & Medicinal Chemistry Letters 2012, 22, 2110 – 2113.

[32] Kwak S – Y, Yang J – K, Choi H – R, et al. Synthesis and dual biological effects of hydroxycinnamoyl phenylalanyl /prolyl hydroxamic acid derivatives as tyrosinase inhibitor and antioxidant [J]. Bioorganic & Medicinal Chemistry Letters, 2013, 23, 1136 – 1142.

[33] Lee J – Y, Choi H – J, Chung T – W, et al. Caffeic acid phenethyl ester inhibits alpha – melanocyte stimulating hormone – induced melanin synthesis through suppressing transactivation activity of microphthalmia – associated transcription factor [J]. J. Nat. Prod., 2013, 76, 1399 – 1405.

[34] 张春乐. 抑制剂对蘑菇酪氨酸酶的抑制效应及抗菌活性 [D]. 厦门大学硕士论文,2007.

[35] Monteiro A S, Almeida J, Cabral G, et al. Synthesis and evaluation of N – acylamino acids derivatives of triazenes. Activation by tyrosinase in human melanoma cell lines [J]. European Journal of Medicinal Chemistry, 2013, 70, 1 – 9.

[36] Balakrishna C, Payili N, Yennam S, et al. Synthesis of new kojic acid based unnaturala – amino acid derivatives [J]. Bioorganic & Medicinal Chemistry Letters, 2015, 25, 4753 – 4756.

[37] 宋长伟,熊丽丹,王裕军,等. 新型龙胆酸衍生物的合成及其抑制酪氨酸酶活性研究[J]. 有机化学,2012,32,1753 – 1758.

[38] Kashima Y, Miyazawa M. Synthesis, antioxidant capacity, and structure – activity relationships of tri – O – methylnorbergenin analogues on tyrosinase inhibition [J]. Bioorganic & Medicinal Chemistry Letters, 2013, 23, 6580 – 6584.

[39] Khan H, Saeed M, Khan M A, et al. Isolation of long - chain esters from the rhizome of Polygonatum verticillatumby potent tyrosinase inhibition [J]. Med. Chem. Res., 2013, 22, 2088 - 2092.

[40] Jia Y - L, Zheng J, Yu F, et al. Anti - tyrosinase kinetics and antibacterial process of caffeic acid N - nonyl ester in Chinese Olive (Canarium album) postharvest [J]. International Journal of Biological Macromolecules, 2016, 91, 486 - 495.

[41] Yi W, Wu X, Cao R, et al. Biological evaluations of novel vitamin C esters as mushroom tyrosinase inhibitors and antioxidants [J]. Food Chemistry, 2009, 117, 381 - 386.

[42] 曹春晓. 维生素 E 琥珀酸酯诱导黑色素瘤 B16 细胞分化的体外实验研究[D]. 2010, 河北医科大学硕士论文.

3.4 编者相关论文

3.4.1 Letters in Drug Design & Discovery (2013, 10, 529 - 534)

Inhibitory effects ofsubstituted cinnamic acid esters on mushroom tyrosinase

Zhenghua Zhang, Jinbing Liu*, Fengyan Wu, Liangzhong Zhao

Department of Biology and Chemical Engineering, Shaoyang University, Shao Shui Xi Road, Shaoyang 422100, *PRC*

* Corresponding author: Department of Biology and Chemical Engineering, Shaoyang University, Shao Shui Xi Road, Shaoyang 422100, PRC

Phone: +86 - 739 - 5431768; Fax: +86 - 739 - 5431768

E - mail address: xtliujb@ yahoo. com. cn

Abstract

A series of substituted cinnamic acid esters were synthesized and their inhibitory effects on the diphenolase activity of mushroom tyrosinase were evaluated. Compound 8 was found to be the most potent inhibitor with IC_{50} value of 5. 60μmol/L. Preliminary structure activity relationships (SARs) were concluded. The inhibition kinetics analyzed by Lineweaver - Burk plots revealed that compound 8 was anti - competitive inhibitor.

Keywords: substituted cinnamic acid esters, tyrosinase inhibitor, inhibition mechanism.

INTRODUCTION

Tyrosinase (monophenol o - diphenol, oxygen oxidoreductase, EC 1. 14. 18. 1, syn. polyphenol oxidase) is a copper - containing enzyme that is present in micro - organisms, animals and plants [1]. It is well known that tyrosinase can catalyzes two distinct reactions of melanin synthesis, the hydroxylation of monophenols to o - diphenols (monophenolase activity) and the oxidation of *o* - diphenols to *o* - quinones (diphenolase activity) [2]. Due to the high reactivity, quinines could polymerize spontaneously to form high molecular weight brown - pigments (melanins) or react with amino acids and proteins to enhance brown color of the pigment produced [3, 4]. Hyperpigmentations, such as senile lentigo, melasma, freckles, and pigmented acne scars, are of particular concern to women. The treatment usually involves the use of medicines or medicinal cosmetics containing depigmenting agents or skin whitening agents [5, 6]. In clinical usage, tyrosinase inhibitors are used for treatments of dermatological disorders related to melanin hyperaccumulation and are essential in cosmetics for depigmentation [7 - 11], such as age spots and freckle, were caused by the accumulation of an excessive level of epidermal pigmentation [12, 13].

Previous reports confirmed that tyrosinase not only was involved in melanizing in animals, but alsowas one of the main causes of most fruits and vegetables quality loss during post harvest handling and processing, leading to faster degradation and shorter shelf life [14]. Tyrosinase has also been linked to Parkinson's and other neurodegenerative diseases [15]. In insects, tyrosinase is uniquely associated with three different biochemical processes, including sclerotization of cuticle, defensive encapsulation and melanization of foreign organism, and wound healing [16]. These processes provide potential targets for developing safer and effective tyrosinase inhibitors as insecticides and ultimately for insect control. Thus, the development of safe and effective tyrosinase inhibitors is of great concern in the medical, agricultural, and cosmetic industries. However, only a few such as arbutin, tropolone, and 1 - phenyl - 2 - thiourea (PTU) are used as therapeutic agents and cosmetic products [12,17].

Caffeic acid is one of the most widely distributed hydroxycinnamate and phenylpropanoid metabolites in plant tissues [18]. It is usually found as various simple derivatives including amides, esters, sugar esters, and glycosides. In addition, substituted cinnamic acid esters and amide derivatives exhibit a broad spectrum of biological activities, including anti - oxidative properties, and they have been shown to scavenge a number of reactive species, including DPPH radicals, and peroxyl and hydroxyl radicals [19 - 23]. Caffeic acid and it's derivatives also show antimicrobial, anti - infammatory, antineoplastic and antioxidant activity [24]. In addition, Caffeic acid phenethyl ester is a potent inhibitor of ornithine decarboxylase, cyclooxygenase, lipoxygenases and HIV - 1 integrase. In the last decade, much attention has been focused on the substituted cinnamic acid and it's derivatives [20]. However, the tyrosinase inhibitor activities of this kind of compounds have seldom appeared in the literature. Stimulated by these results, in the present investigation, we synthesized a series of substituted cinnamic acid ester compounds, their inhibitory activities against mushroom tyrosinase were evaluated using kojic acid as comparing substance. Meanwhile, the structure - activity relationships of these compounds were also primarily discussed.

MATERIALS AND METHODS

Materials

Melting points (m. p.) were determined with WRS - 1B melting point apparatus and the thermometer was uncorrected. NMR spectra were recorded on Bruker 400 spectrometers or Bruker 500 spectrometers at25℃ in $CDCl_3$ or DMSO - d_6. All chemical shifts (δ) are quoted in parts per million downfield from TMS and coupling constants (J) are given in hertz. Abbreviations used in the splitting pattern were an follows: s = singlet , d = doublet, t = triplet, q = quintet, m = multiplet, LC - MS spectra were recorded using the LCMS - 2010A. All reactions were monitored by TLC (Merck Kieselgel 60 F254) and the spots were visualized under UV light. Infrared (IR) spectra were recorded as potassium bromide pellets on VECTOR 22 spectrometer.

Tyrosinase, L - 3, 4 - dhydroxyphenylalanine (L - DOPA) andkojic acid were purchased from Sigma - Aldrich Chemical Co. Other chemicals were purchased from commercial suppliers and were dried and purified when necessary.

General procedures for the synthesis ofsubstituted cinnamic acid ester compounds

The appropriate substitutedcinnamic acid (30 mmol) and 70mL dichloromethane were placed in a 250mL 3 - neck round bottom flask equipped with a magnetic stirring bar. The mixture was stirred at room temperature, and the thionyl chloride (45 mmol) was added dropwise to the reaction system over 0.5 h. The reaction mixture was heated to reflux. The reaction was completed under reflux for about 2 h. Appropriate alcohol (1.5 equiv) was added over 1 h, and the reaction mixture was stirred for 3 h.

The reaction mixture was then concentrated in vacuo, and the residue was purified by column chromatography on silica gel (petroleum ether (60 - 90) - AcOEt) to give the substitutedcinnamic acid esters.

methyl 3 - (4 - hydroxyphenyl) acrylate (1)

Yield 95.2%. White solid, mp 133.1 ~ 133.6 ℃; IR(KBr): 3390, 2961, 1727, 1680, 1653, 1632, 1469, 1269, 1199, 832cm^{-1}. ^{1}HNMR (500MHz, $CDCl_3$) δ 7.66(d,J =16.0Hz, 1H, =CH), 7.43 - 7.40(m, 2H, Ph - H), 6.88 - 6.86(m,2H, Ph - H), 6.31(d,J =16.0Hz, 1H, =CH), 5.45(s, 1H, OH), 3.81 (s, 3H, OCH_3), ^{13}CNMR (125MHz, $CDCl_3$) δ 169.43, 158.13, 145.07, 130.06, 126.94, 115.96, 114.88, 51.82. MS (ESI): m/z(100%) 177 (M - 1).

methyl 3 - (4 - ethoxyphenyl) acrylate (2)

Yield 87.5%. White solid, mp 115.4 ~ 116.0 ℃; IR(KBr): 3189, 2975, 2853, 1727, 1470, 1392, 1285, 1252, 1172, 833, 807cm^{-1}. ^{1}HNMR(400MHz, $CDCl_3$) δ 7.67(d,J =16.0Hz, 1H, =CH), 7.47(d, J =8.0Hz, 2H, Ph - H), 6.90(d, J =8.0Hz, 2H, Ph - H), 6.33(d,J =16.0Hz, 1H, =CH), 4.09(q, J =6.8Hz, 2H, OCH_2), 3.79(s, 3H, OCH_3), 1.45(d, J =6.8Hz, 3H, CH_3), ^{13}CNMR (100MHz, $CDCl_3$) δ 167.75, 160.87, 144.58, 129.72, 127.07, 115.26, 114.90, 63.66, 51.50, 14.72.

methyl 3 - (4 - nitrophenyl) acrylate (3)

Yield 89.4%. Yellow solid, mp 95.6 ~ 96.2 ℃; IR(KBr): 3092, 2987, 1722, 1695, 1635, 1558, 1513, 1344, 1312, 1172, 847cm^{-1}. ^{1}HNMR(400MHz, $CDCl_3$) δ 8.26(d,J =8.0Hz, 2H, Ph - H), 7.74(d,J =16.0Hz, 1H, =CH), 7.68(d,J =8.0Hz, 2H, Ph - H), 6.58(d,J =16.0Hz, 1H, =CH), 3.80(s, 3H, OCH_3), ^{13}CNMR(100MHz, $CDCl_3$) δ 166.45, 148.54, 141.89, 140.49, 128.65, 124.17, 122.11, 52.06. MS (ESI): m/z(100%) 208 (M + 1).

methyl 3 – (4 – hydroxy – 3 – methoxyphenyl) acrylate (4)

Yield 63.3%. Yellow oil; IR(KBr): 3381, 2954, 1728, 1693, 1643, 1453, 1187, 817cm^{-1}. ^{1}HNMR(400MHz, DMSO – d_6) δ 7.86(d, *J* = 16.0Hz, 1H, = CH), 7.75 – 7.73(m, 2H, Ph – H), 7.25(d, *J* = 8.4Hz, 1H, Ph – H), 6.85(d, *J* = 16.0Hz, 1H, = CH), 5.75(s, 1H, OH), 3.85(s, 3H, OCH_3), 3.81(s, 3H, OCH_3), ^{13}CNMR(100MHz, DMSO – d_6) δ 166.79, 154.78, 146.73, 141.28, 127.25, 123.67, 115.58, 113.63, 113.55, 56.10, 52.12.

methyl 3 – (3 – hydroxy – 4 – methoxyphenyl) acrylate (5)

Yield 67.2%. White solid, mp 75.3 ~ 75.8 ℃; IR (KBr): 3375, 2951, 1718, 1636, 1431, 1264, 1167, 1128, 1022, 805cm^{-1}. ^{1}HNMR (400MHz, $CDCl_3$) δ 7.63(d, *J* = 16.0Hz, 1H, = CH), 7.11(d, *J* = 8.0Hz, 1H, Ph – H), 7.06(s, 1H, Ph – H), 6.87(d, *J* = 8.0Hz, 1H, Ph – H), 6.31(d, *J* = 16.0Hz, 1H, = CH), 5.14(s, 1H, OH), 3.98(s, 3H, OCH_3), 3.81(s, 3H, OCH_3), ^{13}CNMR (100MHz, $CDCl_3$) δ 167.76, 148.58, 145.88, 144.72, 128.03, 121.83, 115.82, 113.04, 110.56, 56.00, 51.61. MS (ESI): m/z(100%) 207 (M – 1).

propyl 3 – (3 – hydroxy – 4 – methoxyphenyl) acrylate (6)

Yield 39.9%. Yellow solid, mp 281.2 ~ 283.6 ℃; IR(KBr): 2936, 2838, 1727, 1631, 1602, 1507, 1436, 1262, 1017, 806cm^{-1}. ^{1}HNMR(400MHz, DMSO – d_6) δ 7.86(d, *J* = 16.0Hz, 1H, = CH), 7.75 – 7.73(m, 2H, Ph – H), 7.25 (d, *J* = 8.4Hz, 1H, Ph – H), 6.85(d, *J* = 16.0Hz, 1H, = CH), 5.75(s, 1H, OH), 4.11(t, *J* = 6.8Hz, 2H, CH_2), 3.85(s, 3H, OCH_3), 1.68 – 1.64(m, 2H, CH_2), 0.93(t, *J* = 6.8Hz, 3H, CH_3). ^{13}CNMR(100MHz, DMSO – d_6) δ 164.89, 153.78, 147.73, 140.20, 127.35, 123.27, 115.53, 113.48, 113.40, 65.85, 56.66, 22.14, 10.77. MS (ESI): m/z(100%) 235 (M – 1).

isopropyl 3 – (3 – hydroxy – 4 – methoxyphenyl) acrylate (7)

Yield 52.4%. White solid, mp 267 ~ 268℃; IR(KBr): 2977, 2837, 1703, 1629, 1602, 1510, 1429, 1259, 1020, 804cm^{-1}. ^{1}HNMR(400MHz, DMSO – d_6) δ 7.86(d, *J* = 16.0Hz, 1H, = CH), 7.75 – 7.73(m, 2H, Ph – H), 7.25(d, *J* = 8.4Hz, 1H, Ph – H), 6.85(d, *J* = 16.0Hz, 1H, = CH), 5.75(s, 1H, OH), 5.04 – 4.96(m, 1H, CH), 3.85(s, 3H, OCH_3), 1.26(d, 6H, 2CH_3), ^{13}CNMR (100MHz, DMSO – d_6) δ 164.89, 153.78, 147.38, 140.19, 127.33, 123.27, 115.55, 113.48, 113.38, 67.62, 56.61, 22.21. MS (ESI): m/z(100%) 235 (M – 1).

pentyl 3 – (3 – hydroxy – 4 – methoxyphenyl) acrylate (8)

Yield 50.4%. Yellow solid, mp 253 ~ 254℃; IR(KBr): 2941, 2836, 1730, 1602, 1502, 1435, 1020, 804cm^{-1}. ^{1}HNMR(400MHz, DMSO – d_6) δ 7.86(d,J = 16.0Hz, 1H, = CH), 7.75 – 7.74(m, 2H, Ph – H), 7.25(d,J = 8.4Hz, 1H, Ph – H), 6.85(d,J = 16.0Hz, 1H, = CH), 5.75(s, 1H, OH), 4.14(t,J = 6.4Hz, 2H, OCH_2), 3.85(s, 3H, OCH_3), 1.65 – 1.59(m, 2H, CH_2), 1.33 – 1.28(m, 4H, 2CH_2), 0.88((t,J = 4.2Hz, 3H, CH_3). ^{13}CNMR(100MHz, DMSO – d_6) δ 164.88, 153.78, 147.38, 140.20, 127.34, 123.28, 115.53, 113.45, 113.47, 65.41, 56.69, 29.10, 28.10, 22.27, 14.31. MS (ESI): m/z(100%) 263 (M – 1).

benzyl 3 – (3 – hydroxy – 4 – methoxyphenyl) acrylate (9)

Yield 60.4%. Yellow solid, mp 258.7 ~ 259.2℃; IR(KBr): 2977, 1720, 1631, 1602, 1511, 1432, 1258, 1200, 1117, 1022, 802cm^{-1}. ^{1}HNMR(400MHz, DMSO – d_6) δ 7.86(d,J = 16.0Hz, 1H, = CH), 7.75 – 7.73(m, 2H, Ph – H), 7.41(d,J = 9.6Hz, 2H, Ph – H), 7.32 (d, J = 8.4Hz, 1H, Ph – H), 7.25 – 7.22(m, 3H, Ph – H), 6.85(d,J = 16.0Hz, 1H, = CH), 5.75(s, 1H, OH), 5.22(s,2H, CH_2), 3.85(s, 3H, OCH_3), ^{13}CNMR(100MHz, DMSO – d_6) δ 164.88, 153.77, 146.29, 146.25, 140.19, 129.10, 128.93, 128.45, 127.35, 126.89, 123.27, 115.54, 113.49, 63.39, 56.60. MS (ESI): m/z(100%) 283 (M – 1).

methyl 3 – (3,4 – dihydroxyphenyl) acrylate (10)

Yield 56.6%, Yellow solid, mp 163.0 ~ 163.8℃; IR(KBr): 3474, 3093, 2855,1728, 1677, 1604, 1444, 1279, 1241, 1190, 1044, 818cm^{-1}. ^{1}HNMR (400MHz, $CDCl_3$) δ 7.60(d,J = 16.0Hz, 1H, = CH), 7.07(s, 1H, Ph – H), 7.03 (d, J = 8.4Hz, 1H, Ph – H), 6.88(d, J = 8.4Hz, 1H, Ph – H), 6.29(d,J = 16.0Hz, 1H, = CH), 5.45(s, 1H, OH), 5.30(s, 1H, OH), 3.79(s, 3H, OCH_3).

methyl 3 – (3,4 – dimethoxyphenyl) acrylate (11)

Yield 55.6%, Yellow solid, mp 68.3 ~ 68.8℃; IR(KBr): 3098, 2935, 1708, 1631, 1464, 1387, 1306, 1254, 1157, 807cm^{-1}. ^{1}HNMR (400MHz, $CDCl_3$) δ 7.66(d,J = 16.0Hz, 1H, = CH), 7.12(d, J = 8.0Hz, 1H, Ph – H), 7.05(s, 1H, Ph – H), 6.88(d, J = 8.0Hz, 1H, Ph – H),6.33(d,J = 16.0Hz, 1H, = CH), 3.92(s, 6H, 2OCH_3), 3.80(s, 3H, OCH_3),^{13}CNMR (100MHz, $CDCl_3$) δ 167.66, 151.15, 149.22, 144.78, 127.37, 122.59, 115.49, 111.05,

109.67, 55.96, 55.88, 51.60.

ethyl 3 - (3,4 - dimethoxyphenyl) acrylate (12)

Yield 92.5%, Yellow oil; IR(KBr): 3099, 2931, 1712, 1633, 1505, 1471, 1309, 1259, 1164, 807cm^{-1}. ^{1}HNMR(400MHz, $CDCl_3$) δ 7.65(d, J = 16.0Hz, 1H, = CH), 7.11(d, J = 8.0Hz, 1H, Ph - H), 7.05(s, 1H, Ph - H), 6.87(d, J = 8.0Hz, 1H, Ph - H), 6.33(d, J = 16.0Hz, 1H, = CH), 4.28(q, J = 7.2Hz, 2H, OCH_2), 3.91(s, 6H, $2OCH_3$), 1.35(t, J = 7.2Hz, 3H, CH_3), ^{13}CNMR (100MHz, $CDCl_3$) δ 167.19, 151.07, 149.20, 144.48, 127.44, 122.55, 115.96, 111.03, 109.60, 60.34, 55.94, 55.85, 14.34.

propyl 3 - (3,4 - dimethoxyphenyl) acrylate (13)

Yield 90.6%, Yellow oil; IR(KBr): 3021, 2935, 1708, 1631, 1511, 1464, 1387, 1254, 1157, 807cm^{-1}. ^{1}HNMR(400MHz, $CDCl_3$) δ 7.65(d, J = 16.0Hz, 1H, = CH), 7.12(d, J = 8.0Hz, 1H, Ph - H), 7.06(s, 1H, Ph - H), 6.88(d, J = 8.0Hz, 1H, Ph - H), 6.34(d, J = 16.0Hz, 1H, = CH), 4.23(t, J = 6.0Hz, 2H, OCH_2), 3.92(s, 6H, $2OCH_3$), 1.76 - 1.67(m, 2H, CH_2), 1.02(t, J = 7.2Hz, 3H, CH_3), ^{13}CNMR (100MHz, $CDCl_3$) δ 167.41, 151.15, 149.29, 144.55, 127.55, 122.65, 116.06, 111.12, 109.69, 66.11, 56.04, 55.96, 22.20, 10.54.

isopropyl 3 - (3,4 - dimethoxyphenyl) acrylate (14)

Yield 93.3%, Yellow oil; IR(KBr): 2976, 2835, 1701, 1638, 1512, 1308, 1261, 1175, 1024, 808cm^{-1}. ^{1}HNMR(400MHz, $CDCl_3$) δ 7.61(d, J = 16.0Hz, 1H, = CH), 7.14(s, 1H, Ph - H), 7.04(d, J = 8.0Hz, 1H, Ph - H), 6.85(d, J = 8.0Hz, 1H, Ph - H), 6.31(d, J = 16.0Hz, 1H, = CH), 4.28 - 4.22(m, 1H, CH), 3.93(s, 6H, $2OCH_3$), 1.35(d, J = 7.2Hz, 6H, $2CH_3$), ^{13}CNMR (100MHz, $CDCl_3$) δ 166.59, 151.11, 149.31, 144.11, 127.61, 122.41, 116.59, 111.23, 109.89, 67.50, 55.91, 55.86, 23.52, 21.91.

Tyrosinase assay

The spectrophotometric assay for tyrosinase was performed according to the method reported by our group with some slight modifications. Briefly, all the synthesized compounds were screened for the diphenolase inhibitory activity of tyrosinase using L - DOPA as substrate. All the active inhibitors from the preliminary screening were subjected to IC_{50} studies. All the compounds were dissolved in

DMSO. The final concentration of DMSO in the test solution was 2.0%. Phosphate buffer, pH 6.8, was used to dilute the DMSO stock solution of test compounds. Thirty units of mushroom tyrosinase (0.5 mg/mL) were first pre-incubated with the compounds, in 50 mmol/L phosphate buffer (pH 6.8), for 10 min at 25°C. Then the L-DOPA (0.5 mmol/L) was added to the reaction mixture and the enzyme reaction was monitored by measuring the change in absorbance at 475 nm of formation of the DOPA chrome for 1 min. The measurement was performed in triplicate for each concentration and averaged before further calculation. IC_{50} value, a concentration giving 50% inhibition of tyrosinase activity, was determined by interpolation of the dose-response curves. The percent of inhibition of tyrosinase reaction was calculated as following:

$$\text{Inhibition rate}(\%) = [(B-S)/B] \times 100$$

Here, the B and S are the absorbances for the blank and samples. Kojic acid was used as reference standard inhibitors for comparison.

RESULTS AND DISCUSSION

Chemistry

According to the general procedure shown in Scheme 1. The appropriate substitutedcinnamic acid was dissolved in dichloromethane. The mixture was stirred at room temperature, and the thionyl chloride was added dropwise to the reaction system. The reaction mixture was heated to reflux. The reaction was completed under refluxing for about 2 h. Appropriate alcohol was added to the mixture, and the reaction mixture was stirred for about 3 h. The reaction mixture was then concentrated in vacuo, and the residue was purified by column chromatography on silica gel to give the substitutedcinnamic acid esters [25].

Biological evaluation

For evaluating the tyrosinase inhibitory activity, all the synthesized compounds were subjected to tyrosinase inhibition assay with L-DOPA as substrate, according tothe method reported by our group with some slight modifications [14, 26]. The tyrosinase inhibitory activity of kojic acid was ever reported, therefore, it was selected as comparing substance. The IC_{50} values of cinnamic derivatives against tyrosinase were summarized in Table 1, and IC_{50} values of all these compounds were determined from logarithmic concentration-inhibition curves and given as means of three experiments.

Our results showed that compounds1, 3, 5 – 8 exhibited potent inhibitory on mushroom tyrosinase with IC_{50} values ranged from 5. 60μmol/L to 94. 42μmol/L. Especially, compounds 5 – 8 showed more potent inhibitory activities than the other compounds. Compounds 11 – 14 showed low or no inhibitory activities. This suggested that hydroxyl group at the 3 – position of benzene ring and methoxyl group at the 4 – position of benzene ring might play an important role in determining this kind of compounds inhibitory activities. The hydroxyl group and methoxyl group might be more beneficial to inhibit the tyrosinase enzyme by participating in metal – ligand binding interaction with the dicopper nucleus. Interestingly, compound 8 was found to be the most potent inhibitor with an IC_{50} value of 5. 60 μmol/L. Comparing to the tyrosinase inhibitory activities of compounds 5 – 8, with the increase of lipophilicity of compounds, the inhibitory activities increased gradually. These results showed that the inhibitory activities of compounds with hydroxyl group at the 3 – position and methoxyl group at the 4 – position at benzene ring might be relate to the lipophilicity. Because lipophilic groups can affect both the inhibition potency, as well as the ability of the tyrosine to compete with the inhibitors. Minor bulky lipophilic group may interact with the enzyme hydrophobic pocket and augment binding affinity [27]. However, compound 9 exhibited no inhibitory activity. It suggested that benzyl ester may hinder the complex formation between inhibitor unit and the metallic center of the active site in enzyme due to the steric hindrance. Because compound has the benzene of rigid structure and the other compounds have the fatty chain of non – rigid structure. Except for the compound 9, the synthesized 3 – (3 – hydroxy – 4 – methoxyphenyl) acrylic acid derivatives exhibited more potent inhibitory activities than 3 – (3 – hydroxy – 4 – methoxyphenyl) acrylic acid (Fig. 1). In addition, compound 3 showed more potent inhibitory activity on mushroom tyrosinase than 3 – (4 – nitrophenyl) acrylic acid (Fig. 1), and thc compound 1 also showed more potent inhibitory activity than 3 – (4 – hydroxyphenyl) acrylic acid (Fig. 1). The results suggested that ester group of these compounds might be facilitated their inhibitory effects on the diphenolase activity of mushroom tyrosinase.

Kinetic analysis

Among the testedcinnamic derivatives, compound 8 showed the highest inhibitory activity, and hence, we carried out the kinetic analysis of compound 8 for tyrosinase inhibition with respect to L – DOPA as a substrate. The inhibition mechanism of compound 8 on mushroom tyrosinase for the oxidation of L – DOPA was determined. Figure 2 showed the relationship of enzyme activity with its concentration in the presence of different concentrations of compound 8. The results showed that the plots of V versus [E] gave a family of straight lines with different slopes but they intersected origin of coordinates. It demonstrated that the inhibitory effect of compound 8 on the tyrosinase was reversible. The kinetic behaviour of compound 8 on the mushroom tyrosinase, during the oxidation of L – DOPA, was determined from Lineweaver – Burk double reciprocal plots. The result (Fig. 3a) showed that the plots of 1/V versus 1/[S] gave a family of straight lines with same slopes. The values of Vmax and Km decreased with increasing concentrations of the inhibitor 8, which indicated that compound 8 was anti – competitive inhibitor of mushroom tyrosinase. The result showed that compound 8 exhibited inhibitory activity only when substrate could bind with enzyme to form complex. The inhibition constant for the inhibitor binding with the substrate – enzyme complex (KIS) was obtained from the secondary plot (Fig. 3b) as 38.65μmol/L.

Conclusions

The present investigation reported that somesubstituted cinnamic acid esters showed potent inhibitory effects on the diphenolase activity of mushroom tyrosinase. Interestingly, compound 8 was found to be the most potent inhibitor with IC_{50} value of 5.60μmol/L. Preliminary structure activity relationships (SARs) analysis indicated that (1) substituted group position and substituted group kind might have effect on the inhibitory activities, especially, hydroxyl group at the 3 – position of benzene ring and methoxyl group at the 4 – position of benzene might play an important role in determining their inhibitory activities; (2) to the compounds 5 – 8, the increase of lipophilicity of compound might improve the inhibitory activity; (3) non – rigid ester group of these compounds might be play an important role in improvement of inhibitory activities. The inhibition mechanism analysis of compound 8 demonstrated that the inhibitory effect of the compound on the tyrosinase was reversible, and the inhibitory type of compound 8 on the diphenolase activity was anti – competitive.

Acknowledgements

This work wasfinancially supported by the Foundation of Education Department of

Hunan Province, China (09B091), and the Key Subject Foundation of Shaoyang University (2008XK201).

REFERENCES

[1] Claus H, Decker H. Copper - Containing Oxidases: Occurrence in Soil Microorganisms, Properties, and Applications. *Syst. Appl. Microbiol* . 2006, *29* , 3 - 14.

[2] Fang, Y. ; Chen, Y. ; Feng, G. Benzyl benzoates: New phlorizin analogs as mushroom tyrosinase inhibitors. *Bioorg. Med. Chem* . 2011, *19* , 1167 - 1171.

[3] Kubo, I. ; Nihei, K. I. ; Shimizu, K. Oxidation products of quercetin catalyzed by mushroom tyrosinase. *Bioorg. Med. Chem . Lett* . 2004, *12* , 5343 - 5347.

[4] Matsuura, R. ; Ukeda, H. ; Sawamura, M. Tyrosinase inhibitory activity of citrus essential oils. *J. Agric. Food Chem.* 2006, *54* , 2309 - 2313.

[5] Wang, H. M. ; Chen, C. Y. ; Chen, C. Y. (-) - *N* - Formylanonaine from *Michelia alba* as human tyrosinase inhibitor and antioxidant. *Bioorg. Med. Chem* . 2010, *18* , 5241 - 5247.

[6] Tripathi, R. K. ; Hearing, V. J. ; Urabe, K. ; Aroca, P. ; Spritz, R. A. Mutational mapping of the catalytic activities of human tyrosinase. *J. Biol. Chem* . 1992, *267* , 23707 - 23712.

[7] Gillbro, J. M. ; Marles, L. K. ; Hibberts, N. A. ; Schallreuter, K. U. Autocrine catecholamine biosynthesis and the beta - adrenoceptor signal promote pigmentation in human epidermal melanocytes. *J. Invest. Dermatol* . 2004, *123* , 346 - 353.

[8] Shiino, M. ; Watanabe, Y. ; Umezawa, K. Synthesis of *N* - substituted *N* - nitroso - hydroxylamines as inhibitors of mushroom tyrosinase. *Bioorg. Med. Chem.* 2001, *9* , 1233 - 1240.

[9] Spencer, J. D. ; Chavan, B. ; Marles, L. K. ; Kauser, S. ; Rokos, H. ; Schallreuter, K. U. A novel mechanism in control of human pigmentation by {beta} - melanocyte - stimulating hormone and 7 - tetrahydrobiopterin. *J. Endocrinol* . 2005, *187* , 293 - 302.

[10] Schallreuter, K. U. ; Hasse, S. ; Rokos, H. ; Chavan, B. ; Shalbaf, M. ; Spencer, J. D. ; Wood, J. M. Cholesterol regulates melanogenesis in human epidermal melanocytes and melanoma cells. *Exp. Dermatol* . 2009, *18* , 680 - 688.

[11] Wood, J. M.; Decker, H.; Hartmann, H.; Chavan, B.; Rokos, H.; Spencer, J. D.; Hasse, S.; Thornton, M. J.; Shalbaf, M.; Paus, R.; Schallreuter, K. U. Senile hair graying: H_2O_2 – mediated oxidative stress affects human hair color by blunting methionine sulfoxide repair. *FASEB J*. 2009, *23*, 2065 – 2075.

[12] Gupta, A. K.; Gover, M. D.; Nouri, K. N.; Taylor, S. The treatment of melasma: a review of clinical trials. *J. Am. Acad. Dermatol*. 2006, *55*, 1048 – 1065.

[13] Thanigaimalai, P.; Hoang, T. A. L. Structural requirement(s) of *N* – phenylthioureas and benzaldehyde thiosemicarbazones as inhibitors of melanogenesis in melanoma B 16 cells. *Bioorg. Med. Chem. Lett*, 2010, *20*, 2991 – 2993.

[14] Yi, W.; Cao, R. Synthesis and biological evaluation of novel 4 – hydroxybenzaldehyde derivatives as tyrosinase inhibitors. *Eur. J. Med. Chem.* 2010, *45*, 639 – 646.

[15] Zhu, Y. J.; Zhou, H. T.; Hu, Y. H. Antityrosinase and antimicrobial activities of 2 – phenylethanol, 2 – phenylacetaldehyde and 2 – phenylacetic acid. *Food Chem*. 2011, *124*, 298 – 320.

[16] Ashida, M.; Brey, P. Role of the integument in insect defense: pro – phenol oxidase cascade in the cuticular matrix. *P. Natl Acad. Sci. USA*. 1995, *92*, 10698 – 10702.

[17] Battaini, G.; Monzani, E.; Casella, L.; Santagostini, L.; Pagliarin, R. Inhibition of the catecholase activity of biomimetic dinuclear copper complexes by acid kojic. *J. Biol. Inorg. Chem*. 2000, *5*, 262 – 268.

[18] Nardini, M.; Pisu, P.; Gentili, V.; Natella, F.; Piccolella, E.; Scaccini, C. Effect of caffeic acid on tert – butyl hydroperoxide – induced oxidative stress in U937. *Free Radical Biol. Med*. 1998, *25*, 1098 – 1105.

[19] Buzzi, F. C.; Franzoi, C. L.; Antonini, G.; Fracasso, M.; Filho, V. C. Antinociceptive properties of caffeic acid derivatives in mice. *Eur. J. Med. Chem.* 2009, *44*, 4596 – 4602.

[20] Rajan, P.; Vedernikova, I.; Cos, P.; Berghe, D. V.; Augustynsa, K.; Haemersa, A. Synthesis and evaluation of caffeic acid amides as antioxidants. *Bioorg. Med. Chem. Lett*. 2001, *11*, 215 – 217.

[21] Son, S.; Lewis, B. A. Free radical scavenging and antioxidative activity

of caffeic acid amide and ester analogues: Structureactivity relationship. *J. Agric. Food Chem.* 2002, *50* , 468 – 472.

[22] Kikuzaki, H. ; Hisamoto, M. ; Hirose, K. ; Akiyama, K. ; Taniguchi, H. Antioxidant properties of ferulic acid and its related compounds. *J. Agric. Food Chem.* 2002, *50* , 2161 – 2168.

[23] Hung, C. ; Tsai, W. ; Kuo, L. Y. Evaluation of caffeic acid amide analogues as anti – platelet aggregation and anti – oxidative agents. *Bioorg. Med. Chem.* 2005, *13* , 1791 – 1797.

[24] Fu, J. ; Cheng, K. ; Zhang, Z. ; Fang, R. Synthesis, structure and structure – activity relationship analysis of caffeic acid amides as potential antimicrobials. *Eur. J. Med. Chem.* 2010, *45* , 2638 – 2643.

[25] Yang, J. ; Marriner, G. A. ; Wang, X. Synthesis of a series of caffeic acid phenethyl amide (CAPA) fluorinated derivatives: Comparison of cytoprotective effects to caffeic acid phenethyl ester (CAPE). *Bioorg. Med. Chem.* 2010, *18* , 5032 – 5038.

[26] Liu, J. B. ; Cao, R. H. ; Yi, W. ; Ma, C. M. ; Wan, Y. Q. ; Zhou, B. H. ; Ma, L. ; Song, H. C. A class of potent tyrosinase in hibitors: Alkylidenethiosemicarbazide compounds. *Eur. J. Med. Chem.* 2009, *44* , 1773 – 1778.

[27] Khatib, S. ; Nerya, O. ; Musa, R. ; Tamir, S. ; Peter, T. ; Vaya, J. Enhanced substituted resorcinol hydrophobicity augments tyrosinase inhibition potency. *J. Med. Chem.* 2007, *50* , 2676 – 2681.

Figure captions

Fig. 1. Chemical structure of substituted cinnamic acid.

Fig. 2. The effect of concentrations of tyrosinase on its activity for the catalysis of L – DOPA at different concentration of compound 8. The concentrations of compound 8 for curves 1 – 3 are 0, 5 and 10 μmol/L, respectively.

Fig. 3. Determination of the inhibitory type and inhibition constants of compound 8 on the diphenolase activity of mushroom tyrosinase. (a) Lineweaver – Burk plot of mushroom tyrosinase inhibition by compound 8. Enzyme activity was measured at 475 nm The concentrations of compound 8 for curves 1 – 3 are 0, 5 and 10 μmol/L, respectively. (b) represents the secondary plot of Km versus concentrations of compound 8 to determine the inhibition constant.

1. R_1 = H, R_2 = OH, R = CH_3
2. R_1 = H, R_2 = OCH_2CH_3, R = CH_3
3. R_1 = H, R_2 = NO_2, R = CH_3
4. R_1 = OCH_3, R_2 = OH, R = CH_3
5. R_1 = OH, R_2 = OCH_3, R = CH_3
6. R_1 = OH, R_2 = OCH_3, R = $CH_2CH_2CH_3$
7. R_1 = OH, R_2 = OCH_3, R = i-pr
8. R_1 = OH, R_2 = OCH_3, R = n-pentyl
9. R_1 = OH, R_2 = OCH_3, R = benzyl
10. R_1 = OH, R_2 = OH, R = CH_3
11. R_1 = OCH_3, R_2 = OCH_3, R = CH_3
12. R_1 = OCH_3, R_2 = OCH_3, R = CH_2CH_3
13. R_1 = OCH_3, R_2 = OCH_3, R = $CH_2CH_2CH_3$
14. R_1 = OCH_3, R_2 = OCH_3, R = i-pr

Scheme 1　Synthesis of substituted cinnamic acid ester compounds

Reagents and conditions: i) CH_2Cl_2, RT to Reflux, 2h. ii) CH_2Cl_2, Reflux, 3h.

Table 1　Tyrosinase inhibitory activities and yields of the synthesized compounds

Compounds	CLog*P* [d]	IC_{50} (μmol/L) [a]
1	1.798	94.42
2	2.913	NA[b]
3	2.208	82.07
4	1.6472	4.5% [c]
5	1.6472	60.87
6	2.1762	47.06
7	2.4852	12.29
8	3.7632	5.60
9	3.4152	NA[b]
10	1.201	NA[b]
11	2.123	NA[b]
12	2.652	NA[b]
13	3.181	1.52% [c]
14	2.961	NA[b]
3 - (4 - hydroxyphenyl) acrylic acid	1.572	23.2% [c]
3 - (3 - hydroxy - 4 - methoxyphenyl) acrylic acid	1.4212	138.06
3 - (4 - nitrophenyl) acrylic acid	1.982	NA[b]
Kojic acid		23.67

a. Values were determined from logarithmic concentration - inhibition curves (at least eight points) and are given as means of three experiments

b. not active at 200μmol/L concentration.

c. percent of inhibition of tyrosinase reaction at the 200μmol/L

d. Value of CLog *P* was obtained by ChemBioDraw Ultra 12.0

Fig. 1.

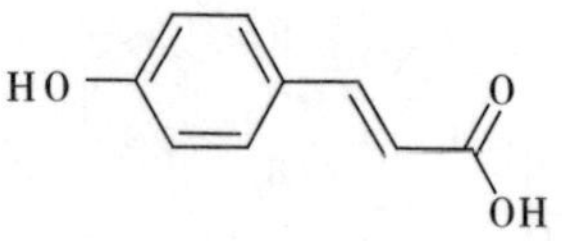

HO

O

OH

3-(4-hydroxy phenyl) acrylic acid

O

O_2N

OH

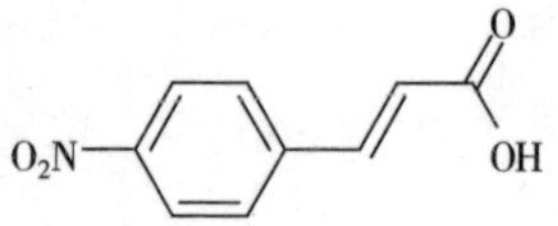

3-(4-nitrophenyl) acrylic acid

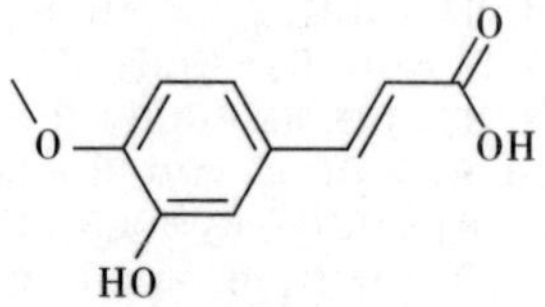

O

O

OH

HO

3-(3-hydroxy-4-methoxyphenyl) acrylic acid

Fig. 2.

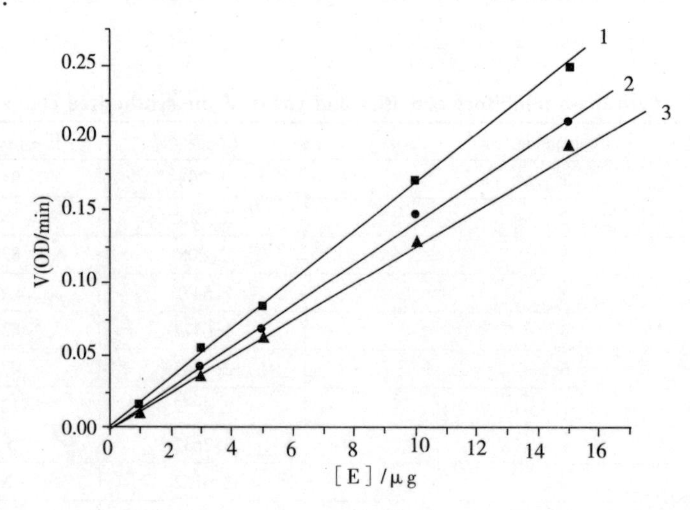

Fig. 3.

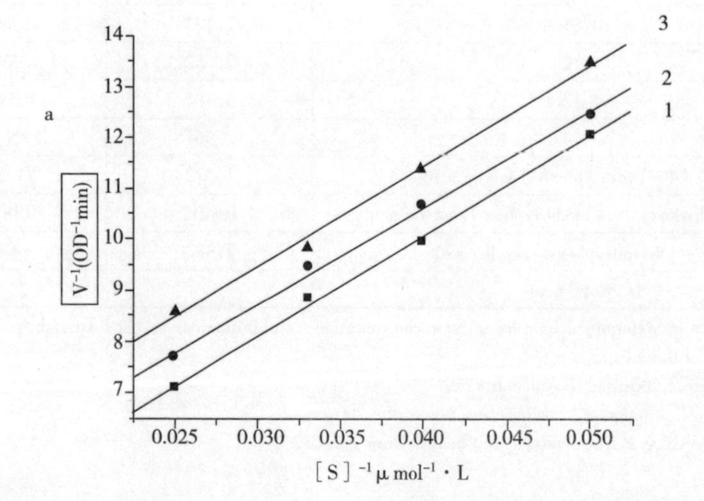

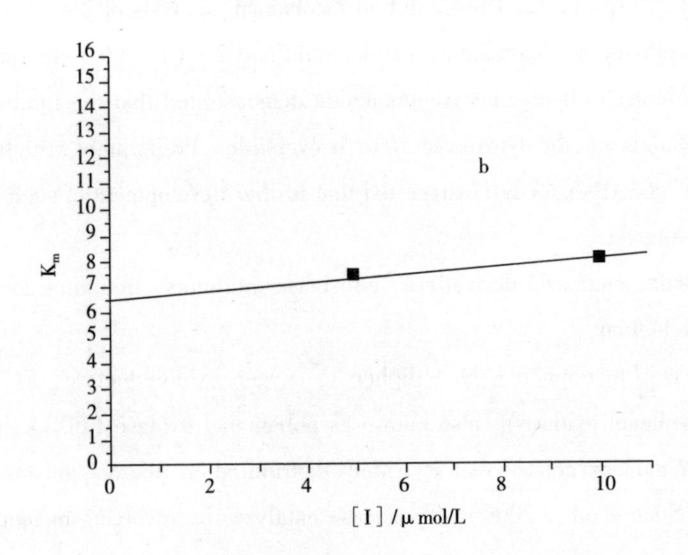

3.4.2 Food Chemistry (2012, 135, 2872 – 2878)

Biological evaluation ofcoumarin derivatives as mushroom tyrosinase inhibitors

Jinbing Liu *[a], Fengyan Wu[a], Lingjuan Chen[b], Liangzhong Zhao[a],

Zibing Zhao[a], Min Wang,[a] Sulan Lei[a]

[a]*Department of Biology and Chemical Engineering*, *Shaoyang University*, *Shao Shui Xi Road*, *Shaoyang* 422100, *PRC*

[b]*College of Food Science and Engineering*, *Central South University of Forestry and Technology*, *Changsha*, *Hunan* 410004, *PRC*

* Corresponding author: Department of Biology and Chemical Engineering, Shaoyang University, Shao Shui Xi Road, Shaoyang 422100, PRC

Phone: +86 – 739 – 5431768; Fax: +86 – 739 – 5431768

E – mail address: xtliujb@ yahoo. com. cn

Abstract

A series of coumarin derivatives were synthesized and their inhibitory effects on the diphenolase activity of mushroom tyrosinase were evaluated. The results showed that some of the synthesized compounds exhibited significant inhibitory activities. Especially, 2 – (1 – (coumarin – 3 – yl) ethylidene) hydrazinecarbothioamide bearing thiose – micarbazide group exhibited the most potent tyrosinase inhibitory activity with

IC_{50} value of 3.44μmol/L. The inhibition mechanism analysis of 2 – (1 – (coumarin – 3 – yl) – ethylidene) hydrazinecarbothioamide and 2 – (1 – (6 – chlorocoumarin – 3 – yl) ethylidene) – hydrazinecarbothioamide demonstrated that the inhibitory effects of the compounds on the tyrosinase were irreversible. Preliminary structure activity relationships' (SARs) analysis suggested that further development of such compounds might be of interest.

Keywords: coumarin derivatives, tyrosinase inhibitors, inhibition mechanism.

1. Introduction

Tyrosinase (monophenol oro – diphenol, oxygen oxidoreductase, EC 1.14.18.1, syn. polyphenol oxidase), also known as polyphenol oxidase (PPO), is a copper – containing monooxygenase that is widely distributed in microorganisms, animals, and plants (Song et al., 2006). Tyrosinase catalyzes by involving molecular oxygen in two distinct reactions: in the hydroxylation of monophenols too – diphenols (monophenolase) and in the oxidation of o – diphenols to o – quinones (diphenolase) (Chen et al., 2003). Due to the high reactivity, quinines could polymerize spontaneously to form high molecular weight brown – pigments (melanins) or react with amino acids and proteins to enhance brown color of the pigment produced (Matsuura et al., 2006). Hyperpigmentations, such as *senile lentigo*, melasma, freckles, and pigmented acne scars are of particular concern to women. The treatment usually involves the use of medicines or medicinal cosmetics containing depigmenting agents or skin whitening agents (Tripathi et al., 1992). In clinical usage, tyrosinase inhibitors are used for treatments of dermatological disorders related to melanin hyperaccumulation and are essential in cosmetics for depigmentation (Schallreuter et al., 2009; Wood et al., 2009). For example, age spots and freckle were caused by the accumulation of an excessive level of epidermal pigmentation (Thanigaimalai et al., 2010).

Previous reports confirmed that tyrosinasewas one of the main causes of most fruits and vegetables quality loss during post harvest handling and processing, leading to faster degradation and shorter shelf life (Yi et al., 2010). Tyrosinase has also been linked to Parkinson's and other neurodegenerative diseases (Zhu et al., 2011). In insects, tyrosinase is uniquely associated with three different biochemical processes, including sclerotization of cuticle, defensive encapsulation and melanization of foreign organism, and wound healing (Ashida et al., 1995). These processes provide potential targets for developing safer and effective tyrosinase inhibitors as insecticides and ultimately for insect control. Thus, the development of safe and effective tyrosinase inhibitors is of great concern in the medical, agricultural, and cosmetic industries. However, only a few such as kojic acid, arbutin, tropolone, and 1 - phenyl - 2 - thiourea (PTU) (Fig. 1) are used as therapeutic agents and cosmetic products (Battaini et al., 2000).

Coumarin derivatives are an important class of compounds, widely present in plants, including edible vegetables and fruits (Rai et al., 2010; Curini et al., 2006). Coumarin derivatives are of great interest due to their diverse structural features and versatile biological properties, such as anti - inflammatory, antioxidant, vasorelaxant, cytotoxic, anti - HIV, antitubercular and antimicrobial (Belluti et al., 2010; Roussaki et al., 2010; Ostrov et al., 2007; Neyts et al., 2009; Kostova et al., 2006; Chimenti et al., 2010; Upadhyay et al., 2010). In particular, their antibacterial, antifungal and anticancer activities make the compounds attractive for further derivatization and screening as novel therapeutic agents (Khode et al., 2009). The literature survey revealed that compounds with thiourea moieties have been reported to demonstrate a wide range of pharmacological activities, which include antibacterial, antifungal, anticonvulsant and tyrosinase inhibitory activity. Such as phenylthioureas, alkylthioureas and 1,3 - bis - (5 - methanesulfonylbutyl) thiourea, displayed weak or moderate tyrosinase inhibitory activity (Ley et al., 2001). More recently, our investigations also demonstrated that thiosemicarbazide derivatives exhibited potent inhibitory activities against mushroom tyrosinase (Liu et al., 2008). During the last years, extensive studies on the pharmacology of coumarin derivatives have been reported, but the tyrosinase inhibitor activities of this kind of compounds have hardly ever appeared in the literature. Stimulated by these results, in the present investigation, we synthesized a series of coumarin derivatives bearing thiosemicarbazide moieties or ester groups, their inhibitory activities against mushroom tyrosinase were evaluated using kojic acid as comparing substance. Meanwhile, the structure - activity relationships of these compounds were also primarily discussed. The aim of the present study was the discovery of safe and efficient compounds as food additives or food preservatives, which can offer a clue to the design and synthesis of novel tyrosinase inhibitors.

2. Materials and methods

2.1 Chemical reagents and instruments

Melting points (m. p.) were determined with WRS – 1B melting point apparatus and the thermometer was uncorrected. NMR spectra were recorded on Bruker 400 spectrometersat25℃ in $CDCl_3$ or DMSO – d_6. All chemical shifts (δ) are quoted in parts per million downfield from TMS and coupling constants (J) are given in hertz. Abbreviations used in the splitting pattern were an follows: s = singlet, d = doublet, t = triplet, q = quintet, m = multiplet, LC – MS spectra were recorded using the LCMS – 2010A. All reactions were monitored by TLC (Merck Kieselgel 60 F254) and the spots were visualized under UV light. Infrared (IR) spectra were recorded as potassium bromide pellets on VECTOR 22 spectrometer.

Tyrosinase, L – 3, 4 – dhydroxyphenylalanine (L – DOPA) andkojic acid were purchased from Sigma – Aldrich Chemical Co. Other chemicals were purchased from commercial suppliers and were dried and purified when necessary.

2.2 General procedures for the synthesis ofsubstituted 3 – acetylcoumarin

Piperidine (5 mol %) was added to the mixture of substituted salicylaldehyde (1 mmol) and ethylacetoacetate (1.1 mmol) in dry CH_3CN (10 mL), then the reaction mixture was stirred for about 4 h at the room temperature. The progress of the reaction was monitored by TLC. After completion of the reaction, formed precipitate was collected by filtration and washed with cold CH_3CN, then dried under vacuum to provide the substituted 3 – acetylcoumarin.

2.2.1 3 – acetylcoumarin (1)

Yield 76%. Yellow solid, mp 120.7 ~ 122.1℃. IR (KBr): 3078, 3026, 1723, 1675, 1598, 1556, 1441, 1406, 1351, 1297, 1220, 1198, 1162, 1098, 967, 762 cm^{-1}. ^{1}HNMR (400 MHz, $CDCl_3$): δ 8.49 (s, 1H, C = CH), 7.65 – 7.63 (m, 2H, Ph – H), 7.37 – 7.32 (m, 2H, Ph – H), 2.71 (s, 3H, CH_3); ^{13}CNMR (100 MHz, $CDCl_3$) δ 195.41 (1C), 159.17 (1C), 155.27 (1C), 147.39 (1C), 134.33 (1C), 130.17 (1C), 124.92 (1C), 124.48 (1C), 118.20 (1C), 116.63 (1C), 30.49 (1C).

2.2.2 6 – chloro – 3 – acetylcoumarin (2)

Yield 89.87%. Yellow solid, mp 213.1 ~ 213.9℃. IR (KBr): 2946, 1726,

1611, 1469, 1412, 1355, 1278, 1134, 1082, 970, 829, 765 cm^{-1}. ^{1}HNMR (400 MHz, $CDCl_3$): δ 8.40(s, 1H, C=CH), 7.63 (dd, H, J =8.5Hz, 2.5Hz, Ph-H), 7.60 (d, 1H, J =2.5Hz, Ph-H), 7.33 (d, 1H, J =8.5Hz, Ph-H), 2.72 (s, 3H, CH_3); ^{13}CNMR (100 MHz, $CDCl_3$) δ 195.41 (1C), 158.57 (1C), 152.23 (1C), 145.39 (1C), 141.08 (1C), 132.12 (1C), 128.65 (1C), 127.56 (1C), 126.94 (1C), 118.02 (1C), 16.14 (1C).

2.2.3 6-bromo-3-acetylcoumarin (3)

Yield 94.10%. Yellow solid, mp 229.1~229.9℃. IR (KBr): 3039, 1732, 1675, 1608, 1550, 1415, 1351, 1233, 1201, 1063, 980, 832, 768 cm^{-1}. ^{1}HNMR (400 MHz, $CDCl_3$): δ 8.40 (s, 1H, C=CH), 7.78 (dd, H, J =8.5Hz, 2.5Hz, Ph-H), 7.74 (d, 1H, J =2.5Hz, Ph-H), 7.27 (d, 1H, J =8.5Hz, Ph-H), 2.72 (s, 3H, CH_3); ^{13}CNMR (100 MHz, $CDCl_3$) δ 195.41 (1C), 158.62 (1C), 152.71 (1C), 145.32 (1C), 140.77 (1C), 134.69 (1C), 131.12 (1C), 126.65 (1C), 120.88 (1C), 118.67 (1C), 116.18 (1C), 16.16 (1C).

2.3 The general procedure for the synthesis of substituted 2-(1-(coumarin-3-yl) ethylidene)hydrazinecarbothioamide compound.

The appropriate substituted 3-acetylcoumarin (10 mmol) was dissolved in anhydrous ethanol (20 mL), thiosemicarbazide (10 mmol) and acetic acid (0.5 mL) were added to the above system. The reaction mixture was refluxed for 6 h. The reaction was monitored by TLC. After completion of the reaction, the reaction mixture was cooled to room temperature. The appearing precipitate was filtered and recrystallized from 95% alcohol to obtain the corresponding substituted 2-(1-(coumarin-3-yl) ethylidene) hydrazinecarbothioamide compound.

2.3.1 2-(1-(coumarin-3-yl)ethylidene)hydrazinecarbothioamide (4)

Yield 85.38%. Yellow solid, mp 214.2~216.7℃. IR (KBr): 3385, 3235, 3155, 1716, 1598, 1499, 1428, 1367, 1290, 1236, 1111, 1069, 970, 861, 765 cm^{-1}. ^{1}HNMR (400 MHz, DMSO-d_6) δ 10.61 (s, 1H, NH), 8.46 (s, 1H, C=CH), 8.39 (s, 1H, NH_2), 7.94 (s, 1H, NH_2), 7.76 (d, J =8.5Hz, H, Ph-H), 7.65 (t, 1H, J =8.5Hz, Ph-H), 7.44 (d, 1H, J =8.5Hz, Ph-H), 7.38 (t, 1H, J =8.5Hz, Ph-H), 2.38 (s, 3H, CH_3); ^{13}CNMR (100 MHz, DMSO-d_6) δ 179.72 (1C), 159.56 (1C), 154.07 (1C), 146.39 (1C), 142.55 (1C), 132.83 (1C), 129.56 (1C), 126.25 (1C), 125.20 (1C), 119.39

(1C), 116.41 (1C), 16.47 (1C).

2.3.2 2 – (1 – (6 – chlorocoumarin – 3 – yl)ethylidene)hydrazinecarbothioamide (5)

Yield 87.68%. Yellow solid, mp 232.0 ~ 232.9℃. IR (KBr): 3417, 3235, 3138, 1732, 1707, 1588, 1502, 1479, 1451, 1428, 1287, 1233, 1204, 1130, 1092, 954, 925, 871, 832, 772 cm^{-1}. ^{1}HNMR (400 MHz, DMSO – d_6) δ 10.49 (s, 1H, NH), 8.44 (s, 1H, NH_2), 8.41 (s, 1H, C = CH), 8.44 (s, 1H, NH_2), 7.82 (s, H, Ph – H), 7.68 (dd, 1H, J = 8.5Hz, 2.5Hz, Ph – H), 7.47 (d, 1H, J = 8.5Hz, Ph – H), 2.45 (s, 3H, CH_3); ^{13}CNMR (100 MHz, DMSO – d_6) δ 179.31 (1C), 158.64 (1C), 151.96 (1C), 145.37 (1C), 140.63 (1C), 131.79 (1C), 128.35 (1C), 127.63 (1C), 126.82 (1C), 120.40 (1C), 117.97 (1C), 15.87 (1C).

2.3.3 2 – (1 – (6 – bromocoumarin – 3 – yl)ethylidene)hydrazinecarbothioamide (6)

Yield 89.15%. Yellow solid, mp 231.9 ~ 232.9℃. IR (KBr): 3404, 3222, 3135, 1720, 1588, 1495, 1467, 1422, 1236, 1233, 1204, 1101, 1060, 957, 861, 826, 775 cm^{-1}. ^{1}HNMR (400 MHz, DMSO – d_6) δ 10.60 (s, 1H, NH), 8.44 (s, 1H, NH_2), 8.40 (s, 1H, C = CH), 7.90 (s, 1H, NH_2), 7.80 (s, H, Ph – H), 7.68 (dd, 1H, J = 8.5Hz, 2.5Hz, Ph – H), 7.41 (d, 1H, J = 8.5Hz, Ph – H), 2.24 (s, 3H, CH_3); ^{13}CNMR (100 MHz, DMSO – d_6) δ 179.30 (1C), 158.58 (1C), 152.63 (1C), 145.32 (1C), 140.71 (1C), 134.53 (1C), 130.93 (1C), 126.73 (1C), 120.80 (1C), 118.35 (1C), 116.14 (1C), 15.86 (1C).

2.4 The general procedures for the synthesis of substituted coumarin – 3 – carboxylic acid ester compounds.

The substituted coumarin – 3 – carboxylic acid ethyl esters (Scheme 2) were prepared by Knoevenagel reaction. To a mixture of diethyl malonate (1 mmol) and the appropriate salicylaldehyde (1 mmol) in ethanol (10 mL) was added piperidine (5 mol %) and the reaction mixture was stirred at the room temperature. The progress of the reaction was monitored by TLC. After completion of the reaction, the solvent was removed under vacuum and the residue was purified by chromatography. The obtained substituted coumarin – 3 – carboxylic acid ethyl ester (2 mmol) was dissolved in 10%

NaOH (50 mL), then 3 N HCl (50 mL) was added the mixture. The suspension was filtered and dried under vacuum to provide substituted coumarin – 3 – carboxylic acid. The coumarin – 3 – carboxylic acid (2 mmol) was added to thionyl chloride (30 mL), the reaction mixture was refluxed for about 2 hours, the thionyl chloride was removed under vacuum. The desired substituted coumarin – 3 – carbonyl chloride was obtained. To a mixture of substituted coumarin – 3 – carbonyl chloride (2 mmol) and toluene (20 mL) or ethyl ether (20 mL) was added dropwise appropriate alcohol (2 mmol) and then the reaction mixture was refluxed for about 12h. The solvent was removed under vacuum and the residue was purified by chromatography.

2.4.1 ethylcoumarin – 3 – carboxylate (7)

Yield 85.06%. Yellow solid, mp 93.1 ~ 96.2℃. IR (KBr): 3055, 2972, 1764, 1697, 1608, 1559, 1479, 1447, 1371, 1297, 1242, 1204, 1130, 1028, 983, 916, 797 cm^{-1}. ^{1}HNMR (400 MHz, $CDCl_3$) δ 8.52 (s, 1H, C = CH), 7.66 – 7.60 (m, 2H, Ph – H), 7.37 – 7.30 (m, 2H, Ph – H), 4.44 (q, J = 7.2Hz, 2H, CH_2), 1.43 (t, J = 7.2Hz, 3H, CH_3); ^{13}CNMR (100 MHz, $CDCl_3$) δ 162.97 (1C), 157.94 (1C), 155.08 (1C), 148.50 (1C), 134.30 (1C), 129.46 (1C), 124.80 (1C), 118.28 (1C), 117.82 (1C), 116.71 (1C), 61.91 (1C), 14.15 (1C).

2.4.2 ethyl 6 – chlorocoumarin – 3 – carboxylate (8)

Yield 88.00%. White solid, mp 173.7 ~ 176.5℃. IR (KBr): 3068, 2972, 1752, 1697, 1614, 1556, 1470, 1364, 1287, 1239, 1207, 1021, 996, 839, 791 cm^{-1}. ^{1}HNMR (400 MHz, $CDCl_3$) δ 8.43 (s, 1H, C = CH), 7.59 (s, 1H, Ph – H), 7.57 (d, J = 8.4Hz, 1H, Ph – H), 7.32 (d, J = 8.4Hz, 1H, Ph – H), 4.44 (q, J = 7.2Hz, 2H, CH_2), 1.42 (t, J = 7.2Hz, 3H, CH_3); ^{13}CNMR (100 MHz, $CDCl_3$) δ 162.62 (1C), 156.00 (1C), 153.44 (1C), 147.08 (1C), 134.11 (1C), 130.10 (1C), 128.41 (1C), 119.50 (1C), 118.79 (1C), 118.23 (1C), 62.18 (1C), 14.16 (1C).

2.4.3 ethyl 6 – bromocoumarin – 3 – carboxylate (9)

Yield 67.45%. White solid, mp 171.0 ~ 172.3 ℃. IR (KBr): 3068, 2972, 1752, 1710, 1614, 1595, 1550, 1467, 1409, 1364, 1284, 1236, 1207, 1018, 989, 858, 791 cm^{-1}. ^{1}HNMR (400 MHz, $CDCl_3$) δ 8.43 (s, 1H, C = CH), 7.76 – 7.71 (m, 2H, Ph – H), 7.27 (d, J = 8.8Hz, 1H, Ph – H), 4.45 (q, J =

7.2Hz, 2H, CH_2), 1.43 (t, J = 7.2Hz, 3H, CH_3); ^{13}CNMR (100 MHz, $CDCl_3$) δ 162.59 (1C), 155.89 (1C), 153.92 (1C), 146.94 (1C), 136.88 (1C), 131.49 (1C), 119.49 (1C), 119.32 (1C), 118.50 (1C), 117.30 (1C), 62.18 (1C), 14.17 (1C).

2.4.4 ethyl 7-hydroxycoumarin-3-carboxylate (10)

Yield 51.04%. Yellow solid, mp 177.5~179.9 ℃. IR (KBr): 3462, 2991, 1745, 1678, 1614, 1444, 1380, 1297, 1242, 1217, 1140, 1034, 970, 842, 797 cm^{-1}. ^{1}HNMR (400 MHz, $CDCl_3$) δ 11.06 (s, 1H, OH), 8.51 (s, 1H, C=CH), 8.08 (d, J = 8.8Hz, 1H, Ph-H), 6.87-6.84 (m, 2H, Ph-H), 4.43 (t, J = 7.2Hz, 2H, CH_2), 1.42 (t, J = 7.2Hz, 3H, CH_3); ^{13}CNMR (100 MHz, $CDCl_3$) δ 162.81 (1C), 156.00 (1C), 153.62 (1C), 147.83 (1C), 134.30 (1C), 129.55 (1C), 128.94 (1C), 119.63 (1C), 119.27 (1C), 118.65 (1C), 61.87 (1C), 14.51 (1C).

2.4.5 coumarin-3-carboxylic acid (11)

Yield 69.76%. White solid, mp 203.6~204.3℃. IR (KBr): 3052, 1745, 1691, 1614, 1489, 1431, 1371, 1223, 1207, 1037, 983, 826, 800, 768 cm^{-1}. ^{1}HNMR (400 MHz, $CDCl_3$) δ 12.32 (s, 1H, COOH), 9.00 (s, 1H, C=CH), 7.82-7.76 (m, 2H, Ph-H), 7.51-7.47 (m, 2H, Ph-H); ^{13}CNMR (100 MHz, $CDCl_3$) δ 164.09 (1C), 162.39 (1C), 154.56 (1C), 151.48 (1C), 135.76 (1C), 130.49 (1C), 126.25 (1C), 118.45 (1C), 117.20 (1C), 114.89 (1C).

2.4.6 6-chlorocoumarin-3-carboxylic acid (12)

Yield 93.18%. White solid, mp >300℃. IR (KBr): 3110, 3046, 1755, 1726, 1617, 1582, 1479, 1409, 1374, 1342, 1249, 1210, 1150, 1079, 967, 880, 791 cm^{-1}. ^{1}HNMR (400 MHz, DMSO-d_6) δ 13.41 (s, 1H, COOH), 8.70 (s, 1H, C=CH), 8.05 (s, 1H, Ph-H), 7.78-7.75 (m, 1H, Ph-H), 7.50 (d, J = 9.2Hz, 1H, Ph-H); ^{13}CNMR (100 MHz, DMSO-d_6) δ 164.71 (1C), 164.52 (1C), 158.09 (1C), 146.88 (1C), 134.18 (1C), 132.57 (1C), 116.47 (1C), 113.25 (1C), 111.48 (1C), 102.19 (1C).

2.4.7 6-bromocoumarin-3-carboxylic acid (13)

Yield 67.45%. White solid, mp 208.3~210.3 ℃. IR (KBr): 3436, 3049, 1764, 1681, 1611, 1553, 1476, 1415, 1364, 1242, 1204, 1140, 1066, 964,

874, 804 cm^{-1}. ^{1}HNMR (400 MHz, DMSO – d_6) δ 13.41 (s, 1H, COOH), 8.70 (s, 1H, C = CH), 8.05 (s, 1H, Ph – H), 7.78 – 7.75 (m, 1H, Ph – H), 7.50 (d, J = 9.2Hz, 1H, Ph – H); ^{13}CNMR (100 MHz, DMSO – d_6) δ 164.71 (1C), 164.52 (1C), 158.09 (1C), 146.88 (1C), 134.18 (1C), 132.57 (1C), 116.47 (1C), 113.25 (1C), 111.48 (1C), 102.19 (1C).

2.4.8 7 – hydroxycoumarin – 3 – carboxylic acid (14)

Yield 93.58%. Yellow solid, mp 263.4 ~ 265.3 ℃. IR (KBr): 3158, 2834, 1726, 1710, 1611, 1572, 1499, 1454, 1396, 1367, 1322, 1278, 1220, 1137, 1028, 993, 852, 794 cm^{-1}. ^{1}HNMR (400 MHz, DMSO – d_6) δ 12.79 (s, 1H, COOH), 11.22 (s, 1H, OH), 8.69 (s, 1H, C = CH), 7.76 (d, J = 8.8Hz, 1H, Ph – H), 6.88 (dd, J = 8.8Hz, 1H, Ph – H), 6.77 (d, J = 8.8Hz, 1H, Ph – H); ^{13}CNMR (100 MHz, DMSO – d_6) δ 164.69 (1C), 164.48 (1C), 158.08 (1C), 157.46 (1C), 149.88 (1C), 132.49 (1C), 114.53 (1C), 112.95 (1C), 111.08 (1C), 102.09 (1C).

2.4.9 methylcoumarin – 3 – carboxylate (15)

Yield 89.65%. White solid, mp 156.8 ~ 158.7℃. IR (KBr): 3046, 2946, 1758, 1694, 1614, 1559, 1470, 1431, 1355, 1300, 1242, 1204, 1085, 1005, 832, 794 cm^{-1}. ^{1}HNMR (400 MHz, $CDCl_3$) δ 8.47 (s, 1H, C = CH), 7.61 – 7.58 (m, 2H, Ph – H), 7.33 – 7.26 (m, 2H, Ph – H), 3.96 (s, 3H, CH_3); ^{13}CNMR (100 MHz, $CDCl_3$) δ 169.25 (1C), 155.8 (1C), 147.69 (1C), 147.63 (1C), 134.29 (1C), 134.26 (1C), 130.19 (1C), 128.49 (1C), 118.75 (1C), 118.26 (1C), 53.04 (1C).

2.4.10 methyl 6 – bromocoumarin – 3 – carboxylate (16)

Yield 86.50%. Yellow solid, mp 176.8 ~ 178.9℃. IR (KBr): 3039, 2972, 1748, 1697, 1617, 1588, 1556, 1473, 1435, 1377, 1287, 1236, 1204, 999, 880, 816, 794 cm^{-1}. ^{1}HNMR (400 MHz, $CDCl_3$) δ 8.47 (s, 1H, C = CH), 7.76 – 7.20 (m, 2H, Ph – H), 7.27 (d, J = 8.8Hz, 1H, Ph – H), 3.97 (s, 3H, CH_3); ^{13}CNMR (100 MHz, $CDCl_3$) δ 163.26 (1C), 155.91 (1C), 154.00 (1C), 147.50 (1C), 137.06 (1C), 131.56 (1C), 119.29 (1C), 119.13 (1C), 118.54 (1C), 117.40 (1C), 53.06 (1C).

2.4.11 methyl 7 – hydroxycoumarin – 3 – carboxylate (17)

Yield 93.58%. White solid, mp 116.2 ~ 117.2 ℃. IR (KBr): 3055, 2956,

1745, 1694, 1614, 1563, 1451, 1361, 1313, 1268, 1246, 1217, 1153, 996, 919, 797 cm^{-1}. ^1HNMR (400 MHz, $CDCl_3$) δ 9.96 (s, 1H, OH), 8.55 (s, 1H, C = CH), 7.66 – 7.60 (m, 2H, Ph – H), 7.36 – 7.32 (m, H, Ph – H), 3.95 (s, 3H, CH_3); $^{13}CNMR$ (100 MHz, $CDCl_3$) δ 163.69 (1C), 156.64 (1C), 155.23 (1C), 149.04 (1C), 134.42 (1C), 129.52 (1C), 124.85 (1C), 118.00 (1C), 117.85 (1C), 116.78 (1C), 52.86 (1C).

2.4.12 pentylcoumarin – 3 – carboxylate (18)

Yield 53.28%. Yellow solid, mp 133.4 ~ 134.5℃. IR (KBr): 3055, 2956, 2927, 1764, 1691, 1620, 1563, 1470, 1415, 1383, 1351, 1294, 1246, 1207, 1085, 989, 829, 788 cm^{-1}. ^1HNMR (400 MHz, DMSO – d_6) δ 8.71 (s, 1H, C = CH), 8.08 (d, J = 8.8Hz, 1H, Ph – H), 7.79 – 7.58 (m, 2H, Ph – H), 7.50 (d, J = 8.8Hz, 1H, Ph – H), 4.72 (t, J = 6.8Hz, 2H, CH_2), 2.52 – 2,50 (m, 2H, CH_2), 1.73 – 1.66 (m, 2H, CH_2), 1.37 – 1.31 (m, 2H, CH_2), 0.92 (t, J = 6.8Hz, 3H, CH_3); $^{13}CNMR$ (100 MHz, DMSO – d_6) δ 162.95 (1C), 156.05 (1C), 153.65 (1C), 147.81 (1C), 134.31 (1C), 129.61 (1C), 128.93 (1C), 119.67 (1C), 119.37 (1C), 118.69 (1C), 65.81 (1C), 28.25 (1C), 27.95 (1C), 22.24 (1C), 14.31 (1C).

2.4.13 isopropyl 6 – chlorocoumarin – 3 – carboxylate (19)

Yield 67.45%. Yellow solid, mp 82.1 ~ 82.6 ℃. IR (KBr): 3046, 2982, 2959, 1748, 1604, 1563, 1473, 1457, 1374, 1297, 1249, 1210, 1137, 1101, 1005, 922, 864, 797 cm^{-1}. ^1HNMR (400 MHz, $CDCl_3$) δ 8.38 (s, 1H, C = CH), 7.60 – 7.57 (m, 2H, Ph – H), 7.32 (d, J = 8.8Hz, 1H, Ph – H), 5.30 – 5.24 (m, 1H, CH), 1.40 (d, J = 6.4Hz, 6H, $2CH_3$); $^{13}CNMR$ (100 MHz, $CDCl_3$) δ 162.00 (1C), 155.97 (1C), 153.41 (1C), 146.56 (1C), 134.09 (1C), 130.08 (1C), 128.47 (1C), 120.00 (1C), 118.91 (1C), 118.23 (1C), 69.99 (1C), 21.76 (2C).

2.4.14 isopropyl 6 – bromocoumarin – 3 – carboxylate (20)

Yield 69.75%. White solid, mp 217.3 ~ 218.8 ℃. IR (KBr): 3065, 2972, 1745, 1710, 1611, 1563, 1473, 1409, 1367, 1290, 1246, 1207, 1101, 1005, 970, 877, 829, 794 cm^{-1}. ^1HNMR (400 MHz, $CDCl_3$) δ 8.37 (s, 1H, C = CH), 7.75 – 7.70 (m, 2H, Ph – H), 7.27 (d, J = 8.8Hz, 1H, Ph – H), 5.30 – 5.24 (m, 1H, CH), 1.40 (d, J = 6.4Hz, 6H, $2CH_3$); $^{13}CNMR$ (100 MHz, $CDCl_3$)

δ 161.99 (1C), 155.93 (1C), 153.90 (1C), 146.41 (1C), 136.76 (1C), 131.45 (1C), 119.95 (1C), 119.38 (1C), 118.25 (1C), 117.30 (1C), 70.02 (1C), 21.78 (2C).

2.4.15 isopropyl 7 – hydroxycoumarin – 3 – carboxylate (21)

Yield 61.36%. White solid, mp 204.7 ~ 205.9 ℃. IR (KBr): 3062, 2975, 1742, 1704, 1620, 1563, 1473, 1419, 1364, 1290, 1246, 1207, 1105, 1009, 967, 832, 791 cm^{-1}. ^{1}HNMR (400 MHz, $CDCl_3$) δ 9.98 (s, 1H, OH), 8.44 (s, 1H, C = CH), 7.64 – 7.59 (m, 2H, Ph – H), 7.34 – 7.30 (m, 1H, Ph – H), 5.28 – 5.23 (m, 1H, CH), 1.39 (d, J = 6.5Hz, 6H, $2CH_3$); ^{13}CNMR (100 MHz, $CDCl_3$) δ 162.34 (1C), 156.70 (1C), 155.05 (1C), 148.03 (1C), 134.21 (1C), 129.48 (1C), 124.82 (1C), 118.66 (1C), 117.87 (1C), 116.69 (1C), 69.63 (1C), 21.80 (2C).

2.4.16 pentyl 7 – hydroxycoumarin – 3 – carboxylate (22)

Yield 56.18%. Yellow solid, mp 53.2 ~ 55.0 ℃. IR (KBr): 3046, 2949, 1764, 1723, 1604, 1447, 1367, 1294, 1242, 1207, 1124, 1009, 919, 861, 797 cm^{-1}. ^{1}HNMR (500 MHz, $CDCl_3$) δ 9.87 (s, 1H, OH), 8.49 (s, 1H, C = CH), 7.64 – 7.59 (m, 2H, Ph – H), 7.35 – 7.30 (m, 1H, Ph – H), 4.43 (t, J = 6.5Hz, 2H, CH_2), 1.80 – 1.74 (m, 2H, CH_2), 1.41 – 1.36 (m, 4H, $2CH_2$), 0.93 (t, J = 7.0Hz, 3H, CH_3); ^{13}CNMR (125 MHz, $CDCl_3$) δ 163.15 (1C), 156.59 (1C), 155.18 (1C), 148.33 (1C), 134.23 (1C), 129.46 (1C), 124.77 (1C), 118.51 (1C), 117.90 (1C), 116.76 (1C), 66.08 (1C), 28.26 (1C), 28.01 (1C), 22.43 (1C), 13.96 (1C).

2.5 Tyrosinase assay

The spectrophotometric assay for tyrosinase was performed according to the method reported by our groups with some slight modifications (Yi et al., 2010; Liu et al., 2008). Briefly, all the synthesized compounds were screened for the diphenolase inhibitory activity of tyrosinase using L – DOPA as substrate. All the compounds were dissolved in DMSO. The final concentration of DMSO in the test solution was 2.0%. Phosphate buffer, pH 6.8, was used to dilute the DMSO stock solution of test compounds. Thirty units of mushroom tyrosinase (0.5 mg/mL) were first pre – incubated with the compounds, in 50 mmol/L phosphate buffer (pH 6.8), for 10 min at 25°C. Then the L – DOPA (0.5 mmol/L) was added to the reaction mixture and the enzyme reaction was monitored by measuring the change in absorbance at 475 nm of formation of the DOPA chrome for 1 min. The measurement was performed in triplicate for each concentration and averaged before further calculation. IC_{50} value, a concentration giving 50% inhibition of tyrosinase activity, was determined by interpolation of the dose - response curves. The percent of inhibition of tyrosinase reaction was calculated as following:

$$\text{Inhibition rate}(\%) = [(B - S)/B] \times 100$$

Here, the B and S are the absorbances for the blank and samples. Kojic acid was used as reference standard inhibitors for comparison.

3. Results and discussion

3.1 Synthesis

According to the general procedure shown inScheme 1, substituted coumarin was synthesized by Knoevenagel condensation. In presence of piperidine as catalyst, the reaction of substituted salicylaldehyde and ethyl acetoacetate was carried out at room temperature. In present of acetic acid as catalyst, the obtained substituted coumarin was further converted into it ' s thiosemicarbazide derivative. The yields of these compounds were from moderate to good (Chimenti et al., 2010; Liu et al., 2008; Wanare et al., 2010; Chimenti et al., 2007). According to the general procedure shown in Scheme 2, the substituted coumarin - 3 - carboxylate ethyl ester was obtained from substituted salicylaldehyde and diethyl malonate by the Knoevenagel condensation. The substituted coumarin - 3 - carboxylate ethyl ester was further converted into the target compound by the hydrolysis reaction, chlorination reaction, esterification reaction in turn (Khode et al., 2009; Chimenti et al., 2009).

3.2 Tyrosinase inhibitory activity

For evaluating the tyrosinase inhibitory activity, all the synthesized compoundswere subjected to tyrosinase inhibition assay with L - DOPA as substrate, according to the method reported by our groups with some slight modifications. The tyrosinase inhibitory activities of kojic acid was ever reported, therefore, it was selected as comparing substance. The IC_{50} values of coumarin derivatives against tyrosinase were summarized in Table 1, and IC_{50} values of all these compounds were determined from logarithmic concentration - inhibition curves and given as means of three experiments.

Our results showed that compounds3 – 6 and 20 exhibited potent inhibition on mushroom tyrosinase with IC_{50} values ranged from 3. 44 to 195. 03 μmol/L. Especially, compounds 4 – 6 bearing thiosemicarbazide showed more potent inhibitory activities than the other compounds. In addition, compound 4 demonstrated more potent inhibitory activity than the reference standard inhibitor kojic acid. Most of the coumarin – 3 – carboxylic acid analogues showed low or no activities against tyrosinase. The results may be related to the structure of tyrosinase contained a type – 3 copper center with a coupled dinuclear copper active site in the catalytic core. Tyrosinase inhibition of compounds 4 – 6 depended on the competency of the sulfur atom to chelate with the dicopper nucleus in the active site, and tyrosinase would lose its catalyzing ability after forming complex (Gerdemann et al., 2002).

Comparing to the tyrosinase inhibitory activities of substituted 3 – acetylcoumarin, compounds with halogen atom exhibited more potent inhibitory activities than the 3 – acetylcoumarin, with the increase of lipophilicity of compound, the inhibitory activities increased gradually. These results showed that the inhibitory activities of 3 – acetylcoumarin and it's halogen substituted analogues ralated to the lipophilicity. However, 2 – (1 – (coumarin – 3 – yl) ethylidene) hydrazinecarbothioamide showed the most potent inhibitory activity in all their homologues, and with the increase of the radius of halogen atom, the inhibitory activities decreased gradually. In addition, the inhibitory activities of the 2 – (1 – (coumarin – 3 – yl) ethylidene) hydrazinecarbothioamide and it's homologues were more potent than the inhibitory activities of 3 – acetylchroman – 2 – one and it's homologues. This result might be related to different inhibitory mechanism. Because of the 2 – (1 – (coumarin – 3 – yl) ethylidene) hydrazinecarbothioamide and it's homologues mainly depended on chelation of the sulfur atom with the active center of tyrosinase, the increase of the the radius of halogen atom might cause stereohindrance for the inhibitors approaching the active site of the enzyme. However, substituted 3 – acetylcoumarins are similar to benzaldehyde – type inhibitors, the tyrosinase inhibitory machanism of this type inhibitors come from ability to form a Schiff base with a primary amino group in the enzyme (Kubo et al., 1999). The increase of lipophilicity might be benefited to the formation of Schiff base.

As shown in table 1, coumarin – 3 – carboxylic acid and all of the substituted coumarin – 3 – carboxylic acids exhibited certain inhibitory activities against tyrosinase at the concentration of 200μmol/L. However some of the coumarin – 3 – carboxylic acid esters and the substituted coumarin – 3 – carboxylic acid esters lost their inhibitory activities against tyrosinase. The results indicated that the carboxyl group might be effective group to the interaction of compound with the active site of tyrosinase. From table 1, comparing the ethyl esters of coumarin – 3 – carboxylic acid and their substituted analogues, only the ethyl esters of coumarin – 3 – carboxylic acid with halogen atom showed weak inhibitory activities at the concentration of 200μmol/L. Among the methyl esters of coumarin – 3 – carboxylic acid, only the methyl coumarin – 3 – carboxylate exhibited inhibitory activity at the concentration of 200μmol/L. The isopropyl ester of coumarin – 3 – carboxylic acid and it's substituted analogues had certain inhibitory activity against tyrosinse, especially, isopropyl 6 – bromocoumarin – 3 – carboxylate with the IC_{50} value of 131.05μmol/L showed the most potent activity than the other ester compounds. For the pentyl ester compounds, the inhibitory percent of pentyl coumarin – 3 – carboxylate is 33.32% at the concentration of 200μmol/L, and pentyl 7 – hydroxycoumarin – 3 – carboxylate exhibited no inhibitory activity at the same concentration. These results suggested that the ester group structure might be an important factor for the improvement of inhibitory activity.

3.3 Inhibitory mechanism

The inhibitory mechanisms of the selected compounds4, 5 on mushroom tyrosinase for the oxidation of L – DOPA were determined. Figure 2, 3 showed the relationship between enzyme activity and concentration in the presence of different concentrations of the above – mentioned compounds. The results displayed that the plots of V versus $[E]$ gave a family of parallel straight lines with the same slopes. These results demonstrated that the inhibitory effects of compound 4, 5 on the tyrosinase were irreversible, suggesting that substituted 2 – (1 – (coumarin – 3 – yl) ethylidene) hydrazinecarbothioamide compounds effectively inhibited the enzyme by binding to its binuclear active site irreversibly. The result may be related to the structure of tyrosinase, within the structure, there are two copper ions in the active centre of tyrosinase and a lipophilic long – narrow gorge near to the active centre. Compound 4 or 5 could exhibit strong affinity for copper ions in the active centre and form a reversible non – covalent complex with the tyrosinase, and this then reacts to produce the covalently modified "dead - end complex" (Tsou, 1988).

The present investigation reported thatcoumarin derivatives had potent inhibitory effects on the diphenolase activity of mushroom tyrosinase. Interestingly, compound 4 was found to be the most potent inhibitor with IC_{50} value of 3.44μmol/L. Preliminary structure activity relationships (SARs) analysis indicated that (1) thiosemicarbzide moiety might play an important role in determining their inhibitory activities, because of the sulfur atom could chelate with the dicopper nucleus in the active site, and tyrosinase would lose its catalyzing ability after forming complex; (2) carboxyl group might be effective group to the interaction of compound with the active site of tyrosinase; (3) the ester group structure might be an important factor for the improvement of inhibitory activity. The inhibition mechanism analysis of compounds 4, 5 demonstrated that the inhibitory effects of the two compounds on the tyrosinase were irreversible.

Acknowledgment

This work wasfinancially supported by the Foundation of Education Department of Hunan Province, China (09B091), and the Key Subject Foundation of Shaoyang University (2008XK201).

References

Ashida, M., Brey, P., et al. (1995). Role of the integument in insect defense: pro - phenol oxidase cascade in the cuticular matrix. *Proceeding of the National Academy of Sciences of the United States of America*, *92*, 10698 - 10702.

Battaini, G., Monzani, E., Casella, L., Santagostini, L., et al. (2000). Inhibition of the catecholase activity of biomimetic dinuclear copper complexes by acid kojic. *Journal of Biological Inorganic Chemistry*, *5*, 262 - 268.

Belluti, F., Fontana, G., Bo, L., Carenini, N., Giommarelli, C., et al. (2010). Design, synthesis and anticancer activities of stilbene - coumarin hybrid compounds: Identification of novel proapoptotic agents. *Bioorganic and Medicinal Chemistry*, 18, 3543 - 3550.

Chen, Q. X., Liu, X. D., Huang, H., et al. (2003). Inactivation Kinetics of Mushroom Tyrosinase in the Dimethyl Sulfoxide Solution. *Biochemistry (Mosc)*, *68*, 644 - 649. Matsuura, R., Ukeda, H., Sawamura, M., et al. (2006). Tyrosinase inhibitory activity of citrus essential oils. *Journal of Agricultural Food Chemistry*, *54*,

2309 – 2313.

Chimenti, F., Bizzarri, B., Bolasco, A., Secci, D. (2010). Synthesis and Anti – Helicobacter pylori Activity of 4 – (Coumarin – 3 – yl) thiazol – 2 – ylhydrazone Derivatives. *Journal of Heterocyclic Chemistry*, 47, 1269 – 1274.

Chimenti, F., Bizzarri, B., Bolasco, A., et al. (2007). A novel class of selective anti – *Helicobacter pylori* agents 2 – oxo – 2*H* – chromene – 3 – carboxamide derivatives

Bioorganic and Medicinal Chemistry Letters, *17*, 3065 – 3071.

Chimenti, F., Secci, D., Bolasco, A., Chimenti, P., et al. (2009). Synthesis, Molecular Modeling, and Selective Inhibitory Activity against Human Monoamine Oxidases of 3 – Carboxamido – 7 – Substituted Coumarins. *Journal Medicinal Chemistry*, *52*, 1935 – 1942.

Curini, M., Cravotto, G., Epifano, F., Giannone, G., et al. (2006). Chemistry and biological activity of natural and synthetic prenyloxycoumarins. *Current Medicinal Chemistry*, 13, 199 – 222.

Gerdemann, C., Eicken, C., Krebs, B. (2002). The crystal structure of catechol oxidase: new insight into the function of type – 3 copper proteins. *Accounts of Chemical Research*, 35, 183 – 191.

Khode, S., Maddi, V., Aragade, P., et al. (2009). Synthesis and pharmacological evaluation of a novel series of 5 – (substituted) aryl – 3 – (3 – coumarinyl) – 1 – phenyl – 2 – pyrazolines as novel anti – inflammatory and analgesic agents. *European Journal Medicinal Chemistry*, 44, 1682 – 1688.

Kostova, I. (2006). coumarins as inhibitors of HIV reverse transcriptase. *Current HIV Research*, *4*, 347 – 363.

Kubo, I., Kinst – Hori, I., et al. (1999). Tyrosinase inhibitory activity of the olive oil flavor compounds. *Journal of Agricultural Food Chemistry*, *47*, 4574 – 4578

Ley, J. P., Bertram, H., et al. (2001). Hydroxy – or Methoxy – Substituted Benzaldoximes and Benzaldehyde – *O* – alkyloximes as Tyrosinase Inhibitors. *Bioorganic and Medicinal Chemistry*, *9*, 1879 – 1885.

Liu, J. B., Yi, W., Wan, Y. Q., Ma, L., Song, H. C. (2008). 1 – (1 – Arylethylidene) thiosemicarbazide derivatives: A new class of tyrosinase inhibitors. *Bioorganic and Medicinal Chemistry*, *14*, 1096 – 1102.

Neyts, J., De Clercq, E., Singha, R., Chang, Y. H., Das, A. R., Chakraborty, S. K., Hong, S. C., Tsay, S. – C., Hsu, M. – H., Hwu, J. R. (2009). Synthesis and Evaluation of Vancomycin Aglycon Analogues That Bear Modifications in the N – Terminal d – Leucyl Amino Acid. *Journal Medicinal Chemistry*, *52*, 1486 – 1490.

Ostrov, D. A., Hernάndez Prada, J. A., Corsino, P. E., Finton, K. A., Le, N., Rowe, T. C. (2007). Discovery of Novel DNA Gyrase Inhibitors by High – Throughput Virtual Screening. *Antimicrobial Agents and Chemotherapy*, *51*, 3688 – 3698.

Rai, U. S., Isloor, A. M., Shetty, P., et al. (2010). Novel chromeno [2,3 –*b*] – pyrimidine derivatives as potential anti – microbial agents. *European Journal Medicinal Chemistry*, *45*, 2695 – 2699.

Roussaki, M., Kontogiorgis, C., Hadjipavlou – Litina, D. J., Hamilakis, S., et al. (2010). A novel synthesis of 3 – aryl coumarins and evaluation of their antioxidant and lipoxygenase inhibitory activity. *Bioorganic and Medicinal Chemistry Letters*, 20, 3889 – 3892.

Schallreuter, K. U., Hasse, S., Rokos, H., Chavan, B., Shalbaf, M., Spencer, J. D.,

Wood, J. M. (2009). Cholesterol regulates melanogenesis in human epidermal melanocytes and melanoma cells. *Experimental Dermatology*, *18*, 680 – 688.

Song, K. K., Huang, H., Han, P., Zhang, C. L., Shi, Y., Chen, Q. X. (2006). Inhibitory effects of cis – and trans – isomers of 3,5 – dihydroxystilbene on the activity of mushroom tyrosinase. *Biochemical Biophysical Research Communication*, *342*, 1147 – 1151.

Thanigaimalai, P., Hoang, T. A. L., et al. (2010). Structural requirement (s) of *N* – phenylthioureas and benzaldehyde thiosemicarbazones as inhibitors of melanogenesis in melanoma B 16 cells. *Bioorganic and Medicinal Chemistry Letters*, *20*, 2991 – 2993.

Tripathi, R. K., Hearing, V. J., Urabe, K., Aroca, P., Spritz, R. A., et al. (1992). Mutational mapping of the catalytic activities of human tyrosinase. *Journal of Biological Chemistry*, *267*, 23707 – 23712.

Tsou, C. L., et al. (1988). Kinetics of substrate reaction during irreversible modification of enzyme activity. *Advances in Enzymology and Related Areas of Molecular Biology*, 61, 381 - 436.

Upadhyay, K. K., Mishra, R. K., et al. (2010). Zn^{2+} Specific Colorimetric Receptor Based on Coumarin. *Bulletin of the Chemical Society of Japan*, 83, 1211 -1215.

Wanare, G., Aher, R., Kawathekar, N., et al. (2010). Synthesis of novel α - pyranochalcones and pyrazoline derivatives as *Plasmodium falciparum* growth inhibitors. *Bioorganic and Medicinal Chemistry Letters*, *20*, 4675 -4678.

Wood, J. M., Decker, H., Hartmann, H., Chavan, B., Rokos, H., Spencer, J. D., et al. (2009). Senile hair graying: H_2O_2 - mediated oxidative stress affects human hair color by blunting methionine sulfoxide repair. *FASEB Journal*, *23*, 2065 -2075.

Yi, W., Cao, R., et al. (2010). Synthesis and biological evaluation of novel 4 - hydroxybenzaldehyde derivatives as tyrosinase inhibitors. *European Journal Medicinal Chemistry*, 45, 639 -646.

Zhu, Y. J., Zhou, H. T., Hu, Y. H., et al. (2011). Antityrosinase and antimicrobial activities of 2 - phenylethanol, 2 - phenylacetaldehyde and 2 - phenylacetic acid. *Food Chemistry*, *124*, 298 -302.

Figure captions

Figure 1 Chemical structure of known tyrosinase inhibitors

Figure 2 The effect of concentrations of tyrosinase on its activity for the catalysis of L - DOPA at different concentration of compound 4. The concentrations of compound 4 for curves 1 - 4 are 0, 25, 50 and 100 μmol/L, respectively.

Figure 3The effect of concentrations of tyrosinase on its activity for the catalysis of L - DOPA at different concentration of compound 5. The concentrations of compound 5 for curves 1 - 4 are 0, 25, 50 and 100 μmol/L, respectively.

Scheme 1

Synthesis of coumarin compounds bearing thiosemicarbazide moieties.

Reagents and conditions: I) Piperidine, Acetonitrile, RT, 1 - 4h. II) Acetic acid, Ethanol, Reflux, 6h.

1. R=H;   2. R=6-chloro;   3. R=6-bromo
4. R=H;   5. R=6-chloro;   6. R=6-bromo

Scheme 2

7. R = H;   8. R = 6-chloro;   9. R = 6-bromo;
10. R = 7-hydroxy;   11. R = H;   12. R = 6-chloro;
13. R = 6-bromo;   14. R = 7-hydroxy;
15. R = H, R' = methyl;   16. R = 6-bromo, R' = methyl;
17. R = 7-hydroxy, R' = methyl;   18. R = H, R' = pentyl;
19. R = 6-chloro, R' = isopropyl;   20. R = 6-bromo , R' = isopropy
21. R = 7-hydroxy, R' = isopropyl;   22. R = 7-hydroxy , R' = pentyl

Synthesis of the substituted 2 - oxo - 2H - coumarin - 3 - carboxylate esters.

Reagents and conditions: i) Piperidine, Ethanol, RT. ii) 10% NaOH /3 N HCl.

iii) thionyl chloride, Reflux. iv) Appropriate alcohol/ Toluene or ethyl ether, Reflux.

Table 1 Tyrosinase inhibitory activities and yields of the synthesized compounds

Compounds	Yield (%)	CLog*P* [d]	IC_{50} (μmol/L) [a]	percent of inhibition[c]
1	76%	0.9114	- -	8.3%
2	89.87%	1.7644	- -	42.45%
3	94.10%	1.9144	195.03	
4	85.38%	1.5506	3.44	
5	87.68%	2.4036	34.32	
6	89.15%	2.5536	114.68	
7	85.06%	1.91	- -	NA[b]
8	88.00%	2.623	- -	23.2%
9	67.45%	2.773	- -	12.88%
10	51.04%	2.17114	- -	NA[b]
11	69.76%	1.54175	- -	43.38%
12	93.18%	2.26375	- -	38.57%
13	67.45%	2.41375	- -	29.68%
14	93.58%	1.43245	- -	48.4%
15	89.65%	1.381	- -	33.33%
16	86.50%	2.244	- -	NA[b]
17	93.58%	1.64214	- -	NA[b]
18	53.28%	3.497	- -	33.32%
19	67.45%	2.932	- -	15.62%
20	69.75%	3.082	131.05	
21	61.36%	2.48014	- -	13.9%
22	56.18%	3.75814	- -	NA[b]
Kojic acid			23	

a. Values were determined from logarithmic concentration - inhibition curves (at least eight points) and are given as means of three experiments

b. not active at 200μmol/L concentration.

c. percent of inhibition of tyrosinase reaction at the 200μmol/L

d. Value of CLog*P* was obtained by ChemBioDraw Ultra 12.0

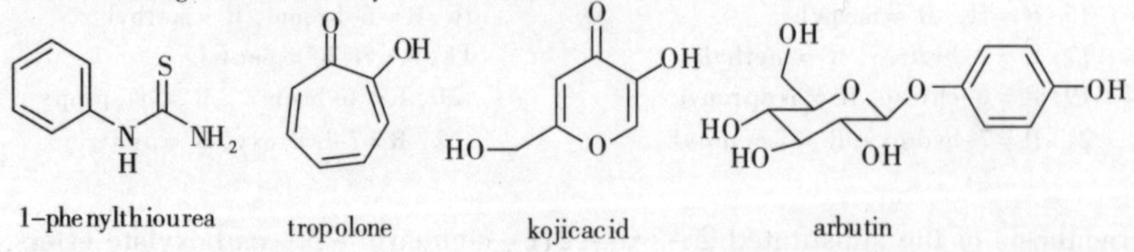

S N NH2 H O OH O OH HO— O OH O O OH HO HO OH

1-phenylthiourea tropolone kojicacid arbutin

Figure 1

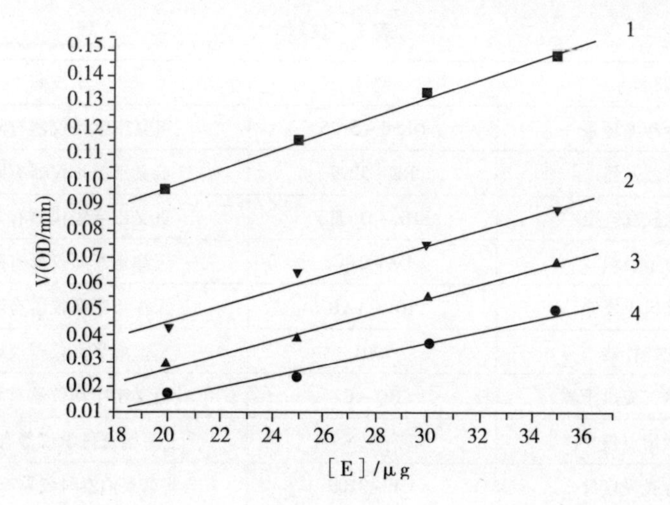

Figure 2

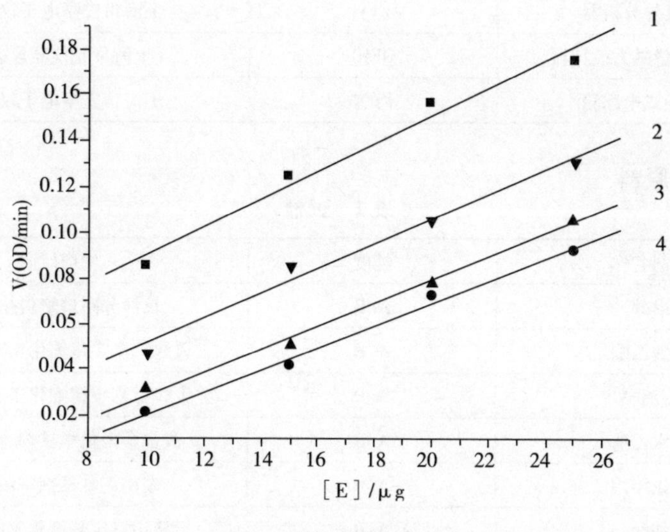

Figure 3

3.4.3　硕士陈昌红毕业论文部分内容

3.4.3.1　苯乙酸酯衍生物的合成研究

仪器与药品

仪器

表1　仪器

仪器名称	型号	生产商
低温冷却循环泵	DLSB－5/25	巩义市宇华仪器有限公司
旋转蒸发器	RE－5299	巩义市宇华仪器有限公司
循环水式真空泵	SHZ－D(Ⅲ)	巩义市宇华仪器有限公司
真空干燥箱	DZ－2BC	天津市泰斯仪器有限公司
电热鼓风干燥箱	101－1AB	天津市泰斯仪器有限公司
电磁搅拌器	7901	上海华光仪器仪表厂
玻璃仪器气流烘干器	RQ－C	巩义市宇华仪器有限公司
显微熔点仪	SGW® X－4	上海精密科学仪器有限公司
紫外分光光度计	UV－2100	北京莱伯泰科仪器有限公司
红外光谱分析仪	impact410	美国尼高力公司
三用紫外分析器	ZF_7	上海市安亭电子仪器厂
集热式恒稳磁力搅拌器	DF10	上海华光仪器仪表厂
海尔柜式电冰箱	海尔	上海市安亭电子仪器厂

试剂和原料

表2　试剂

试剂名称	纯度	生产厂商
甲醇	A. R	长沙分路口塑料化工厂
乙酸乙酯	A. R	天津市富宇精细化工有限公司
石油醚	A. R	天津市富宇精细化工有限公司
无水乙醇	A. R	成都金山化学试剂有限公司
二氯甲烷	A. R	湖南汇虹试剂有限公司
盐酸	A. R	天津市富宇精细化工有限公司
固体氢氧化钠	A. R	天津市永大化学试剂有限公司
甲氧胺盐酸盐	≥98%	上海达瑞精细化学品有限公司
氨基脲盐酸盐	≥98%	上海达瑞精细化学品有限公司
浓硫酸	≥98%	邵阳市方华化工有限公司
95%乙醇	A. R	湖南汇虹试剂有限公司
乙氧胺盐酸盐	≥98%	上海达瑞精细化学品有限公司
无水硫酸钠	A. R	上海强顺化学试剂有限公司
硫代氨基脲	≥98%	上海达瑞精细化学品有限公司

续表

试剂名称	纯度	生产厂商
苯乙酸甲酯	≥98%	上海达瑞精细化学品有限公司
对甲基苯乙酸	≥98%	上海达瑞精细化学品有限公司
对甲氧基苯乙酸	≥98%	上海达瑞精细化学品有限公司
对羟基苯乙酸甲酯	≥98%	上海达瑞精细化学品有限公司
对三氟甲基苯乙酸	≥98%	上海达瑞精细化学品有限公司
对二甲氨基吡啶	≥98%	上海达瑞精细化学品有限公司
环二己基碳二亚胺	≥98%	上海达瑞精细化学品有限公司

苯乙酸酯的合成

α－甲酰基苯乙酸甲酯的合成

合成路线：

$COOCH_3$　CHO　甲酸乙酯　室温搅拌　$COOCH_3$

150.17　178.18

将3g苯乙酸甲酯加入到100mL反应瓶中，加入3mL甲酸乙酯和50mL甲苯，加入0.55g金属钠，在室温下搅拌反应。反应12h后。金属钠完全消失，体系呈橙红色。加入5mL无水乙醇，搅拌20min后，加入60mL水，搅拌5min，分层。有机层以2×50mL水洗，合并水相，以3×25mL二氯甲烷洗。水层以稀盐酸酸化至pH＝3，以3×30mL二氯甲烷萃取，以2×30mL饱和碳酸氢钠溶液洗，无水硫酸钠干燥，减压蒸除溶剂，得2.352g橙红色油状物，产物记为C1。IR(cm－1，KBr)：2981，1733，1684，1653，1597，1450，1375，1168；1HNMR：7.47－7.26(m，5H，Ph－H)，6.40（d，1H，J＝7.2Hz，CH)，3.85(s，3H，CH_3)；13CNMR：175.0，167.3，129.3，128.6，128.2，128.1，61.1，52.2。

α－羟甲基苯乙酸甲酯的合成

合成路线：

OH　CHO　$COOCH_3$　硼氢化钠　冰浴　$COOCH_3$

178.18　180.2

将2.352g α－甲酰基苯乙酸甲酯加入到100mL反应瓶中。加入70mL甲醇搅拌、冰浴冷却。加入硼氢化钠，加入瞬间产生大量气泡。加完体系呈淡黄色，保温反应10h。减压蒸除甲醇。加入60mL二氯甲烷，充分搅拌。分别经20mL水洗，20mL饱和食盐水洗，无水碳酸钠干燥。减压蒸除溶剂，剩余物柱层析，得到产物1.756g，产物记为C2。IR(cm－1，KBr)：2972，2950，2879，1732，1601，1454，1249，1163，1048. 1HNMR：7.35－7.25(m，5H，Ph－H)，4.14－4.09(m，2H，CH_2)，3.86－3.82(m，1H，CH)，3.91 (s，3H)，2.86(s，1H，OH)；13CNMR：173.6，135.7，128.8，128.2，127.7，64.6，54.2，52.1。

α－甲酰基苯乙酸酯缩氨基硫胺席夫碱的合成

合成路线：

CHO; $COOCH_3$ —乙醇/回流→ $HC{=}N{-}NHCSNH2$; $COOCH_3$

178.18　　251.3

将3.1g α－甲酰基苯乙酸甲酯置于50mL反应瓶当中，加入30mL无水乙醇，加入1.6g硫代氨基脲，在80℃下回流反应。反应3h后，减压蒸除溶剂。剩余物用15mL 95%的乙醇重结晶，抽滤，滤饼，自然干燥。得到白色固体3.38g，MP＝124～126℃，产物记为C3。1HNMR：10.134(d，J＝13.6Hz，1H，＝CH)，7.63～7.28 (m，6H，Ph－H，NH)，7.02(s，1H，NH)，6.67(s，1H，NH)，4.50(dd，J＝7.2Hz，2.4Hz，1H)，4.23－3.74(CH_3，3H). 13CNMR：178.6，170.7，144.1，143.8，129.1，128.2，128.1，61.8，52.7. IR(cm－1，KBr)：2971，1724，1641，1600，1586，1500，1452，1083。

α－羟甲基苯乙酸甲酯苯甲酸酯的合成

合成路线：

CH_2OH; $COOCH_3$ —DCC/DMAP/低温→ CH_2OCOPh; $COOCH_3$

180.2　　284.31

将1.0g α－羟甲基苯乙酸酯，1.13g环二己基碳二亚胺(DCC)，0.67g苯甲酸，0.02g对二甲氨基吡啶(DMAP)加入到反应瓶当中，加入50mL二氯甲烷，搅拌，于－10℃保温反应。反应10h后，抽滤。滤液减压蒸馏，剩余物柱层析分离(硅胶100～200目)，洗脱剂乙酸乙酯：石油醚＝1：4，得0.8g淡黄色黏稠物体，产物记为 C4。[1] HNMR：7.99(d, J =8.4Hz, 1H, Ph－H), 7.56(t, 1H, J = 7.6Hz, Ph－H), 7.43－7.31 (m, 7H, Ph－H), 4.86(t, J =9.2Hz, 1H, CH), 4.62－4.58(m, CH_2, 1H), 4.21－4.08 (m, CH_2 1H), 3.71(s, 2H, CH_3). [13] CNMR：171.96, 166.07, 135.03, 133.0, 129.56, 128.89, 128.84, 128.32, 128.12, 128.01, 65.63, 52.27, 50.88。

α－羟甲基苯乙酸甲酯乙酰水杨酸酯的合成

合成路线：

CH_2OH / $COOCH_3$ 180.2 —DCC/DMAP, 低温→ CH_2OCO / $OCOCH_3$ / $COOCH_3$ 342.34

将1.017g α－羟甲基苯乙酸酯，1.3g 环二己基碳二亚胺(DCC)，1.017g 乙酰水杨酸，0.021g 对二甲氨基吡啶(DMAP)到反应瓶当中，加入60mL 二氯甲烷，搅拌，与－10℃保温反应。反应7h 后，抽滤。滤液减压蒸馏。剩余物柱层析分离(硅胶100～200目)，洗脱剂乙酸乙酯：石油醚＝1：3，得1.152g 无色油状物，产物记为 C5。1HNMR：7.94(d, J＝10.8Hz, 1H, Ph－H), 7.59(t, J＝7.6Hz, 1H, Ph－H), 7.39－7.28(m, 6H, Ph－H), 7.21(d, J＝8.4Hz, 1H, Ph－H), 5.30, 3.73(s, 3H, CH_3), 4.84(t, 1H, J＝9.2Hz, CH), 4.24－4.07(m, 2H, CH2), 2.35(s, 3H, CH_3). 13CNMR：171.9, 169.6, 164.0, 150.8, 134.9, 134.0, 131.8, 129.2, 128.9, 128.1, 126.0, 123.9, 122.9, 65.7, 52.3, 50.6, 21.0. IR(cm^{-1}, KBr)：2943, 2898, 1733, 1695, 1652, 1601, 1505, 1456。

α－羟甲基苯乙酸甲酯3,4－二甲氧基肉桂酸酯的合成

合成路线：

CH_2OH / $COOCH_3$ 180.2 —DCC/DMAP, 低温→ $CH_2OCCHCH$ (C=O) / $COOCH_3$ 370.4

将1.8g α－羟甲基苯乙酸酯，2.3g 环二己基碳二亚胺(DCC)，2.08g3,4－二甲氧基肉桂酸，0.037g 对二甲氨基吡啶(DMAP)加入到反应瓶当中，加入80mL

二氯甲烷,搅拌,与 - 10℃保温反应。反应4h后,抽滤。滤液减压蒸馏,剩余物柱层析分离(硅胶100~200目),洗脱剂乙酸乙酯:石油醚=1:3,得1.822g淡黄色黏稠物体,产物记为C6。1HNMR:7.62(d, J=16Hz, 1H, =CH),7.35-7.29(m, 5H, Ph-H), 7.08(d, 1H, J=8.4Hz, Ph-H), 7.02(s, 1H, Ph-H), 6.85(d, J=8.4Hz, 1H, Ph-H), 6.29(d, J=16Hz, =CH, 1H), 4.74(t, J=9.6Hz, CH, 1H), 4.49(dd, J=5.6Hz, 9.6Hz, 1H, CH_2), 4.05(dd, J=5.6Hz, 9.6Hz, 1H, CH_2), 3.89(s, 6H, OCH_3), 3.71(s, 3H, CH_3)。13CNMR:171.5, 166.7, 151.1, 149.1, 145.2, 135.0, 128.8, 128.0, 127.8, 127.2, 122.7, 115.1, 110.9, 109.5, 65.0, 61.1, 55.9, 50.9. IR(cm^{-1},KBr):2966, 2845, 1711, 1693, 1631, 1598, 1555, 1467, 1385, 852, 816, 762。

α-甲酰基对甲基苯乙酸甲酯的合成

合成路线:

COOH 浓硫酸 甲醇 150.17 → $COOCH_3$ 164.2 $CHOOCH_2CH_3$ 室温搅拌 → CHO $COOCH_3$ 192.21

将10.0g对甲基苯乙酸加入到反应瓶中,加入40mL无水甲醇,加入3mL浓硫酸,室温搅拌7h后,减压蒸除溶剂。剩余物加入40mL水,搅拌,3×20mL二氯甲烷萃取,30mL饱和氯化钠溶液洗涤,无水硫酸钠干燥,得无色液体9.4g。称取3.28g对甲基苯乙酸甲酯加入到100mL的反应瓶当中,加入2.5mL甲酸乙酯和40mL甲苯,加入0.5g金属钠。在RT下搅拌保温反应。21h,金属钠完全消失。加入5mL无水乙醇,搅拌10min,加入40mL水,搅拌5min,分层。有机层以2×30mL水洗,合并水相,以2×25mL二氯甲烷洗。水层以稀盐酸酸化至pH=3,以3×25mL二氯甲烷萃取,合并有机相。以2×30mL饱和NaHCO3洗,无水硫酸钠干燥。减压蒸出溶剂,得1.77g淡红色液体,产物记为C7。^{1}HNMR:7.89-7.92(d, 1H, *J* =6.3Hz, -CHO), 7.14-7.32(m, 4H, Ph-H), 4.41-4.48(q, 1H, *J* =7.2Hz, $-CH_2$), 4.1-4.17(q, 2H, *J* =7.2Hz, -CH), 2.34(s, 3H, CH_3), 1.4-1.44(t, 3H, *J* =7.2Hz, CH_3). IR(cm^{-1}, KBr):2959, 2927, 2866, 1729, 1661, 1601, 1511, 1444, 819, 717。

α-羟甲基对甲基苯乙酸甲酯的合成

合成路线:

OH

CHO

COOCH$_3$　硼氢化钠　COOCH$_3$

冰浴

192.21　194.23

将2.7g α－甲酰基苯乙酸甲酯加入到100mL反应瓶当中,加入50mL甲醇溶解,冰浴冷却至－10℃。分批加入0.6g硼氢化钠,加入瞬间产生大量的气泡。保温,搅拌反应。5.5 h后,蒸出甲醇。剩余物加入40mL水,以2×30mL二氯甲烷萃取,2×25mL饱和食盐水洗,无水硫酸钠干燥,减压蒸出溶剂,得橙黄色液体2.15g,产物记为C8。^{1}HNMR:7.15－7.12(m, 4H, Ph－H), 4.09－4.2(m, 3H, 2CH$_2$,CH), 3.71－3.86(m, 2H, CH$_2$), 2.334(s, 3H, CH$_3$), 1.73－1.74(s, 1H,OH), 1.19－1.24(t, 3H, J＝7.2Hz, CH$_3$).IR(cm^{-1}, KBr):3419, 2978, 2932, 2875, 1731, 1513, 1457, 1371, 1340, 1249, 1175, 1041, 817。

α－甲酰基对甲基苯乙酸甲酯缩氨基硫脲席夫碱的合成

合成路线:

CHO　HC=N-NHCSNH2

COOCH$_3$　乙醇　COOCH$_3$

回流

192.21　265.33

将1.11g α－甲酰基对甲基苯乙酸甲酯和0.6g氨基硫脲置于100mL反应瓶当中,加入30mL无水乙醇,在常温下搅拌2min.置于56℃回流反应,原料在加热过程中,全部溶解。76min后,蒸出溶剂。用20mL 95%的乙醇重结晶,得到1.25g白色晶体,产物记为C9。1HNMR:9.12(d, J＝6.3Hz, 1H, ＝CH), 7.16－7.48(m, 4H, Ph－H), 6.27(s, 1H, NH), 4.21－4.25 (m, 2H, CH$_2$), 3.77 (m, 1H, CH), 2.36－2.37(m, 3H, CH$_3$), 1.62(s, 2H, NH$_2$), 1.26－1.29 (m, CH$_3$, 3H). IR (cm－1,KBr):1721, 1707, 1599, 1517, 1463, 1363, 1291, 1201, 1110。

α－甲酰基对甲氧基苯乙酸甲酯的合成

合成路线:

称取7.8g对甲氧基苯乙酸甲酯加入到100mL反应瓶当中,加入8mL甲酸乙酯和50mL甲苯,加入1.15g金属钠。在RT下搅拌保温反应。体系逐渐由土黄

CHO
COOCH3 甲酸乙酯 室温搅拌 COOCH3
O O
180.2 208.21

色变成淡棕黄色。反应4.5h后,金属钠完全消失。加入10mL无水乙醇,搅拌10min,加入40mL水,搅拌5min,体系呈乳浊液,分层。有机层以2×30mL水洗,合并水相,以3×20mL二氯甲烷洗。水相以稀盐酸酸化至pH=3,以3×20mL二氯甲烷萃取,合并有机相。无水硫酸钠干燥,减压蒸出溶剂,得金棕色液体5g,产物记为C10。1HNMR:7.99-8.0(d, J=2.1Hz, 1H, -CHO),6.84-7.26(m, 4H, Ph-H),4.4-4.47(q, 2H, J=6.9Hz, $-CH_2$),4.10-4.17(q, 1H, J=7.2Hz, CH), 3.79(s, 3H, $-CH_3$), 3.9(s, 3H, $-CH_3$)。IR(cm-1, KBr): 2980, 2932, 2902, 2836, 1732, 1652, 1610, 1513, 1245, 1175, 1032, 834。

α-羟甲基对甲氧基苯乙酸甲酯的合成

合成路线:

OH
CHO
COOCH3 硼氢化钠 冰浴 COOCH3
O O
208.2 210.23

将3g α-甲酰基对甲氧基苯乙酸甲酯加入到100mL反应瓶当中,加入40mL甲醇溶解,冰浴冷却至-10℃。分批加入0.7g硼氢化钠,加入瞬间产生大量的气泡。保温,搅拌。10 h后,蒸出甲醇。剩余物加入40mL水,以2×30mL二氯甲烷萃取,2×25mL饱和食盐水洗,无水硫酸钠干燥,减压蒸出溶剂,得棕色液体1.33g,产物记为C11。1HNMR:7.17-7.26(q, 2H, J=2.1Hz, Ph-H), 6.85-6.88(q, 2H, J=2.1Hz,Ph-H), 4.05-4.22 (m, 3H, -CH, $-CH_2$), 3.76-3.8(m, 5H, CH_2, $-OCH_3$), 2.06(d, 1H, 5.1Hz, OH), 1.20-1.25(t, 6H, J=1.2Hz, $-CH_3$)。IR(cm-1, KBr):3419,2978,2929,2881,1731,1513,1456,1249,1175,1041。

α-羟甲基对甲氧基苯乙酸甲酯苯乙酸酯的合成

合成路线:

将0.49g对甲氧基苯乙酸,0.54g α-羟甲基对甲氧基苯乙酸甲酯,0.7g环二己基碳二亚胺(DCC),0.01g对二甲氨基吡啶(DMAP)置于100mL反应瓶当

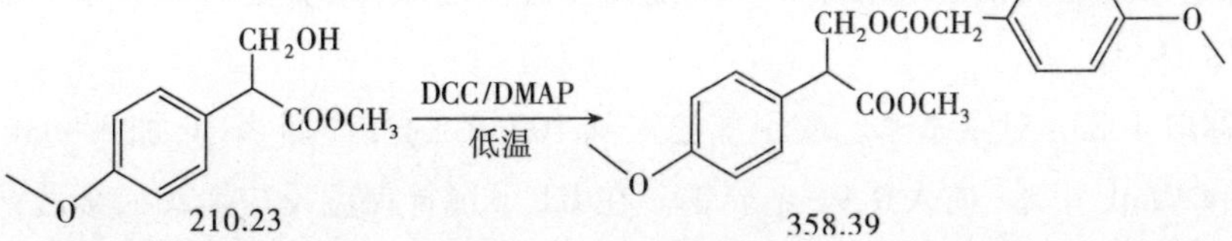

中,加入30mL二氯甲烷。于-10℃下保温搅拌反应。11h后,过滤,蒸除溶剂。剩余物柱层析(乙酸乙酯：石油醚=1：5),得到青色透明黏液0.39g,产物记为C12。^{1}HNMR:6.86-6.89,7.08-7.10(m, 4H, Ph-H), 7.26-7.29(m, 4H, Ph-H), 4.01-4.20(m, 3H, CH, CH_2), 4.47-4.57, 4.74-4.78(m, 2H, CH_2), 3.70(s, 6H, 2-OCH_3), 3.46(m, 2H, CH_2), 1.19-1.22(d, 3H, J = 7.2Hz, CH_3). IR(cm^{-1}, KBr):2933, 2851, 1769, 1743, 1709, 1650, 1608, 1513, 1451, 1368, 1246, 1133, 1078, 1033。

α-甲酰基对甲氧基苯乙酸甲酯缩氨基硫脲席夫碱的合成

合成路线：

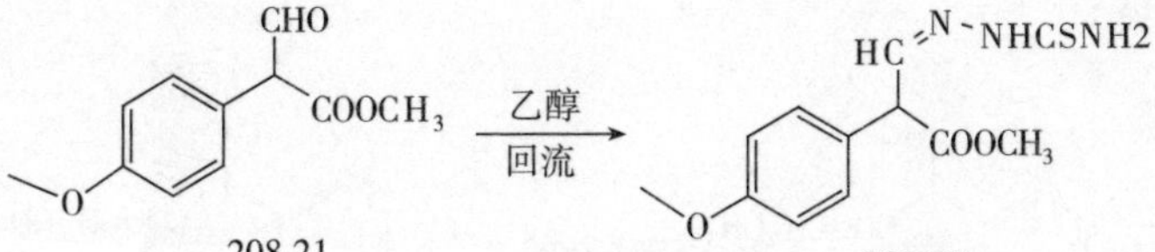

将0.62g α-甲酰基对甲氧基苯乙酸甲酯和0.27g氨基硫脲置于50mL反应瓶当中,加入15mL无水乙醇,在常温下搅拌2min。置于70℃回流反应,原料在加热过程中,全部溶解。3h后,蒸出溶剂。用8mL 90%的乙醇溶液洗涤,干燥。得到0.4g白色固体,MP=141℃-142℃,产物记为C13。^{1}HNMR:9.74(m, 1H, CH), 7.5(m, 1H, CH), 7.2-7.28(m, 2H, Ph-H), 6.9-6.92(m, 2H, Ph-H), 6.36(s, 1H, NH), 4.21-4.44(m, 2H, CH_2), 3.76-3.83(m, 3H, CH_3), 1.675(s, 2H, NH_2), 1.26-1.29(t, 3H, J =5.2Hz, CH_3)。

α-甲酰基对氟基苯乙酸乙酯的合成

合成路线：

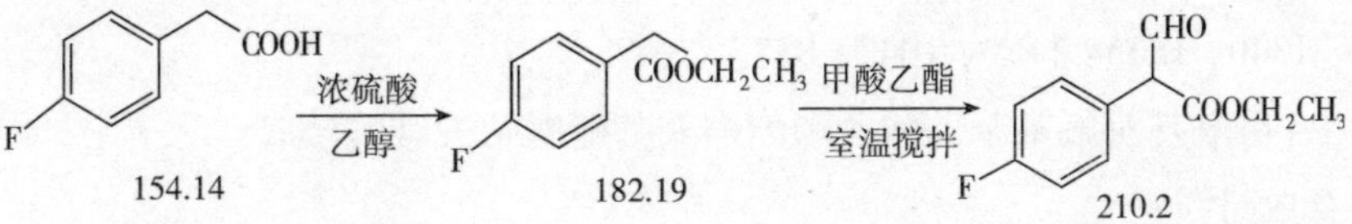

加入15.4g对氟苯乙酸置于反应瓶当中,加入40mL无水乙醇,6mL浓硫酸,室温搅拌反应10h,蒸除大部分溶剂,加入40mL水,用3×20mL二氯甲烷萃取,

20mL 饱和碳酸氢钠溶液洗涤,30mL 饱和食盐水洗涤,无水硫酸钠干燥。得到无水液体 18.4g。

称取 4.83g 对氟基苯乙酸甲酯加入到 100mL 的反应瓶当中,加入 4mL 甲酸乙酯和 40mL 甲苯,加入 0.9g 金属钠。在 RT 下搅拌保温反应。20h,金属钠完全消失。加入 5mL 无水乙醇,搅拌 10min,加入 40mL 水,搅拌 5min,分液。有机层以 2×30mL 水洗,合并水相,以 2×25mL 二氯甲烷洗。水层以稀盐酸酸化至 pH =2,瓶底出现黄色油状物。用 2×20mL 二氯甲烷萃取,合并有机相。减压蒸出溶剂,柱层析,得 3.8g 红色液体,产物记为 C14。^{1}HNMR:8.39 - 8.43(d, 1H, J = 5.4Hz, CHO), 7.16 - 7.26(m, 4H, Ph - H), 4.42 - 4.49(q, 2H, J = 7.2Hz, CH_2), 3.76 - 3.98(m, 1H, CH), 1.4 - 1.45(t, 3H, J = 7.2Hz, CH_3). IR (cm^{-1}, KBr): 2984, 2950, 1732, 1660, 1510, 1444, 1418, 1375, 1340, 1275, 1224, 1157, 1023, 838。

α - 羟甲基对氟基苯乙酸乙酯的合成

合成路线:

CHO / $COOCH_2CH_3$ / F ——硼氧化钠 / 冰浴——> OH / $COOCH_2CH_3$ / F

210.2 212.23

将 4.36g α - 甲酰基对氟基苯乙酸乙酯加入到 100mL 反应瓶当中,加入 40mL 甲醇溶解,冰浴冷却至 -10℃。分批加入 0.86g 硼氢化钠,加入瞬间产生大量的气泡。保温,搅拌反应。9h 后,蒸出甲醇。剩余物加入 40mL 水,以 2×30mL 二氯甲烷萃取,2×25mL 饱和食盐水洗,无水硫酸钠干燥,减压蒸出溶剂,得无色透明液体 2.65g,产物记为 C15。1HNMR:7.00 - 7.05(t, 2H, J = 8.4Hz, 9Hz, Ph - H), 7.22 - 7.27(q, 2H, J = 5.4Hz, Ph - H), 4.05 - 4.22(m, 3H, CH, CH_2), 3.71 - 3.84(m, 2H, CH_2), 2.12(s, 1H, OH), 1.2 - 1.24(t, 3H, J = 7.2Hz, CH_3); IR(cm - 1, KBr): 3398, 2977, 2952, 2881, 1731, 1604, 1509, 1372, 1340, 1225, 1176, 1042, 837。

α - 甲酰基对氟基苯乙酸乙酯酯缩氨基硫脲席夫碱的合成

合成路线:

将 0.67g α - 甲酰基对氟基苯乙酸乙酯和 0.36g 氨基硫脲置于 100mL 反应瓶当中,加入 30mL 无水乙醇,在常温下搅拌 2min。置于 56℃ 回流反应,原料在

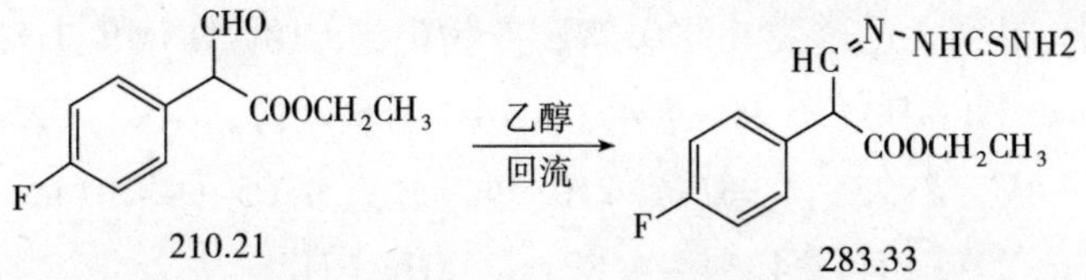

CHO COOCH$_2$CH$_3$ F 210.21 —乙醇 回流→ HC=N-NHCSNH2 COOCH$_2$CH$_3$ F 283.33

加热过程中,全部溶解。3h 后,蒸出溶剂。柱层析,得到 0.82g 白色固体,产物记为 C16。1HNMR：7.48 – 7.50(d, 1H, J = 6.9Hz, CH), 7.23 – 7.27(m, 2H, Ph – H), 7.04 – 7.09 (m, 2H, Ph – H), 6.34(s, 1H, NH), 4.45 – 4.5(t, 1H, J = 7.2Hz, CH), 4.19 – 4.23(q, 2H, J = 4.2Hz, CH$_2$), 1.64(s, 2H, NH$_2$), 1.23 (t, 3H, J = 7.2Hz, CH$_3$)。

α – 甲酰基对氟基苯乙酸乙酯缩甲氧胺席夫碱的合成

合成路线：

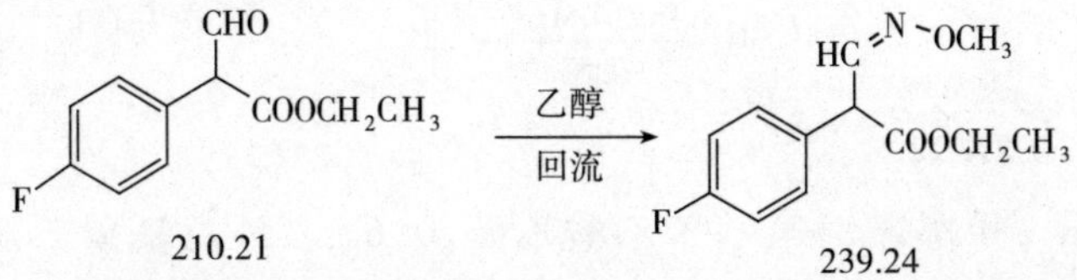

CHO COOCH$_2$CH$_3$ F 210.21 —乙醇 回流→ HC=N-OCH$_3$ COOCH$_2$CH$_3$ F 239.24

将 0.77g α – 甲酰基对氟基苯乙酸甲酯和 0.33g 甲氧胺盐酸盐置于 100mL 反应瓶当中,加入 0.16g 氢氧化钠,加入 30mL 无水乙醇,在常温下搅拌反应 10h。50℃反应 2h 后,蒸出溶剂,柱层析,得到透明液体 0.65g,产物记为 C17。^{1}HNMR：7.64 – 7.66(d, 1H, J = 8.1Hz, CH), 7.03 – 7.30(m, 4H, Ph – H), 4.15 – 4.23(m, 2H, CH$_2$), 4.44 – 4.47, 4.9 – 5.0(dd, J = 8.1Hz, 2.4Hz, 1H, CH), 3.85 – 3.88(d, 3H, J = 11.4Hz, CH$_3$), 1.21 – 1.27(q, 3H, J = 5.7Hz, CH$_3$). IR (cm^{-1}, KBr)：2979, 2940, 2901, 1737, 1603, 1509, 1464, 1437, 1330, 1227, 1157, 1044, 839。

α – 羟甲基对氟基苯乙酸乙酯苯甲酸酯的合成

合成路线：

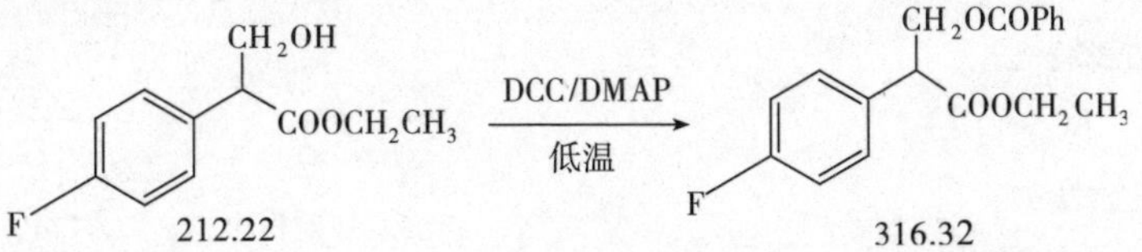

CH$_2$OH COOCH$_2$CH$_3$ F 212.22 —DCC/DMAP 低温→ CH$_2$OCOPh COOCH$_2$CH$_3$ F 316.32

将 0.36g 苯甲酸, 0.6g 环二己基碳二亚胺(DCC), 0.6g α – 羟甲基对氟基苯乙酸乙酯,置于 100mL 反应瓶当中。加入催化量对二甲氨基吡啶(DMAP),加入 30mL 二氯甲烷。于 – 10℃下保温搅拌反应。10h 后,抽滤。蒸出滤液中的溶剂。

剩余物柱层析，得到无色透明液体0.27g，产物记为C18。^{1}HNMR：7.02－7.08（t，2H，J ＝8.7Hz，Ph－H），7.34－7.45（m，4H，Ph－H），7.54－7.56（m，1H，J ＝1.2Hz，Ph－H），7.95－7.98（m，2H，Ph－H），4.05－4.10（m，2H，CH_2），4.13－4.21（m，1H，CH），4.54－4.82（m，2H，CH_2），1.2－1.24（t，3H，J ＝6.9Hz，CH_3）。IR（cm^{-1}，KBr）：2928，1724，1712，1601，1510，1274，1225，1175，1158，1109，1069，1025。

α－羟甲基基对氟基苯乙酸乙酯乙酰水杨酸酯的合成

合成路线：

CH_2OH / $COOCH_2CH_3$ / F（212.22） $\xrightarrow[\text{低温}]{\text{DCC/DMAP}}$ CH_2OCO / $OCOCH_3$ / $COOCH_2CH_3$ / F（374.36）

将0.6g α－甲酰基对氟基苯乙酸乙酯，0.6g 二环乙基碳二亚胺（DCC），0.54g阿司匹林和0.01g 4－二甲氨基吡啶（DMAP），加入到100mL 反应瓶当中，加入30mL 二氯甲烷，搅拌，于－10℃下保温反应。13h 后停止反应。抽滤，滤液减压蒸馏。剩余物柱层析分离（硅胶 100～200 目，洗脱剂 乙酸乙酯：石油醚＝1：3）。得透明黏液0.1g，产物记为C19。^{1}HNMR：7.02－7.11（m，3h，Ph－H），7.26－7.33（m，3H，Ph－H），7.53－7.57（q，2H，J ＝1.8Hz，Ph－H），7.87－7.91（q，1H，J ＝4.5Hz，Ph－H），4.49－4.55，4.71－4.78（qq，2H，J ＝6Hz，9.3Hz，CH_2），4.16－4.20（t，1H，J ＝6.9Hz，CH_2），4.02－4.05（m，1H，CH），2.336（s，3H，CH_3），1.19（t，3H，J ＝7.2Hz，CH_3）. IR（cm^{-1}，KBr）：2980，2932，2854，1769，1732，1700，1650，1605，1538，1509，1450，1368，1224，1197，1078，1040，1014。

α－羟甲基对氟基苯乙酸乙酯3，4－二甲氧基肉桂酸酯的合成

合成路线：

CH_2OH / $COOCH_2CH_3$ / F（212.22） $\xrightarrow[\text{低温}]{\text{DCC/DMAP}}$ $CH_2OCCHCH$（O） / $COOCH_2CH_3$ / F（402.41）

将0.6g α－羟甲基对氟基苯乙酸乙酯，0.62g 3，4－二甲氧基肉桂酸，0.6g环二己基碳二亚胺（DCC），催化量对二甲氨基吡啶（DMAP）加入到100mL反应瓶当中，加入50mL二氯甲烷，于－10℃下保温搅拌反应。16h后停止反应。抽滤，滤液减压蒸馏除去溶剂。剩余物柱层析，乙酸乙酯：石油醚＝1：3的洗脱剂。得到白色固体0.78g，产物记为C20。^{1}HNMR：7.58－7.63（d，1H，J＝15.9Hz，Ph－H），7.31－7.34（m，2H，Ph－H），7.02－7.08（m，4H，Ph－H），6.85－6.88（d，1H，J＝8.4Hz，＝CH），6.24－6.29（d，1H，J＝15.9Hz，＝CH），3.92（s，6H，2CH_3），4.00－4.03（m，1H，CH），4.16－4.21（m，2H，CH_2），4.43－4.49，4.64－4.71（qq，2H，J＝6Hz，12hz，CH_2），1.21－1.26（t，3H，J＝7.2Hz，CH_3）．IR（cm^{-1}，KBr）：3241，3155，2971，1726，1595，1507，1465，1363，1302，1221，1199，1159，1092。

α－甲酰基对三氟甲基苯乙酸乙酯的合成

合成路线：

F_3C–C$_6$H$_4$–CH$_2$COOH（204.15） —浓硫酸/乙醇→ —甲酸乙酯/室温搅拌→ F_3C–C$_6$H$_4$–CH(CHO)$COOCH_2CH_3$（260.3）

加入7g对三氟甲基苯乙酸到反应瓶当中，加入35mL无水乙醇，加入4mL浓硫酸，在室温下搅拌反应。12h后，蒸除部分溶剂，加入30mL水。用3×20mL二氯甲烷萃取，2×15mL饱和碳酸氢钠溶液洗涤，2×20mL饱和氯化钠溶液洗涤，无水硫酸钠干燥，得到金黄色液体6.57g。

称取2.33g对三氟甲基苯乙酸乙酯加入到100mL的反应瓶当中，加入4mL甲酸乙酯和20mL甲苯，加入0.4g金属钠。在RT下搅拌保温反应。10h后，金属钠完全消失。加入30mL水，搅拌3min，分出有机层。有机层以2×20mL水洗，合并水相，以2×15mL二氯甲烷洗。水层以稀盐酸酸化至pH＝3，以3×20mL二氯甲烷萃取，合并有机相。以2×15mL饱和碳酸氢钠溶液洗，2×15mL氯化钠溶液洗涤，无水硫酸钠干燥。得红色液体1.2g，产物记为C21。^{1}HNMR：8.15－8.18（d，1H，J＝8.1Hz，CHO），7.39－7.82（m，4H，Ph－H），4.25－4.51（q，2H，J＝7.2Hz，CH_2），4.12－4.27（m，1H，CH），1.42－1.46（t，3H，J＝7.2Hz，CH_3）．IR（cm^{-1}，KBr）2985，2938，2875，1736，1662，1615，1467，1446，1420，1395，1325，1278，1180，1125，1069，1020，884。

α－羟甲基对三氟甲基苯乙酸乙酯的合成

合成路线：

$$\text{F}_3\text{C-C}_6\text{H}_4\text{-CH(CHO)COOCH}_2\text{CH}_3 \ (260.2) \xrightarrow[\text{冰浴}]{\text{硼氢化钠}} \text{F}_3\text{C-C}_6\text{H}_4\text{-CH(CH}_2\text{OH)COOCH}_2\text{CH}_3 \ (262.23)$$

将3.7g α－甲酰基对三氟甲基苯乙酸乙酯加入到100mL反应瓶当中，加入40mL甲醇溶解，冰浴冷却至－10℃。分批加入0.68g硼氢化钠，加入瞬间产生大量的气泡。保温，搅拌反应。21h后，蒸出甲醇。剩余物加入50mL水，以2×30mL二氯甲烷萃取，2×20mL饱和碳酸氢钠溶于洗涤，柱层析(先用乙酸乙酯：比石油醚＝1：4，后用乙酸乙酯：石油醚＝1：1洗涤)得到淡黄色液体2.2g，产物记为C22。^{1}HNMR：7.39－7.41，7.59－7.62(dd，4H，J＝8.1Hz，7.8Hz，Ph－H)，4.10－4.22(m，3H，CH_2，CH)，3.84－3.92(m，2H，CH_2)，3.73(s，1H，OH)，1.21－1.25(t，3H，J＝7.2Hz，CH_3)．IR(cm^{-1}，KBr) 3444，2986，2955，2887，1735，1619，1437，1420，1731，1420，1372，1328，1166，1125。

α－羟甲基对三氟甲基苯乙酸乙酯苯甲酸酯的合成

合成路线：

$$\text{F}_3\text{C-C}_6\text{H}_4\text{-CH(CH}_2\text{OH)COOCH}_2\text{CH}_3 \ (262.23) \xrightarrow[\text{低温}]{\text{DCC/DMAP}} \text{F}_3\text{C-C}_6\text{H}_4\text{-CH(CH}_2\text{OCOPh)COOCH}_2\text{CH}_3 \ (366.33)$$

将0.4g苯甲酸，0.64g环二已基碳二亚胺(DCC)，0.75g α－羟甲基对三氟甲基苯乙酸乙酯，置于100mL反应瓶当中。加入催化量二甲氨基吡啶(DMAP)，加入40mL二氯甲烷。于－10℃下保温搅拌反应。20h后，抽滤，蒸出滤液中的溶剂。剩余物柱层析，得到淡黄色液体0.16g，产物记为C23。^{1}HNMR：7.94－7.97(t，2H，J＝7.2Hz，Ph－H)，7.64－7.61(m，2H，Ph－H)，7.56－7.5(m，2H，Ph－H)，7.45－7.40(m，3H，Ph－H)，4.14－4.23(m，3H，CH，CH_2)，4.59－4.56，4.80－4.87(m，2H，CH_2)，1.20－1.25(t，3H，J＝7.2Hz，CH_3)．

α－羟甲基对三氟甲基苯乙酸乙酯乙酰水杨酸酯的合成

合成路线：

将0.75g α－羟甲基对三氟苯乙酸乙酯，0.64g二环乙基碳二亚胺(DCC)，0.54g阿司匹林和0.015g 4－二甲氨基吡啶(DMAP)，加入到100mL反应瓶当中，

CH_2OH $COOCH_2CH_3$ F_3C 262.23 —DCC/DMAP, 低温→ CH_2OCO $OCOCH_3$ $COOCH_2CH_3$ F_3C 424.37

加入35mL二氯甲烷，搅拌，于-10℃下保温反应。20h后停止反应。抽滤，滤液减压蒸馏。剩余物柱层析分离（硅胶100～200目，洗脱剂 乙酸乙酯：石油醚=1：4）。得透明黏液0.16g，产物记为C24。^{1}HNMR：7.89-7.86（q，2H，J=1.2Hz，Ph-H），7.08-7.11（d，2H，J=8.4Hz，Ph-H），7.26-7.31（m，1H，Ph-H），7.47-7.39（m，1H，Ph-H），7.61-7.48（m，1H，Ph-H），7.86-7.89（m，1H，Ph-H），4.11-4.22（m，3H，CH，CH_2），4.54-4.59，4.76-4.82（m，2H，CH_2），2.32（s，3H，CH_3），1.20-1.25（t，3H，J=7.2Hz，CH_3）．IR（cm^{-1}，KBr）：2981，1770，1731，1650，1618，1606，1581，1452，1422，1369，1326，1292，1253，1197，1162，1124，1069。

α-羟甲基对三氟甲基苯乙酸乙酯3，4-二甲氧基肉桂酸的合成

合成路线：

CH_2OH $COOCH_2CH_3$ F_3C 262.23 —DCC/DMAP, 低温→ $CH_2OCCHCH$ (O) $COOCH_2CH_3$ F_3C 452.42

将0.46g α-羟甲基对三氟甲基苯乙酸乙酯，0.38g 3，4-二甲氧基肉桂酸，0.38g环二己基碳二亚胺（DCC），催化量二甲氨基吡啶（DMAP）加入到100mL反应瓶当中，加入25mL二氯甲烷，于-10℃下保温搅拌反应。20h后停止反应。抽滤，滤液减压蒸馏除去溶剂。剩余物柱层析，（乙酸乙酯：石油醚=1：6的洗脱剂）。得到淡黄色固体0.25g，产物记为C25。^{1}HNMR：7.57-7.64（m，3H，Ph-H），7.03-7.10（m，2H，Ph-H），7.39-7.41（d，1H，J=5.1Hz，Ph-H），7.47-7.50（d，1H，J=7.8Hz，Ph-H），6.85-6.88（d，1H，J=8.1Hz，CH），6.23-6.29（d，1H，J=15.9Hz，=CH），4.09-4.25（m，3H，CH，CH_2），4.48-4.69，4.72-4.76（qq，J=6Hz，9Hz，2H，CH_2），3.91（s，6H，CH_3），1.21-1.26（m，3H，CH_3）．IR（cm^{-1}，KBr）：2968，2941，2899，1729，1700，1627，1596，1514，1465，1327，1260，1161，1138，1121。

α-甲酰基对三氟甲基苯乙酸乙酯缩氨基硫脲席夫碱的合成

合成路线：

CHO / COOCH$_2$CH$_3$ / F$_3$C —(乙醇，回流)→ HC=N-NHCSNH2 / COOCH$_2$CH$_3$ / F$_3$C

260.21　　333.33

将0.52g α－甲酰基对三氟甲基苯乙酸甲酯和0.2g氨基硫脲置于100mL反应瓶当中，加入10mL无水乙醇，在常温下搅拌反应。14h后蒸除溶剂。柱层析（乙：石＝1∶3），得到白色固体0.35g，产物记为C26。^{1}HNMR：7.25－7.65（m，4H，Ph－H），6.98（s，1H，NH），4.54（d，1H，J＝6.6Hz，CH），4.18－4.27（m，2H，CH$_2$），3.76－3.77（d，1H，J＝1.8Hz，CH），1.64（s，2H，NH$_2$），1.23－1.29（m，3H，CH$_3$）. IR（cm^{-1}，KBr）：3423，3251，3160，2980，1725，1615，1596，1515，1333，1306，1163，1109，1068。

α－甲酰基对三氟甲基苯乙酸乙酯缩甲氧胺席夫碱的合成

合成路线：

CHO / COOCH$_2$CH$_3$ / F$_3$C —(乙醇，回流)→ HC=N-OCH$_3$ / COOCH$_2$CH$_3$ / F$_3$C

260.21　　289.25

将0.74g α－甲酰基对三氟甲基苯乙酸甲酯和0.25g甲氧胺盐酸盐置于100mL反应瓶当中，加入0.12g氢氧化钠，加入20mL无水乙醇，在常温下搅拌反应22h，50℃反应2h后，蒸出溶剂，柱层析得黄色液体0.46g，产物记为C27。^{1}HNMR：7.59－7.69（m，1H，CH），7.12－7.14（q，1H，Ph－H），7.42－7.47（t，2H，J＝7.2Hz，Ph－H），4.16－4.23，5.0－5.1（tt，2H，J＝7.2Hz，6.6Hz，CH$_2$），3.88－3.92（d，3H，J＝7.2Hz，CH$_3$），3.67－3.75（m，1H，CH），1.22－1.26（m，3H，－CH$_3$）。IR（cm^{-1}，KBr）：2986，2942，2904，1738，1618，1464，1437，1420，1327，1166。

α－取代对四氢吡喃醚基苯乙酸甲酯的合成

合成路线：

HO / COOH + (二氢吡喃) —(对甲苯磺酸吡啶盐)→ —(甲酸乙酯，室内搅拌)→ CHO / COOCH$_2$CH$_3$

152.15

加入1.66g对羟基苯乙酸甲酯到反应瓶当中，加入15mL二氯甲烷，加入1g 3,4－二氢吡喃，加入催化量的对甲苯磺酸吡啶盐。室温搅拌反应，30h后，蒸除溶剂，柱层析，得无色液体1.24g。

加入1.24g对四氢吡喃醚基苯乙酸甲酯到反应瓶当中，加入7mL甲酸乙酯，加入0.15g金属钠，室温搅拌反应。加料完毕，有气体冒出，体系逐渐变得浑浊。反应5h后，蒸除未反应的甲酸乙酯，加入20mL水，加入稀盐酸调pH＝5，搅拌5min，2×20mL二氯甲烷萃取，10mL碳酸氢钠溶液洗涤，20mL饱和氯化钠溶液洗涤，无水硫酸钠干燥，得金黄色液体0.72g，收率52.04%，产物记为C28。^{1}HNMR：8.02(m, 1H, CH), 7.92－7.95 (d, 2H, J ＝8.7Hz, PH－H), 6.9－6.93(d, 2H J ＝9Hz, Ph－H), 4.47(q, 2H, J ＝7.2Hz, CH_2), 4.95－4.99(m, 1H, CH), 3.85－3.96(m, 1H, CH), 3.55－3.58(m, 2H, CH_2), 1.71－1.86 (m, 6H, $3CH_2$), 1.39－1.43(t, 3H, J ＝7.2Hz, CH_3). IR(cm^{-1}, KBr)：2945, 1729, 1660, 1600, 1514, 1443, 1374, 1273, 1235, 1161, 1022, 965。

α－羟甲基对四氢吡喃醚基苯乙酸甲酯的合成

合成路线：

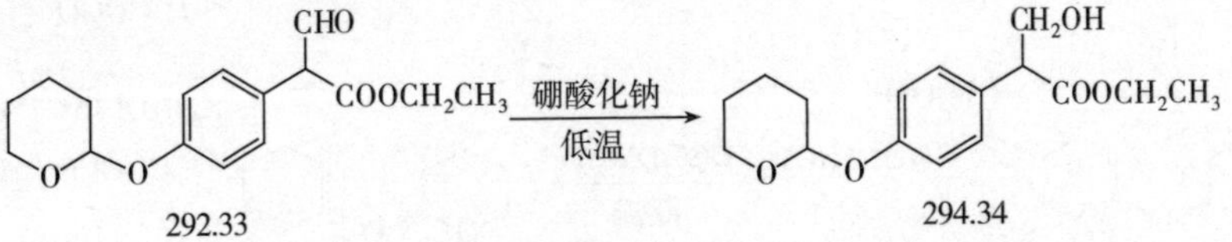

CHO　COOCH$_2$CH$_3$　—硼酸化钠/低温→　CH$_2$OH　COOCH$_2$CH$_3$

292.33　　294.34

加入2.1g α－甲酰基对四氢吡喃醚基苯乙酸甲酯到反应瓶中，加入25mL无水乙醇，－10℃保温搅拌，分批加入0.37g硼氢化钠，保温反应7h后，减压蒸除大部分溶剂，加入25mL水搅拌10min，2×25mL二氯甲烷萃取，2×20mL饱和氯化钠溶液洗涤，无水硫酸钠干燥，蒸除溶剂，得到黄色固体1.09g，产物记为C29。^{1}HNMR：7.12－7.26(m, 2H, Ph－H), 6.74－6.79(m, 2H, Ph－H), 4.65－4.66(m, 2H, CH_2), 4.12－4.2(m, 3H, CH, CH_2), 3.53－3.55(t, 2H, J ＝5.1Hz, CH_2), 3.69－3.81(d, 1H, J ＝3Hz, CH), 1.53－1.85(m, 6H, $3CH_2$), 1.21－1.25(m, 3H, CH_3). IR(cm^{-1}, KBr)：3348, 2940, 2866, 1512, 1611, 1517, 1370, 1239, 1176, 1109, 1036, 965, 832。

α－羟甲基对四氢吡喃醚基苯乙酸甲酯苯甲酸酯的合成

合成路线：

0.84g α－羟甲基对四氢吡喃基苯乙酸甲酯加入到反应瓶当中，加入0.36g苯甲酸，0.64g环二己基碳二亚胺和0.073g对二甲氨基吡啶，加入30mL二氯甲

CH_2OH $COOCH_2CH_3$ 硼氢化钠 低温 CH_2OCOPh $COOCH_2CH_3$

294.33 398.45

烷做溶剂。在 -10℃下保温搅拌反应,12h 后,过滤,蒸除溶剂。剩余物柱层析,得淡黄色黏状物 0.53g,产物记为 C30。1HNMR: 8.18 (d, 1H, J = 1.5Hz ,Ph - H), 7.96 - 8.00(m, 2H, Ph - H), 7.42 - 7.55(m, 2H, Ph - H), 7.21 - 7.32 (m, 2H, Ph - H), 7.02 - 7.05(m, 2H, Ph - H), 5.40 - 5.42 (t, 1H, J = 3.3Hz, CH), 4.03 - 4.23(m, 3H, CH_2, CH), 4.52 - 4.58(m, 2H, CH_2), 4.58 - 4.82(m, 2H, CH_2), 3.89(t, 1H, CH), 3.63 - 3.71(m, 2H, CH_2), 1.26, 1.62 - 1.66, 1.83 - 1.86(m, 6H, CH_2), 1.19 - 1.24(t, 3H, J = 7.2Hz, CH_3). IR (cm^{-1}, KBr): 2933, 2853, 1740, 1718, 1601, 1509, 1451, 1368, 1270, 1202, 1175, 1110, 1063, 1024, 965。

α - 羟甲基对四氢吡喃醚基苯乙酸甲酯乙酰水杨酸酯的合成

合成路线:

H_3COCO CH_2OH $COOCH_2CH_3$ DCC/DMAP 低温 CH_2OCO $COOCH_2CH_3$

294.33 456.49

加入 0.84g α - 羟甲基对四氢吡喃基苯乙酸甲酯加入到反应瓶当中,加入 0.54g乙酰水杨酸,0.64g 环二已基碳二亚胺(DCC)和 0.073g 对二甲氨基吡啶(DMAP),加入 30mL 二氯甲烷做溶剂。在 -10℃下保温搅拌反应,12h 后,过滤,蒸除溶剂。剩余物柱层析,得淡黄色黏状物 0.68g,产物记为 C31。^{1}HNMR: 7.90 - 7.94(m, 1H, Ph - H), 7.38 - 7.66(m, 1H, Ph - H), 7.10 - 7.26(m, 2H, Ph - H) 6.99 - 7.04(m, 2H, Ph - H), 6.77 - 6.81(m, 2H, Ph - H), 5.40 - 5.41(d, 1H, J = 3Hz, CH), 4.71 - 4.74(m, 2H, CH_2), 4.12 - 4.17(m, 1H, CH), 4.32 - 4.54(m, 2H, CH_2), 3.56 - 3.64(m, 2H, CH_2), 2.33(s, 3H, CH_3), 1.53 - 2.03(m, 6H, $2CH_2$), 1.21 - 1.25(t, 3H, J = 3.9Hz, CH_3). IR (cm^{-1}, KBr): 2945, 2872, 1763, 1739, 1652, 1607, 1581, 1509, 1452, 1369, 1241, 1123, 1077, 1037, 964。

α - 羟甲基对四氢吡喃醚基苯乙酸甲酯 3,4 - 二甲氧基肉桂酸酯的合成

合成路线:

DCC/DMAP
低温

294.33

484.54

加入0.84g α-羟甲基对四氢吡喃基苯乙酸甲酯加入到反应瓶当中,加入0.64g 3,4-二甲氧基肉桂酸,0.64g 环二已基碳二亚胺和0.073g 对二甲氨基吡啶,加入30mL 二氯甲烷做溶剂。在-10℃下保温搅拌反应,11h 后,过滤,蒸除溶剂。剩余物柱层析,得0.7g 白色固体,产物记为C32。^{1}HNMR:7.63-7.58(d, 1H, J =15.9Hz, Ph-H), 7.23-7.28(m, 2H, Ph-H), 7.01-7.10(m, 2H, Ph-H), 7.20(m, 1H, =CH), 6.80-6.87(m, 2H, Ph-H) 6.25-6.31(d, 1H, J =15.9Hz, =CH), 4.15-4.20(t, 1H, J =7.2Hz, CH), 4.42-4.46(q, 1H, J =5.7Hz, CH_2), 4.63-4.71(q, 1H, J =2.7Hz, CH_2), 3.71(m, 2H, CH), 3.91(s, 6H, 2CH_3), 1.84-1.85(m, 2H, CH_2), 1.62-1.67(m, 2H, CH_2), 1.54-1.48(m, 2H, CH_2), 1.20-1.25(t, 3H, J =6.9Hz, CH_3); IR (cm^{-1}, KBr): 3427, 2945, 1728, 1710, 1631, 1598, 1513, 1466, 1452, 1239, 1201, 1156, 1021, 963。

合成结果分析与讨论

以各种苯乙酸酯与甲酸乙酯为起始原料,合成α-甲酰基取代苯乙酸酯化合物,再同氨基硫脲或甲氧胺盐酸盐缩合得到α-甲酰基取代苯乙酸酯缩氨基硫脲或α-甲酰基取代苯乙酸酯缩甲氧胺。α-甲酰基取代苯乙酸酯经硼氢化钠还原得到α-羟甲基取代苯乙酸酯,再同苯甲酸、取代苯乙酸、肉桂酸、水杨酸在DCC 和 DMAP 的作用下脱水酯化,得到α-羟甲基取代苯乙酸酯苯甲酸酯、α-羟甲基取代苯乙酸酯苯乙酸酯、α-羟甲基取代苯乙酸酯水杨酸酯、α-羟甲基取代苯乙酸酯3,4-二甲氧基肉桂酸酯。合成了32 个α-取代苯乙酸酯衍生物,得到了这些化合物相关的合成工艺参数,理化性质。所有化合物用溴化钾为载体进行了红外表征,也用氘代氯仿或者氘代 DMSO 为溶剂进行核磁共振氢谱检测,部分化合物还进行了核磁共振碳谱检测。

α-甲酰基苯乙酸酯衍生物合成分析与讨论

表3 α-甲酰基苯乙酸酯衍生物的合成结果

序号	编号	中文名	摩尔收率
1	C1	α-甲酰基苯乙酸甲酯	66.06%
2	C7	α-甲酰基对甲基苯乙酸甲酯	46.1%
3	C10	α-甲酰基对甲氧基苯乙酸甲酯	55.42%
4	C14	α-甲酰基对氟基苯乙酸乙酯	68.19%
5	C21	α-甲酰基对三氟甲基苯乙酸乙酯	46.1%
6	C28	α-甲酰基对四氢吡喃醚基苯乙酸甲酯	52.04%

备注:摩尔收率=(产物的物质的量/投料较小的原料的物质的量)×100%

α-甲酰基苯乙酸酯衍生物(见表3)合成的摩尔收率在46.1%~73.82%。苯环4号位被氟基取代的α-甲酰基苯乙酸乙酯收率最高为68.19%,其次是没有被取代的α-甲酰基苯乙酸甲酯为66.06%,α-甲酰基对甲氧基苯乙酸甲酯、α-甲酰基对四氢吡喃醚基苯乙酸甲酯的收率相差不大分别为55.42%和52.04%,α-甲酰基对甲基苯乙酸甲酯和α-甲酰基对三氟甲基苯乙酸甲酯都为46.1%,说明苯环4号位被给电子和吸电子能力强的基团取代后,收率会略有下降,被大体积的给电子基团取代后,对收率影响不大。

红外图谱上最大的特征是1653 cm^{-1} -1662 cm^{-1}处有比较强烈的吸收峰,这是甲酰基在这里的伸缩振动引起的吸收峰。α-甲酰基对三氟甲基苯乙酸甲酯的三氟甲基的吸电子诱导效应使甲酰基特征吸收峰为1662 cm^{-1}比α-甲酰基苯乙酸甲酯的1653 cm^{-1}向高波数移动了9。α-甲酰基对甲氧基苯乙酸甲酯的甲酰基吸收峰为1652 cm^{-1},给电子共轭效应和吸电子诱导效应使甲酰基特征吸收峰向低波数移动了1。苯环4号位被氟基、四氢吡喃醚基、甲基取代的化合物的甲酰基特征吸收峰为1660 cm^{-1}、1660 cm^{-1}、1661 cm^{-1},说明氟基、四氢吡喃醚基、甲基产生的电子效应作用于甲酰基产生的效果一样。

由于碳氧双键较强的去屏蔽作用,甲酰基上的氢的化学位移出现在低场7.92~8.43。甲酰基上的氢与其相邻的一个叔碳氢发生偶合,产生双重分裂峰。α-甲酰基对甲氧基苯乙酸甲酯的甲酰基上氢的化学位移是7.99-8.00,偶合常数为3Hz。吸电子基团的三氟甲基取代的α-甲酰基对三氟甲基苯乙酸乙酯,其甲酰基上氢的化学位移为8.15-8.18,向低场移了0.16。被氟基取代后的α-甲酰基对氟基苯乙酸甲酯乙酰基上氢的位移为8.39-8.43,向低场移了0.4。

α－羟甲基苯乙酸酯衍生物合成分析与讨论

表4　α－羟甲基苯乙酸酯衍生物的合成结果

序号	编号	中文名	摩尔收率
1	C8	α－羟甲基对甲基苯乙酸甲酯	78.8%
2	C11	α－羟甲基对甲氧基苯乙酸甲酯	43.91%
3	C15	α－羟甲基对氟基苯乙酸乙酯	60.2%
4	C22	α－羟甲基对三氟甲基苯乙酸乙酯	58.87%
5	C29	α－羟甲基对四氢吡喃醚基苯乙酸甲酯	51.49%

备注:摩尔收率＝(产物的物质的量/投料较小的原料的物质的量)×100%

α－羟甲基苯乙酸酯衍生物(见表4)合成的摩尔收率在43.91%～78.8%。α－羟甲基苯乙酸甲酯C2的收率为73.82%，其余的同类物α－羟甲基对甲氧基苯乙酸甲酯C11、α－羟甲基对甲基苯乙酸甲酯C8、α－羟甲基对氟基苯乙酸乙酯C15、α－羟甲基对三氟甲基苯乙酸乙酯C22、α－羟甲基对四氢吡喃醚基苯乙酸甲酯C29分别为43.91%,78.8%,60.2%,58.87%,51.49%。苯环4号位被甲基C8和未被取代的C2合成的收率较高,甲氧基取代的C11,给电子诱导能力增强强,导致α位的氢活性减弱,反应难度增加,收率降低。氟取代的苯乙酸乙酯,三氟甲基取代的苯乙酸乙酯,四氢吡喃醚基取代的苯乙酸甲酯反应活性略有降低,但收率变化不是很大。

这类化合物在红外谱图上最大特征是3348 cm^{-1}～3444 cm^{-1}之间有宽而长的吸收带,这是络合羟基或游离羟基的特征吸收峰。α－羟甲基对甲基苯乙酸乙酯的醇羟基在红外谱图上的吸收峰出现在3419 cm^{-1}处,甲氧基取代的α－羟甲基对甲氧基苯乙酸甲酯未发生红移。氟基取代的α－羟甲基对氟基苯乙酸甲酯吸收为3398 cm^{-1}向低波处移动了21,α－羟甲基对三氟甲基苯乙酸甲酯吸收向高波处移动了25,这是由于三氟甲基强吸电子的作用。四氢吡喃醚基取代的α－羟甲基对四氢吡喃醚基苯乙酸甲酯的醇羟基在红外谱图上的吸收峰出现在3348处,相比于甲基取代后的化合物向低波处移动了71。

α－羟甲基对甲基苯乙酸乙酯在氢谱上,1.19－1.24出现三重峰,积分面积为三个氢为甲基,与其相连的亚甲基化学位移出现在3.7－3.8。2.33出现的单峰为苯环上甲基的化学位移,7.15－7.26出现的多重峰为苯环上四个氢的化学位移,1.74处出现的单峰为羟基氢的化学位移。α－羟甲基对氟基苯乙酸乙酯,受氟取代基的影响,苯环四个氢的化学位移出现在7.00－7.05和7.22－7.27

处,呈对称的多重峰。4.02－4.22出现的多重峰,积分面积为3,为一个亚甲基和一个叔碳氢的化学位移。3.71－3.84出现的多重峰,积分面积为2,是亚甲基的吸收峰。2.12出现的单峰是羟基的化学位移,比甲基取代的α－羟甲基对甲基苯乙酸乙酯向低场多位移了0.38。1.2－1.24出现的三重峰积分面积为3,是甲基的吸收峰。

α－甲酰基苯乙酸酯席夫碱衍生物合成分析与讨论

表5　α－甲酰基苯乙酸酯席夫碱衍生物的合成结果

序号	编号	中文名	摩尔收率
1	C3	α－甲酰基苯乙酸酯缩氨基硫胺席夫碱	75.1%
2	C9	α－甲酰基对甲基苯乙酸甲酯缩氨基硫脲席夫碱	81.57%
3	C13	α－甲酰基对甲氧基苯乙酸甲酯缩氨基硫脲席夫碱	47.75%
4	C16	α－甲酰基对氟基苯乙酸乙酯酯缩氨基硫脲席夫碱	90.8%
5	C17	α－甲酰基对氟基苯乙酸乙酯缩甲氧胺席夫碱	74.17%
6	C26	α－甲酰基对三氟甲基苯乙酸乙酯缩氨基硫脲席夫碱	52.54%
7	C27	α－甲酰基对三氟甲基苯乙酸乙酯缩甲氧胺席夫碱	55.9%

备注:摩尔收率＝(产物的物质的量/投料较小的原料的物质的量)×100%

α－甲酰基对三氟甲基苯乙酸乙酯缩甲氧胺席夫碱和α－甲酰基对氟基苯乙酸乙酯缩甲氧胺席夫碱合成的摩尔收率分别为55.90%和74.17%。亚胺的红外吸收峰不强,在1634 cm^{-1}－1657 cm^{-1}的特征峰不是特别明显。亚胺碳上的氢的化学位移在7.12－7.66之间,是双重峰。

α－甲酰基取代苯乙酸酯和氨基硫脲发生缩合反应的活性很高(见表5),反应比较彻底,收率比较高,在47.75%～90.8%。α－甲酰基苯乙酸酯缩氨基硫脲C3的摩尔收率为75.1%,甲基取代的α－甲酰基对甲基苯乙酸乙酯缩氨基硫脲是81.57%,甲氧基取代的α－甲酰基对甲氧基苯乙酸乙酯缩氨基硫脲是47.57%,氟基取代的α－甲酰基对氟基苯乙酸乙酯缩氨基硫脲收率是90.8%,三氟甲基取代的α－甲酰基对三氟甲基苯乙酸乙酯缩氨基硫脲收率是52.54%。甲氧基给电子诱导能力较强,减弱酰羰基的亲电性,不利于脱水。三氟甲基吸电子诱导能力太强,甲酰基主要以醇式存在为主,不利于亲电试剂的进攻。

红外图谱上这些化合物在1733 cm^{-1}有比较强烈的酯羰基的吸收峰,1652 cm^{-1}有比较弱的碳氮双键的吸收峰。α－甲酰基对三氟甲基苯乙酸乙酯缩氨基硫脲,3423 cm^{-1}和3250 cm^{-1}出现比较尖的强的吸收峰伯胺和仲胺的伸缩振动吸收峰，2978 cm^{-1} 和 2933 cm^{-1} 是甲基和亚甲基的伸缩振动吸收峰，1725 cm^{-1} 是酯羰基的吸收峰，1646 cm^{-1} 是亚胺的振动吸收峰，1515 cm^{-1} 和 1596 cm^{-1} 是苯环骨架振动吸收峰，1110 cm^{-1} 是碳硫双键的振动吸收峰。α－甲酰基对甲氧基苯乙酸乙酯缩氨基硫脲在红外图谱上，伯胺的伸缩振动吸收峰出现在 3438 cm^{-1}，甲基的吸收峰在 2927 cm^{-1}，1718 cm^{-1} 出现比较强烈的吸收峰，是酯羰基伸缩振动的吸收峰，1652 比较弱的吸收峰是碳氮双键的振动吸收峰，1594 cm^{-1} 和 1508 cm^{-1} 为苯环骨架的振动吸收峰。因为甲氧基总的电子效应使其它基团的电子云密度增强，三氟甲基侧减弱，导致羰基和苯环骨架略向低波数移动，而伯胺和亚胺的吸收向高波数移动。

氢的核磁共振图谱上，这类化合物碳氮双键上的氢受双键的影响，化学位移在低场 9.12～10.13，受与其相邻的一个叔氢的影响，为双峰。α－甲酰基对三氟甲基苯乙酸乙酯缩氨基硫脲，与亚胺连接的氢的化学位移在 9.59 处。7.65－7.62和 7.40－7.42 为对称的多重峰，积分面积都为 2，分别是苯环四个氢的化学位移。伯胺上单有两个孤对电子，导致碳氮键不能自由转动，氮上的两个氢化学不等价，它们化学位移分别在 6.98 和 6.35。伯胺的化学位移在 1.64 处。4.18－4.27 的多重峰，积分面积为 2，是亚甲基的化学位移。4.54－4.56 的双峰是叔氢的化学位移，1.23－1.29 是甲基的化学位移。α－甲酰基对氟基苯乙酸乙酯缩氨基硫脲的氢谱上，亚胺上的氢的化学位移为 9.52－9.55 的双重峰，叔氢的化学位移为 4.45－4.50 的多重峰，与碳相连的三个氢的化学位移分别在 7.04、6.34、3.73 处。α－甲酰基苯乙酸乙酯缩氨基硫脲亚胺上的氢的化学位移为 10.13，叔氢的化学位移在 4.5 的多重峰，伯胺上氢的化学位移为 7.02、6.67 的单峰。苯环上被吸电子的三氟甲基和给电子的氟基取代后，与亚胺连接的氢的化学位移向高场移动。

α－羟甲基苯乙酸酯交酯衍生物合成分析与讨论

表 6　α－羟甲基苯乙酸酯交酯衍生物的合成结果

序号	编号	中文名	摩尔收率
1	C4	α－羟甲基苯乙酸甲酯苯甲酸酯	51.2%
2	C5	α－羟甲基苯乙酸甲酯乙酰水杨酸酯	59.62%
3	C6	α－羟甲基苯乙酸甲酯 3,4－二甲氧基肉桂酸酯	49.2%
4	C19	α－羟甲基基对氟基苯乙酸乙酯乙酰水杨酸酯	9.43%
5	C20	α－羟甲基基对氟基苯乙酸乙酯 3,4－二甲氧基肉桂酸酯	68.56%

续表

序号	编号	中文名	摩尔收率
6	C23	α-羟甲基对三氟甲基苯乙酸乙酯苯甲酸酯	15.25%
7	C24	α-羟甲基对三氟甲基苯乙酸乙酯乙酰水杨酸酯	13.18%
8	C25	α-羟甲基对三氟甲基苯乙酸乙酯3,4-二甲氧基肉桂酸	31.47%
9	C30	α-羟甲基对四氢吡喃醚基苯乙酸甲酯苯甲酸酯	46.55%
10	C31	α-羟甲基对四氢吡喃醚基苯乙酸甲酯乙酰水杨酸酯	52.14%
11	C32	α-羟甲基对四氢吡喃醚基苯乙酸甲酯3,4-二甲氧基肉桂酸酯	50.62%

备注:摩尔收率=(产物的物质的量/投料较小的原料的物质的量)×100%

α-羟甲基取代苯乙酸酯与苯甲酸、水杨酸、3,4-二甲氧基肉桂酸在环二己基碳二亚胺(DCC)的脱水作用和对二甲氨基吡啶(DMAP)的催化作用下发生酯化,但受位阻的影响比较大。α-羟甲基取代苯乙酸乙酯乙酰水杨酸酯衍生物合成的摩尔收率在9.43%~59.62%,α-羟甲基基对氟基苯乙酸乙酯乙酰水杨酸酯的合成收率比较低只有9.43%。α-羟甲基苯乙酸酯苯甲酸酯衍生物衍生物合成的摩尔收率在15.25%~51.2%。α-羟甲基取代苯乙酸酯与肉桂酸反应的位阻相比于同水杨酸和苯甲酸较小,这些衍生物的收率在31.47%~68.56%,比其余两类的收率要好。

在红外图谱上,这四个系列的化合物都有两个不同环境中酯羰基,所以具有两个大而强烈的酯羰基的特征吸收峰。α-羟甲基苯乙酸酯苯甲酸酯衍生物最大的特征是1724 cm^{-1}-1740 cm^{-1}和1712 cm^{-1}-1718 cm^{-1}有两个比较强烈的吸收峰,这是两个酯羰基的特征吸收峰。α-羟甲基苯乙酸酯水杨酸酯衍生物最大特征是1763 cm^{-1}-1770 cm^{-1}、1731 cm^{-1}-1739 cm^{-1}、1652 cm^{-1}-1650 cm^{-1}三处有比较强烈的吸收峰,有些在1695 cm^{-1}-1700 cm^{-1}处也有一个吸收峰,这些事三个酯羰基伸缩振动的特征吸收峰。α-羟甲基苯乙酸酯肉桂酸酯衍生物在红外谱图上,这个系列的化合物在1711 cm^{-1}-1728 cm^{-1}、1693 cm^{-1}-1710 cm^{-1}、1627 cm^{-1}-1631 cm^{-1}有中强度的吸收峰,前面两处是两个酯羰基的吸收峰,后面一处是碳碳双键的伸缩振动吸收峰。α-羟甲基对甲氧基苯乙酸甲酯苯乙酸酯在红外谱图上2933 cm^{-1},2851 cm^{-1}是甲基和亚甲基的特征吸收峰,1769 cm^{-1},1709 cm^{-1}是两个酯羰基的特征吸收峰。

α-羟甲基对氟基苯乙酸乙酯苯甲酸酯在核磁共振的氢谱上,两个苯环上的9个氢的化学位移在7.02-7.98,叔氢的化学位移为4.05-4.10的多重峰。与叔碳相连的亚甲基上的两个氢是化学不等价的,其化学位移分别是4.54-4.60

和4.76－4.82的多重峰。甲基的化学位移为1.2－1.24的三重峰，与其相连的亚甲基的化学位移4.13－4.21的多重峰。α－羟甲基对氟基苯乙酸乙酯乙酰水杨酸酯的两个苯环的8个氢的化学位移在7.02－7.91之间，叔氢的化学位移为4.02－4.05的三重峰，与叔碳相连的亚甲基上的氢的化学位移分别为4.49－4.55和4.71－4.78，与亚甲基相连的甲基氢的化学位移为1.19－1.24的三重峰，亚甲基的化学位移为4.16－4.20的多重峰，乙酰基上的甲基氢的化学位移为2.34的单峰。α－羟甲基对氟基苯乙酸乙酯3’,4’－二甲氧基肉桂酸酯在核磁共振的氢谱上，两个苯环上的7个氢的化学位移在7.02－7.63之间，叔氢的化学位移为4.00－4.03的三重峰，与叔碳相连的亚甲基上的氢的化学位移分别为4.43－4.49和4.64－4.71，与亚甲基相连的甲基氢的化学位移为1.21－1.26的三重峰，亚甲基的化学位移为4.16－4.21的多重峰，与碳碳双键相连的两个氢的化学位移分别为6.85－6.83和6.24－6.29的双峰，两个甲氧基上的氢的化学位移都在3.92处。

小结

通过实验合成了32个苯乙酸酯类化合物，得到了合成这些化合物的工艺条件和工艺参数，基本的物理化学性质。取代苯乙酸酯发生克莱森缩合时，苯环上有强吸电子的取代基不利于缩合的发生，强给电子取代基不利于α碳负离子的形成，这都容易造成α－甲酰基苯乙酸酯衍生物的收率偏低。α－甲酰基苯乙酸酯与甲氧胺和氨基硫脲反应容易进行，收率较高。α－羟甲基苯乙酸酯与苯甲酸、水杨酸、肉桂酸反应，能够顺利进行，但因为位阻较大，收率不是很高。这些化合物通过1HNMR、13CNMR、IR表征，确定了其化学结构是目标化合物。

3.4.3.2　苯乙酸酯抗肿瘤活性筛选

实验原理

3－(4,5－二甲基－2－噻唑)－2,5－二苯基溴化四唑(MTT)[53]是一种能接受氢原子的染料。活细胞线粒体中与NADP相关的脱氢酶在细胞内可将黄色的MTT转化成不溶性的蓝紫色的甲臜，而死细胞则无此功能。用DMSO溶解甲臜后，在一定波长下用酶标仪测定光密度值，既可定量测出细胞的存活率。

材料与方法

实验材料

细胞：人结肠癌细胞(HCT－116)、人肝癌细胞(HepG2)、胃腺癌细胞(BGC－823)、人肺支气管癌细胞(NCI－H1650、1650)，卵巢癌细胞(A2780、SK－OV－3)来自北药物所药物：泰素 北京乐博生物科技有限公司，批号为970617－2;注射用M T T购自S I G M A公司,其余试剂为市售分析纯。

待测样品:32个α－取代苯乙酸化合物,邵阳学院药物合成研究实验室提供,编号C1－C32。

实验方法

(1)接种细胞:选用对数生长期的贴壁肿瘤细胞,用胰酶消化后,用10%小牛血清的RPMI1640培养液配成40000个/mL的细胞悬液,接种在96孔培养板中,每孔接种100ul,37℃,5% CO_2 培养24h。

(2)加药:受试药物以10倍被稀释,作5个稀释度。实验组加样品10 μL,每孔终体积为200 μL,用1640培养液补足。37℃,5% CO_2 培养3d。

(3)结果测试:弃上清液,每孔加入100ul新鲜配制的0.5mg/mL MTT的无血清培养液,37℃继续培养4h。小心弃上清,并加入150 μL DMSO溶解甲臜沉淀,用微型超声振荡器混匀,在酶标仪上测定波长540 nm处的光密度值。

(4)结果评定:肿瘤细胞生长抑制率＝$[(OD_{对照}-OD_{实验})/(OD_{对照}-OD_{空白})]\times100\%$

实验结果

用人结肠癌细胞(HCT－116)、人肝癌细胞(HepG2)、胃腺癌细胞(BGC－823)、人肺支气管癌细胞(NCI－H1650、1650),卵巢癌细胞(A2780、SK－OV－3)培养到对数期后,加入不同浓度化合物,3d后加入四氮唑(MTT),用酶标仪,检测540nm处的吸光度,间接得到活的癌细胞的数量。利用公式肿瘤细胞生长抑制率＝$[(OD_{对照}-OD_{实验})/(OD_{对照}-OD_{空白})]\times100\%$,得到一定浓度化合物对癌细胞的抑制能力。作抑制率－浓度(化合物)的曲线,便得到这种化合物对特定癌细胞的半抑制浓度IC_{50},其结果见表7。

表7 抗肿瘤活性测试结果

序号	编号	IC_{50}(mg/mL)				
		HCT－116	HepG2	BGC－823	NCI－H1650	A2780
1	C13	8.8×10^{-6}	3.77×10^{-6}	$>10\times10^{-6}$	2.5×10^{-6}	3.40×10^{-6}
2	C12	$>10\times10^{-6}$	$>10\times10^{-6}$	$>10\times10^{-6}$	$>10\times10^{-6}$	$>10\times10^{-6}$
3	C11	$>10\times10^{-6}$	$>10\times10^{-6}$	$>10\times10^{-6}$	$>10\times10^{-6}$	$>10\times10^{-6}$
4	C10	$>10\times10^{-6}$	$>10\times10^{-6}$	$>10\times10^{-6}$	$>10\times10^{-6}$	$>10\times10^{-6}$
5	C16	1.91×10^{-6}	2.18×10^{-6}	7.27×10^{-6}	1.98×10^{-6}	2.11×10^{-6}

续表

序号	编号	IC_{50}（mg/mL）				
		HCT－116	HepG2	BGC－823	NCI－H1650	A2780
6	C20	$>10\times10^{-6}$	$>10\times10^{-6}$	$>10\times10^{-6}$	$>10\times10^{-6}$	$>10\times10^{-6}$
7	C18	$>10\times10^{-6}$	$>10\times10^{-6}$	$>10\times10^{-6}$	$>10\times10^{-6}$	$>10\times10^{-6}$
8	C14	$>10\times10^{-6}$	$>10\times10^{-6}$	$>10\times10^{-6}$	$>10\times10^{-6}$	$>10\times10^{-6}$
9	C19	$>10\times10^{-6}$	$>10\times10^{-6}$	$>10\times10^{-6}$	$>10\times10^{-6}$	$>10\times10^{-6}$
10	C15	$>10\times10^{-6}$	$>10\times10^{-6}$	$>10\times10^{-6}$	$>10\times10^{-6}$	$>10\times10^{-6}$
11	C17	$>10\times10^{-6}$	$>10\times10^{-6}$	$>10\times10^{-6}$	$>10\times10^{-6}$	$>10\times10^{-6}$
12	C7	$>10\times10^{-6}$	$>10\times10^{-6}$	$>10\times10^{-6}$	$>10\times10^{-6}$	$>10\times10^{-6}$
13	C8	$>10\times10^{-6}$	$>10\times10^{-6}$	$>10\times10^{-6}$	$>10\times10^{-6}$	$>10\times10^{-6}$
14	C29	$>10\times10^{-6}$	$>10\times10^{-6}$	$>10\times10^{-6}$	$>10\times10^{-6}$	$>10\times10^{-6}$
15	C28	$>10\times10^{-6}$	$>10\times10^{-6}$	$>10\times10^{-6}$	$>10\times10^{-6}$	$>10\times10^{-6}$
16	C32	$>10\times10^{-6}$	$>10\times10^{-6}$	$>10\times10^{-6}$	$>10\times10^{-6}$	$>10\times10^{-6}$
17	C31	$>10\times10^{-6}$	$>10\times10^{-6}$	$>10\times10^{-6}$	$>10\times10^{-6}$	$>10\times10^{-6}$
18	C30	$>10\times10^{-6}$	$>10\times10^{-6}$	$>10\times10^{-6}$	$>10\times10^{-6}$	$>10\times10^{-6}$
19	C22	$>10\times10^{-6}$	$>10\times10^{-6}$	$>10\times10^{-6}$	$>10\times10^{-6}$	$>10\times10^{-6}$
20	C26	3.67×10^{-6}	5.19×10^{-6}	$>10\times10^{-6}$	8.05×10^{-6}	2.59×10^{-6}
21	C25	$>10\times10^{-6}$	$>10\times10^{-6}$	$>10\times10^{-6}$	$>10\times10^{-6}$	$>10\times10^{-6}$
22	C21	$>10\times10^{-6}$	$>10\times10^{-6}$	$>10\times10^{-6}$	$>10\times10^{-6}$	$>10\times10^{-6}$
23	C24	$>10\times10^{-6}$	$>10\times10^{-6}$	$>10\times10^{-6}$	$>10\times10^{-6}$	$>10\times10^{-6}$
24	C23	$>10\times10^{-6}$	$>10\times10^{-6}$	$>10\times10^{-6}$	$>10\times10^{-6}$	$>10\times10^{-6}$
25	C27	$>10\times10^{-6}$	$>10\times10^{-6}$	$>10\times10^{-6}$	$>10\times10^{-6}$	$>10\times10^{-6}$
26	C9	$>10\times10^{-6}$	$>10\times10^{-6}$	5.22×10^{-6}	5.34×10^{-6}	5.92×10^{-6}
27	C1	$>10\times10^{-6}$	$>10\times10^{-6}$	$>10\times10^{-6}$	$>10\times10^{-6}$	$>10\times10^{-6}$
28	C2	$>10\times10^{-6}$	$>10\times10^{-6}$	$>10\times10^{-6}$	$>10\times10^{-6}$	$>10\times10^{-6}$
29	C3	3.21×10^{-6}	8.00×10^{-6}	8.18×10^{-6}	3.25×10^{-6}	$>10\times10^{-6}$
30	C5	$>10\times10^{-6}$	$>10\times10^{-6}$	$>10\times10^{-6}$	$>10\times10^{-6}$	$>10\times10^{-6}$
31	C6	$>10\times10^{-6}$	$>10\times10^{-6}$	$>10\times10^{-6}$	$>10\times10^{-6}$	$>10\times10^{-6}$
32	C4	$>10\times10^{-6}$	$>10\times10^{-6}$	$>10\times10^{-6}$	$>10\times10^{-6}$	$>10\times10^{-6}$
Taxol		1.95×10^{-8}	1.06×10^{-8}	2.56×10^{-8}	3.17×10^{-8}	5.87×10^{-7}

结果与讨论

如表7所知,32个化合物中,有5个化合物:α－甲酰基苯乙酸甲酯缩氨基硫脲(C3)、α－甲酰基对甲基苯乙酸甲酯缩氨基硫脲(C9)、α－甲酰基对甲氧基苯乙酸甲酯缩氨基硫脲(C13)、α－甲酰基对氟基苯乙酸甲酯缩氨基硫脲(C16)、α－甲酰基对三氟甲基苯乙酸甲酯缩氨基硫脲(C26),对癌细胞有明显的抑制作用。

表8　抗人结肠癌(HCT－116)筛选结果

化合物 / IC_{50} mg/mL	C13	C16	C26	C9	C3	Taxol
HCT－116	8.8×10^{-6}	1.91×10^{-6}	3.67×10^{-6}	$>10\times10^{-6}$	3.21×10^{-6}	1.95×10^{-8}

如表8所知,α－甲酰基对氟基苯乙酸乙酯缩氨基硫脲(C16)对人结肠癌细胞(HCT－116)的抑制活性最好其半抑制率为1.91×10^{-6}。但α－甲酰基对甲基苯乙酸甲酯缩氨基硫脲(C9)对结肠癌细胞(HCT－116)却没有抑制活性;苯环4号位被甲氧基取代的C13对结肠癌细胞(HCT－116)抑制作用相比于C16差了4.6倍。苯环4号位未被取代的α－甲酰基苯乙酸甲酯缩氨基硫脲(C9)和被强吸电子的三氟甲基取代的α－甲酰基对三氟甲基苯乙酸甲酯缩氨基硫脲(C26)对结肠癌细胞(HCT－116)的抑制作用相差不大。说明了4号位被给电子的基团取代后,会减弱对结肠癌细胞(HCT－116)的抑制作用。苯环4号位被强吸电子的官能团取代后,对结肠癌细胞(HCT－116)抑制作用没有明显的提高。

表9　抗肝癌细胞(HepG2)筛选结果

化合物 / IC_{50} mg/mL	C13	C16	C26	C9	C3	Taxol
HepG2	3.77×10^{-6}	2.18×10^{-6}	5.19×10^{-6}	$>10\times10^{-6}$	8.00×10^{-6}	1.06×10^{-8}

如表9已知:对肝癌细胞(HepG2)的抑制活性中,α－甲酰基对氟基苯乙酸酯缩氨基硫脲席夫碱(C16)的抑制活性最好,其IC_{50}是2.18×10^{-6}。4号位被甲基取代的α－甲酰基对甲基苯乙酸甲酯缩氨基硫脲(C9)对肝癌细胞(HepG2)的没有抑制活性,α－甲酰基苯乙酸甲酯缩氨基硫脲(C3)、α－甲酰基对甲氧基苯乙酸甲酯缩氨基硫脲(C13)、α－甲酰基对三氟甲基苯乙酸甲酯缩氨基硫脲(C26)对肝癌细胞(HepG2)的抑制作用差别不是很大。没有与氨基硫对脲结合的α－甲酰基对氟基苯乙酸甲酯和同甲氧胺结合的α－甲酰基对氟基苯乙酸甲酯缩甲氧胺席夫碱肝癌细胞(HepG2)没有抑制作用。说明氨基硫脲可能是一个

活性基团。

表 10　抗胃腺癌细胞(BGC－823)筛选结果

化合物 / IC_{50} mg/mL	C13	C16	C26	C9	C3	Taxol
BGC－823	$>10\times10^{-6}$	7.27×10^{-6}	$>10\times10^{-6}$	5.22×10^{-6}	8.18×10^{-6}	2.56×10^{-8}

如表 10 已知：α－甲酰基对甲氧基苯乙酸甲酯缩氨基硫脲(C13)和 α－甲酰基对三氟甲基苯乙酸甲酯缩氨基硫脲(C26)对胃腺癌细胞(BGC－823)没有抑制作用。α－甲酰基苯乙酸甲酯缩氨基硫脲(C3)，α－甲酰基对甲基苯乙酸甲酯缩氨基硫脲(C9)和 α－甲酰基对氟基苯乙酸甲酯缩氨基硫脲(C16)对胃腺癌细胞(BGC－823)的抑制作用差别不大，其抑制的 IC_{50} 分别为 8.18×10^{-6} mg/mL，5.22×10^{-6} mg/mL，7.27×10^{-6} mg/mL。

表 11　抗人肺支气管癌细胞(NCI－H1650)筛选结果

化合物 / IC_{50} mg/mL	C13	C16	C26	C9	C3	Taxol
NCI－H1650	2.5×10^{-6}	1.98×10^{-6}	8.05×10^{-6}	5.34×10^{-6}	3.25×10^{-6}	3.17×10^{-8}

如表 11 已知：对人肺支气管癌细胞(NCI－H1650)的抗肿瘤筛选中，α－甲酰基对氟基苯乙酸乙酯缩氨基硫脲(C16)对其的抑制效果最好，半抑制率为 1.98×10^{-6}。苯环对位被甲氧基取代的 α－甲酰基对甲氧基苯乙酸甲酯缩氨基硫脲对人肺支气管癌细胞(NCI－H1650)的抑制活性与 α－甲酰基对氟基苯乙酸乙酯缩氨基硫脲接近。苯环对位被强吸电子的三氟甲基取代的 α－甲酰基对三氟甲基苯乙酸甲酯缩氨基硫脲和给电子的甲基取代的 α－甲酰基对甲基苯乙酸甲酯缩氨基硫脲，它们对人肺支气管癌细胞(NCI－H1650)的抑制活性相比于 α－甲酰基对甲氧基苯乙酸甲酯缩氨基硫脲和 α－甲酰基对氟基苯乙酸乙酯缩氨基硫脲有所减弱。说明苯环 4 号位被吸电子或给电子基团取代产生的电子效应对抗人肺支气管癌细胞有一定影响，但不是很大。

表 12　抗卵巢癌细胞(A2780)筛选结果

化合物 / IC_{50} mg/mL	C13	C16	C26	C9	C3	Taxol
A2780	3.40×10^{-6}	2.11×10^{-6}	2.59×10^{-6}	5.92×10^{-6}	$>10\times10^{-6}$	5.87×10^{-7}

如表 12 已知：α－甲酰基对甲氧基苯乙酸甲酯缩氨基硫脲(C13)、α－甲酰基对氟基苯乙酸乙酯缩氨基硫脲(C16)和 α－甲酰基对三氟甲基苯乙酸甲酯缩

氨基硫脲(C26)对卵巢癌细胞(A2780)的抑制活性比较接近,IC_{50}分别为3.40×10^{-6}mg/mL,2.11×10^{-6}mg/mL,2.59×10^{-6}mg/mL。苯环4号位被甲基取代的α-甲酰基对甲基苯乙酸甲酯缩氨基硫脲对卵巢癌细胞(A2780)的抑制作用有所减弱,其抑制的IC_{50}为5.92×10^{-6}mg/mL。苯环未被取代的α-甲酰基苯乙酸酯缩氨基硫脲对卵巢癌细胞(A2780)没有抑制作用。说明苯环4号位被取代是发挥抗卵巢癌活性的重要条件之一。

对照组Taxol(泰素,紫杉醇)对人结肠癌细胞(HCT-116)、人肝癌细胞(HepG2)、胃腺癌细胞(BGC-823)、人肺癌细胞(1650),卵巢癌细胞(SK-OV-3)都有很强的抑制作用,其半抑制率分别为1.95×10^{-8}mg/mL,1.06×10^{-8}mg/mL,2.56×10^{-8}mg/mL,3.17×10^{-8}mg/mL,5.87×10^{-7}mg/mL,其抑制活性比α-甲酰基苯乙酸酯缩氨基硫脲衍生物强了1-2个数量级。α-苯乙酸酯衍生物的抗肿瘤的构效关系需要深入讨论,继续优化设计新的化合物。

α-甲酰基苯乙酸酯衍生物,α-羟甲基苯乙酸酯衍生物,α-甲酰基苯乙酸缩甲氧胺衍生物,α-羟甲基苯乙酸酯苯甲酸酯衍生物,α-羟甲基苯乙酸酯水杨酸酯衍生物,α-羟甲基苯乙酸酯肉桂酸酯衍生物,α-羟甲基苯乙酸酯苯乙酸酯衍生物对人结肠癌细胞(HCT-116)、人肝癌细胞(HepG2)、胃腺癌细胞(BGC-823)、人肺支气管癌细胞(NCI-H1650、1650),卵巢癌细胞(A2780、SK-OV-3)都没有明显的抑制作用。说明硫脲基结构是发挥抗癌细胞的重要活性基团。

小结

32个苯乙酸酯衍生物进行抗肿瘤筛选,有5个化合物对肿瘤细胞有抑制作用。5个化合物都是α-甲酰基苯乙酸酯缩氨基硫脲类衍生物,对抗人结肠癌细胞(HCT-116)、人肝癌细胞(HepG2)、胃腺癌细胞(BGC-823)、人肺支气管癌细胞(NCI-H1650、1650),卵巢癌细胞(A2780、SK-OV-3)有抑制作用。α-甲酰基苯乙酸甲酯缩氨基硫脲(C3)、α-甲酰基对甲氧基苯乙酸甲酯缩氨基硫脲(C13)、α-甲酰基对氟基苯乙酸甲酯缩氨基硫脲(C16)、α-甲酰基对三氟甲基苯乙酸甲酯缩氨基硫脲(C26)对人结肠癌细胞(HCT-116)的半抑制率在1.91×10^{-6}mg/mL到8.8×10^{-6}mg/mL之间,α-甲酰基对氟基苯乙酸乙酯缩氨基硫脲C16对人结肠癌细胞(HCT-116)的抑制活性最好其半抑制率为1.91×10^{-6}mg/mL。抑制活性结果分析发现硫脲基可能是发挥抗肿瘤活性必不可少的基团之一。α-甲酰基苯乙酸酯缩氨基硫脲类衍生物相比于对照药紫杉醇的抗肿瘤活性低了1-2个数量级,因此,α-甲酰基苯乙酸酯缩氨基硫脲类衍生物的抗肿瘤活性有待提高。

3.4.3.3　苯乙酸酯α-葡萄糖苷酶抑制活性筛选

α-葡萄糖苷酶主要包括麦芽糖酶、蔗糖酶、异构麦芽糖酶、乳糖酶等酶类，能够在低聚糖的非还原末端切开α-1,4糖苷键，释放出葡萄糖。它能将对硝基苯基-α-D-吡喃葡萄糖苷催化还原为葡糖糖和对硝基苯酚。用对硝基苯基-α-D-吡喃葡萄糖苷为底物，在紫外可见分光光度仪上，于400nm处读取开始反应的1min内吸光度的变化值。此值相当于葡萄糖的生成速度，从而检测出化合物对α-糖苷酶的抑制活性。

生化试剂

29个α-取代苯乙酸酯衍生物

α-糖苷酶　　Sigma化学公司，活力为179 U/mg

磷酸氢二钾　　天津市恒兴化学试剂制造有限公司

磷酸二氢钾　　天津市大茂化学试剂厂

对硝基苯基α-D-吡喃葡萄糖苷(301)　　Johnson Matthey化学公司

二甲基亚砜(Dimethyl sulfoxide, DMSO)　　Sigma化学公司

仪器设备

可见/紫外分光光度计	UV4802	UNICO公司
冰箱	BCD-256KF A	海尔公司
pH计	FE20型	梅特勒-托利多仪器公司
低温冰柜	BD-198E	海尔公司

苯乙酸酯α-葡萄糖苷酶抑制效果测试

α-糖苷酶抑制剂的活性测定方法是根据以上原理并结合陈玲娟[55]等的方法改编而来的，在1.5mL的离心管中加入总量为1mL的反应体系，其中10μL样品溶液、30μL L-DOPA溶液、10μL α-糖苷酶溶液和950μL PBS，样品在反应体系中的浓度为100μmol/L。100μmol/L样品抑制率越高则抑制效果越好，反之则抑制效果越差，其计算公式[55]如下：

100μmol/L样品抑制率=[(OD标-*OD*样)/ OD标]×100%——公式1

式中OD标为对硝基苯基-α-D-吡喃葡萄糖苷和α-糖苷酶，没加样品时每分钟的吸光度变化值；OD样为对硝基苯基-α-D-吡喃葡萄糖苷、α-糖苷酶加样品时每分钟的吸光度。

实验结果

优选30个α-取代的苯乙酸酯衍生物进行抑制α-糖苷酶活性筛选。将30

个样品配置成10mmol/L的DMSO溶液，用对硝基苯基 α－D－吡喃葡萄糖苷为底物，磷酸氢二钾和磷酸二氢钾配置缓冲溶液为环境体系，加入适量的样品溶液，使样品在体系中的浓度为100μmol/L，25℃下均质3min，加入 α－糖苷酶后利用紫外可见光分光光度计，扫描一分钟内在400nm下光度值的变化量。利用公式1，得到了100μmol/L样品对 α－糖苷酶抑制率，其结果如表13。

表13　α－糖苷酶抑制筛选结果

序号	中文名字	编号	100μmol/L 抑制率
1	α－羟甲基对甲氧基苯乙酸甲酯	C11	无活性
2	α－甲酰基对甲氧基苯乙酸甲酯	C10	8.42%
3	α－羟甲基对甲基苯乙酸甲酯	C8	4.63%
4	α－甲酰基对甲基苯乙酸甲酯	C7	19.29%
5	α－羟甲基对氟基苯乙酸甲酯	C15	2.92%
6	α－甲酰基对氟基苯乙酸甲酯	C14	10.38%
7	α－羟甲基对三氟甲基苯乙酸甲酯	C22	无活性
8	α－甲酰基对三氟甲基苯乙酸甲酯	C21	无活性
9	α－羟甲基对四氢吡喃醚基苯乙酸甲酯	C29	4.91%
10	α－甲酰基对四氢吡喃醚基苯乙酸甲酯	C28	无活性
11	α－甲酰基对氟基苯乙酸甲酯缩甲氧胺席夫碱	C17	4.44%
12	α－羟甲基对四氢吡喃醚基苯乙酸甲酯苯甲酸酯	C30	6.27%
13	α－羟甲基对甲氧基苯乙酸甲酯苯乙酸酯	C12	0.91%
14	α－羟甲基对三氟甲基苯乙酸甲酯3,4－二甲氧基肉桂酸酯	C25	19.9%
15	α－甲酰基对三氟甲基苯乙酸甲酯缩氨基硫脲席夫碱	C26	24.05%
16	α－羟甲基对氟基苯乙酸甲酯3,4－二甲氧基肉桂酸酯	C20	37.84%
17	α－甲酰基对氟基苯乙酸酯缩氨基硫脲席夫碱	C16	55.27%
18	α－羟甲基对三氟甲基苯乙酸甲酯苯甲酸酯	C23	6.81%
19	α－羟甲基对三氟甲基苯乙酸乙酰水杨酸酯	C24	4.09%
20	α－甲酰基对甲基苯乙酸甲酯缩氨基硫脲席夫碱	C9	10.76%
21	α－甲酰基对甲氧基苯乙酸乙酯缩氨基硫脲席夫碱	C13	无活性
22	α－羟甲基对氟基苯乙酸乙酯苯甲酸酯	C18	无活性
23	α－羟甲基对氟基苯乙酸乙酰水杨酸酯	C19	无活性

续表

序号	中文名字	编号	100μmol/L 抑制率
24	α-甲酰基对三氟甲基苯乙酸甲酯缩甲氧胺席夫碱	C27	无活性
25	α-羟甲基对四氢吡喃醚基苯乙酸甲酯水杨酸酯	C31	无活性
26	α-羟甲基对四氢吡喃醚基苯乙酸甲酯肉桂酸酯	C32	无活性
27	α-羟甲基苯乙酸甲酯	C2	18.66%
28	α-羟甲基苯乙酸乙酯苯甲酸酯	C4	2.86%
29	α-羟甲基苯乙酸乙酰水杨酸酯	C5	3.68%
30	α-羟甲基苯乙酸甲酯肉桂酸酯	C6	1.1%
31	脱氧野尻霉素		52.52%

如表13所示，α-甲酰基对氟基苯乙酸酯缩氨基硫脲席夫碱(C16)对α-糖苷酶的抑制活性最好，100μmol/L 的样品浓度抑制率为55.27%，比阳性对照药脱氧野尻霉素的抑制活性要好。100μmol/L C2，C7，C9，C14，C20，C25，C26的样品浓度对α-糖苷酶抑制率在10%-50%之间。C4，C5，C6，C8，C10，C12，C15，C17，C23，C24，C29，C30对α-糖苷酶的抑制作用不强，100μmol/L 的浓度的抑制率在0%-10%之间。C11，C13，C18，C19，C21，C22，C27，C28，C31，C32对α-糖苷酶没有抑制作用。所以，对α-糖苷酶抑制有较强抑制活性的个数占所筛选α-苯乙酸酯衍生物的26.6%，抑制活性较弱的占40%，没有抑制活性的占33.4%。

结果与讨论

30个α-取代苯乙酸酯衍生物中，α-甲酰基苯乙酸酯缩氨基硫脲衍生物、α-甲酰基苯乙酸酯衍生物、α-羟甲基苯乙酸酯衍生物(C2)、α-羟甲基苯乙酸酯肉桂酸酯衍生物对α-糖苷酶有抑制活性对α-糖苷酶有明显的抑制活性。α-甲酰基苯乙酸缩甲氧胺衍生物，α-羟甲基苯乙酸酯苯甲酸酯衍生物，α-羟甲基苯乙酸酯水杨酸酯衍生物，α-羟甲基苯乙酸酯苯乙酸酯衍生物(C12)对α-糖苷酶的抑制活性比较弱(0~10%)，或者没有抑制作用。

α-甲酰基对氟基苯乙酸酯缩氨基硫脲席夫碱(C16)对α-糖苷酶的抑制活性最好，100μmol/L 的样品浓度抑制率为55.27%。苯环4号位被甲基取代的α-甲酰基对甲基苯乙酸酯缩氨基硫脲席夫碱(C9)，当浓度为100μmol/L时，对α-糖苷酶的抑制率为10.76%。苯环4号位被三氟甲基取代的α-甲酰基对三氟甲基苯乙酸酯缩氨基硫脲席夫碱(C26)，在100μmol/L的浓度时，对α-糖苷酶的抑制率为24.05%。苯环4号位被甲氧基取代的α-甲酰基对甲氧基苯乙酸酯缩氨基硫脲席夫碱(C13),在100μmol/L的浓度时,对α-糖苷酶的抑制率-8.04%。以上说明苯环的4号位给电子能力太强的基团取代对α-糖苷酶的抑制活性没有提高的作用,吸电子和体积比较大的基团取代苯环对位后,会减弱对α-糖苷酶的抑制作用。

100μmol/L的α-甲酰基对甲基苯乙酸甲酯(C7)对α-糖苷酶的抑制率为19.29%,α-甲酰基对氟基苯乙酸乙酯(C14)对α-糖苷酶的抑制率为10.38%,苯环4号位被甲氧基代的α-甲酰基对甲氧基苯乙酸甲酯(C10)的抑制率为8.42%,被三氟甲基取代的α-甲酰基对三氟甲基苯乙酸乙酯(C21)是-7.45%,被四氢吡喃醚基取代的α-甲酰基对四氢吡喃醚基苯乙酸甲酯(C28)的抑制率为-6.14%。以上说明在α-甲酰基苯乙酸酯系列中,苯环4号位被吸电子太强的官能团和体积太大的官能团取代对α-糖苷酶反而起促进作用。苯环4号位被甲氧基、氟基、甲基给电子基团取代的C10,C14,C7对α-糖苷酶的抑制活性依次增强.

100μmol/L的α-羟甲基苯乙酸甲酯(C2)对α-糖苷酶的抑制活性为18.66%,其余的同类物α-羟甲基对甲氧基苯乙酸甲酯(C11)、α-羟甲基对甲基苯乙酸甲酯(C8)、α-羟甲基对氟基苯乙酸乙酯(C15)、α-羟甲基对三氟甲基苯乙酸乙酯(C22)、α-羟甲基对四氢吡喃醚基苯乙酸甲酯(C29)对α-糖苷酶的抑制率在1.95%~4.91%,抑制效果不理想。

100μmol/L的α-羟甲基对三氟甲基苯乙酸乙酯3,4-二甲氧基肉桂酸酯C25和α-羟甲基对氟基苯乙酸甲酯3,4-二甲氧基肉桂酸酯C20对α-糖苷酶有抑制率分别为19.9%和37.84%。α-羟甲基苯乙酸甲酯3,4-二甲氧基肉桂酸酯C6对α-糖苷酶有抑制率为1.1%。以上说明在这个体系中苯环4号位被吸电子的基团取代能够增强抑制α-糖苷酶的活性。

3.4.3.4 抗肿瘤活性和α-糖苷酶抑制活性比较

表14 苯乙酸酯衍生物α-糖苷酶抑制活性和抗肿瘤活性对比结果

序号	编号	中文名字	100μmol/L抑制率	是否有抗肿瘤活性
1	C7	α-甲酰基对甲基苯乙酸甲酯	19.29%	无活性
2	C14	α-甲酰基对氟基苯乙酸甲酯	10.38%	无活性
3	C25	α-羟甲基对三氟甲基苯乙酸甲酯3,4-二甲氧基肉桂酸酯	19.9%	无活性
4	C26	α-甲酰基对三氟甲基苯乙酸甲酯缩氨基硫脲席夫碱	24.05%	有

续表

序号	编号	中文名字	100μmol/L 抑制率	是否有抗肿瘤活性
5	C20	α－羟甲基对氟基苯乙酸甲酯 3,4－二甲氧基肉桂酸酯	37.84%	无活性
6	C16	α－甲酰基对氟基苯乙酸酯缩氨基硫脲席夫碱	55.27%	有
7	C9	α－甲酰基对甲基苯乙酸甲酯缩氨基硫脲席夫碱	10.76%	有
8	C13	α－甲酰基对甲氧基苯乙酸乙酯缩氨基硫脲席夫碱	无活性	有
9	C3	α－甲酰基对苯乙酸乙酯缩氨基硫脲席夫碱	未测	有
10	C2	α－羟甲基苯乙酸甲酯	18.66%	无活性

如表 14,8 个 α－取代苯乙酸酯化合物对 α－糖苷酶有较强的抑制作用,5 个 α－取代苯乙酸酯化合物具有抗肿瘤活性。其中 3 个化合物既有抑制 α－糖苷酶的活性,又有抗肿瘤的活性;1 个有抗肿瘤活性,但未进行 α－糖苷酶抑制活性的筛选,1 个化合物只有抗肿瘤活性,5 个化合物只有 α－糖苷酶抑制活性。

5 个具有抗肿瘤活性的化合物,都是同一类型的化合物,属于 α－甲酰基苯乙酸乙酯缩氨基硫脲类化合物。它们的区别只是苯环 4 号位的取代基不同。8 个具有较强 α－糖苷酶抑制活性的化合物,是属于 4 类不同类型的苯乙酸酯衍生物,3 个 α－甲酰基苯乙酸乙酯缩氨基硫脲类化合物,1 个 α－羟甲基苯乙酸甲酯,2 个 α－甲酰基苯乙酸乙酯,2 个 α－羟甲基苯乙酸乙酯 3’,4’－二甲氧基肉桂酸酯,每个类型的不同之处只是苯环的取代基不同。

抗肿瘤的药物的毒副作用相比于其它药物,其毒副作用比较大。鉴于 α－甲酰基苯乙酸乙酯缩氨基硫脲类化合物既有抗肿瘤活性,又有 α－糖苷酶抑制活性,所以这类化合物不适合于作为 α－糖苷酶抑制剂药物进行研究。但作为新型的抗肿瘤的先导化合物,进行药物开发,还是具有一定的研究价值。

α－羟甲基苯乙酸甲酯类化合物,α－甲酰基苯乙酸乙酯类化合物,α－羟甲基苯乙酸乙酯 3’,4’－二甲氧基肉桂酸酯类化合物没有抗肿瘤活性,也没有这 3 类的 α－糖苷酶抑制剂的研究报道,适合于作为新型的 α－糖苷酶抑制剂进行研究。特别是 α－羟甲基苯乙酸乙酯 3’,4’－二甲氧基肉桂酸酯类化合物,对 α－糖苷酶抑制活性都比较好。已经筛选的两个这类化合物对 α－糖苷酶的抑制活性为 37.85% 和 19.9%,只比 α－甲酰基苯乙酸乙酯缩氨基硫脲类化合物略差,比阳性对照药去氧野艽霉素的抑制活性相差不是很大。

小结

抑制 α－糖苷酶体外检测结果:α－甲酰基对氟基苯乙酸酯缩氨基硫脲席夫

碱(C16)对 α-糖苷酶的抑制活性最好,100μmol/L 的样品浓度抑制率为 55.27%。7 个 α-取代苯乙酸酯衍生物,在 100μmol/L 的浓度时,对 α-糖苷酶的抑制率在 10% ~50%。α-甲酰基对氟基苯乙酸酯缩氨基硫脲席夫碱(C16)对 α-糖苷酶的抑制活性比阳性对照药脱氧野尻霉素的抑制活性稍好,α-羟甲基对氟基苯乙酸甲酯 3,4-二甲氧基肉桂酸酯 C20 对 α-糖苷酶的抑制活性比阳性对照药脱氧野尻霉素的抑制活性(37.84%)相差不大。α-甲酰基苯乙酸乙酯缩氨基硫脲类化合物既有抗肿瘤活性,不适合于作为 α-糖苷酶抑制剂药物进行研究。α-羟甲基苯乙酸甲酯类化合物,α-甲酰基苯乙酸乙酯类化合物,α-羟甲基苯乙酸乙酯 3',4'-二甲氧基肉桂酸酯类化合物适合于作为新型的 α-糖苷酶抑制剂进行研究。

3.4.3.5 总结与展望

总结

合成了 32 个 α-取代苯乙酸酯衍生物,得到了相关工艺及工艺参数,并通过了 1HNMR、13CNMR、IR 表征,确定了其化学结构。从各种苯乙酸酯经过克莱森缩合,得到 α-甲酰基取代苯乙酸酯席夫碱类的化合物,收率在 46.1% ~90.8%。α-甲酰基取代苯乙酸酯衍生物经硼氢化钠还原、酯化得到不同种类的 α-羟甲基苯乙酸酯交酯类衍生物,收率在 9.43% -68.56%。

α-甲酰基苯乙酸酯缩氨基硫脲衍生物对抗人结肠癌细胞(HCT-116)、人肝癌细胞(HepG2)、胃腺癌细胞(BGC-823)、人肺支气管癌细胞(NCI-H1650、1650),卵巢癌细胞(A2780、SK-OV-3)有抑制作用。α-甲酰基苯乙酸甲酯缩氨基硫脲(C3)、α-甲酰基对甲氧基苯乙酸甲酯缩氨基硫脲(C13)、α-甲酰基对氟基苯乙酸甲酯缩氨基硫脲(C16)、α-甲酰基对三氟甲基苯乙酸甲酯缩氨基硫脲(C26)对人结肠癌细胞(HCT-116)的半抑制率在 1.91×10^{-6}mg/mL 到 8.8×10^{-6}mg/mL 之间,α-甲酰基对氟基苯乙酸乙酯缩氨基硫脲 C16 对人结肠癌细胞(HCT-116)的抑制活性最好其半抑制率为 1.91×10^{-6}mg/mL。

抑制 α-糖苷酶体外检测结果:α-甲酰基对氟基苯乙酸酯缩氨基硫脲席夫碱(C16)对 α-糖苷酶的抑制活性最好,100μmol/L 的样品浓度抑制率为 55.27%。7 个 100μmol/L 的样品浓度对 α-糖苷酶抑制率在 10% ~50%。α-甲酰基对氟基苯乙酸酯缩氨基硫脲席夫碱(C16)对 α-糖苷酶的抑制活性比阳性对照药脱氧野尻霉素的抑制活性稍好,α-羟甲基对氟基苯乙酸甲酯 3,4-二甲氧基肉桂酸酯 C20 对 α-糖苷酶的抑制活性比阳性对照药脱氧野尻霉素的抑制活性(37.84%)相差不大。α-羟甲基苯乙酸甲酯类化合物,α-甲酰基苯乙酸乙酯类化合物，α－羟甲基苯乙酸乙酯 3’，4’－二甲氧基肉桂酸酯类化合物适合于作为新型的 α－糖苷酶抑制剂，具有一定的研究价值。

展望

通过体外抑制 α－糖苷酶抑制活性和抗肿瘤活性的筛选，发现了抗肿瘤的 α－甲酰基取代苯乙酸酯缩氨基硫脲类的苗头化合物，为设计高效低毒的抗肿瘤药物提供了分子骨架。α－甲酰基对氟基苯乙酸酯缩氨基硫脲席夫碱（C16）对 α－糖苷酶的抑制活性也比较好的抑制作用，比阳性对照药脱氧野尻霉素的抑制活性稍强；羟甲基对氟基苯乙酸甲酯 3，4－二甲氧基肉桂酸酯 C20 对 α－糖苷酶的抑制活性比阳性对照药脱氧野尻霉素的抑制活性稍弱。这两个化合物作为先导化合物，优化其结构，研究其抑制机理，毒理活性等，具有一定的研究价值。

第4章　杂环、酚类酪氨酸酶抑制剂

4.1　杂环类酪氨酶抑制剂

2015年M. J. Matos基于香豆素结构，设计并合成了一类3-芳基、3-杂环芳基香豆素（图4-1），为了发现这类化合物对于酪氨酸酶抑制活性的结构特征，进行药理特征和分子对接评价[1]。所有合成的化合物都进行了酪氨酸酶抑制活性测试，其中5,7-二羟基-3-(3-噻吩基)香豆素（12b）是这一系列化合物中抑制活性是最好的，其IC_{50}值为0.19μmol/L，大约比阳性对照曲酸的抑制活

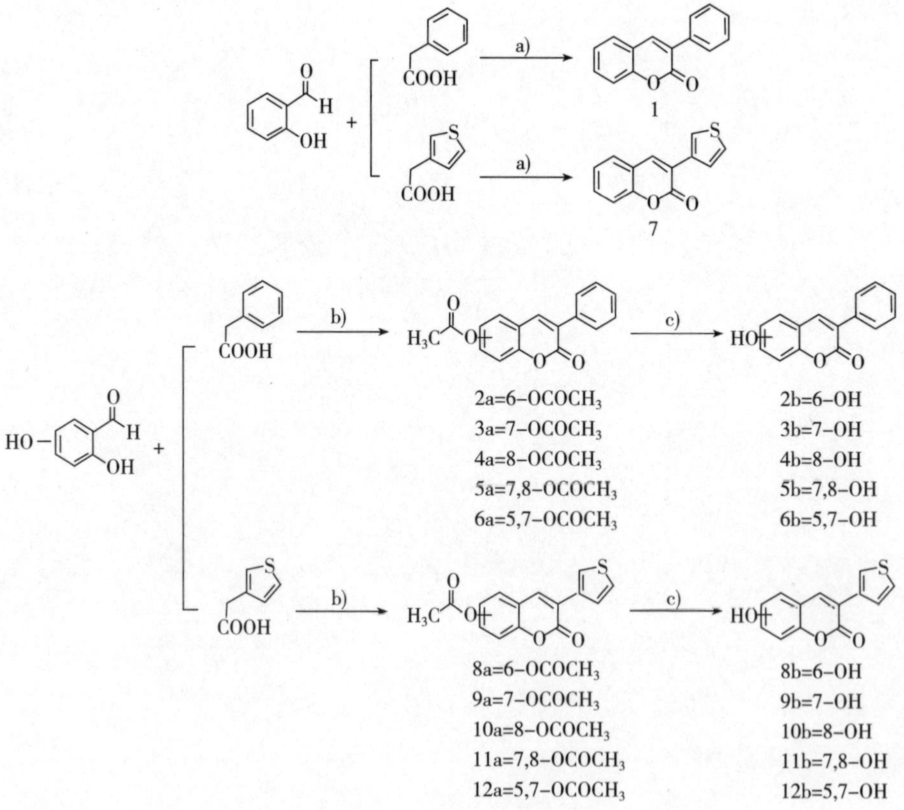

图4-1　3-芳基、3-杂环芳基香豆素合成路线

性强 100 倍。抑制动力学研究揭示以 L－多巴为底物，化合物 5，7－二羟基－3－（3－噻吩基）香豆素（12b）是酪氨酸酶的竞争型抑制剂，该化合物 B16F10 黑色素细胞的细胞毒性的缺失性也进行了评价，所有化合物抗氧化能力同样得到了测试，为了更好地讨论构效关系，进行了分子对接研究。

2015 年 Shailendra Asthana 等人研究了四个羟基香豆素（图 4－2）酪氨酸酶抑制活性的构效关系，其中两个化合物的羟基在苯环上，另外两个是羟基取代了吡喃酮环上的氢原子，对这些基团与酪氨酸酶相互作用方式进行了研究[2]。这些化合物对酶的作用呈现出不同行为特征，6－羟基香豆素、7－羟基香豆素对酪氨酸酶的抑制活性较弱。有趣的是催化产物是 6，7－二羟基香豆素，并未发现有 5，6－和 7，8－两种异构体的存在，而且当酪氨酸酶作用于典型底物 L－酪氨酸时，两类化合物都能够减少多巴色素形成。虽然没有在吡喃酮环上含有羟基的香豆素类化合物可以作为酪氨酸酶的底物的报道，但是 3－羟基香豆素却是酪氨酸酶的有效抑制剂，4－羟基香豆素不是酪氨酸酶的抑制剂。这些结果与分子对接模拟预测进行了比较，获得了一些关于新合成的具有香豆素结构化合物潜在的有用信息，如类似于 3－羟基香豆素结构。

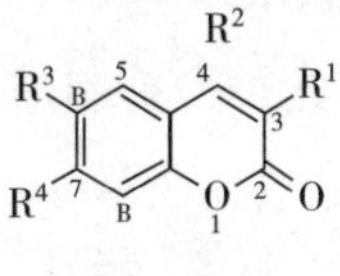

3-Hydroxycoumarin	R^1=-OH　R^2=-H	R^3=-H　R^4=-H
4-Hydroxycoumarin	R^1=-H　R^2=-OH	R^3=-H　R^4=-H
6-Hydroxycoumarin	R^1=-H　R^2=-H	R^3=-OH　R^4=-H
7-Hydroxycoumarin	R^1=-H　R^2=-H	R^3=-H　R^4=-OH

图 4－2　羟基香豆素结构

2016 年 Marwa Gardelly 等人合成了一类二氮磷酰胺（图 4－3），具体合成路线如下：一系列的 α－氨基腈通过 4－羟基香豆素和丙二睛及一系列芳香醛发生缩合反应得到，将所得的中间体和劳森试剂反应得到二氮磷酰胺及其异构体[3]。所有的目标化合物都经过了核磁共振氢谱、碳谱、磷谱及质谱进行表征。所有合成化合物都进行了细胞毒性和酪氨酸酶抑制活性考察，结果发现所有化合物都是潜在的酪氨酸酶抑制剂。

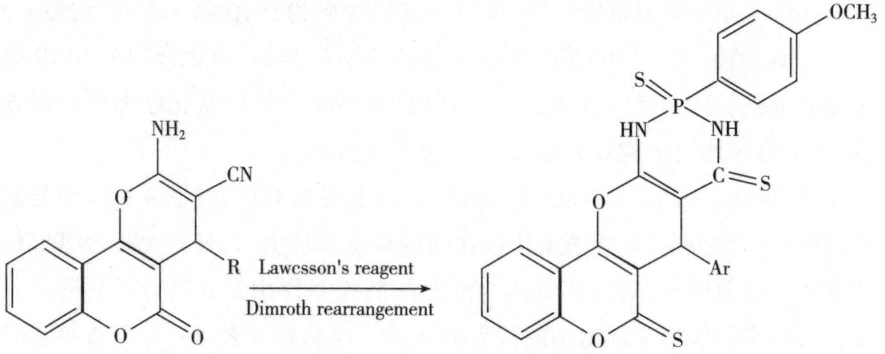

NH2 CN O R Lawcsson's reagent Dimroth rearrangement O O OCH3 S P HN NH C S O Ar O S

图4-3 二氮磷酰胺的合成

2016年Pingping Liu等人为了找到高效低毒的酪氨酸酶抑制剂,设计并合成了一系列N-芳基-N'-苯基硫脲衍生物(图4-4),并测试了所合成化合物对于酪氨酸酶二酚酶的抑制活性[4]。测试结果表明,4,5,6,7-氢-2-[{(苯氨基)硫甲基}氨基]-苯[b]硫苯基-3-甲酸衍生物(3a-i)表现出中等抑制酪氨酸酶二酚酶的活性。当骨架4,5,6,7-四氢苯基[b]硫苯基-3-甲酸被2-(1,3,4-三氮唑-2-基)硫代羧酸(6a-k)取代后,抑制活性明显加强,特别是化合物N-[3-(三氟甲基)苯基-N'-[5-(羰甲硫基)-1,3,4-三唑-2-基]硫脲(6h)对酪氨酸酶抑制活性最强,其半抑制浓度为6.13μmol/L,明显高于阳性对照曲酸,抑制机理研究表明该化合物是非竞争型抑制剂。

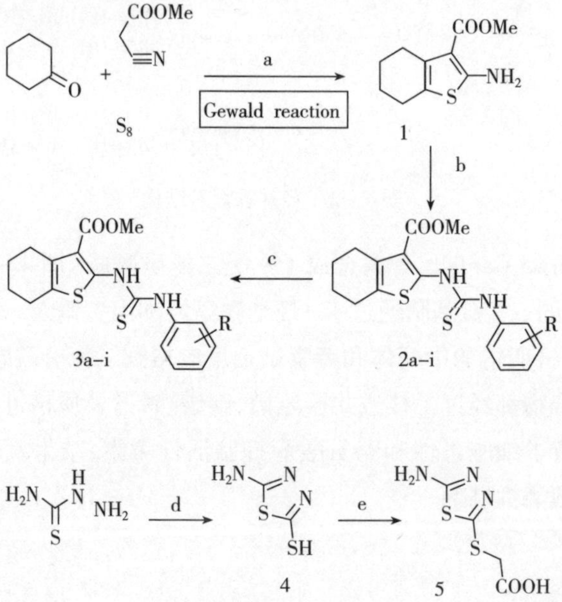

COOMe O + N S_8 a Gewald reaction COOMe NH2 S 1 b COOMe S NH NH S R 2a–i c COOMe S NH NH S R 3a–i

H H2N N NH2 S d H2N N S N SH 4 e H2N N S N S COOH 5

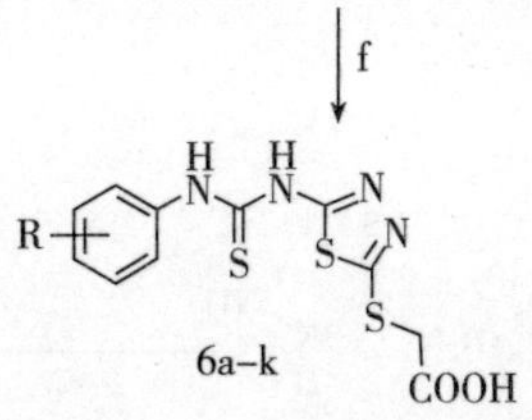

图 4－4　N－芳基－N'－苯基硫脲衍生物的合成

2015 年 Wenlin Xie 等人通过 5－羟基－2－氯甲基 －4H－吡喃－4－酮和 5－取代－3－巯基－4－氨基－1,2,4－三氮唑发生亲核取代反应合成 5－取代－3－[5－羟基－4－吡喃－2－基甲基巯基]－4－氨基－1,2,4－三唑(图 4－5)[5],并考察了目标化合物对于酪氨酸酶的抑制活性,结果显示所有合成的化合物对酪氨酸酶具有明显的抑制活性,特别是化合物 5－(4－氯苯基) －3－[5－羟基－4－吡喃－2－基－甲巯基]－4－氨基－1,2,4－三唑具有最好的酪氨酸酶抑制活性,其半抑制浓度为(4.50 ±0.34)μmol/L。5－(4－氯苯基) －3－[5－羟基－4－吡喃－2－基－甲巯基]－4－氨基－1,2,4－三唑的抑制动力学证实该化合物对酪氨酸酶的抑制作用属于竞争型抑制,同时也讨论了该类化合物的构效关系。

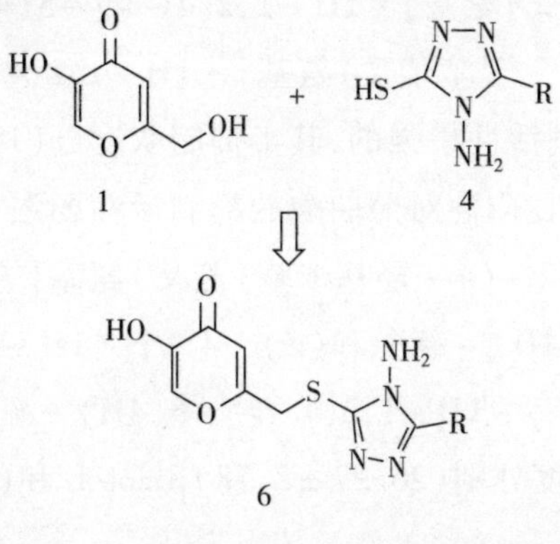

图 4－5　5－取代－3－[5－羟基－4－吡喃－2－基甲基巯基]－4－氨基－1,2,4－三唑的合成

发现新型高效低毒酪氨酸酶抑制剂及开发较好细胞毒试剂在药物开发和研究中仍然是非常重要的。2015 年 Hua－Li Qin 等人合成了一类含有查尔酮结构的吡唑啉衍生物(图 4－6),探讨了这类化合物的酪氨酸酶二酚酶的抑制活性[6]。结果表明,这类化合物具有明显的酪氨酸酶抑制活性,其中 4 个化合物对酪氨酸酶的抑制活性比阳性对照曲酸还要强。采用分子对接法研究了所合成化合物与酪氨酸酶相互作用的分子机制,构效关系和分子模拟结果具有高度的一致性。

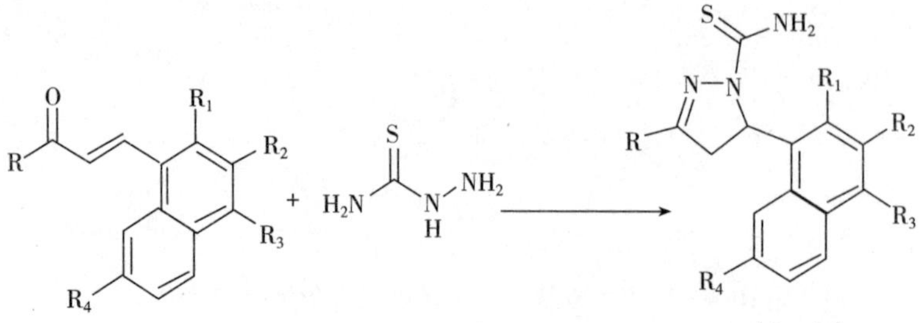

O R R_1 R_2 R_3 R_4 + S H_2N N H NH_2 → S NH_2 N－N R R_1 R_2 R_3 R_4

（1-6） （1a-6a）

1a:R=Ra,R_1=H,R_2=H,R_3=$N(CH_3)_2$,R_4=H
2a:R=Ra,R_1=H,R_2=H,R_3=OCH_3,R_4=OCH_3
3a:R=Ra,R_1=OCH_3,R_2=OCH_3,R_3=H_5,R_4=H
4a:R=Rb,R_1=H,R_2=H,R_3=$N(CH_3)_2$,R_4=H_1
5a:R=Rb,R_1=H,R_2=H,R_3=OCH_3,R_4=$OCHN_3$
6a:R=Rb,R_1=OCH_3,R_2=OCH_3,R_3=H,R_4=H

图4－6 吡唑啉衍生物的合成路线

2015年Muhammad Rafiq等人通过微波辅助合成法合成了10个三唑席夫碱类化合物（图4－7），亚胺是通过氨基取代的三唑和不同取代的醛类化合物反应。所有合成的化合物评价了其对酪氨酸酶的抑制活性，其中（Z）－3－（2－氟苄基）－4－［（4－氟苄叉）氨基］－1H－1，2，4－唑－5（4H）－硫酮和（Z）－3－（2－氟苄基）－4－［（4－羟基苄叉）氨基］－1H－1，2，4－唑－5（4H）－硫酮是所有合成化合物中抑制活性最强的，其半抑制浓度分（10.09 ±1.03）μmol/L和（6.23 ±0.85）μmol/L，比阳性对照曲酸的抑制活性要强[7]。与阳性对照曲酸相比，化合物（Z）－4－｛［2－（4－羟基苯基）乙叉］氨基｝－3－（2－甲氧苄基）－1H－1，2，4－三唑－5（4H）－硫酮和（Z）－4－｛［（1H－吡咯－2－基）甲叉］氨基｝－3－（4－羟基苄基）－1H－1，2，4－唑－5（4H）－硫酮具有中等的酪氨酸酶抑制活性，其半抑制浓度分别（20.27 ±2.78）μmol/L和（26.02 ±4.14）μmol/L。

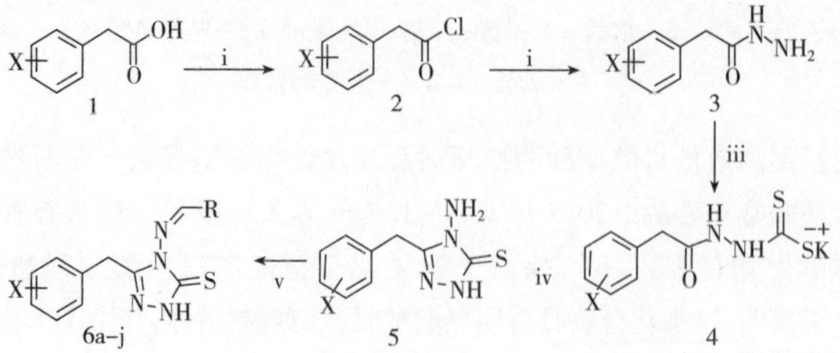

X OH O 1 —i→ X Cl O 2 —i→ X H N NH_2 O 3 —iii→ X H N NH S $\overset{-+}{SK}$ O 4 —iv→ X NH_2 N S N－NH 5 —v→ X N R N S N－NH 6a–j

图4－7

采用双倒数曲线分析了(Z)-3-(2-氟苄基)-4-[(4-氟苄叉)氨基]-1H-1,2,4-唑-5(4H)-硫酮和(Z)-3-(2-氟苄基)-4-[(4-羟基苄叉)氨基]-1H-1,2,4-唑-5(4H)-硫酮的抑制动力学,这两个化合物属于非竞争型抑制剂,其抑制常数分别为0.023 mmol/L和0.022 mmol/L。

2014年景临林等人以叠氮糖炔丙基溴和香草醛为原料,通过醚化"Click chemistry"、氧化、水解4步反应得到一系糖基化三氮唑香草酸衍生物(图4-8),产物结构经^{1}HNM、IR、EI-MS和元素分析确认[8]。通过对目标化合物进行酪氨酸

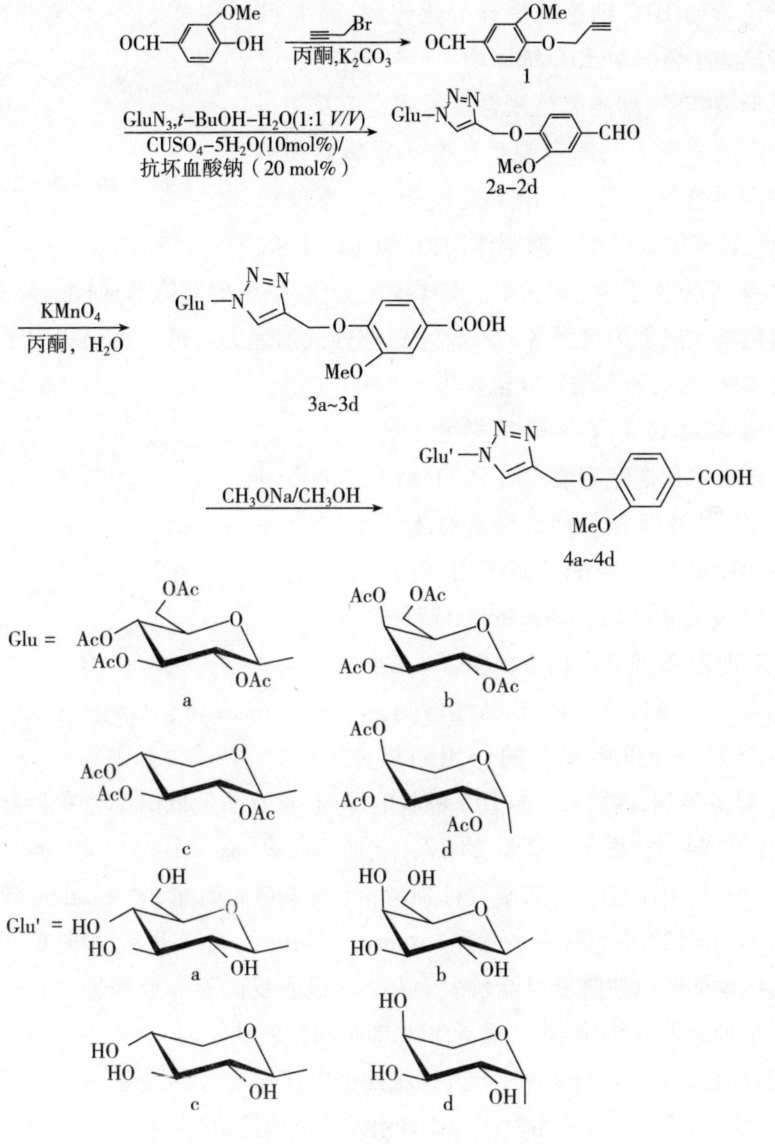

OMe
OCH—　—OH
Br
丙酮,K_2CO_3
OMe
OCH—　—O
1

$GluN_3$,t-BuOH-H_2O(1:1 V/V)
$CUSO_4$-$5H_2O$(10mol%)/
抗坏血酸钠(20 mol%)
N=N
Glu-N
O　CHO
MeO
2a-2d

$KMnO_4$
丙酮,H_2O
N=N
Glu—N
O　COOH
MeO
3a~3d

CH_3ONa/CH_3OH
N=N
Glu'—N
O　COOH
MeO
4a~4d

Glu =　OAc　AcO　AcO　OAc　O　a
AcO　OAc　AcO　OAc　O　b
AcO　AcO　OAc　O　c
AcO　AcO　AcO　O　d

Glu' =　OH　HO　HO　OH　O　a
HO　OH　HO　OH　O　b
HO　HO　OH　O　c
HO　HO　OH　O　d

图4-8　糖基化三氮唑香草酸衍生物的合成

酶抑制活性筛选,结果表明,所得目标化合物均具有较强的酪氨酸酶抑制活性,其中3-甲氧基-4-((1-((2R,3R,4S,5S,6R)-3,4,5-三羟基-6-(羟基甲基)四氢-2H-吡喃-2-基)-1H-1,2,3-三唑-4-基)甲氧基)苯甲酸活性最佳,IC_{50}为(0.12 ± 0.04)mmol/L。实验结果证明:糖基化三氮唑基团的引入可以显著提高香草酸对酪氨酸酶的抑制活性,为现有酪氨酸酶抑制剂的结构修饰与改造提供了一条新的途径。

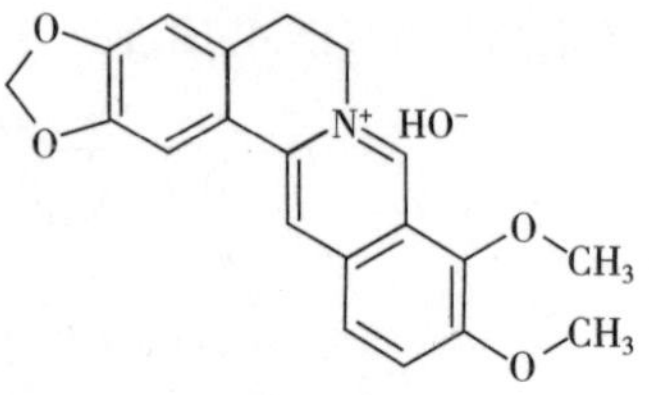

2010年杨菁为了测定盐酸小檗碱(图4-9)对小鼠黑色素瘤B16细胞的诱导分化作用,以MTT法测定盐酸小檗碱对B16细胞增殖的抑制作用,通过克隆形成实验、细胞黑色素含量的测定以及B16细胞体内成瘤能力的测定观察盐酸小檗碱对B16细胞的诱导分化作用[9]。结果发现盐酸小檗碱对B16细胞具有明显的增殖抑制作用;并使B16细胞生长缓慢,平铺不重叠,呈正常上皮样细胞分化,细胞的克隆形成能力降低,细胞内的黑色素生成能力增强,C57/BL小鼠体内成瘤能力降低。结果得出了盐酸小檗碱对B16细胞具有诱导分化作用。

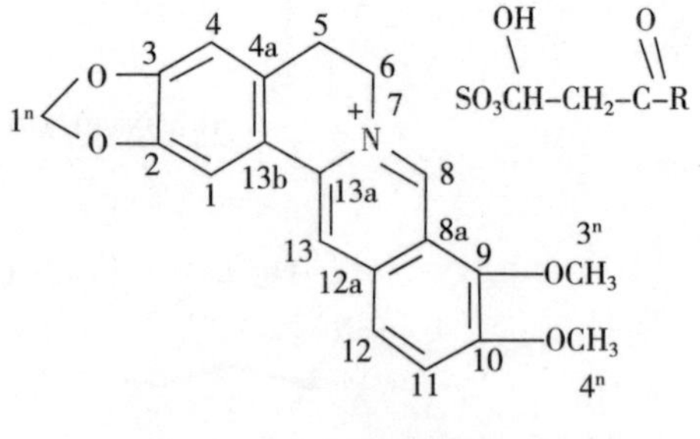

图4-9　盐酸小檗碱的结构式

2010年林菁为了探讨黄癸素(图4-10)诱导小鼠黑色素瘤B16细胞凋亡的作用,应用MTT法检测黄癸素对体外培养的B16细胞增殖的抑制作用,观察量效及时效关系;通过Rhodamine123染色、琼脂糖凝胶电泳、Hoechst33258染色、caspase -3和caspase-8活性检测,观察黄癸素诱导细胞凋亡的作用和机制[10]。结果发现,黄癸素抑制B16细胞的增殖,具有明显的时间依赖性和浓度依赖性,作用24 h、48 h、72 h的IC_{50}分别16.59 mg·L^{-1}、9.29 mg·L^{-1}和6.22 mg.L^{-1};B16细胞经黄癸素处理后,出现染色质固缩、DNALadder等凋亡表现,Rhodamine123染色荧光降低和caspase - 3、caspase - 8活性增强,提示黄癸素引起线粒体膜电位降低可能触发caspases级联反应而导致细胞凋亡。得出了黄癸素可抑制B16细胞增殖,诱导B16细胞凋亡的结论。

图4-10　黄癸素结构式

2014年Chin-Feng Chan等人合成四个甲巯咪唑(2-巯基-1-甲基咪唑,MMI)衍生物(图4-11),考察了这四个化合物的酪氨酸酶抑制活性,探讨了甲基咪唑及其衍生物对酪氨酸酶的抑制动力学[11]。结果表明，三个化合物对酪氨酸酶具有明显的抑制活性，IC_{50}值分别为1.50 mmol/L，4.11 mmol/L和1.43 mmol/L，其他化合物不具有任何酪氨酸酶抑制活性。动力学分析表明，化合物3是非竞争型抑制剂，而化合物1、2是混合型抑制剂，进一步研究表明化合物3表现出在B16/F10黑色素瘤细胞的黑色素合成具有明显的抑制作用，并未表现出细胞毒性，化合物1、2具有同样的特性。

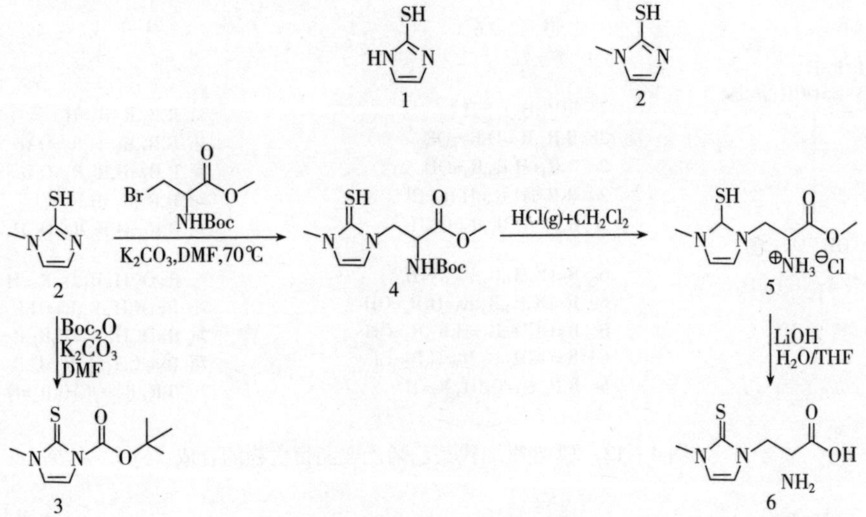

图4-11　甲巯咪唑衍生物的合成

2014年Ibrahim Chaaban等人合成了一系列新的噁唑啉取代萘乙烯乙酸酯衍生物和噁唑啉取代-4-甲氧基萘乙烯乙酸酯（图4-12）[12]。对所合成的化合物通过国家癌症中心进行体外抑制黑色素瘤活性测试，两个含有乙酰氧基在苯环1位或3位表现出很好的抗肿瘤活性，同时通过计算机模拟揭示了该类化合物的药效团。

2014年Mariusz Mojzych等人合成了具有酪氨酸酶抑制活性的西地那非类似物——吡唑[4,3-e][1,2,4]哒嗪磺酰胺（图4-13）[13]。活性测试结果表明，西地那非类似物具有一定的酪氨酸酶抑制活性，但具有很低的PDE5A抑制活性。构效关系研究表明，化合物中的伯氨基或仲氨基是影响抑制活性的重要基团；酚羟基能够与铜离子发生螯合作用而导致酶的抑制和底物络合相竞争，而这类含有伯氨基和仲氨基的化合物的竞争抑制可能是由于氨基的影响，然而其他化合物表现出较弱的酪氨酸酶抑制活性，因此可以推测与一个铜离子初步结合应该能够得到加强，主要是通过抑制剂和酶活性中心的另外一个铜离子或氨基酸残基发生相互作用，从而达到抑制酪氨酸酶活性的目的，分子的柔性和二氨乙基的存在使得化合物能够与酪氨酸酶氨基酸残基产生很强的相互作用，这一点也通过分子对接得到了证实。

1 R=H
5 R=OCH_3,*from*4

2a R,R_1,R_2,R_3=H
2b R,R_1,R_3=H,R_2=OH
2c R,R_1=H,R_2,R_3=OH
2d R,R_1=H,R_3=H,R_2=Cl
2e R,R_2=H,R_1,R_3=OCH_3

6a R=OCH_3,R_1,R_2,R_3=H
6b R=OCH_3,R_1,R_3=H,R_2=OH
6c R=OCH_3,R_1=H,R_2,R_3=OH
6d R=OCH_3,R_1,R_3=H,R_2=Cl
6e R,R_1,R_3=OCH_3,R_2=H

3a R,R_1,R_2,R_3=H
3b R,R_1,R_3=H,R_2=OAc
3c R,R_1=H,R_2,R_3=OAc
3d R,R_1,R_3=H,R_2=Cl
3e R,R_2=H,R_1,R_3=OCH_3

7a R=OCH_3,R_1,R_2,R_3=H
7b R=OCH_3,R_1,R_3=H,R_2=OAc
7c R=OCH_3,R_1=H,R_2,R_3=OAc
7d R=OCH_3,R_1,R_3=H,R_2=Cl
7e R,R_1,R_3=OCH_3,R_2=H

图 4－12　噁唑啉取代萘乙烯乙酸酯衍生物的合成

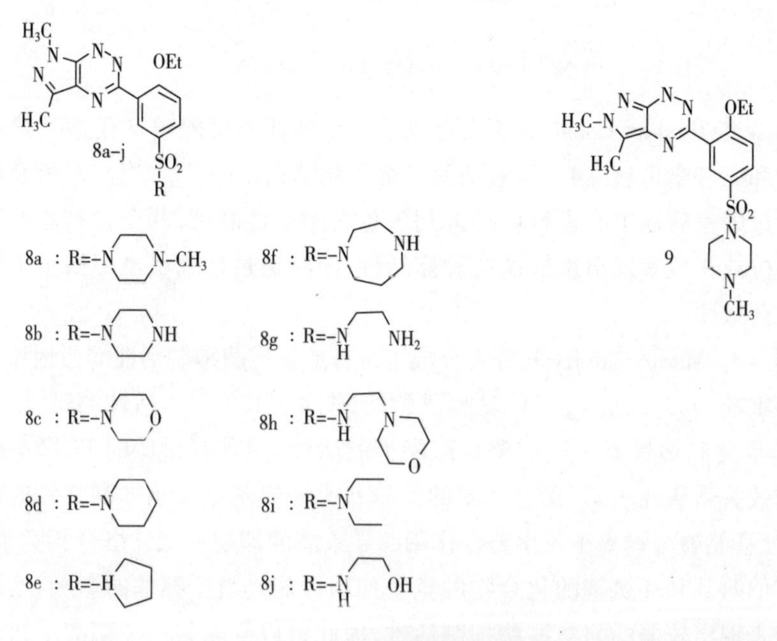

图 4－13　吡唑哒嗪磺酰胺结构式

2014 年 Zhiyong Chen 等人设计并合成了一类羟基或甲氧基取代 5 – 苯乙烯巴比妥或硫代巴比妥，评价了所有化合物对于酪氨酸酶二酚酶的抑制活性[14]。结果表明，部分化合物比阳性对照广泛用作酪氨酸酶抑制剂的曲酸活性要强。特别是 3,4 – 二羟基化的 5 – (3,4 – 二羟基苯乙烯) 嘧啶 – 2,4,6(1H,3H,5H) – 三酮抑制活性最强，其 IC_{50} 值为 1.52μmol/L，抑制动力学分析表明，该化合物对酪氨酸酶的抑制作用属于不可逆抑制，同时进行了构效关系分析，苯乙烯亚结构的引入是抑制酪氨酸酶活性的关键；苯环上羟基的数目对抑制活性有重要影响；苯环上 4 – 位上的羟基是绝对必需的。推测了该类化合物与酪氨酸酶相互作用的模式（图 4 – 14）。

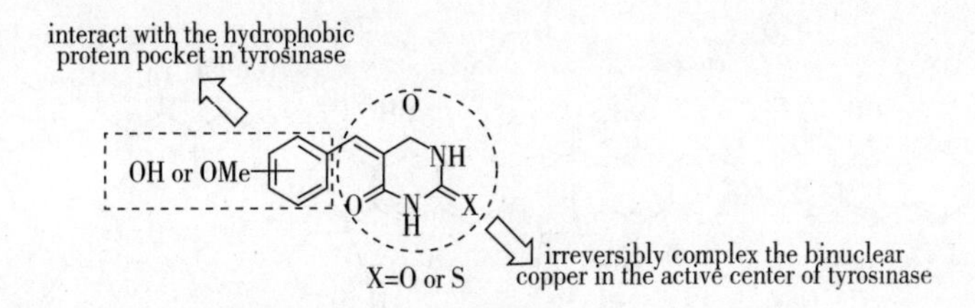

图 4 – 14　与酪氨酸酶相互作用模式

2014 年 Likai Xia 等人以硝酸铈铵为催化剂通过[3 + 2]环加成得到 5 – 羟基 – 4 – 乙酰基 – 2,3 – 二氢萘[1,2 – b]并呋喃（图 4 – 15）[15]。所有合成化合物都进行了酪氨酸酶抑制活性、抗氧化活性、抗菌活性评价。以 L – DOPA 为底物，采用紫外光谱测试了所合成化合物对酪氨酸酶的抑制活性，在所合成的化合物中，具有一个乙酰基和富电子的二氢萘并呋喃结构的化合物对酪氨酸酶抑制活性最强，其半抑制浓度为 8.91μg/mL，与阳性对照曲酸的抑制活性相当，抑制机理研究表明该化合物是可逆竞争型抑制剂，与曲酸抑制机理相似，该化合物的抑制常数为 Ki = 11.2μmol/L。

2013 年 Dong – Fang Li 等人为了找到新型酪氨酸酶抑制剂，以曲酸为起始原料，设计并合成了一类羟基吡啶酮和 L – 苯丙氨酸结合物（图 4 – 16）[16]。对所合成的化合物进行了酪氨酸酶抑制活性评价，结果发现，((S) – (5 – (苄氧基) – 1 – 辛基 – 4 – 氧代 – 1,4 – 二氢吡啶 – 2 – 基) 甲基 – 2 – 氨基 – 3 – 苯基丙酸，5e) 表现出较强的酪氨酸酶抑制活性，对于单酚酶和二酚酶的半抑制浓度分别是 12.6μmol/L 和 4.0μmol/L。抑制机理研究证实了这些结合物是混合型抑制剂，这就说明这类化合物能够同时和游离酶、酶—底物络合物相互作用，MTT 法证实该类化合物无毒性，因此，该类化合物可以用于食品保鲜和化妆品行业。

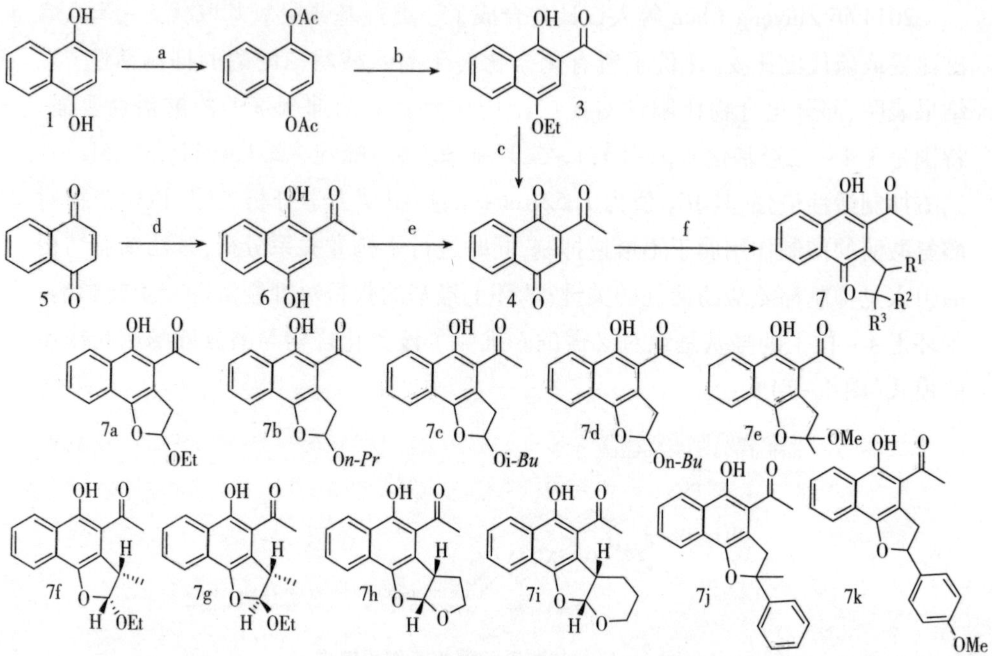

图4－15　二氢萘并呋喃衍生物的合成

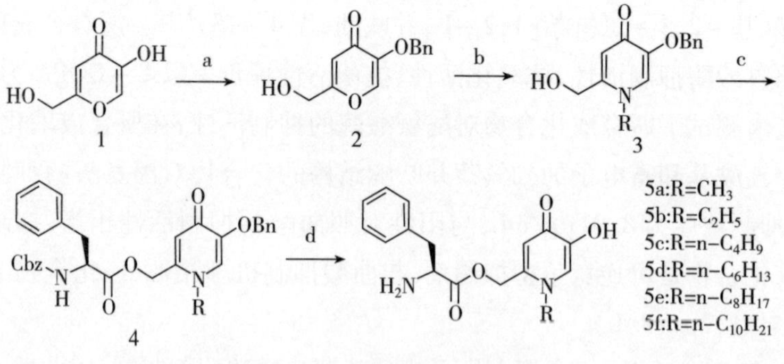

图4－16　吡啶酮衍生物的合成

Artoxanthochromane（图4－17）是一种D－A型4－异丙基苯基邻苯二酚和氧化白藜芦醇的连接产物，2013年Horng－Huey Ko等人从黄果菠萝蜜的心材中分离出来，通过核磁共振一维谱、二维谱及其他光谱对所得化合物进行结构表征，Artoxanthochromane化合物的结构为2－（2，4－羟基苯基）－3－（3，5－二羟基苯基）－7－羟基－4，4－二甲基色满烷[17]，提出了这一化合物生物合成的机理，探讨了其对酪氨酸酶的抑制活性和清除自由基的能力，结果表明，该化合物具有微弱的酪氨酸酶抑制活性。

2013年Mohamed M. El Sadek等人合成了一类新的呋喃基-1,3,4-噁二唑衍生物,并测试了所合成化合物酪氨酸酶抑制活性,考察了呋喃基-1,3,4-噁二唑衍生物不同取代基对酪氨酸酶抑制活性的影响[18]。希望能够找到用于临床治疗由于酪氨酸酶活性失调引起的帕金森病、精神分裂症、孤独症、注意力缺损、多动症及癌症等。构效关系讨论表明,苯环上没有取代基呋喃基-1,3,4-噁二唑衍生物具有最好的酪氨酸酶抑制活性,这可能是由于位阻小而使得其易于和酪氨酸酶活性位点相结合,当苯环上有氯原子或硝基时会使抑制活性明显下降;化合物含有两个1,3,4-噁二唑环或者一个噁二唑和一个三氮唑环对酪氨酸酶抑制活性影响明显,含有两个1,3,4-噁二唑环和一个1,2,3-三氮唑环的化合物抑制活性最好,随着含氮杂环数目的增加,对酪氨酸酶抑制活性也有明显影响。

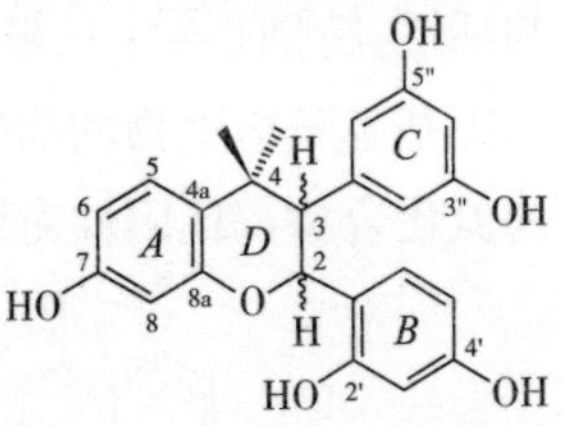

图4-17 Artoxanthochromane结构式

2013年Ramesh L. Sawant等人为了找到潜在的酪氨酸酶抑制剂,在前期研究的基础上,合成了一系列2,5-二取代1,3,4-噁二唑(图4-18),所有的合成化合物都进行了酪氨酸酶抑制活性评价,为了找到这类化合物对于酪氨酸酶抑制活性的结构需求进行了3D QSAR和分子对接研究[19]。结果表明,大部分化合物具有酪氨酸酶抑制活性,3D QSAR和分子对接揭示了电负性和较少大基团的存在会使得酪氨酸酶抑制活性变得更好。

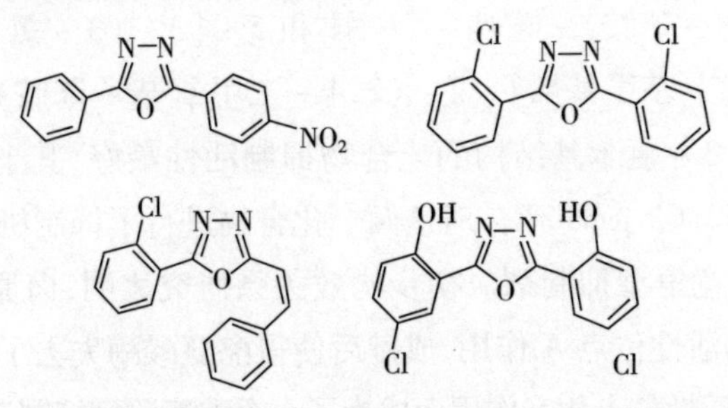

图4-18 2,5-二取代1,3,4-噁二唑结构式

2013年Hooshang Hamidian等人设计并合成了一类新的含有噁唑酮环的偶氮化合物(图4-19),通过4-氨基马尿酸重氮化分别与N,N-二甲基苯胺、1-萘酚、2-萘酚偶合,在与4-氟苯甲醛或4-三氟甲氧基苯甲醛缩合得到六个偶氮染料,所有新合成的化合物通过光谱进行了结构表征[20]。对所合成的化合物进行了酪氨酸酶抑制活性评价,结果表明,所有化合物都具有较高的酪氨酸酶抑

制活性，特别是4－三氟甲氧基衍生物具有最好的抑制酪氨酸酶的活性，N，N－二甲基苯胺衍生物的酪氨酸酶抑制活性强于1－萘酚和2－萘酚衍生物，所有的偶氮化合物都比阳性对照曲酸、L－含羞草氨酸的抑制活性好。

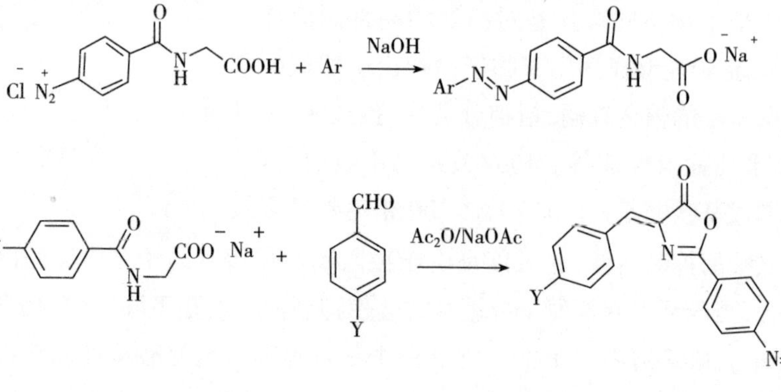

图4－19　噁唑酮偶氮化合物合成路线

2013年Shrikant S. Gawande等人在前期开发新型酪氨酸酶抑制剂的基础上，以吡唑和噻唑烷酮为母核，合成了一类杂环稠合体[21]。一类新型结构的2－(2，4－甲氧基 苯胺)－5甲叉－4－噻唑烷酮衍生物得以合成，通过将2－(2，4－甲氧基－苯胺)－噻唑－4－酮和各种吡唑醛发生缩合得到(图4－20)。酪氨酸酶抑制活性测试结果表明，部分所合成的化合物具有酪氨酸酶抑制活性，特别是化合物5－[3－(2－氯－苯基)－1－苯基－1H－吡唑－4－基甲基－烯]－2－(2，4－二甲氧基－苯胺)－噻唑－4－酮 和5－[3－(3－氯－苯基) －1－苯基－1 H－吡唑－4－基甲基烯]－2－(2，4－二甲氧基－苯胺)－噻唑－4－酮，具有2－氯苯基和3－氯苯基结构的化合物抑制活性最好，其半抑制浓度分别为34.12 μmol/L和52.62 μmol/L。对这两个化合物进行了抑制机理研究，证实了这两个化合物是可逆竞争型抑制剂。初步构效关系研究表明，卤原子可以与活性位点B作用而不是和活性位点A作用，成键后使得酪氨酸酶失去了催化活性；由于羧基能够与酪氨酸酶活性位点相互作用而成为了一个酪氨酸酶抑制活性的有效基团。

2013年Zhixuan Zhou等人合成了一类3，5－芳基－4，5－氢－1H－吡唑化合物(图4－21)，并测试了其酪氨酸酶的抑制活性[22]。二十个含有羟基的3，5－芳基－4，5－氢－1H－吡唑衍生物得以合成。抑制活性测试结果表明，在所有合成的化合物中，化合物1－(5－(3，4－二羟基苯基)－3－(4－羟基苯基)－4，5－二氢－1H－吡唑－1－基)乙基酮具有最好的酪氨酸酶抑制活性，其半抑制浓度为0.301μmol/L。动力学研究表明这些化合物是酪氨酸酶的竞争型抑制剂。

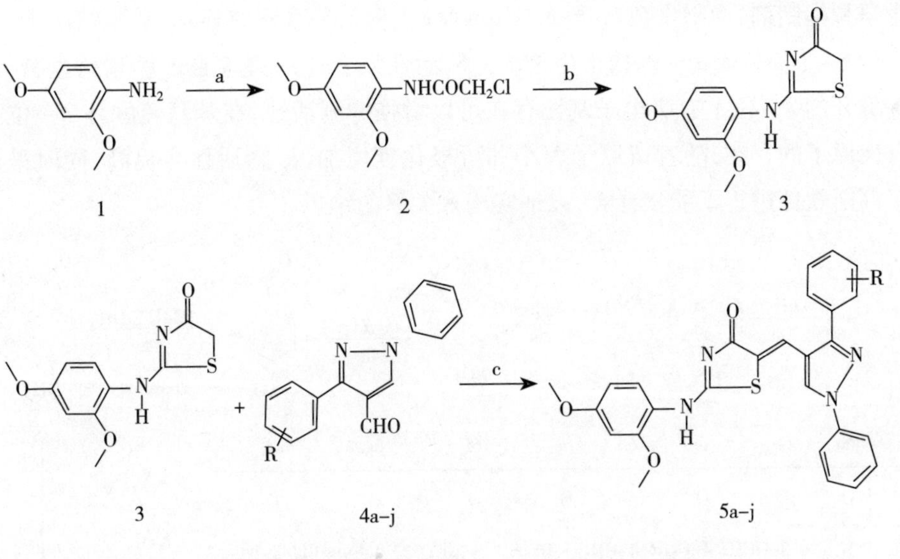

R=H,4-CH_3,+OCH_3,4-Cl,Br,4-F,3-Cl,2-Cl,2,4-Cl,2,4-OCH_3

图 4－20　2－(2,4－甲氧基 苯胺)－5 甲叉－4－噻唑烷酮衍生物合成路线

构效关系研究结果显示，羟基的位置能够明显影响抑制活性，苯环上两个羟基之间的化学基团也能够影响酪氨酸酶的抑制活性，杂环的存在能够增强对酪氨酸酶的抑制活性。这些研究成果为进一步开发新的能够用于临床、食品保鲜及化妆品行业的酪氨酸酶抑制剂奠定了基础。

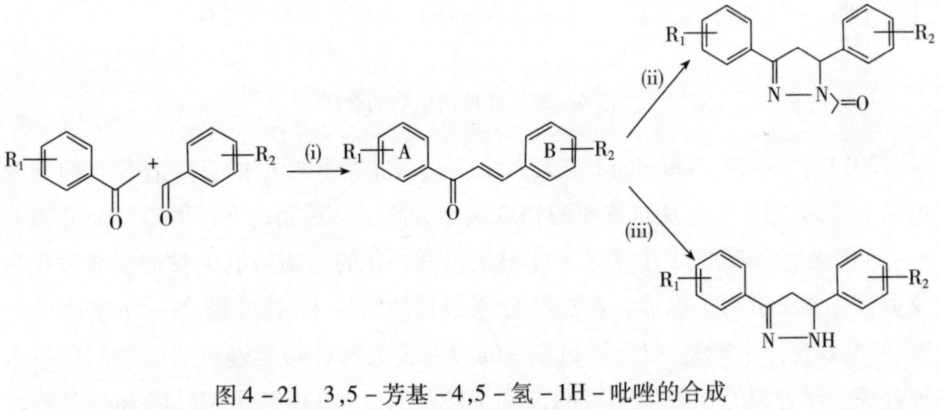

图 4－21　3,5－芳基－4,5－氢－1H－吡唑的合成

2012 年 Nahit Gençer 等人合成了一系列糖精衍生物(图 4－22)，并考察了这些化合物对于香蕉酪氨酸酶二酚酶的抑制活性[23]。结果表明，所有新合成的化合物都具有酪氨酸酶抑制活性，在这些化合物中，6－(3－碘代苯基硫脲）糖精是抑制活性最好的化合物，双倒数曲线分析了该化合物的抑制动力学性质，其为

竞争型抑制剂，抑制常数 K_i = 3.95μmol/L。构效关系研究显示，通常情况下，大多数6－(苯基硫脲) 糖精衍生物的抑制活性强于6－(苯基脲) 糖精衍生物；苯基脲环的3－位上有吸电子基团存在可以增强抑制活性，在苯基硫脲环3－位上有卤原子的存在，随着卤原子大小和可极化度增加，抑制活性会提高，同时采用了高斯软件进行了能量计算，以证实构效关系的结果。

ClSO₃H 60℃,20h；NH₃(eq)–ether 40℃,2h；CrO₃/H₂SO₄ rt,24h；Pd–C,EtOH 2h,reflux；R–C₆H₄–N=C=X, X=O,S, DMF,TEA, 12h,rt → 6a–y

	6a	6b	6c	6d	6e	6f	6g	6h
X	O	O	O	O	O	O	O	O
R	H	3–OCH$_3$	4–OCH$_3$	4–CH$_3$	3–Cl	4–Cl	3,4–di–Cl	3–NO$_2$

	6i	6j	6k	6l	6m	6n	6o	6p
X	O	O	O	O	S	S	S	S
R	4–NO$_2$	2–F	3–F	4–F	H	4–CH$_3$	4–OCH$_3$	2–F

	6q	6r	6s	6t	6u	6v	6y
X	S	S	S	S	S	S	S
R	3–F	4–F	3–I	3–Cl	2,4–di–Cl	3,5–di–Cl	4–NO$_2$

图4－22　糖精衍生物的合成

2013年Seikou Nakamura等人发现从荷花花蕾和叶中的提取液具有抑制黑色素合成的作用[24]。从这些甲醇提取液中分离了一种新的N－甲基降荷叶碱－N－氧化物生物碱和11中苄基异喹啉生物碱。新的生物碱的绝对构型通过化学及生物化学法进行了确证。在这些分离得到的组份中，荷叶碱、N－甲基降荷叶碱、鹅掌秋定、2－羟基－1－甲氧基－6a,7－去氢阿朴啡能够有效的抑制黑色素的合成。对合成的相关生物碱的抑制活性进行比较，得到了阿朴啡和苄基异喹啉生物碱(图4－23)的构效关系。另外(3－30)μmol/L荷叶碱和N－甲基降荷叶碱能够抑制酪氨酸酶的mRNA表达，(3－30)μmol/L N－甲基降荷叶碱能够抑制TRP－1 mRNA表达，(3－30)μmol/L荷叶碱能够抑制TRP－2 mRNA的表达。

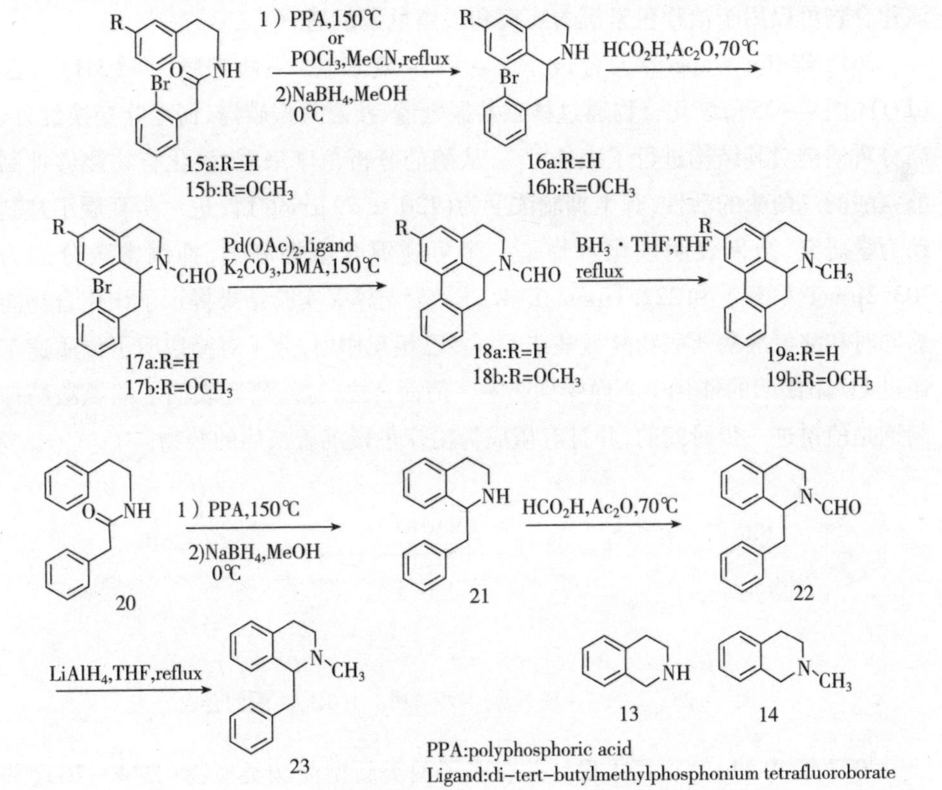

1）PPA,150℃ or $POCl_3$,MeCN,reflux 2)$NaBH_4$,MeOH 0℃

15a:R=H
15b:R=OCH_3

HCO_2H,Ac_2O,70℃

16a:R=H
16b:R=OCH_3

$Pd(OAc)_2$,ligand K_2CO_3,DMA,150℃

17a:R=H
17b:R=OCH_3

$BH_3\cdot THF$,THF reflux

18a:R=H
18b:R=OCH_3

19a:R=H
19b:R=OCH_3

1）PPA,150℃ 2)$NaBH_4$,MeOH 0℃

20　21

HCO_2H,Ac_2O,70℃

22

$LiAlH_4$,THF,reflux

23　13　14

PPA:polyphosphoric acid
Ligand:di-tert-butylmethylphosphonium tetrafluoroborate

图4－23　异喹啉生物碱的合成

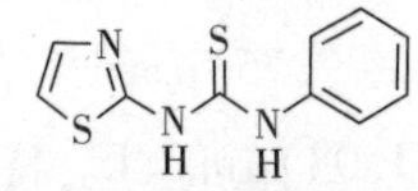

1－苯基－3－(2－噻唑)－2－硫脲(图4－24)是一个很好的多巴胺β－羟基酶抑制剂,能够阻止6－羟基多巴胺诱导的神经衰弱症,但并未见到1－苯基－3－(2－噻唑)－2－硫脲对黑色素合成的影响的报道。2016年Yong Hyun Kim等人研究了1－苯基－3－(2－噻唑)－2－硫脲抑制酪氨酸酶的活性及抑制黑色素合成的机理[25]。结果表明,1－苯基－3－(2－噻唑)－2－硫脲能够降低黑色素的生物合成,同时也能够抑制正常人表皮黑色细胞中酪氨酸酶催化活性。1－苯基－3－(2－噻唑)－2－硫脲的使用可以导致正常人表皮黑色细胞中酪氨酸酶水平的降低,而对酪氨酶相关蛋白1,酪氨酸酶相关蛋白2,小眼相关转录因子没有影响,1－苯基－3－(2－噻唑)－2－硫脲对于酪氨酸酶的mRNA表达无影响。1－苯基－3－(2－噻唑)－2－硫脲能够通过泛素依赖蛋白酶体来加速酪氨酸酶的降解。因此,1－苯基－3－(2－噻唑)－2－硫脲减少黑色素的生物合成是通过降低酶的活性和酪氨酸酶的稳定性实现的,

图4－24　1－苯基－3－(2－噻唑)－2－硫脲结构式

该化合物可以用于治疗色素沉着疾病和去黑色素试剂。

2016 年 Rui Wang 等人合成了 2 -（4 - 氟苯基）- 喹唑啉 - 4(3H) - 酮(FQ)(图 4 - 25)，该化合物通过核磁共振氢谱、核磁共振碳谱、傅里叶变换红外、高分辨质谱对其结构进行了表征[26]。从酶的分析角度来看，该化合物能够抑制酪氨酸酶二酚酶的活性，其半抑制浓度为(120 ± 2)μmol/L。进一步开展了抑制动力学研究，结果表明该化合物是一个可逆混合型抑制剂，抑制常数分别为 703.2μmol/L(K_I) 和 222.1μmol/L(K_{IS})。荧光猝灭实验结果揭示了该化合物能够同时和酪氨酸酶及底物酪氨酸或 L - 多巴相互作用，分子对接说明了传质速率通过 FQ 阻断酶的催化中心而受到影响。简而言之，这一研究表明了酪氨酸酶抑制剂是值得进一步研究的，并且有望成为治疗色素沉着疾病的药物。

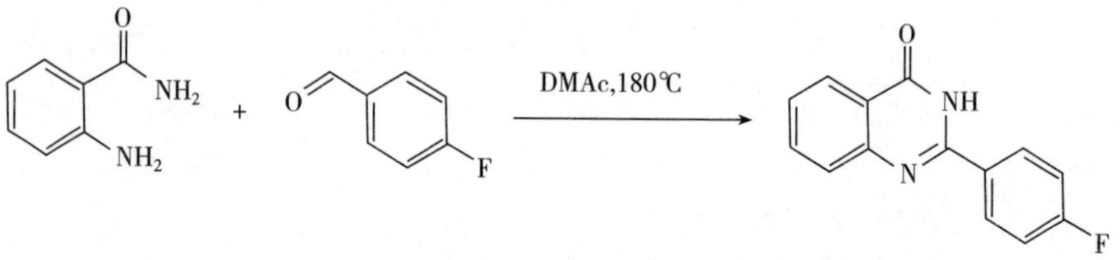

图 4 - 25　2 -（4 - 氟苯基）- 喹唑啉 - 4(3H) - 酮的合成

2017 年 Wenlin Xie 等人设计并合成了两类新型曲酸类似物(图 4 - 26)，并考察了所合成的化合物对酪氨酸酶的抑制活性[27]。结果显示所合成的化合物具有极好的酪氨酸酶抑制活性，其半抑制浓度在(1.35 ± 2.15)μmol/L -（17.50 ± 2.75)μmol/L 范围之间，然而标准对照品曲酸的半抑制浓度只有(20.00 ± 1.08)μmol/L。特别是化合物 5 - 苯基 - 3 - [5 - 羟基 - 4 - 吡喃酮 - 2 - 基 - 甲基巯基] - 4 -（2,4 - 二羟基 - 苄氨基）- 1,2,4 - 三唑 (4f) 表现出最好的酪氨酸酶抑制活性，其半抑制浓度为(1.35 ± 2.15)μmol/L，对该化合物抑制动力学

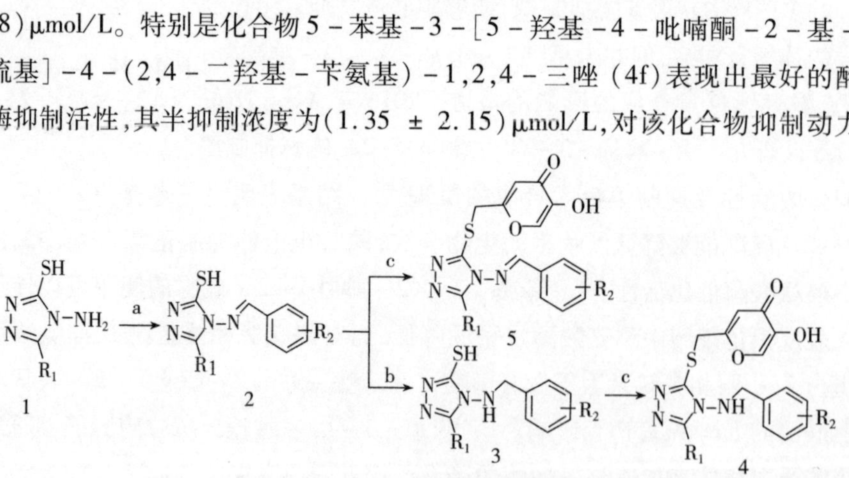

图 4 - 26　曲酸类似物的合成

分析揭示了该化合物对于酪氨酸酶的抑制作用属于竞争型抑制剂。与此同时，构效关系也进行了讨论，在三唑环上的取代基对抑制活性具有重要影响，但是在苯环5－位上存在取代基对酪氨酸酶抑制活性是不利的，有关这类化合物构效关系还要开展进一步深入研究。

酪氨酸酶通过单酚酶羟基化转化为二酚酶二存在于黑色素产物中，为了探明新的阻止皮肤色素沉着和黑色素瘤治疗方法，酪氨酸酶得到了广泛的研究。2017年Stefania Ferro等人合成了一类具有4－(4－氟苄基)哌啶结构的化合物，发现化合物3－(4－苄基哌啶－1－基)－1－(1H－吲哚－3－基)丙－1－酮(1a)是一个潜在的酪氨酸酶抑制剂，其半抑制浓度为252μmol/L。对3－(4－苄基哌啶－1－基)－1－(1H－吲哚－3－基)丙－1－酮的结构进行了修饰，合成了一类新该化合物的类似物(图4－27)，生物活性研究结果表明，部分所合成的化合物对于酪氨酸酶具有抑制作用，其半抑制浓度比先导化合物和阳性对照曲酸还要低[28]。通过双倒数曲线对抑制动力学进行了分析，化合物2a－c和10b是非竞争型抑制剂，而活性最好的化合物2d是混合型抑制剂，其半抑制浓度为7.56μmol/L。更进一步的实验和计算机模拟结构研究进一步阐明了该化合物与酪氨酸酶相互作用的模式。

	R_1
a	H
b	5-OCH_2
c	6-OCH_2
d	5,6-$(OCH_2)_2$

图4－27　3－(4－苄基哌啶－1－基)－1－(1H－吲哚－3－基)丙－1－酮类似物的合成

2016 年 Ming - Jen Chen 等人合成了两个曲酸甲巯咪唑衍生物(图 4 - 28)[29]。5 - 羟基 - 2 - {[(1 - 甲基 - 1H - 咪唑 - 2 - 基)硫代]甲基} - 4H - 吡喃 - 4 - 酮（化合物 4）和 5 - 甲氧基 - 2 - {[(1 - 甲基 - 1H - 咪唑 - 2 - 基)硫代]甲基} - 4H - 吡喃 - 4 - 酮（化合物 5）得到了合成,并考察了其对酪氨酸酶抑制活性及抑制动力学。化合物 4 展现的是一种量效依赖性有效单酚酶抑制剂,其半抑制浓度为 0.03 mmol/L,而该化合物对于二酚酶的抑制活性较弱,其半抑制浓度为 1.29 mmol/L,但仍然比阳性对照曲酸(IC_{50} = 1.80 mmol/L)的抑制活性强。动力学分析表明,该化合物对单酚酶和二酚酶都是非竞争型抑制剂。与此相对,化合物 5 并不具有任何酪氨酸酶抑制活性。化合物 4 具有抑制 DPPH 自由基的活性,其半抑制浓度为 4.09 mmol/L,比酪氨酸酶半抑制浓度要大,这就说明化合物 4 的抗氧化活性部分与其酪氨酸酶的抑制活性有关,化合物 4 也具有抑制 B16/F10 黑色素细胞的黑色素的合成,无细胞毒性。

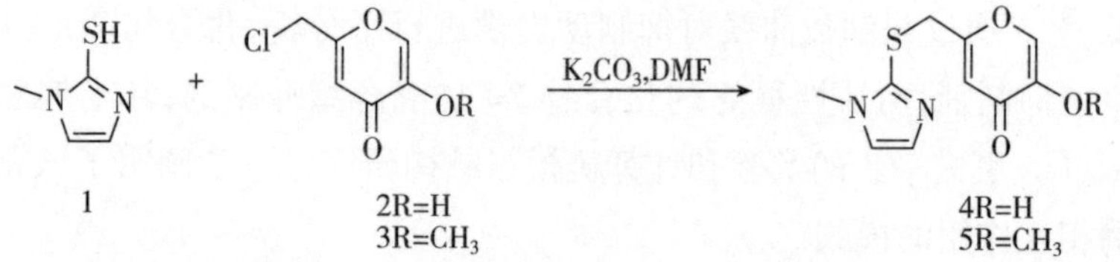

图 4 - 28　曲酸甲巯咪唑衍生物合成路线

黑色素是构成人的皮肤、毛发及眼睛颜色的重要因素,黑色素的存在能够有效防止皮肤受到阳光的损害,但是黑色素过量积累会导致各种皮肤病。2016 年 Mohd Fadhlizil Fasihi Mohd Aluwi 等人合成了一类噁唑衍生物(图 4 - 29),评价了该

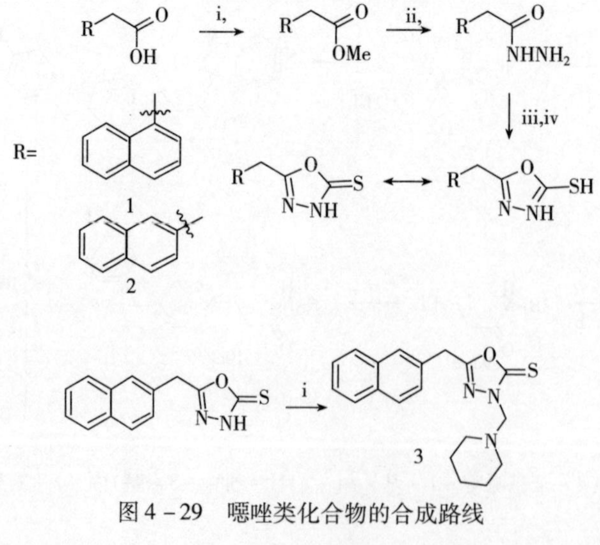

图 4 - 29　噁唑类化合物的合成路线

类化合物对蘑菇酪氨酸酶二酚酶的抑制活性及对于恶性黑色素瘤细胞 SK－MEL－28 的细胞毒性[30]。结果表明化合物1,2,3对蘑菇酪氨酸酶二酚酶具有较好的抑制活性,其半抑制浓度分别为40.46 μmol/L, 27.42μmol/L 和32.51μmol/L,通过双倒数曲线,抑制动力学揭示化合物1,3为混合型抑制剂,其抑制动力学常数 K_i 分别为3.8μmol/L和3.9μmol/L,然而化合物2是竞争型抑制剂,其抑制动力学常数 K_i 为0.7μmol/L。分子对接和动力学模拟进一步证实化合物2和酪氨酸酶活性中心相互作用的行为方式。通过MTT法测试三个化合物对恶性黑色素瘤细胞 SK－MEL－28 的细胞毒性,结果表明该类化合物细胞毒性较低,其浓度在100μmol/L。

2010年 Usman Ghani 等人合成了一系列1,3,4－噻二唑－2(3H)－硫酮 1,3,4－噁二唑－2(3H)－硫酮,4－氨基－1,2,4－三唑－5(4H)－硫酮和取代酰肼(图4－30),化合物具有一定的酪氨酸酶抑制剂。抑制剂的合理设计是基于细菌酪氨酸酶和土豆邻苯二酚氧化酶晶体结构中的活性位点[31]。抑制动力学和活性位点键合作用研究表明,噻二唑,噁二唑和三唑环和酪氨酸酶活性中心单齿状键合,包括疏水性对酪氨酸酶抑制作用的贡献。动力学分析显示所有25个化合物都是混合型抑制剂。构效关系研究揭示了三唑环上3－位,噻二唑/噁二唑环上5－位取代基在该类化合物与酪氨酸酶键合过程中起着重要作用。这一研究成果有利于开发用于治疗色素沉着疾病的新酪氨酸酶抑制剂。

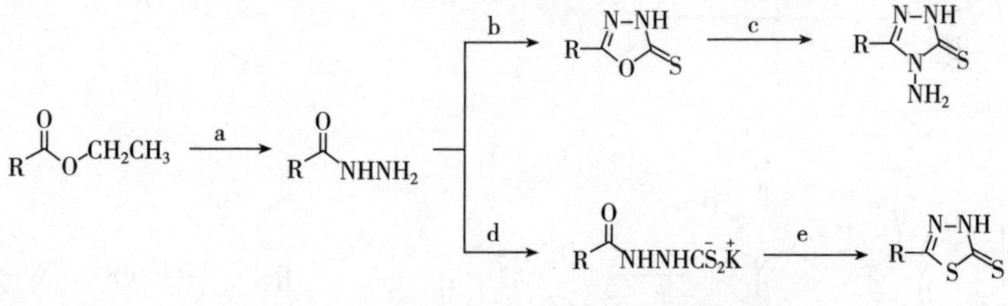

图4－30　噻二唑,噁二唑,三唑类化合物的合成

2016年 De－Yin Zhao 等人设计并合成了两类新型羟基吡啶酮衍生物(图4－31),这两类的化合物的合成都是以曲酸为起始原料,并考察了所合成的化合物酪氨酸酶抑制活性[32]。抑制活性结果表明,这两类化合物具有一定的酪氨酸酶抑制剂,特别是化合物6e和12a对单酚酶的半抑制浓度分别为1.95μmol/L和2.79μmol/L,都低于阳性对照曲酸的半抑制浓(12.50μmol/L),也考察了这两个化合物对二酚酶的抑制活性,半抑制浓度分别为8.97μmol/L和26.20μmol/L。抑制机理显示,这两个化合物是可逆抑制剂,抑制类型是竞争—非竞争的混合型

抑制剂。6e 的抑制动力学常数 K_I 和 K_{IS} 分别为 17.17μmol/L 和 22.09μmol/L，12a 的抑制动力学常数 K_I 和 K_{IS} 分别为 34.41μmol/L 和 79.02μmol/L，化合物 6e 表现出较强与酪氨酸酶螯合作用力。

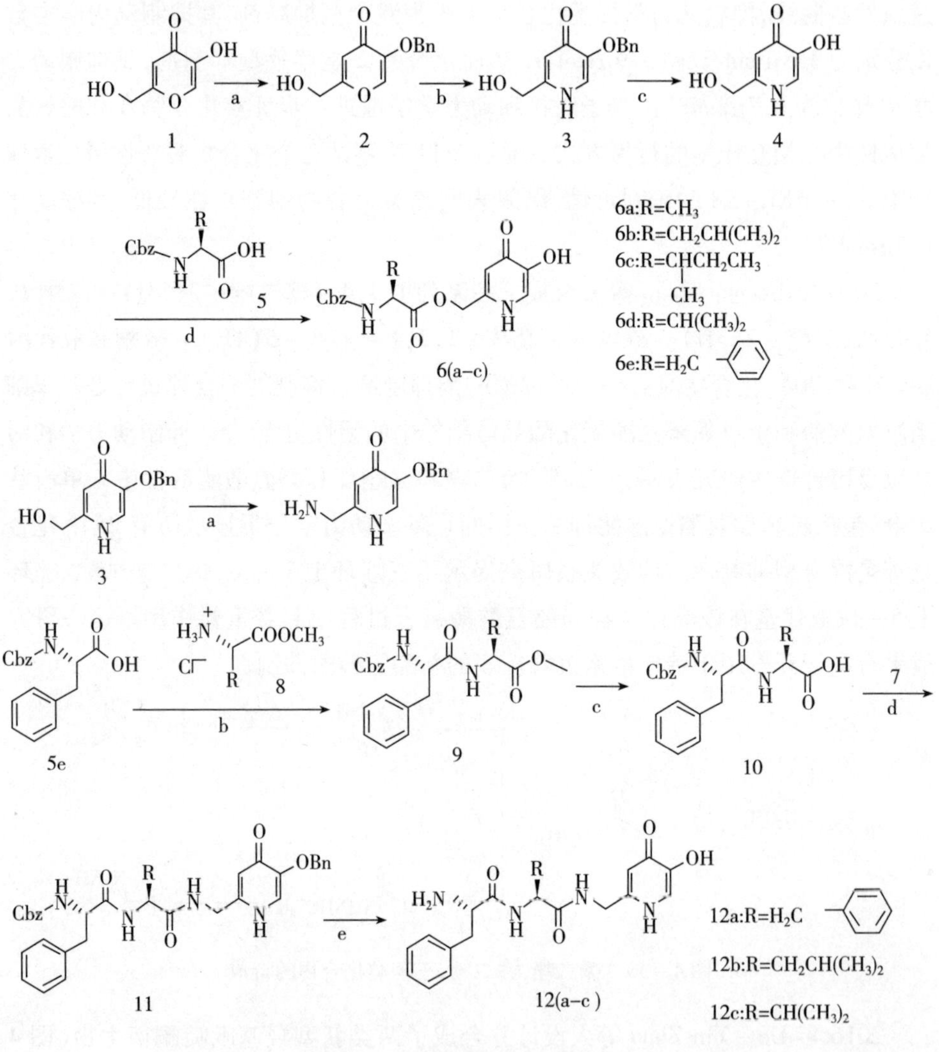

6a:R=CH_3
6b:R=$CH_2CH(CH_3)_2$
6c:R=$CH(CH_3)CH_2CH_3$
6d:R=$CH(CH_3)_2$
6e:R=H_2C–Ph

12a:R=H_2C–Ph
12b:R=$CH_2CH(CH_3)_2$
12c:R=$CH(CH_3)_2$

图 4-31　羟基吡啶酮衍生物的合成

2015 年南昌大学谢娟选取了 2-噻吩甲醛类似物(a、d、e)、2-呋喃甲醛(b)、2-吡咯甲醛(c)作为母体，成功地合成了缩氨基硫脲类化合物（图 4-32）1a-3a，1b-3b，1c-3c，1d 和 1e。采用酶动力学方法研究化合物抑制酪氨酸酶活性，结果表明：芳杂环缩氨基硫脲类化合物比其母体化合物表现出更强的抑制酪氨酸酶的活性；化合物 1a-1c 对酪氨酸酶抑制能力强于对照品曲酸，其中化合物 1a(IC_{50} = 0.43 μmol/L)的抑制能力最强,约是曲酸(IC_{50} = 17.94 μmol/L)的42倍;在硫脲基团中氨基取代的化合物比甲基取代和苯基取代的化合物具有更强的抑制活性;芳杂环中杂原子为 S 的化合物抑制能力强于杂原子为 O 和 N 的化合物[33]。

采用酶动力学方法、紫外光谱法、荧光光谱法、核磁滴定法等手段研究化合物对酪氨酸酶的抑制机理,并揭示其结构与活性之间的关系,结果表明:化合物 1a - 1c 为酪氨酸酶的可逆混合型抑制剂;铜离子螯合实验表明化合物 1a - 1c 与铜离子形成了 1.5∶1 的复合物;在对酪氨酸酶的荧光猝灭实验中,化合物 1a - 1c 对酪氨酸酶的猝灭能力强弱与抑制酪氨酸酶活性相一致,且 1a - Cu^{2+} 复合物与化合物 1a 相比,表现出更弱的猝灭作用。这些结果揭示抑制剂与酪氨酸酶的抑制作用位点可能与酶的活性中心铜离子相关。核磁滴定结果进一步证实这一结论,且硫脲末端氨基上氢原子与酪氨酸酶残基形成氢键更有利于增强酶的抑制活性。此外,ANS 荧光探针揭示了在抑制过程中抑制剂并没有改变酶的疏水环境。

提出了一系列新型、高效、优质的酪氨酸酶抑制剂,总结出的相关构效关系为进一步寻找此类型的酪氨酸酶抑制剂奠定理论基础。对酶抑制剂的抑制机理的研究,为化合物在美白化妆品方面的应用提供了更加科学的理论依据。

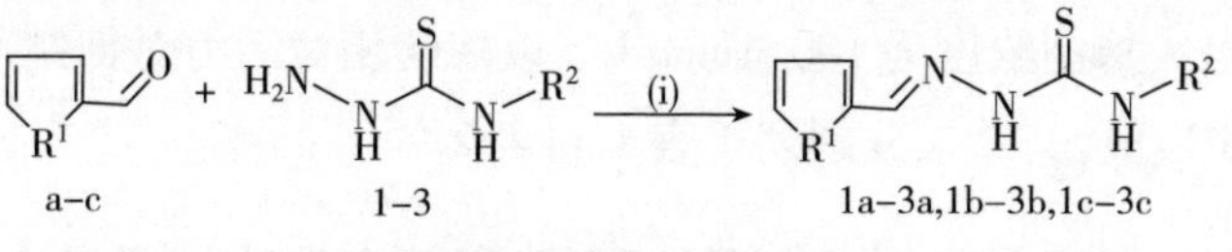

a–c + 1–3 —(i)→ 1a–3a,1b–3b,1c–3c

R^1=S,O or NH;R^2=H,Me or Ph

图4-32 芳杂环氨基硫脲席夫碱的合成路线

关于人类酪氨酸酶抑制剂的研究,开发很好的细胞毒试剂,仍然是当前药物开发和发现中最重要的事情,靶向人类酪氨酸酶多重抑制剂用于治疗黑色素沉积。2016 年 Mubashir Hassan 等人研究了香豆素衍生物、百里香酚衍生物、香兰素衍生物(图4-33)对于人类酪氨酸酶抑制活性,分子对接结果表明,这些化合物与酪氨酸酶的键合能比曲酸与酪氨酸酶的键合能要高[34]。

2010 年阎琴研究了对羟基苯甲醛杂环衍生物对酪氨酸酶的抑制作用[35]。

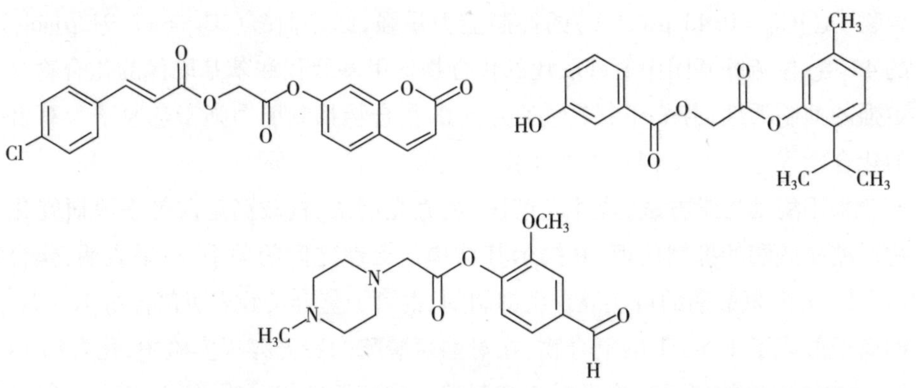

图4－33　香豆素衍生物、百里香酚衍生物、香兰素衍生物结构式

4.1.1　紫外光谱研究羟基取代苯甲醛类衍生物与酪氨酸酶的相互作用

将合成的羟基取代苯甲醛类衍生物进行酶抑制活性初筛，结果表明γ－氨基丁酸（GABA）类衍生物对酪氨酸酶没有明显的抑制作用，3，4－二氢嘧啶－2（1H）－酮（DHPMs）类衍生物抑制作用较弱，而部分巴比妥类衍生物则具有显著的抑制作用。

根据初筛结果豆腐果苷 IC_{50} 为2.60 mmol/L，因此对于糖苷类（硫代）巴比妥类衍生物最大抑制浓度为3.0 mmol/L，而其他（硫代）巴比妥类化合物及DHPM类化合物的最大抑制浓度为100 μmol/L。以酶的相对活力对抑制剂浓度作图得到化合物的 IC_{50} 值，其结果分别列于表1与表2。

表1　糖苷类（硫代）巴比妥对酪氨酸酶的抑制活性研究

Compound	Tyrosinase	
	IC_{50} (mmol/L)	Inhibition rate (%) at 3.0 mmol/L
豆腐果苷	2.62[a]	65.0
E1	0.78	100
E2	>3.0	NA[b]
E3	1.06[c]	78.92
E4	>3.0	NA
E5	>3.0	48.76
E6	>3.0	2.43
E7	0.43	100
F1	1.63	11.20
F2	1.50	100

续表

Compound	Tyrosinase	
	IC_{50} (mmol/L)	Inhibition rate(%) at 3.0 mmol/L
F3	2.26	75.08
F4	1.27	100
F5	>3.0	NA
F6	>3.0	NA
F7	1.73	100
F8	0.13	100
F9	1.16	100
F10	0.34	100
F11	0.47	100
F12	0.23	100
F13	0.43	100
F14	0.87	100
F15	0.28	100
F16	0.05	100
熊果苷	7.3[d]	NA

[a]Values in the literature is 2.54 mmol/L.

[b]No activity at 3.0 mmol/L.

[c]Values in the literature is 0.94 mmol/L.

[d] Values in the literatureis 5.3 – 8.4 mmol/L.

表2　巴比妥及DHPM类衍生物对酪氨酸酶抑制活性研究

Compound	Tyrosinase	
	IC_{50} (μmol/L)	Inhibition rate(%) at 100μmol/L
4–羟基苯甲醛	1220	16.40
2,4–二羟基苯甲醛	113.68	41.23
3,4–二羟基苯甲醛	112.52	43.03
4–甲氧基苯甲醛	262.68	35.76
2,4,6–三羟基苯甲醛	115.25	41.58
2,3,4–三羟基苯甲醛	129.20	38.70
F17	13.98	100
F18	>100	NA
F19	>100	NA
F20	17.15	100

续表

Compound	Tyrosinase	
	IC_{50}(μmol/L)	Inhibition rate(%) at 100μmol/L
F21	1.52	100
F22	>100	NA
F23	5.50	100
F24	>100	6.95
F25	100	50
F26	5.51	100
F27	>100	NA
F28	>100	3.88
F29	14.49	100
F30	>100	27.44
F31	>100	16.86
F32	>100	35.46
F33	6.10	100
F34	>100	17.28
F35	7.80	100
F36	>100	11.0
F37	70.12	70.83
F38	7.14	100
F39	>100	25.71
F40	95.83	51.64
F42	75.42	100
F43	45.45	100
F44	137.29	88.67
F47	28.43	100
F48	107.43	85.88
F49	34.20	100
F50	77.80	100
F57	70.02	100
F58	155.93	95.86
F59	112.76	100
F60	179.31	78.67
F61	132.46	97.49

续表

Compound	Tyrosinase	
	IC_{50}(μmol/L)	Inhibition rate(%) at 100μmol/L
F63	21.10	100
F66	84.74	68.05
F67	37.31	57.86
F72	71.19	72.55
F74	100	50
G16	100	50
G22	3.15	100

GABA类衍生物与DHPM类衍生物大多对酪氨酸酶抑制作用不明显，数据省略未列出。

分析表1的活性筛选数据，对于巴比妥糖苷类酪氨酸酶抑制剂的研究可以得到以下3个结论：

(1) 5－苯亚甲基硫代巴比妥是良好的结构修饰片断。根据初期活性筛选，巴比妥以及硫代巴比妥自身对酪氨酸酶没有抑制作用。但与醛缩合后形成苯亚甲基片断，抑制作用增强；如F1－F16的IC_{50}值大多小于其母体醛豆腐果苷与E1－E7的IC_{50}值。其中硫代巴比妥缩合产物F9－F10抑制活性明显高于相应的巴比妥缩合产物F1－F8。

(2)对于糖配基的亲酯性修饰有利于增强抑制活性。通过分别比较F1/F2，F3/F4，F9/F10，F11/F12，表明无论是巴比妥还是硫代巴比妥类化合物，四乙酰化的糖配基产物活性都高于未保护的糖基化合物。该结论说明，增加糖基的亲酯性有利于增强化合物对酪氨酸酶的抑制活性。

(3)间位羟基有利增强化合物的抑制作用。F8的活性是F4的10倍，而F16的活性是F12的5倍，而在F7/F4与F15/F12的比较中，均未发现此规律，因此可以推断3－羟基取代有利于增强化合物的抑制活性。

根据表2活性数据分析，探讨羟基取代、多醚取代(硫代)巴比妥之间的构效关系，可以得到以下结论：

(1)缩合后的苯亚甲基(硫代)巴比妥片断有利于酶抑制作用，缩合后产物活性与母体醛类化合物相比，具有明显的升高趋势。

(2)对于羟基取代与3,4－多醚取代的(硫代)巴比妥类衍生物，其巴比妥活性较硫代巴比妥高。但对于4－多醚取代的(硫代)巴比妥类衍生物结果刚好相反。

(3) 羟基数目以及取代位置对于酶抑制作用具有重要的影响。单独的增加羟基数目并不能够增强其活性,而保持树脂酚以及儿茶酚的结构片断则可以提高其抑制作用。

(4) 4-羟基对化合物的抑制作用具有决定性意义,对其醚化则可导致活性降低或消失。

(5) 还原亚甲基,增强化合物的柔性结构,则导致活性降低。

为了进一步羟基取代苯甲醛类衍生物的酪氨酸酶抑制机制,选取活性较好的化合物 F16, F21, F29 与 G16 进行酪氨酸酶抑制动力学研究。图 1 为所选化合物对酶抑制类型研究。

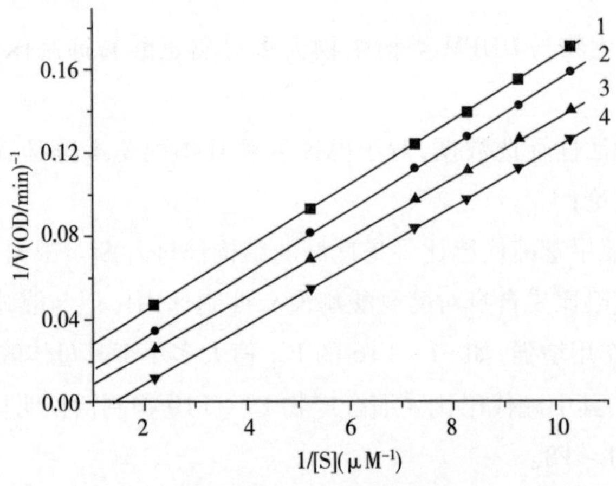

F16 对酪氨酸酶的不可逆抑制作用

曲线 1~4 F16 浓度分别为 0,0.5,0.8 和 1.0mmol/L

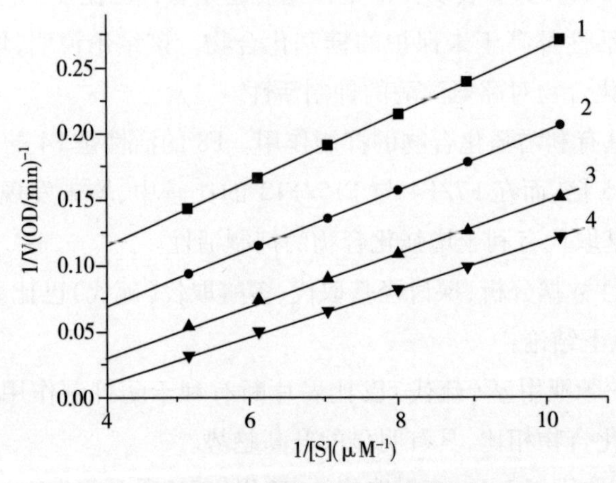

F21 对酪氨酸酶的不可逆抑制作用

曲线 1~4 F21 浓度分别为 0,1.0,1.5 和 2.0μmol/L

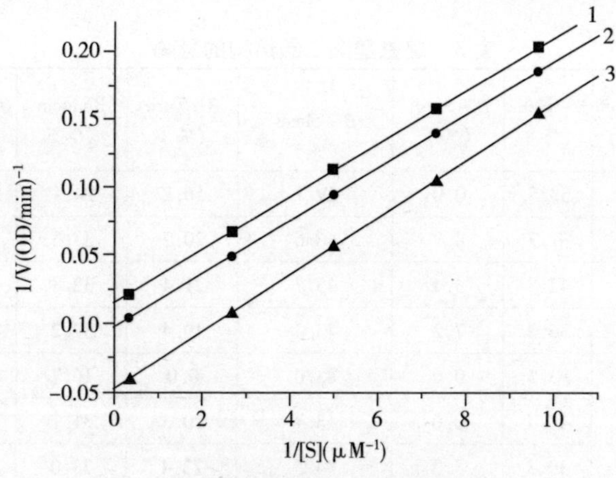

F29 对酪氨酸酶的不可逆抑制作用

曲线 1 ~3 F29 浓度分别为 0,0.5 和 2.0μmol/L

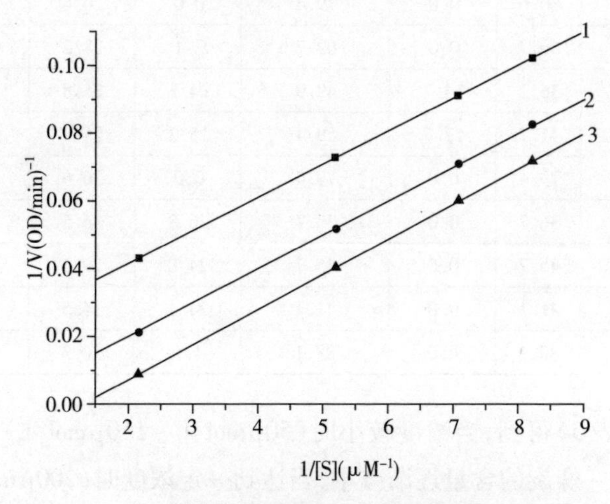

G16 对酪氨酸酶的不可逆抑制作用

曲线 1 ~3 G16 浓度分别为 20,50 和 100mmol/L

图 1F16，F21，F29 与 G16 对酪氨酸酶抑制类型

由图 1 可看出巴比妥类化合物对酪氨酸酶具有不可逆抑制作用,而 3,4 - 二氢嘧啶 - 2(1H) - 酮(DHPM)类化合物结构与巴比妥类似,其抑制类型也为不可逆抑制类型,说明 DHPM 类化合物与巴比妥类似以嘧啶环片断与酪氨酸酶活性中心发生不可逆结合作用。

4.1.2　圆二色谱研究羟基取代苯甲醛类衍生物与酪氨酸酶的相互作用

以活性最好的巴比妥类化合物 F17，F21，F29 与 F33 为例研究巴比妥类化合物对酪氨酸酶二级结构的影响,结果见表 3。

表3 酪氨酸酶二级结构的影响

Compound	Concentration (μmol/L)	α - Helix (%)	β - Sheet (%)	α - Helix + β - Sheet (%)	β - Turn (%)	Random (%)	β - Turn + Random (%)
F17	0	59.7	0.0	59.7	16.2	24.1	40.3
	50	37.7	8.9	46.6	20.2	33.3	53.5
	100	42.3	3.4	45.7	21.4	32.8	54.2
	150	36.2	7.2	43.4	19.4	37.2	56.6
	200	83.3	0.0	83.0	0.0	16.7	16.7
F29	50	48.7	0.0	48.7	20.3	31.1	51.4
	100	39.3	5.2	44.5	22.4	33.0	55.4
	150	45.1	0.0	45.1	20.5	34.5	55.0
	200	80.6	0.0	79.4	0.0	20.6	20.6
F21	50	69.7	0.0	69.7	2.1	28.2	30.3
	100	35.2	14.7	49.9	24.2	25.8	50.0
	150	41.7	17.7	59.4	15.2	25.4	40.6
	200	79.4	0.0	79.4	0.0	20.6	20.6
F33	50	36.7	0.0	36.7	26.8	36.5	53.3
	100	45.7	0.0	45.7	24.8	29.5	54.3
	150	41.1	0.0	41.1	24.4	34.5	58.9
	200	88.1	0.0	88.1	1.6	10.3	11.9

由结果(见表3)得到,当浓度较小时(50μmol/L ~ 150μmol/L),酪氨酸酶的二级结构中的α-螺旋的含量逐渐减小,当达到一定浓度时(200μmol/L),α-螺旋的含量显著升高,Loop(β-TURN)的含量呈先增加后减小的趋势。由表3还可以看出巴比妥和硫代巴比妥类衍生物对酪氨酸酶二级结构的影响表现大致相同,当抑制剂浓度逐渐增大时,α-螺旋的含量逐渐降低,而Loop(β-TURN)的含量逐渐增大,当抑制剂浓度增加到200μmol/L时,α-螺旋的含量迅速增加,而Loop(β-TURN)的含量立即减小。

酪氨酸酶是全α结构模体的酶,因此合适的α-螺旋以及Loop的含量与酶的特定的功能域的形成以及酶的构象变化是相关的,而酶的天然构象以及酶的特定的功能域对酶保持其活性是十分重要的。酶失去活性主要有以下几个原因:第一,由于α-螺旋结构紧密地特殊性,过大的α-螺旋度的构象,使其结构紧密而不利于酶形成活性中心,从而使底物不能与之结合,使酶假象失去活性;

过小的α－螺旋构象，就会彻底破坏酶蛋白质的氢键结构和其他结构而产生酶结构的不可逆破坏。第二，由于Loop是一个柔性极强的构象，Loop越多，它就有越多的机会运动到活性中心或是键合位点的上面，从而覆盖这些有可能与底物结合的位点，导致底物与酶结合的失败，从而使酶失去活性。

从实验结果可以明显地看出，巴比妥类衍生物对酪氨酸酶的作用，先使酶的结构趋为松散，天然构象改变，使酶的活性中心不能形成，此时底物L－DOPA也不能被酶催化反应；但当苯甲醛浓度达到200μmol/L，酶的结构过于紧密，不利于形成酶的活性中心，使酪氨酸酶不能催化多巴发生反应，从而达到抑制效果。

由此可以得出结论，酪氨酸酶的活性与其天然的构象有着重要的关系，天然构象的改变预示着酶活性的改变，而不同的构象改变也预示着抑制剂与酶的相互作用不同。我们认为巴比妥类衍生物通过影响酪氨酸酶的构象发生改变，从而与酶中心的二铜离子发生不可逆的结合，从而导致对酪氨酸酶的不可逆抑制作用。

实验部分

试剂与仪器

酪氨酸酶(Tyrosinase mushroom)，购于Sigma公司，比活力为3900units/mg；L－多巴(L－3－(3,4－Dihydroxyphenyl) alanine，L－DOPA)，购于Alfa Aesar公司；DMSO、KH_2PO_4、K_2HPO_4为分析纯，购于广州化学试剂厂；移液枪为国产SHALON公司产品；紫外吸收用岛津UV－2501PC紫外分光光度计测定。J－810 Jasco圆二色谱仪

实验方法

以L－DOPA为底物的酪氨酸酶活力测定

酪氨酸酶的酶活力测定参考文献：分别加入10μL含不同浓度的抑制剂(DMSO配制)，945μL磷酸缓冲溶液(pH＝6.8)以及5μL的0.45 mg/mL酪氨酸酶的磷酸缓冲溶液(pH＝6.8)于1.5 mL的离心管中，在37℃恒温水浴中保温10min，然后加入40 μL L－DOPA(1.50 mg/mL)溶液启动反应，迅速充分混匀，在37℃恒温条件下测定波长为475 nm的光密度值，由其随时间的增长直线的斜率计算出酶的活力。并用对照扣除缓冲溶液对本试验的影响。

酪氨酸酶的抑制动力学的测定

判断抑制类型：固定底物和样品浓度，改变酶的浓度，得到一系列反应速率值，以V对[En]作图，可得到一条直线。保持底物浓度不变，改变样品浓度，以同样方法得到另外几条直线，根据这一组直线可判断抑制剂的抑制类型。

抑制剂对酶抑制作用的机理是通过 Lineweaver - Burk 双倒数作图法判断。

CD(圆二色谱)的测定

依次在 1mL 的离心管中加入 10μL 含不同浓度的抑制剂的 DMSO 溶液和 40μL 酪氨酸酶水溶液(0.3mg/mL),950μL pH =6.8 磷酸缓冲溶液,摇匀,放置 15 分钟。圆二色谱在 Jasco810 型 CD 光谱仪上进行测定,采用 1mm 液池,在 180 ~260nm 范围内扫描,累加 10 次,按照 chei 的计算方法求各二级结构的比例

小结

本论文以 L - 多巴作为酪氨酸酶催化底物,应用 UV 光谱、CD 光谱研究羟基取代苯甲醛类衍生物对酪氨酸酶的抑制作用机理,现总结如下:

分别测定 GABA 类、巴比妥类和 DHPM 类羟基取代苯甲醛类衍生物以 L - 多巴为酪氨酸酶催化底物时双酚酶抑制活性。结果发现 GABA 类衍生物对酪氨酸酶没有明显的抑制作用,DHPM 类与巴比妥类衍生物均具有一定的活性,其中巴比妥类衍生物的抑制作用较为显著,其抑制类型的测定表明该类衍生物为酶的不可逆抑制剂,同时 DHPM 由于结构与巴比妥类似,也是不可逆抑制类型。

利用圆二色谱研究了巴比妥类化合物与酪氨酸酶的相互作用,及对酪氨酸酶的构象影响,并进一步推测这类化合物对酶作用抑制机理,为后续酪氨酸酶抑制剂的分子设计及实际应用提供了理论基础。

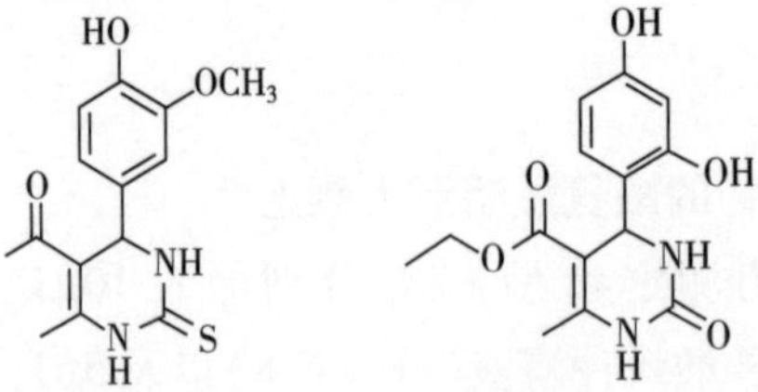

HO OCH3 O NH N H S
G16

OH OH O O NH N H O
G22

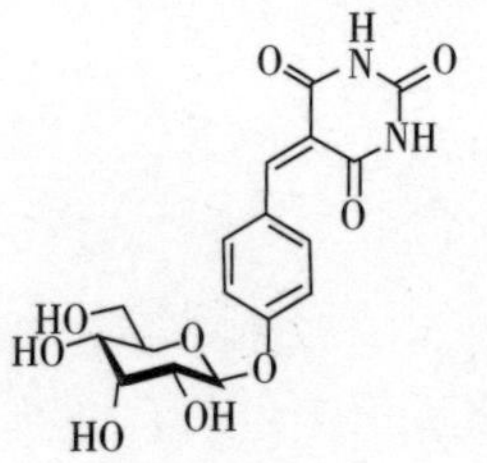

F1 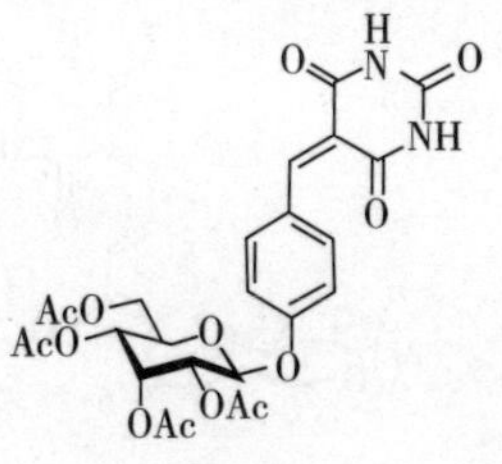 F2 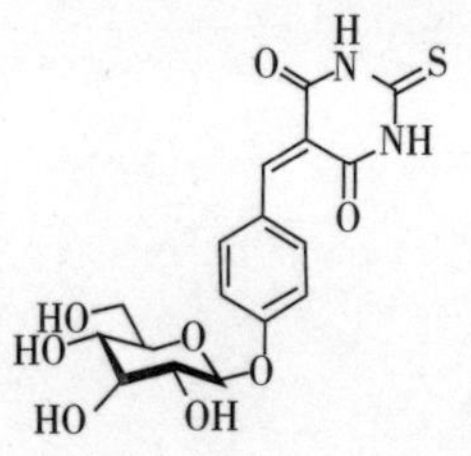 F3

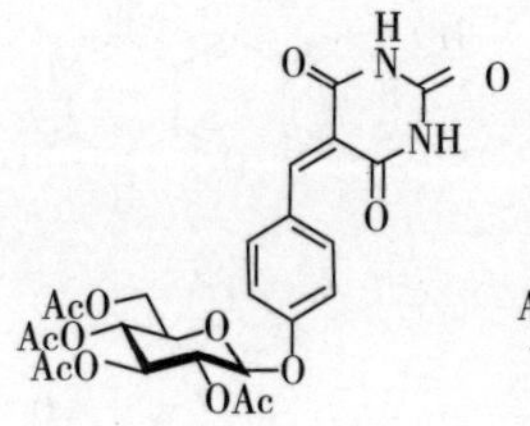

F4 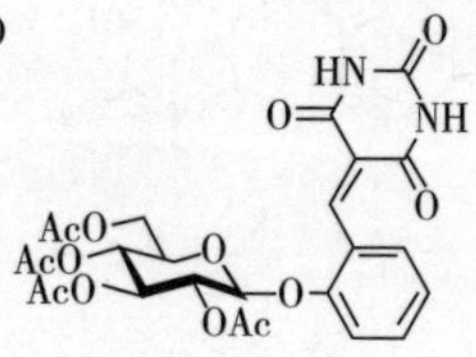 F5 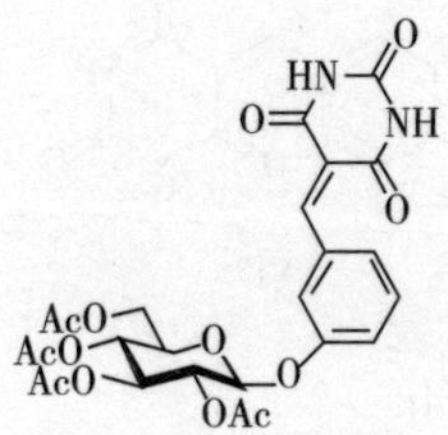 F6

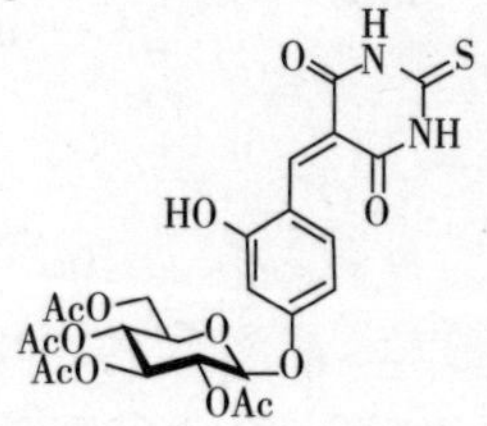

F7 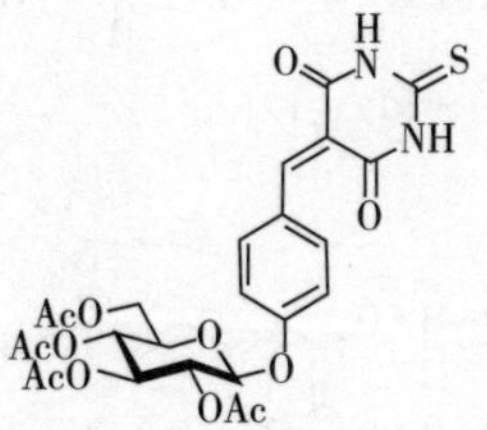 F8 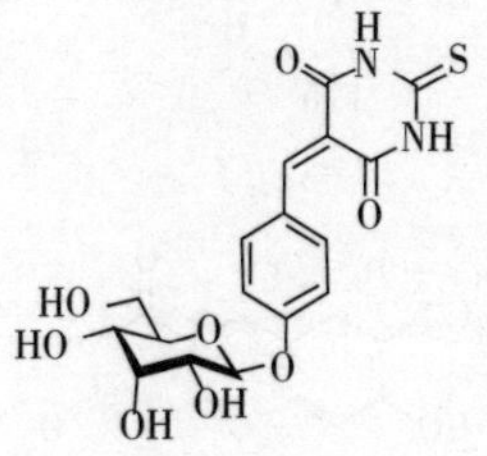 F9

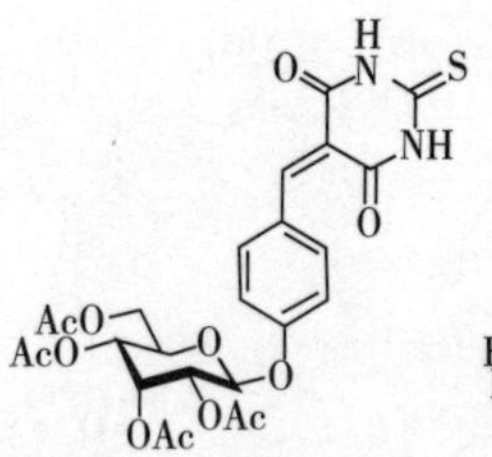

F10 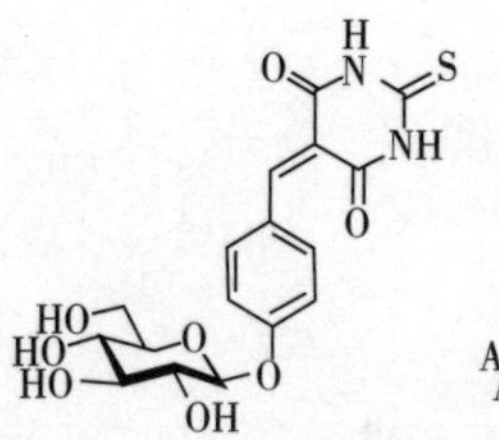 F11 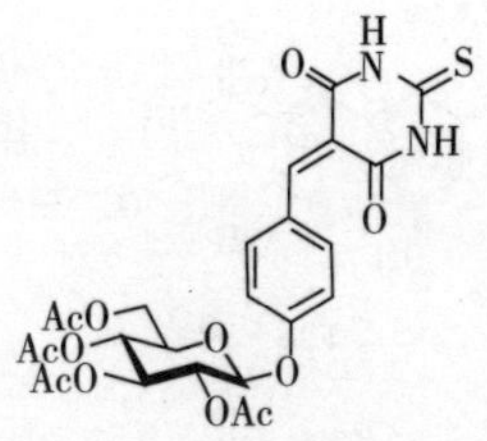 F12

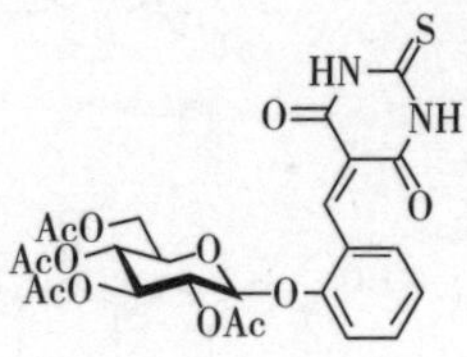

F13 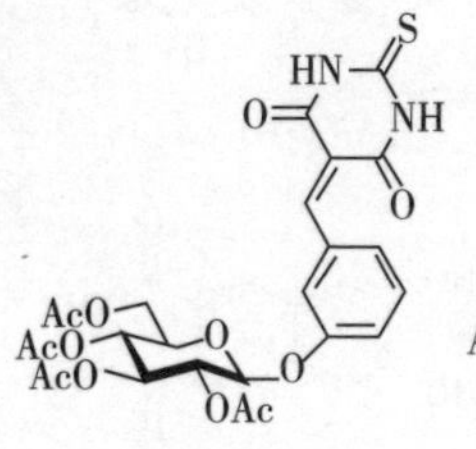 F14 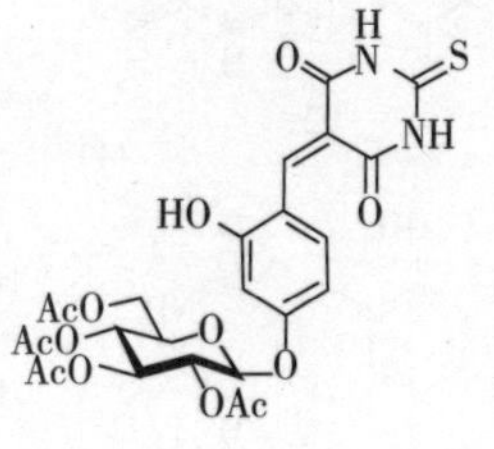 F15

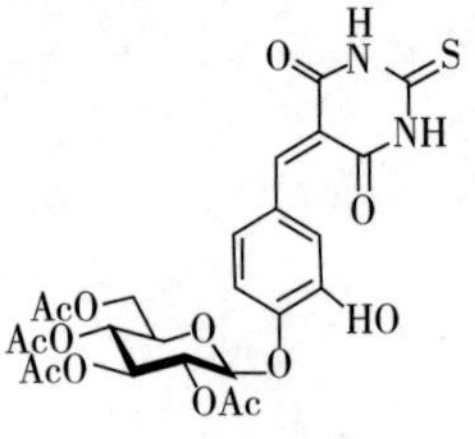

F16

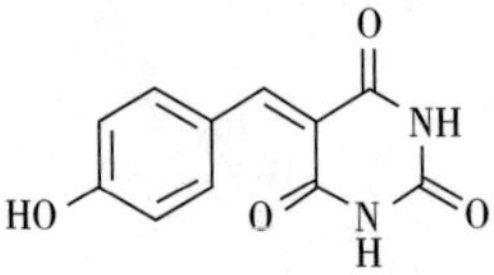

F17 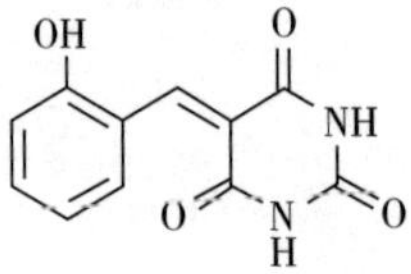 F18 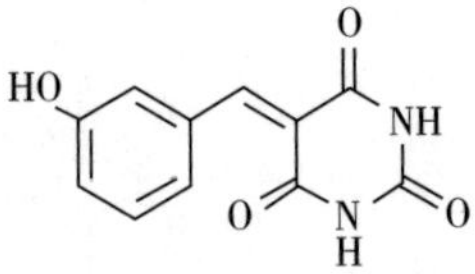 F19

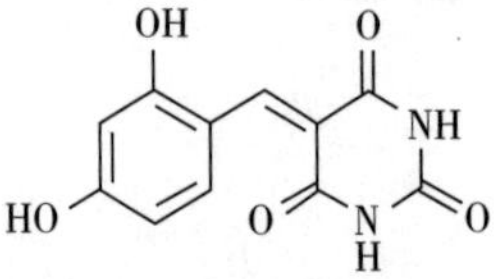

F20 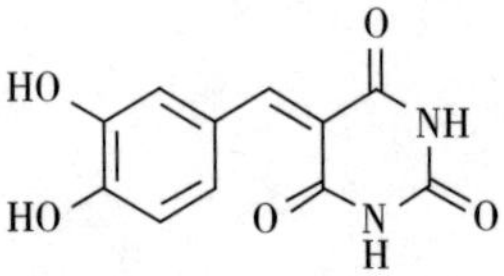 F21 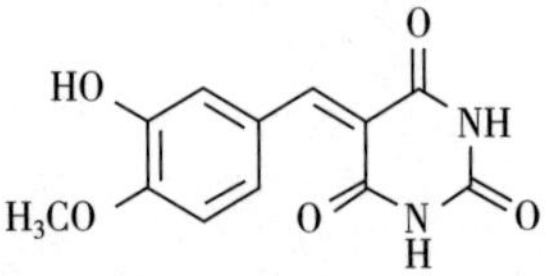 F22

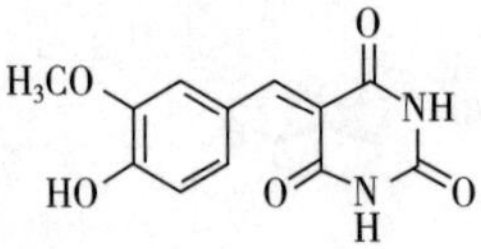

F23 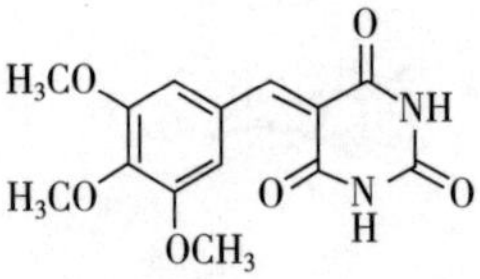 F24 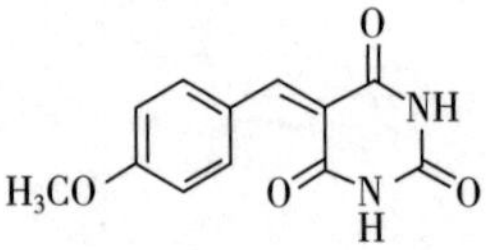 F25

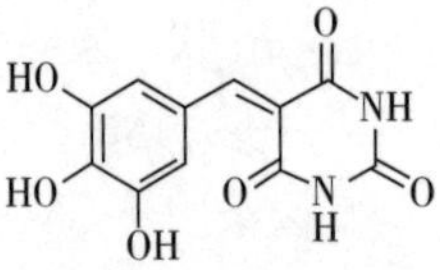

F26 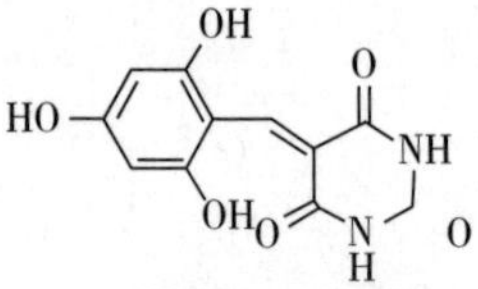 F27 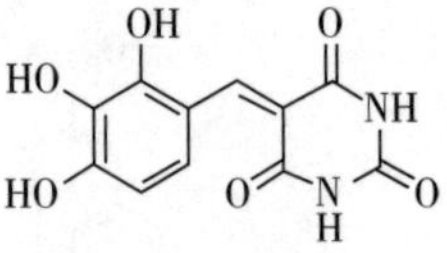 F28

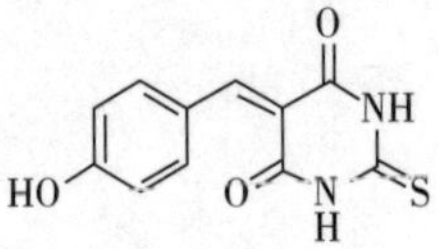

F29 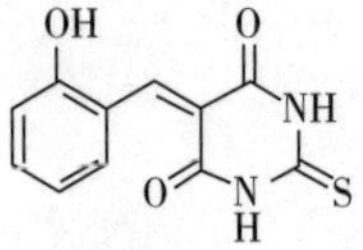 F30 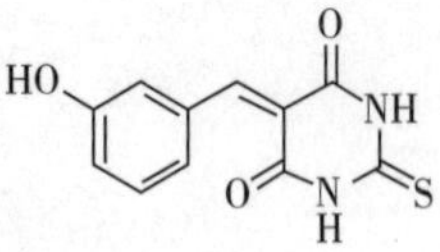 F31

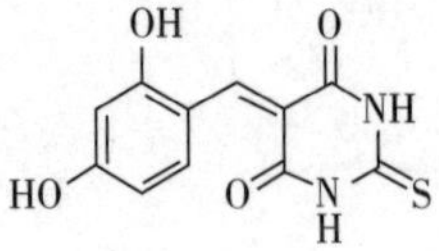

F32 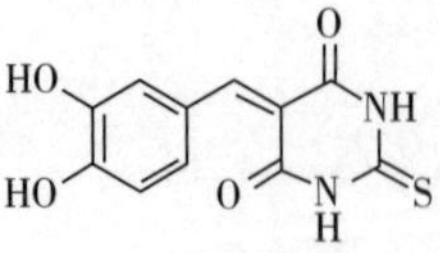 F33 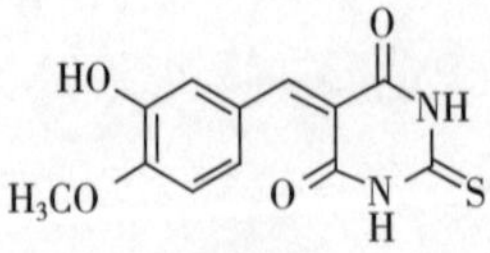 F34

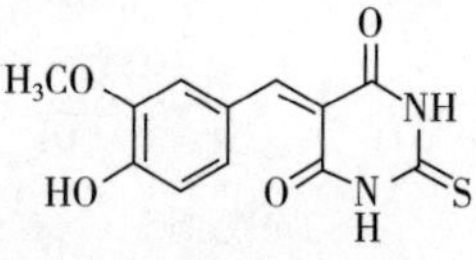

F35 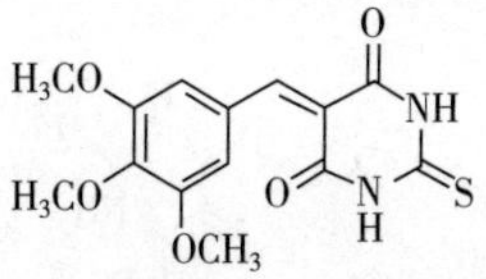 F36 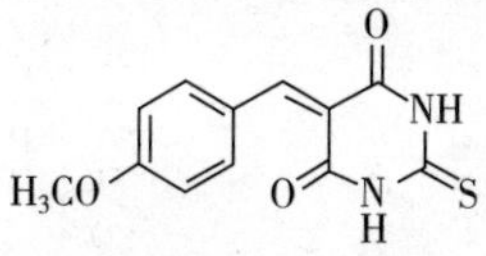 F37

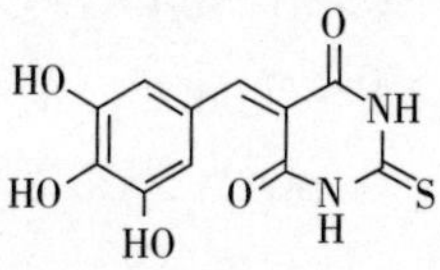

F38 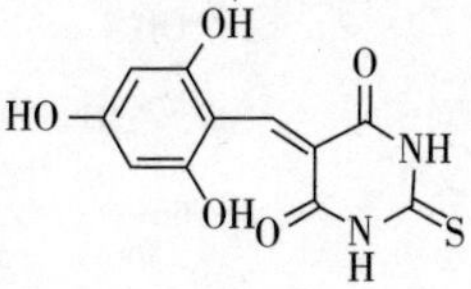 F39 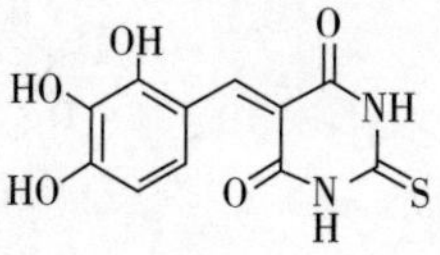 F40

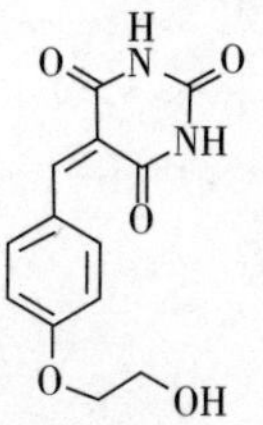

F41 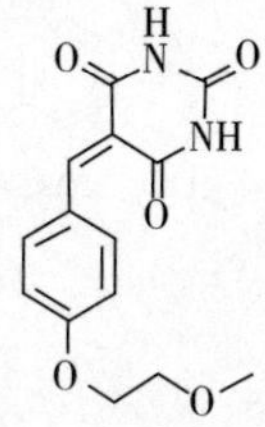 F42 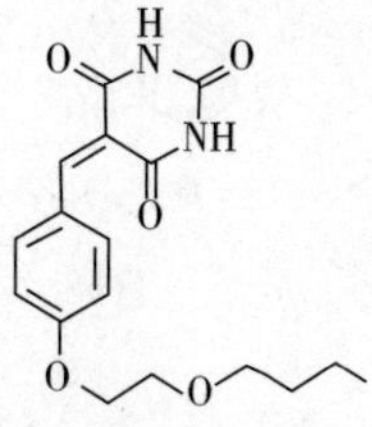 F43 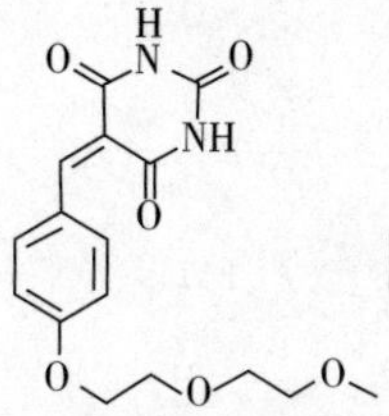 F44 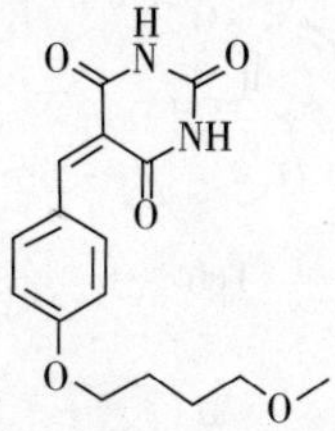 F45

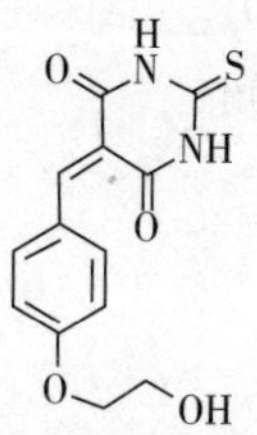

F46 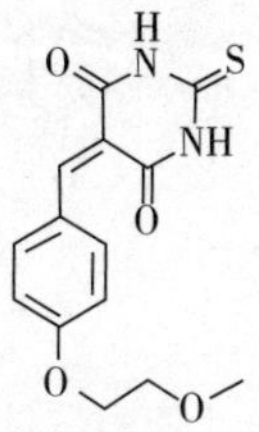 F47 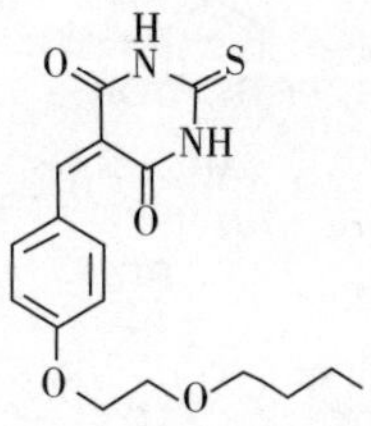 F48 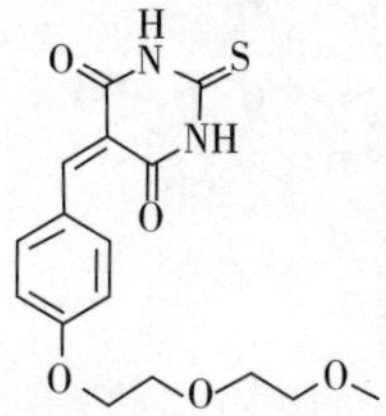 F49 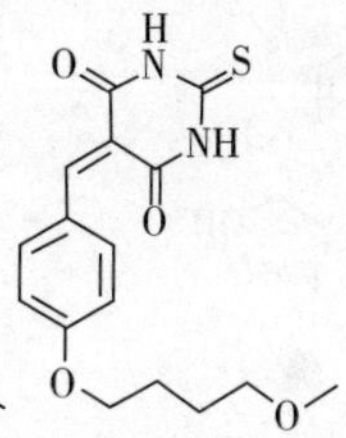 F50

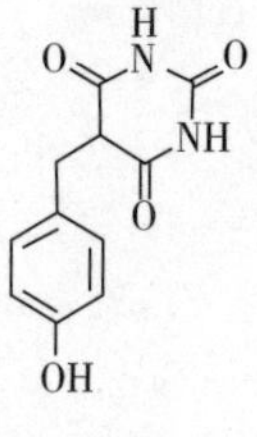

F51 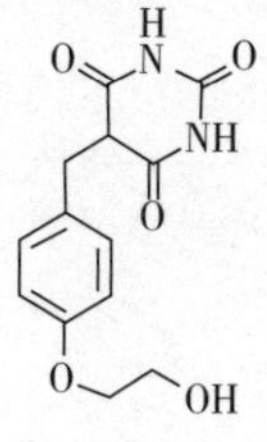 F52 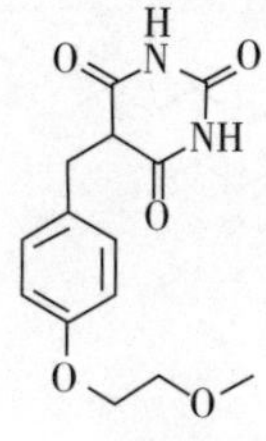 F53 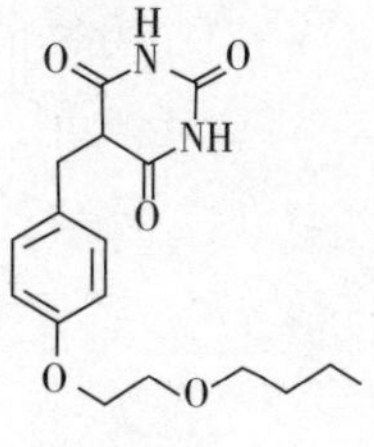 F54 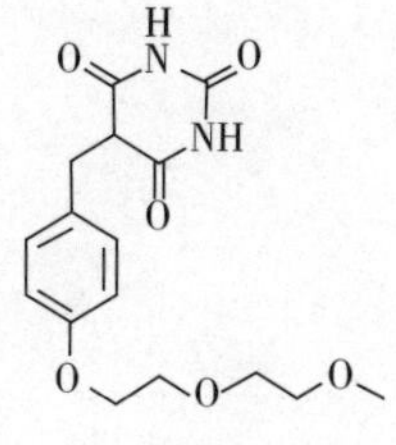 F55

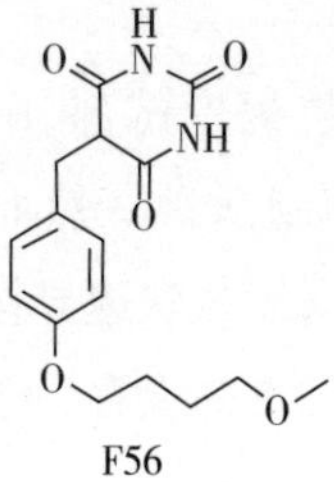

F56 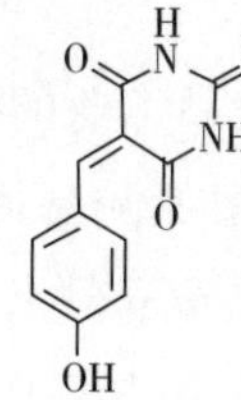 F57 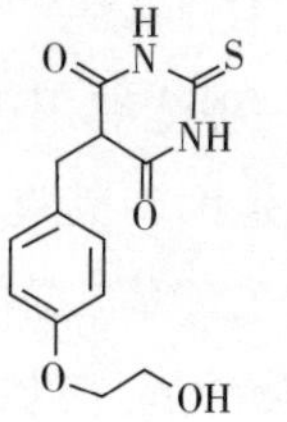 F58 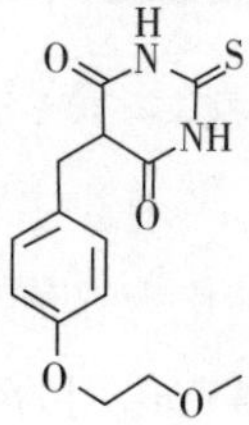 F59 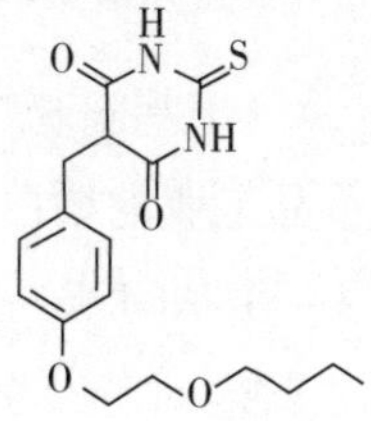 F60

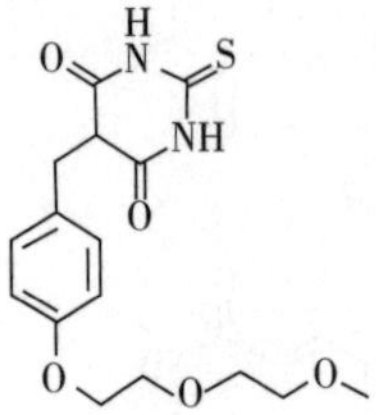

F61 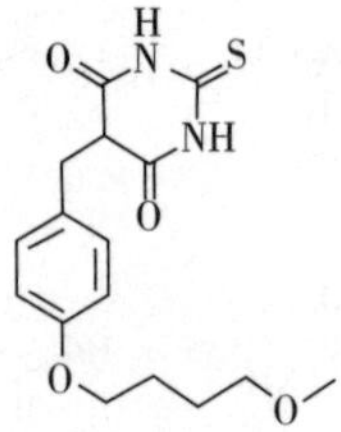 F62 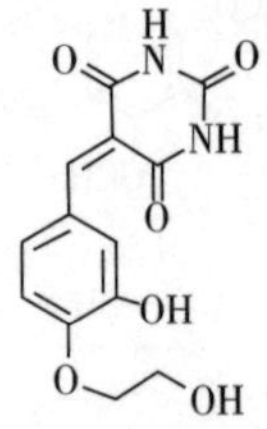 F63 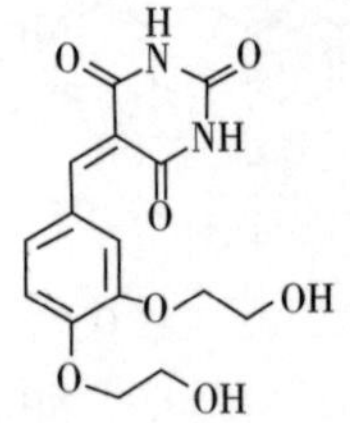 F64 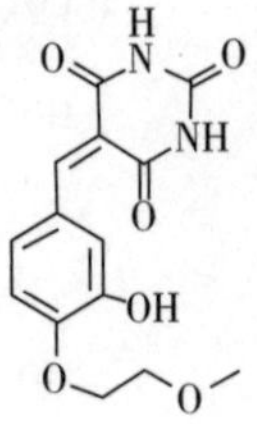 F65

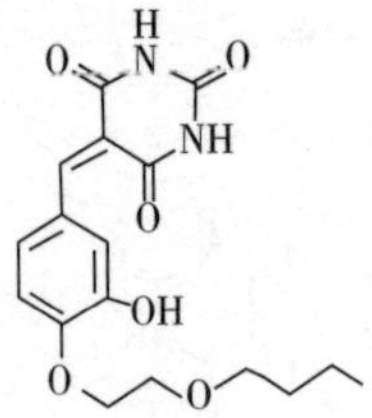

F66 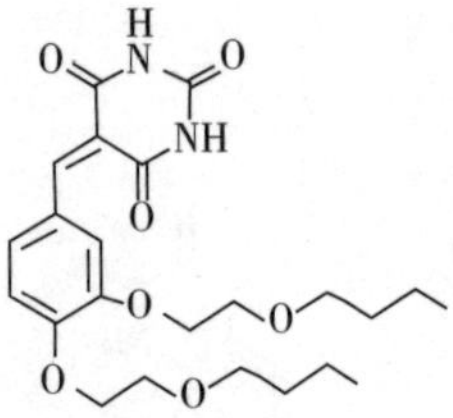 F67 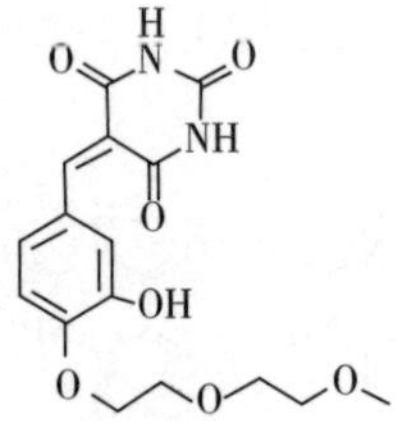 F68 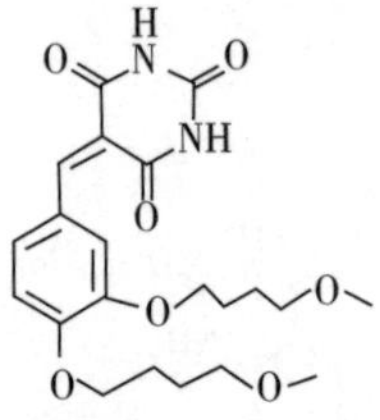 F68

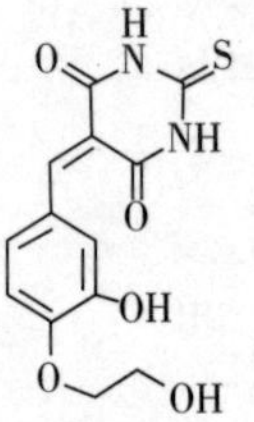

F70 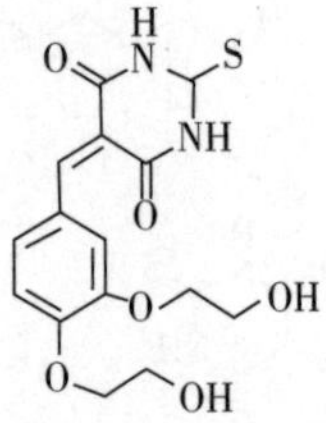 F71 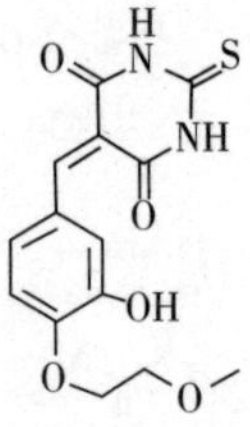 F72 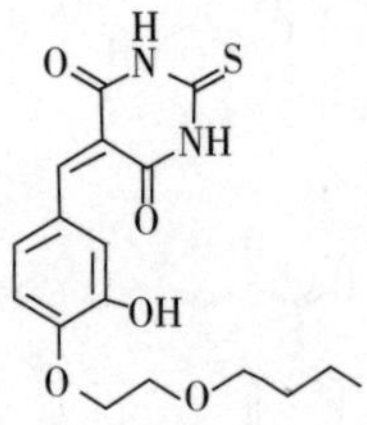 F73

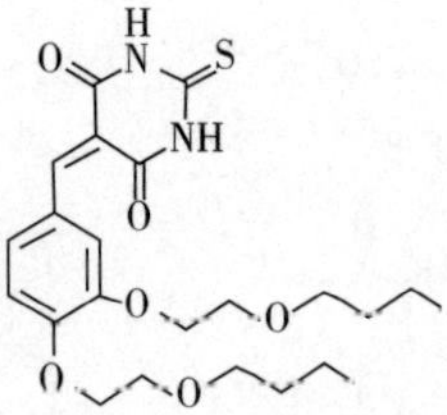

F74

4.2 酚类酪氨酸酶抑制剂

4 - 丁基间苯二酚(图 4 - 34)是一种能够用于食品工业的抗褐变试剂,用于化妆品工业的增白剂,这些用途主要是因为其具有酪氨酸酶抑制活性。然而酶把 4 - 丁基间苯二酚羟基化为邻苯二酚,邻苯二酚氧化成邻苯醌,邻苯醌快速异构化为对苯醌,对于以这种方式起作用的酪氨酸酶,氧化态的酪氨酶(E_{ox})必然会存在于反应中间体中,能够由过氧化氢、维生素C而产生,或者是邻苯二酚和还原剂(NADH)的催化浓度仍然保持为一个常数。2015年Carmen Vanessa Ortiz－Ruiz等人证实了4－丁基间苯二酚是酪氨酸酶的一个底物,探讨了其抑制机理和抑制动力学,动力学特征提高了催化常数为(0.85 ± 0.04)s^{-1},米氏常数为(60.31 ± 6.73)mmol/L[36]。

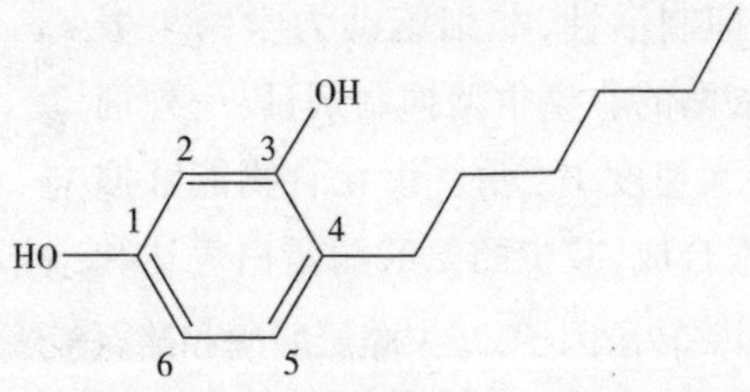

图4－34　4－丁基间苯二酚结构式

2014年周盈利等人研究了茶黄素(Theaflavin,TF1)、茶黄素－3－没食子酸酯(Theaflavins－3－gallate,TF2A)、茶黄素－3'－没食子酸酯(Theaflavins－3'－gallate,TF2B)、茶黄素双没食子酸酯(Theaflavins－3,3'－gallate,TF3)和茶黄酸(Theaflavic acid,TF4)5种单体化合物(图4－35)对酪氨酸酶的抑制作用,并与阳性对照表儿茶素没食子酸酯(Epigallocatechin gallate,EGCG)进行比较[37]。采用L－酪氨酸氧化法测定酪氨酸酶的活性。结果显示,5种茶黄素单体对酪氨酸酶酶活有不同程度的抑制作用,且存在量效关系。除茶黄酸外,其他4种化合物均比表儿茶素没食子酸酯的抑制活性高。说明该类茶黄素单体化合物能有效抑制酪氨酸酶活性,从而有效抑制黑色素的产生,可为研究开发新型有效的酪氨酸酶抑制剂提供重要参考。

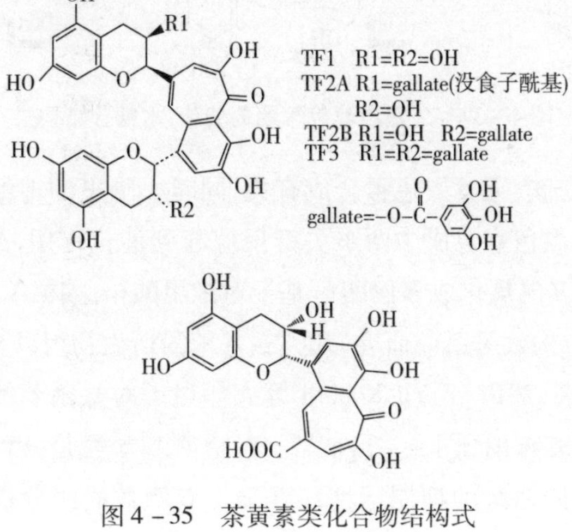

图4－35　茶黄素类化合物结构式

2014 年 Florence Wing – Ki Cheung 等人研究了糖苷化的羟基二苯乙烯的酪氨酸酶抑制活性及抑制特性。2,3,5,4’ – 四羟基二苯乙烯 – 2 – O – β – D – 吡喃糖苷(图 4 – 36)是从何首乌中分离得到的单一化合物,其具有酪氨酸酶抑制活性,非细胞动力学结果表明,该化合物是酪氨酸酶的非竞争型抑制剂;以一种剂量依赖性方式降低最大速度 V_{max} 值。该化合物能够抑制 PKA 诱导的黑色素合成,减少酪氨酸酶蛋白表达和转录因子、小眼畸形相关转录因子,减少酪氨酸酶和酪氨酸酶相关蛋白 1 (TRP – 1)络合物的形成[38]。免疫荧光显微术揭示酪氨酸酶与内质网或溶酶体没有相关性,暗示该化合物酶的成熟过程没有直接联系。2,3,5,4’ – 四羟基二苯乙烯 – 2 – O – β – D – 吡喃糖苷通过对酪氨酸酶的非竞争抑制来调节黑色素合成、下调黑色素蛋白表达、减少酪氨酸酶/TRP – 1 络合物的形成。

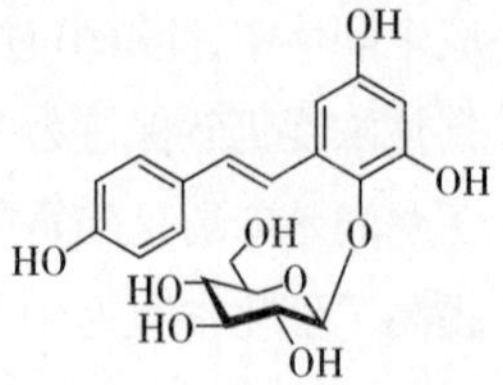

图 4 – 36　2,3,5,4’ – 四羟基二苯乙烯 – 2 – O – β – D – 吡喃糖苷结构式

2014 年 Chisato Oode 等人从油茶中分离得到二氢白藜芦醇葡萄糖苷(图 4 – 37),并首次以三丁基硅烷保护的苯甲醛为起始原料,经过 5 步反应合成了其木糖苷衍生物,二氢白藜芦醇葡萄糖苷是黑色素瘤细胞 B16F0 潜在的黑色素合成抑制剂,比阳性对照曲酸的抑制活性强 40 倍[39]。与此相反,其木糖苷衍生物具备促进黑色素合成的能力,这就说明,在糖苷取代的二氢白藜芦醇的一个单一的羟基甲基是酪氨酸酶抑制剂或促进剂的重要因素。

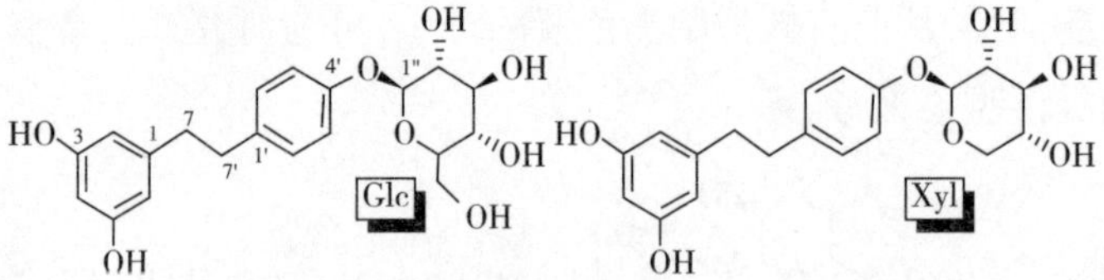

图 4 – 37　二氢白藜芦醇葡萄糖苷、木糖苷结构式

酪氨酸酶与皮肤、毛发黑色素合成有关,同时大脑里的神经黑色也涉及到酪氨酸酶,限速酶在黑色素合成中两步关键反应起到催化作用,酪氨酸羟基化成 L – 多巴,然后 L – 多巴氧化成多巴醌。基于对维甲酸衍生物,N – (4 – 羟基苯基)维酰胺(图 4 – 38)构效关系的研究,几类氨基苯酚衍生物得以合成。为了找到新型酪氨酸酶抑制剂,2014 年 Yu Komori 等人探讨了对氨基苯酚的酪氨酸酶抑制活性,包括对癸氨基苯酚(图 4 – 38)[40]。对癸氨基苯酚是一种有效的酪氨酸酶抑制剂,比阳性对照曲酸的抑制活性还要强。双倒数曲线分析对癸氨基苯酚对

于酪氨酸酶的抑制动力学，结果表明，以多巴胺和酪氨酸酶为底物，对癸氨基苯酚都是酪氨酸酶的非竞争型抑制剂。这些研究成果显示了对癸氨基苯酚是酪氨酸酶抑制剂曲酸的有用替代品。

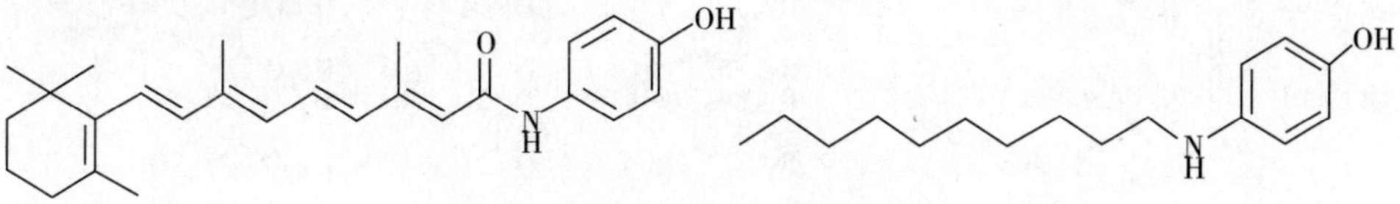

图 4－38　对氨基酚衍生物结构式

腰果壳油从腰果壳中提取得到，是一种丰富天然资源，腰果酚是从腰果壳油中分离得到的主要酚类物质。2016 年 Xiang－Ping Yu 等人第一次报道了腰果酚对酪氨酸酶的抑制活性及抑制机理，腰果酚具有中等的酪氨酸酶抑制活性，研究了腰果酚官能团与腰果酚抑制机理的关系。腰果酚三烯、腰果酚二烯、腰果酚一烯（图 4－39）能够有效降低酪氨酸酶二酚酶活性的稳态速率，这三种腰果酚对酪氨酸酶的半抑制浓度分别为（40.5 ± 3.7）μmol/L，（52.5 ± 3.2）μmol/L 和（56.0 ± 3.6）μmol/L，动力学分析和荧光猝灭结果表明，一个腰果酚能够进入到酪氨酸酶分子中[41]。特征参数进一步证实腰果酚能够和酪氨酸酶相互作用，计算机模拟的分子对接明确了腰果酚能够酪氨酸酶活性位点的氨基酸残基相互作用。

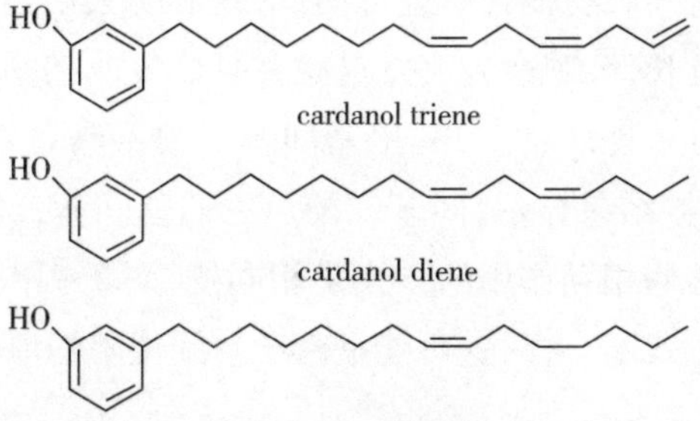

cardanol triene

cardanol diene

cardanol monoene

图 4－39　腰果酚结构式

2017 年意大利学者 Paolo Ruzza 等人采用直接法合成了一类 C_2－对称羟基化的联苯（图 4－40），通过光谱和电化学实验评价了这类化合物对酪氨酸酶的抑制能力[42]。在没有单酚酶活性的酶反应特征滞后时间的情况下，着重研究了其对二酚酶的抑制特性。分别通过分析咖啡酸溶液紫外光谱图的变化和酪氨酸酶－生物传感器阴极的还原性，考察了酪氨酸酶把天然的邻苯二酚氧化成邻苯醌的能力，并对两种方法进行了比较。结果表明，大多数化合物具有较好的酪氨酸酶抑制活性，比已知的羟基化联苯的酪氨酸酶抑制活性强。

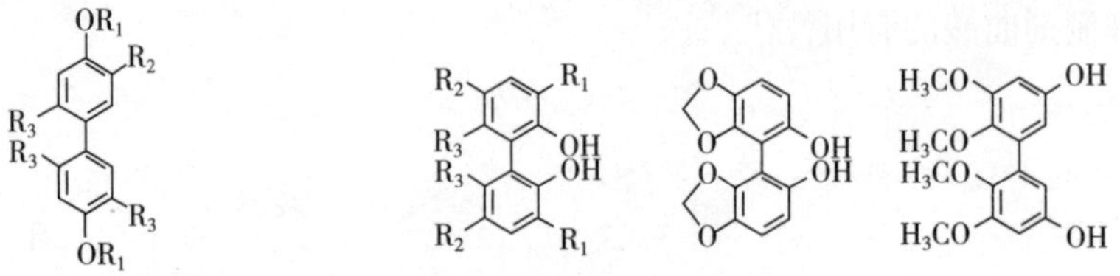

1 $R_1=R_2=R_3=H$,
4 $R_1=R_3=H, R_2=OH$,
5 $R_1=R_3=H$, R_2=prop-2-cn-1-yl,
8 $R_1=R_3=CH_3, R_2=OH$,

2 $R_1=R_2=R_3=H$,
3 $R_1=R_2=H, R_3=OH$,

6

7

图 4－40　羟基化联苯的化学结构

2016 年 Antonio Garcia－Jimenez 等人探讨了间苯二酚及其一些衍生物（4－乙基间苯二酚、2－甲基间苯二酚、4－甲基间苯二酚）（图 4－41）对酪氨酸酶的作用[43]。如果用一个还原剂如抗坏血酸完成催化循环，或者是用邻苯二酚如 4－叔丁基邻苯二酚完成催化循环，在所有情况下，这些化合物是作为酪氨酸酶的底物，把过氧化氢加入到中间体中，此反应同样能够进行。间苯二酚及其衍生物作为酪氨酸酶的底物，研究其特征，同时确定了这些化合物的动力学常数、米氏常数和催化常数。酶的活性是用间苯二酚、4－乙基间苯二酚、4－甲基间苯二酚预孵化至活性在短时间表观损失后进行测试，这和酶失活过程是一致的。然而，如果测试很长时间延长，会观察到突发过程并且恢复酶的活力，证实了这些化合物并非酪氨酸酶的自杀性底物。2－甲基间苯二酚不具备对酪氨酸酶作用方式影响的能力，分子对接结果显示，间苯二酚、4－乙基间苯二酚、4－甲基间苯二酚与酪氨酸酶相互作用的方式是相同的，而 2－甲基间苯二酚与酪氨酸酶作用方式不一样，这可能是由于空间位阻的原因。

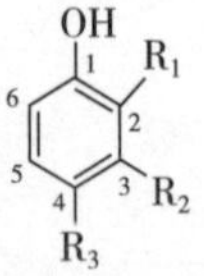

Compound	R_1	R_2	R_3
Resorcinol	$-H$	$-OH$	$-H$
4－Ethylresorcinol	$-H$	$-OH$	$-CH_2CH_3$
2－Methylresorcinol	$-CH_3$	$-OH$	$-H$
4－Methylresorcinol	$-H$	$-OH$	$-CH_3$

图 4－41　间苯二酚及其衍生物结构式

2009 年厦门大学庄江兴从天然植物中分离纯化出对酪氨酸酶具有很强抑制作用的抑制剂并探讨其对单酚酶和二酚酶活性的抑制作用机理。再以小鼠 B16F10 细胞的酪氨酸酶为研究对象，研究其对黑色素生成量的影响及其相关基因的转录调控机制[44]。最后,以果蝇为动物模型研究酪氨酸酶抑制剂对酪氨酸酶代谢途径中相关产物多巴胺生成量的影响。在筛选酪氨酸酶抑制剂过程中,发现了与帕金森病相关的 α - synuclein 蛋白的过表达能显著抑制黑色素的生成量,这预示着酪氨酸酶与帕金森病可能有着某种相关性,因此研究了 α - synuclein 蛋白对小鼠 B16F10 细胞酪氨酸酶转录调控的机制。对酪氨酸酶有抑制作用的化合物:强心酚三烯,强心酚二烯,腰果酚三烯,腰果酚二烯、腰果酚单烯(图 4 - 42),研究了它们对酪氨酸酶的抑制机理及对小鼠 B16F10 细胞的黑色素生成量的影响及其相关基因的转录调控机制;以果蝇为动物模型研究酪氨酸酶抑制剂对酪氨酸酶代谢途径中相关物质含量的影响。建立了动力学反应方程,并求出抑制常数、表观速率常数、微观速率常数,并提出了酶与两分子强心酚三烯结合而失去活性,第一个分子结合为慢反应过程,第二个分子反应为快速结合使酶失活。从抗生素中筛选到了头孢唑啉和头孢地嗪两个蘑菇酪氨酸酶个高效抑制剂,测得了其对蘑菇酪氨酸酶相关的抑制常数。分离得到的化合物对 B16F10 细胞增殖率及黑色素生成的影响。在 10μmol/L 时这些化合物对 B16F10 细胞增殖率基本无影响,而在此浓度下 B16F10 黑色素生成量均增加。分离所得化合物对果蝇多巴胺含量的影响。用 10μmol/L 浓度的化合物喂养果蝇,14 天后果蝇的多巴含量均下降,可能是抑制了酪氨酸酶活性从而抑制多巴生成,进一步转化为多巴胺。开展了 α - synuclein 对酪氨酸酶的调控机制研究,结果表明 α - synuclein 降低了 CAMP 含量、降低 PKA 转录水平、降低 CREB 磷酸化水平、降低 MITF 转录水平、降低了酪氨酸酶的转录表达水平、降低黑色素生成量。用免疫共沉淀法研究 α - synuclein 和酪氨酸酶的相互作用结果表明其并不直接相互作用或者作用较弱。而 α - synuclein 蛋白过量表达可降低酪氨酸酶蛋白表达水平而增加多巴胺对细胞的毒性。

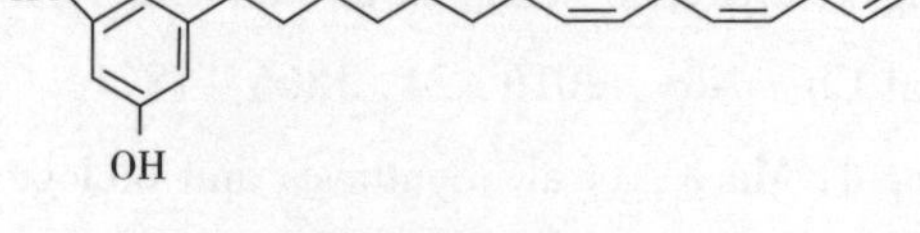

HO

OH

强心酚三烯

HO

OH

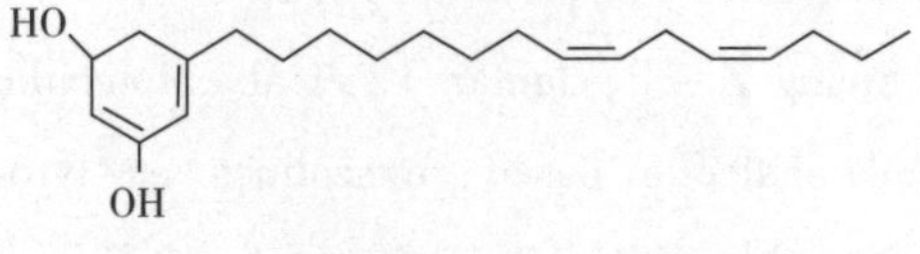

强心酚二烯

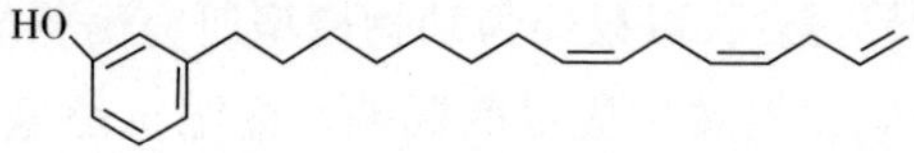

腰果酚三烯

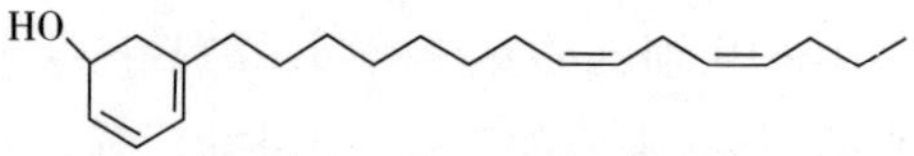

腰果酚二烯

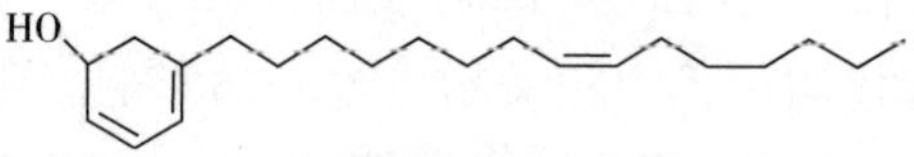

腰果酚单烯

图 4－42　分离得到的酚类化合物结构式

参考文献

[1] Matos M J, Varela C, Vilar S, et al. Design and discovery of tyrosinase inhibitors based on a coumarin scaffold [J]. RSC. Adv., 2015, 5, 94227－94235.

[2] Asthana S, Zucca P, Vargiu A V, et al. Structure－Activity Relationship Study of Hydroxycoumarins and Mushroom Tyrosinase [J]. J. Agric. Food Chem. 2015, 63, 7236－724.

[3] Gardelly M, Trimech B, Belkacem M A, et al. Synthesis of novel diazaphosphinanes coumarin derivatives with promoted cytotoxic and anti－tyrosinase activities [J]. Bioorganic & Medicinal Chemistry Letters, 2016, 26, 2450－2454.

[4] Liu P, Shu C, Liu L, et al. Design and synthesis of thiourea derivatives with sulfur－containing heterocyclic scaffolds as potential tyrosinase inhibitors [J]. Bioorganic & Medicinal Chemistry, 2016, 24, 1866－1871.

[5]Xie W, Zhang J, Ma X, ct al. Synthesis and biological evaluation of Kojic acid derivatives containing 1, 2, 4 － triazole as potent tyrosinase inhibitors [J]. Chem. Biol. Drug. Des., 2015, 86, 1087－1092.

[6] Qin H－L, Shang Z－P, Jantan I, et al. Molecular docking studies and biological evaluation of chalcone based pyrazolines as tyrosinase inhibitors and potential anticancer agents[J]. RSC Adv., 2015, 5, 46330－46338

[7]Rafiq M, Saleem M, Hanif M, et al. Synthesis, structural elucidation and

bioevaluation of 4 - amino - 1,2,4 - triazole - 3 - thione's Schiff base derivatives [J]. Arch. Pharm. Res., 2016, 39, 161 - 171.

[8]景临林,范小飞,贾正平. 糖基化三氮唑香草酸衍生物的合成及酪氨酸酶抑制活性研究[J]. 化学试剂,2014,36,9 - 12;92.

[9]杨菁,林菁. 盐酸小檗碱对 B16 细胞的分化诱导作用[J]. 中国现代应用药学, 2010,27,874 - 877.

[10]林菁,彭华毅. 黄癸素诱导小鼠黑色素瘤 B16 细胞凋亡的研究 [J]. 中国药理学通报, 2010, 26,1630 - 1634.

[11] Chan C - F, Lai S - T, Guo Y - C, et al. Inhibitory effects of novel synthetic methimazole derivatives on mushroom tyrosinase and melanogenesis [J]. Bioorganic & Medicinal Chemistry, 2014, 22, 2809 - 2815.

[12]Chaaban I, Khawass E S M E, Razik H A A E, et al. Synthesis and biological evaluation of new oxadiazoline - substituted naphthalenyl acetates as anticancer agents [J]. European Journal of Medicinal Chemistry, 2014, 87, 805 - 813.

[13]Mojzycha M, Dolashki A, Voelter W. Synthesis of pyrazolo [4,3 - e][1,2,4]triazine sulfonamides, novel Sildenafil analogs with tyrosinase inhibitory activity [J]. Bioorganic & Medicinal Chemistry, 2014, 22, 6616 - 6624.

[14]Chen Z, Cai D, Mou D, et al. Design, synthesis and biological evaluation of hydroxy - or methoxy - substituted 5 - benzylidene (thio) barbiturates as novel tyrosinase inhibitors [J]. Bioorganic & Medicinal Chemistry, 2014, 22, 3279 - 3284.

[15] Xia L, Idhayadhulla A, Lee Y R, et al. Anti - tyrosinase, antioxidant, and antibacterial activities of novel 5 - hydroxy - 4 - acetyl - 2,3 - dihydronaphtho [1,2 - b]furans [J]. European Journal of Medicinal Chemistry, 2014, 86, 605 - 612.

[16] Li D - F, Hu P - P, Liu M - S, et al. Design and Synthesis of Hydroxypyridinone - L - phenylalanine Conjugates as Potential Tyrosinase Inhibitors [J]. J. Agric. Food Chem., 2013, 61, 6597 - 6603.

[17]Ko H - H, Jin Y - J, Lu T - M, et al. A Novel Monoterpene - Stilbene Adduct with a 4,4 - Dimethyl - 2,3 - diphenylchromane Skeleton from *Artocarpus xanthocarpus* [J]. Chemistry biodiversity, 2013, 10, 1269 - 1275.

[18] Mohamed M E S, Seham Y H, Huda E, et al. Synthesis and Bioassay of a New Class of Furanyl - 1,3,4 - Oxadiazole Derivatives [J]. Molecules, 2013, 18, 8550 - 8562.

[19] Ramesh L S, Prashant D L, Jyoti B W. Tyrosinase inhibitory activity, 3D QSAR, and molecular docking study of 2,5 - disubstituted - 1,3,4 - oxadiazoles [J]. Journal of Chemistry, 2013, http://dx.doi.org/10.1155/2013/849782.

[20] Hamidian H, Tagizadeh R, Fozooni S, et al. Synthesis of novel azo compounds containing 5(4H) - oxazolone ring as potent tyrosinase inhibitors [J]. Bioorganic & Medicinal Chemistry, 2013, 21, 2088 - 2092.

[21] Gawande S S, Warangkar S C, Bandgar B P, et al. Synthesis of new heterocyclic hybrids based on pyrazole and thiazolidinone scaffolds as potent inhibitors of tyrosinase [J]. Bioorganic & Medicinal Chemistry, 2013, 21, 2772 - 2777.

[22] Zhou Z, Zhuo J, Yan S, et al. Design and synthesis of 3,5 - diaryl - 4,5 - dihydro - 1H - pyrazoles as new tyrosinase inhibitors [J]. Bioorganic & Medicinal Chemistry, 2013, 21, 2156 - 2162.

[23] Gencer N, Demir D, Sonmez F, et al. New saccharin derivatives as tyrosinase inhibitors [J]. Bioorganic & Medicinal Chemistry, 2012, 20, 2811 - 2821.

[24] Seikou N, Souichi N, Genzo T, et al. Alkaloid constituents from flower buds and leaves of sacred lotus (Nelumbo nucifera, Nymphaeaceae) with melanogenesis inhibitory activity in B16 melanoma cells [J]. Bioorganic & Medicinal Chemistry, 2013, 21, 779 - 787.

[25] Kim Y H, Park J I, Myung C, et al. 1 - Phenyl - 3 - (2 - thiazolyl) - 2 - thiourea inhibits melanogenesis via a dual - action mechanism [J]. Arch Dermatol Res., 2016, 308, 473 - 479.

[26] Wang R, Chai W - M, Yang Q, et al. 2 - (4 - Fluoro - phenyl) - quinazolin - 4 (3H) - one as a novel tyrosinase inhibitor: Synthesis, inhibitory activity, and mechanism [J]. Bioorganic & Medicinal Chemistry, 2016, 24, 4620 - 4625.

[27] Xie W, Zhang H, He J, et al. Synthesis and biological evaluation of novel hydroxybenzaldehydebased kojic acid analogues as inhibitors of mushroom tyrosinase [J]. Bioorganic & Medicinal Chemistry Letters, 2017, 27, 530 - 532.

[28] Stefania F, Laura D L, Maria P G, et al. Chemical exploration of 4 - (4 - fluorobenzyl) piperidine fragment for the development of new tyrosinase inhibitors [J]. European Journal of Medicinal Chemistry, 2017, 125, 992 - 1001.

[29] Chen M - J, Hung C - C, Chen Y - R, et al. Novel synthetic kojic acid - methimazole derivatives inhibit mushroom tyrosinase and melanogenesis [J]. Journal of Bioscience and Bioengineering, 2106, 122, 666 - 672.

[30] Mohd F F M A, Kamal R, Tan H H, et al. Synthesis and effects of oxadiazole derivatives on tyrosinase activity and human SK - MEL - 28 malignant melanoma cells [J]. RSC. Adv. ,2016, 6, 72177 - 72184.

[31] Ghani U, Ullah N. New potent inhibitors of tyrosinase: Novel clues to binding of 1,3,4 - thiadiazole - 2(3H) - thiones, 1,3,4 - oxadiazole - 2(3H) - thiones, 4 - amino - 1,2,4 - triazole - 5(4H) - thiones, and substituted hydrazides to the dicopper active site [J]. Bioorganic & Medicinal Chemistry, 2010, 18, 4042 - 4048.

[32] Zhao D - Y, Zhang M - X, Dong X - W, et al. Design and synthesis of novel hydroxypyridinone derivatives as potential tyrosinase inhibitors [J]. Bioorganic & Medicinal Chemistry Letters, 2016, 26, 3103 - 3108.

[33] 谢娟. 芳杂环缩氨基硫脲类似物的合成及其抑制酪氨酸酶活性研究[D]. 2015,南昌大学硕士论文.

[34] Mubashir H, Zaman A, Qamar A, et al. Exploration of Novel Human Tyrosinase Inhibitors by Molecular Modeling, Docking and Simulation Studies [J]. Interdiscip Sci Comput Life Sci, 2016, DOI 10.1007/s 12539 - 016 - 0171 - x.

[35] 阎琴. 羟基取代苯甲醛类衍生物的设计合成及其生物活性初步研究[D]. 中山大学博士论文,2010年.

[36] Carmen V O - R, Jose B, Jose N R - L, et al. Tyrosinase - Catalyzed Hydroxylation of 4 - Hexylresorcinol, an Antibrowning and Depigmenting Agent: A Kinetic Study [J]. J. Agric. Food Chem. 2015, 63, 7032 - 7040.

[37] 周盈利,刘静,叶琼仙,等. 茶黄素类化合物对酪氨酸酶的抑制作用[J]. 食品科技,2014,39,169 - 172.

[38] Florence W - K C, Albert W - N L, Liu W K, et al. Tyrosinase Inhibitory Activity of a Glucosylated Hydroxystilbene in Mouse Melan - a Melanocytes [J]. J. Nat. Prod., 2014, 77, 1270 - 1274.

[39] Chisato O, Wataru S, Yukiko I, et al. Synthesis of dihydroresveratrol glycosides and evaluation of their activity against melanogenesis in B16F0 melanoma cells [J]. European Journal of Medicinal Chemistry, 2014, 87, 862 – 867.

[40] Komori Y, Imai M, Yamauchi T, et al. Effect of p – aminophenols on tyrosinase activity [J]. Bioorganic & Medicinal Chemistry, 2014, 22, 3994 – 4000.

[41] Yu X – P, Su W – C, Wang Q, et al. Inhibitory mechanism of cardanols on tyrosinase [J]. Process Biochemistry, 2016, 51, 2230 – 2237.

[42] Ruzza P, Serra P A, Fabbri D, et al. Hydroxylated biphenyls as tyrosinase inhibitor: A spectrophotometric and electrochemical study [J]. European Journal of Medicinal Chemistry, 2017, 126, 1034 – 1038.

[43] Garcia – Jimenez A, Teruel – Puche J A, Berna J, et al. Characterization of the action of tyrosinase on resorcinols [J]. Bioorganic & Medicinal Chemistry, 2016, 24, 4434 – 4443.

[44]庄江兴, α – synuclein 蛋白及抑制剂对酪氨酸酶的调控机制[D]. 厦门大学博士论文, 2009.

4.3 编者相关论文

4.3.1 Bioorganic & Medicinal Chemistry Letters 21 (2011) 2376 - 2379

Evaluationof Dihydropyrimidin – (2H) – one analogues and Rhodanine derivatives as tyrosinase inhibitors

Jinbing Liu[*a,1], Fengyan Wu[a,1], Lingjuan Chen[b], Jianming Hu[a], Liangzhong Zhao[a], Changhong Chen[a], Liwang, Peng[a]

[a]*Department of Biology and Chemical Engineering, Shaoyang University, Shao Shui Xi Road, Shaoyang 422100, PRC*

[b]*College of Food Science and Engineering, Central South University of Forestry and Technology, PRC*

* Corresponding author: Department of Biology and Chemical Engineering, Shaoyang University, Shao Shui Xi Road, Shaoyang 422100, PRC

Phone: +86 – 739 – 5431768; Fax: +86 – 739 – 5431768

E – mail address: xtliujb@ yahoo. com. cn

Abstract

A series of dihydropyrimidin – (2H) – one analogues and rhodanine derivatives were synthesized and their inhibitory effects on the diphenolase activity of mushroom tyrosinase were evaluated. The results showed that some of the synthesized compounds exhibited significant inhibitory activities. Especially, compound 15 bearing a hydroxyethoxyl group at position – 4 of phenyl ring exhibited most potent tyrosinase inhibitory activity with IC_{50} value of 0. 56 mmol/L. The inhibition mechanism analysis of compound 15 demonstrated that the inhibitory effect of the compound on the tyrosinase was irreversible. These results suggested that such compounds might be served as lead compounds for further designing new potential tyrosinase inhibitors.

Keywords: Dihydropyrimidin – (2H) – one, Rhodanine, tyrosinase inhibitors , inhibition mechanism.

Dihydropyrimidinones (DHPMs) and their derivatives have attracted interest in medicinal chemistry, exhibiting pharmacological and therapeutic properties. In the past decades, a broad range of biological effects including antiviral, antitumor, antibacterial and anti – inflammatory activities have been described for these compounds [1,2]. Compounds containing the 2 – thioxothiazolidin – 4 – one ring have showed a wide range of pharmacological activities, which includes antimicrobial, antiviral, and anticonvulsant effects [3]. Prominent amongst the resultant hits was a series of derivatives containing a 5 – benzylidenerhodanine substructure (Figure 1). Benzylidenerhodanines have been reported as small molecule inhibitors of numerous targets such as cyclooxygenase and 5 – lipoxygenase, b – lactamase, cathepsin D, HCV NS3 protease, aldose reductase, protein mannosyl transferase and JNK – stimulating phosphatase [4]. Although, during the last years extensive studies on the pharmacology of DHPMs and rhodanine derivatives have been reported, the tyrosinase inhibitor activities of this kind of compounds have never appeared in the literature.

Tyrosinase(monophenol oro – diphenol, oxygen oxidoreductase, EC 1. 14. 18. 1, syn. polyphenol oxidase), also known as polyphenol oxidase (PPO), is a copper – containing monooxygenase that is widely distributed in microorganisms, animals, and plants [5]. Tyrosinase could catalyze two distinct reactions involving molecular oxygen in the hydroxylation of monophenols to o - diphenols (monophenolase) and in the oxidation of o - diphenols to o - quinones (diphenolase) [6]. Due to the high reactivity, quinines could polymerize spontaneously to form high molecular weight brown - pigments (melanins) or react with amino acids and proteins to enhance brown color of the pigment produced [7,8]. Previous reports confirmed that tyrosinase not only was involved in melanizing in animals, but also was one of the main causes of most fruits and vegetables quality loss during post harvest handling and processing, leading to faster degradation and shorter shelf life [9]. Recently, investigation demonstrated that various dermatological disorders, such as age spots and freckle, were caused by the accumulation of an excessive level of epidermal pigmentation [10,11]. Tyrosinase has also been linked to Parkinson's and other neurodegenerative diseases [12]. In insects, tyrosinase is uniquely associated with three different biochemical processes, including sclerotization of cuticle, defensive encapsulation and melanization of foreign organism, and wound healing [13]. These processes provide potential targets for developing safer and effective tyrosinase inhibitors as insecticides and ultimately for insect control. Thus, the development of safe and effective tyrosinase inhibitors is of great concern in the medical, agricultural, and cosmetic industries. However, only a few such as kojic acid, arbutin, tropolone, and 1 - phenyl - 2 - thiourea (PTU) (Figure 2) are used as therapeutic agents and cosmetic products [11,14].

On the other hand, it was reported that phenyl thioureas and alkyl thioureas could exhibit weak to moderate depigmenting activity [15] and Ley and Bertram reported that benzaldoximes and benzaldehyde - o - alkyloximes possess higher tyrosinase inhibitory ability [16]. Tyrosinase belongs with catechol oxidase to the type - 3 copper protein group. Crystal structure of phenyl thioureas bound catechol oxidase was reported. The sulfur atom of this compound binds to both copper ions in the active site of the enzyme [17]. More recently, our investigation also demonstrated that condensation products of aldehyde or ketone and thiosemicarbazidederivatives exhibited potent inhibitory activities against mushroom tyrosinase [18,19]. Stimulated by these results, in the present investigation, we synthesized a series of dihydropyrimidin - (2H) - ones analogues and rhodanine derivatives, and their inhibitory effects on the diphenolase activity of mushroom tyrosinase were investigated.

According to the general procedure shown inScheme 1 [20,21], we carried out the preparation of 3,4 – dihydropyrimidin – 2 – (1H) – ones (DHPMs) and thione analogs from the corresponding aldehydes, pentane –2,4 – dione and urea or thiourea in the presence of magnesium bromide under solvent – free conditions. The synthesis is usually complete within 0.5 – 3 h at 50°C or 100°C, and the target compounds could be purified by recrystallization from 95% alcohol. The yields for these reactions are from moderate to good, and all the target compounds were characterized by IR, NMR and MS. The Schiff base of 3,4 – dihydropyrimidin – 2 (1H) – one was prepared according to Scheme 2[22]. Rhodanine derivatives were synthesised for this study by means of the Knoevenagel condensation of the suitable aldehydes with 2 – thioxo – 4 – thiazolidinone, respectively, in refluxing acetic acid in the presence of sodium acetate (Scheme 3)[23,24]. The target compounds were also characterized by IR, NMR and MS.

For evaluating the tyrosinase inhibitory activity, all the synthesized compounds were subjected to tyrosinase inhibition assay with L – DOPA as substrate, according to the method reported by our groups with some slight modifications[9,19]. The tyrosinase inhibitory activities of arbutin and 4 – methoxycinnamic acid were ever reported, therefore, they were selected as comparing substances. Figure 3 showed that the remaining enzyme activity rapidly decreased with the increasing concentrations of compound 15. The IC_{50} values of dihydropyrimidin – (2H) – one analogues and rhodanine derivatives against tyrosinase were summarized in Table 1, and IC_{50} values of all these compounds were determined from logarithmic concentration – inhibition curves (at least eight points) and are given as means of three experiments .

Our results showed that compounds7 – 16 exhibited potent inhibitory on mushroom tyrosinase with IC_{50} values ranged from 0.56 to 25.11 mmol/L. Some of the synthesized compounds demonstrated more potent inhibitory activities than the reference standard inhibitor arbutin. Compounds 1 – 6 and 17 did not show any significant activities at 50mmol/L concentration. The result may be related to the structure of tyrosinase contained a type – 3 copper center with a coupled dinuclear copper active site in the catalytic core. Tyrosinase inhibition of compounds 7 – 16 depended on the competency of the sulfur atom to chelate with the dicopper nucleus in the active site, and tyrosinase would lose its catalyzing ability after forming complex.

Comparing to the tyrosinase inhibitory activities of 3,4 – dihydropyrimidin – 2 –

(1H) – thione analogs, compounds with hydroxyl group in the benzene ring exhibited more potent inhibitory activities. This suggested that hydroxyl group might play an important role in determining their inhibitory activities. The result was consistent with the Kim's propose that the hydroxyl groups might inhibit the tyrosinase enzyme by participating in a metal – ligand binding interaction with the dicopper nucleus [28]. The introduction of a hydroxy group in the *para* position led to appreciably better inhibitory activity than the hydroxyl group in the *ortho* position, when the substituents introduced in the *meta* position on the phenyl rings may have hindered docking of the inhibitor to tyrosinase, and led to the decrease in inhibitory activities. Interestingly, compound 10 was found to be the most potent inhibitor with an IC_{50} value of 10. 67 mmol/L. These results suggested that benzene ring might cause stereohindrance for the inhibitors approaching the active site of the enzyme.

As shown intable 1, rhodanine derivatives exhibited more potent inhibitor activities than dihydropyrimidin – (2H) – ones analogues, which indicated that the smaller size of the ring in volume would be beneficial to the molecules to approach the active centre of the enzyme. Compound 15 bearing a hydroxyethoxyl group at position – 4 of phenyl ring exhibited most potent tyrosinase inhibitory activity with IC_{50} value of 0. 56 mmol/L. These results suggested that introducing appropriate hydrophobic subunits into position – 4 of benzaldehyde might be facilitated their inhibitory effects on the diphenolase activity of mushroom tyrosinase. Interestingly, hydroxyethoxyl group introduced to the position – 3,4 of phenyl ring (compound 13, IC_{50} = 5. 69mmol/L), led to a dramatic decrease in inhibitory activities. These results further confirmed that the substituents introduced in the *meta* position on the phenyl rings may hinder the complex formation between inhibitors unit and the metallic center of the active site in enzyme due to the steric hindrance.

The inhibitory mechanism of the selected compound15 on mushroom tyrosinase for the oxidation of L – DOPA was determined. Figure 4 showed the relationship between enzyme activity and concentration in the presence of different concentration of the above – mentioned compound. The results displayed that the plots of V versus $[E]$ gave a family of parallel straight lines with the same slopes. These results demonstrated that the inhibitory effect of compound 15 on the tyrosinase was irreversible, suggesting that 2 – thioxothiazolidin – 4 – one moiety of 5 – (4 – (2 –

hydroxyethoxy) – benzylidene) – 2 – thioxothiazolidin – 4 – one effectively inhibited the enzyme by binding to its binuclear active site irreversibly. The result may be related to the structure of tyrosinase, within the structure, there are two copper ions in the active centre of tyrosinase and a lipophilic long – narrow gorge near to the active centre. Hydrophobic subunit and sulfur atom exist in the structure of compound 15, the compound could exhibit strong affinity for copper ions in the active centre and form a reversible non – covalent complex with the tyrosinase, and this then reacts to produce the covalently modified "dead – end complex".

The present investigation reported that dihydropyrimidin – (2H) – one analogues and rhodanine derivatives had potent inhibitory effects on the diphenolase activity of mushroom tyrosinase. Interestingly, compound 15 was found to be the most potent inhibitor with IC_{50} value of 0.56 mmol/L. Preliminary structure activity relationships (SARs) analysis indicated that (1) hydroxyl group in the benzene ring might play an important role in determining their inhibitory activities, big groups might cause stereohindrance for the inhibitors approaching the active site of the enzyme resulting in the decrease of inhibitory activities, the substituents introduced in the *meta* position of the phenyl rings may have hindered docking of the inhibitor to tyrosinase; (2) the smaller size of the ring in volume would be beneficial to the molecules to approach the active centre of the enzyme; (3) Rhodanine derivatives bearing a hydroxyethoxyl group at position – 4 of phenyl ring might be beneficial to increase the inhibitory activities. The inhibition mechanism analysis of compound 15 demonstrated that the inhibitory effect of the compound on the tyrosinase was irreversible.

Acknowledgment

This work wasfinancially supported by the Foundation of Education Department of Hunan Province, China (09B091), and the Key Subject Foundation of Shaoyang University (2008XK201).

References and notes

1. Kappe, C. O. *Eur. J. Med. Chem.* 2000, *35*, 1043.

2. Liang, B.; Wang, X.; Wang, J. X. *Tetrahedron*. 2007, *63*, 1981.

3. Sortino, M..; Delgado, P. *Bioorg. Med. Chem*. 2007, *15*, 484.

4. Irvine, M. W.; Patrick, G. L. *Bioorg. Med. Chem. Lett*. 2008, *18*, 2032.

5. Song, K. K. ; Huang, H. ; Han, P. ; Zhang, C. L. ; Shi, Y. ; Chen, Q. X. *Biochem. Biophys. Res. Commun*. 2006, *342*, 1147.

6. Chen, Q. X. ; Liu, X. D. ; Huang, H. *Biochemistry (Mosc)*. 2003, *68*, 644.

7. Kubo, I. ; Nihei, K. I. ; Shimizu, K. *Bioorg. Med. Chem. Lett*. 2004, *12*, 5343.

8. Matsuura, R. ; Ukeda, H. ; Sawamura, M. *J. Agric. Food. Chem*. 2006, *54*, 2309.

9. Yi, W. ; Cao, R. *Eur. J. Med. Chem*. 2010, *45*, 639.

10. Gupta, A. K. ; Gover, M. D. ; Nouri, K. N. ; Taylor, S. *J. Am. Acad. Dermatol*. 2006, *55*, 1048.

11. Thanigaimalai, P. ; Hoang, T. A. L. *Bioorg. Med. Chem. Lett*. 2010, *20*, 2991.

12. Zhu, Y. J. ; Zhou, H. T. ; Hu, Y. H. *Food Chemistry*. 2011, *124*, 298.

13. Ashida, M. ; Brey, P. *Proc. Natl. Acad. Sci. U. S. A*. 1995, *92*, 10698

14. Battaini, G. ; Monzani, E. ; Casella, L. ; Santagostini, L. ; Pagliarin, R. *J. Biol. Inorg. Chem*. 2000, *5*, 262.

15. Klabunde, T. ; Eicken, C. *Nature Struct. Biol*. 1998, *5*, 1084.

16. Ley, J. P. ; Bertram, H. *J. Bioorg. Med. Chem.* 2001, *9*, 1879.

17. Matsuura, R. ; Ukeda, H. ; Sawamura, M. *J. Agric. Food. Chem*. 2006, *54*, 2309.

18. Liu, J. B. ; Yi, W. ; Wan, Y. Q. ; Ma, L. ; Song, H. C. *Bioorg. Med. Chem*. 2008, *14*,

1096.

19. Liu, J. B. ; Cao, R. H. ; Yi, W. ; Ma, C. M. ; Wan, Y. Q. ; Zhou, B. H. Ma, L. ; Song, H. C. *Eur. J. Med. Chem*. 2009, *44*, 1773.

20. Ahmed, N. ; Lier, J. E. *Tetrahedron Lett*. 2007, *48*, 5407.

21. *General procedures for the synthesis of 3,4 – dihydropyrimidin – 2 – (1H) – ones (DHPMs) compounds*: To a 50 mL flame dried round – bottom flask were added 20 mmol of benzaldehyde, 20 mmol of acetophenone, 20mmol of urea or thiourea and

3.5 mmol of $MgBr_2$. The resulting mixture was heated to 100℃. After completion of the reaction as indicated by TLC, and then the mixture was cooled to room temperature and poured into 50 ml ice - water. The solid product was filtered, washed with ice - water and ethanol (95%), dried, and recrystallized from 95% ethanol. Compound 11: Yellow solid power, m. p. 279.6℃ - 281.3℃; IR (KBr, cm^{-1}) ν: 3267, 2998, 2889, 1608, 1566, 1505, 1463, 1377, 1329, 1252, 1191, 832; ^1H NMR (300 MHz, DMSO - d_6): δ 10.54(bs, 1H, NH), 9.38(bs, 1H, NH), 9.16(s, 1H, OH), 7.10(d,*J* =8.4Hz, 2H, Ph), 6.71(d,*J* =8.4Hz, 2H, Ph), 5.19(d, *J* =2.7Hz, 1H, CH), 2.31(s, 3H, CH_3), 2.24(s, 3H, CH_3); ^{13}C NMR (75 MHz, DMSO - d_6): δ 196.5, 175.1, 158.9, 149.2, 135.8, 128.5, 115.8, 108.1, 55.8, 31.2, 19.1; MS (ESI): m/z(100%) 263 (M+1).

22. Liu, J. B.; Cao, R. H.; Wang, Z. H.; Peng, W. L.; Song, H. C. *Eur. J. Med. Chem.* 2009, *44*, 1737.

23. Irvine, M. W.; Patrick, G. L.; Kewney, J. *Bioorg. Med. Chem. Lett*. 2008, *18*, 2032.

24. *General procedures for the synthesis of Rhodanine derivatives*: The mixture of 2 - thioxo - 4 - thiazolidinone (2.48 mmol), suitable aldehydes (2.54 mmol) and sodium acetate (7.32mmol) in acetic acid (30 mL) was heated to reflux for 3 - 6 h. The reaction was monitored by TLC. After completion of the reaction, upon cooling of the mixture to room temperature, a precipitate formed which was collected by filtration, washed with cold ethanol, and dried under vacuum to provide the target compound. Compound 15: Yellow solid power, M. p. 197.7℃ - 198.4℃; IR (KBr, cm^{-1}) ν: 3068, 2837, 1726, 1684, 1579, 1502, 1438, 1230, 1201, 1172, 1060, 800; ^1H NMR (300 MHz, $CDCl_3$): δ 7.63(s, 1H, =CH), 7.46(d, *J* =8.7Hz, 2H, Ph), 7.03(d,*J* =8.7Hz, 2H, Ph), 4.75(t, *J* =4.5Hz, 2H, OCH_2), 4.26(t, *J* =4.5Hz, 2H, OCH_2); ^{13}C NMR (75 MHz, $CDCl_3$): δ 196.2, 170.0, 161.8, 143.7, 129.6, 126.4, 117.0, 114.8, 70.3, 60.8; MS (ESI): m/z(100%) 282 (M+1).

25. Shi, Y.; Chen, Q. X.; Wang, Q.; Song, K. K.; Qiu, L. *Food Chem.* 2005, *92*, 707.

26. Lee, H. S. *J. Agric. Food. Chem.* 2002, *50*, 1400.

27. Kazuhisa, S.; Koji, N.; Takahisa, N.; Taro, K.; Kenji, S. *J. Biosci.*

Bioeng. 2005, *99*, 272.

28. Kim, D.; Park, J.; Kim, J.; Han, C.; Yoon, J.; Kim, N.; Seo, J.; Lee, C. *J. Agric. Food. Chem*. 2006, *54*, 935.

Figure captions

Figure 1 The substructure of 5 – benzylidenerhodanine

Figure 2 Chemical structure of known tyrosinase inhibitors

Figure 3 Effect of compound 15 on the diphenolase activity of mushroom tyrosinase for the catalysis of L – DOPA at 25℃.

Figure 4 The effect of concentrations of tyrosinase on its activity for the catalysis of L – DOPA at different concentration of compound 15. The concentrations of compound 15 for curves 1 – 4 are 0, 0.1, 0.2 and.3 0 mmol/L, respectively.

Scheme 1

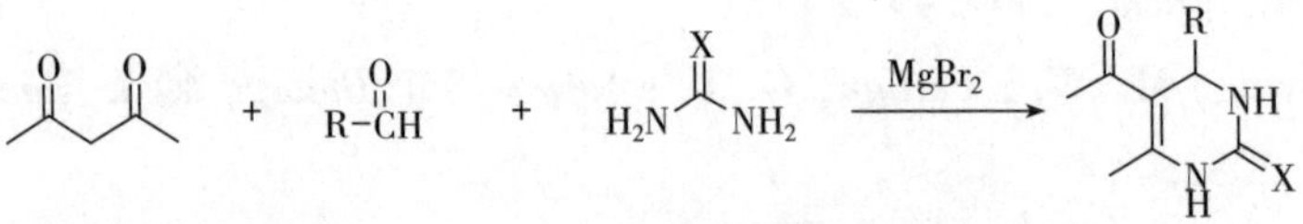

R–CH(=O) + H_2N–C(=X)–NH_2 —$MgBr_2$→

1.R=Ph,X=O	2.R=4–OHPh,X=O
3.R=4–MeOPh,X=O	4.R=2–OHPh,X=O
5.R=4–MeOPh–3–MeOPh,X=O	6.R=Et,X=O
7.R=2–MeOPh,X=S	8.R=4–OH–3–MeOPh,X=S
9.R=4–MeOPh,X=S	10.R=Et,X=S
11.R=4–OPh,X=S	

Synthesis of3,4 – dihydropyrimidin – 2 – (1H) – ones (DHPMs) and thione analogs. Reagents and conditions: $MgBr_2$, 50°C or 100°C, 0.5 – 3 h.

Scheme 2

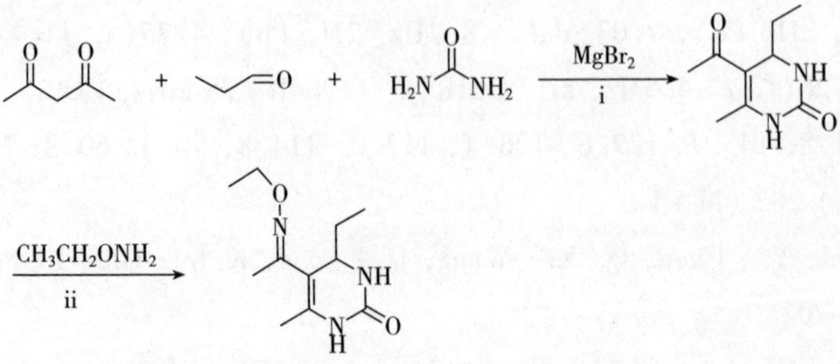

+ H_2N–C(=O)–NH_2 —$MgBr_2$, i→

—$CH_3CH_2ONH_2$, ii→

Synthesis ofthe Schiff base of 3, 4 - dihydropyrimidin - 2 - (1H) - ones (DHPMs). Reagents and conditions: (i) $MgBr_2$, 50° C, 3 h. (ii) anhydrous ethanol, room temp., 24 h.

Scheme 3

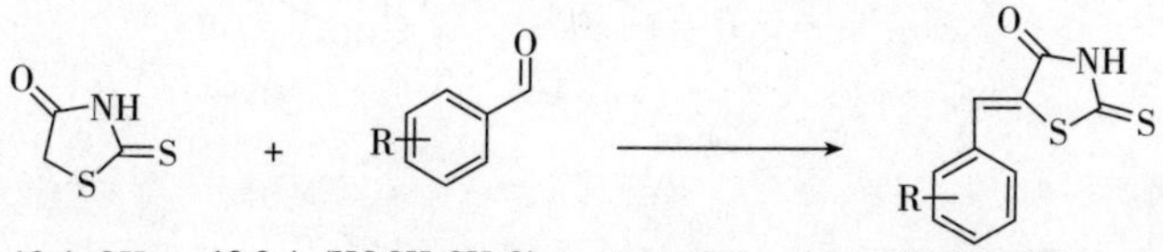

12.4-OH; 13.3,4-$(HOCH_2CH_2O)_2$; 14.2-OH; 15.4-$HOCH_2CH_2O$; 16.4-OH-3-Me;

Synthesis of Rhodanine derivatives. Reagents and conditions: acetic acid, sodium acetate, reflux.

Table 1 Tyrosinase inhibitory activities and yields of the synthesized compounds

Compounds	Yield (%)	IC_{50} (mmol/L)[a]
1	58.5	NA[b]
2	48.9	NA[b]
3	55.0	NA[b]
4	53.7	NA[b]
5	87.2	NA[b]
6	57.0	NA[b]
7	51.2	16.25
8	52.6	20.12
9	58.4	25.11
10	47.2	10.67
11	70.1	12.25
12	33.2	10.06
13	35.8	5.69
14	23.6	9.55
15	35.3	0.56
16	37.2	11.08
17	77.7	NA[b]
4 - Methoxycinnamic acid		0.41 (mmol/L)[c]
Arbutin		10.40 (mmol/L)[d]

a. Values were determined from logarithmic concentration - inhibition curves (at least eight points) and are given as means of three experiments

b. not active at 50mmol/L concentration.

c. IC_{50} values in the literature is 0.34 - 0.43 mmol/L[25,26].

d. The concentration of 10.4 mmol/L corresponding to inhibition percentage, determined in this work, is 30%. The reported IC_{50} value of arbutin was more than 30 mmol/L[27].

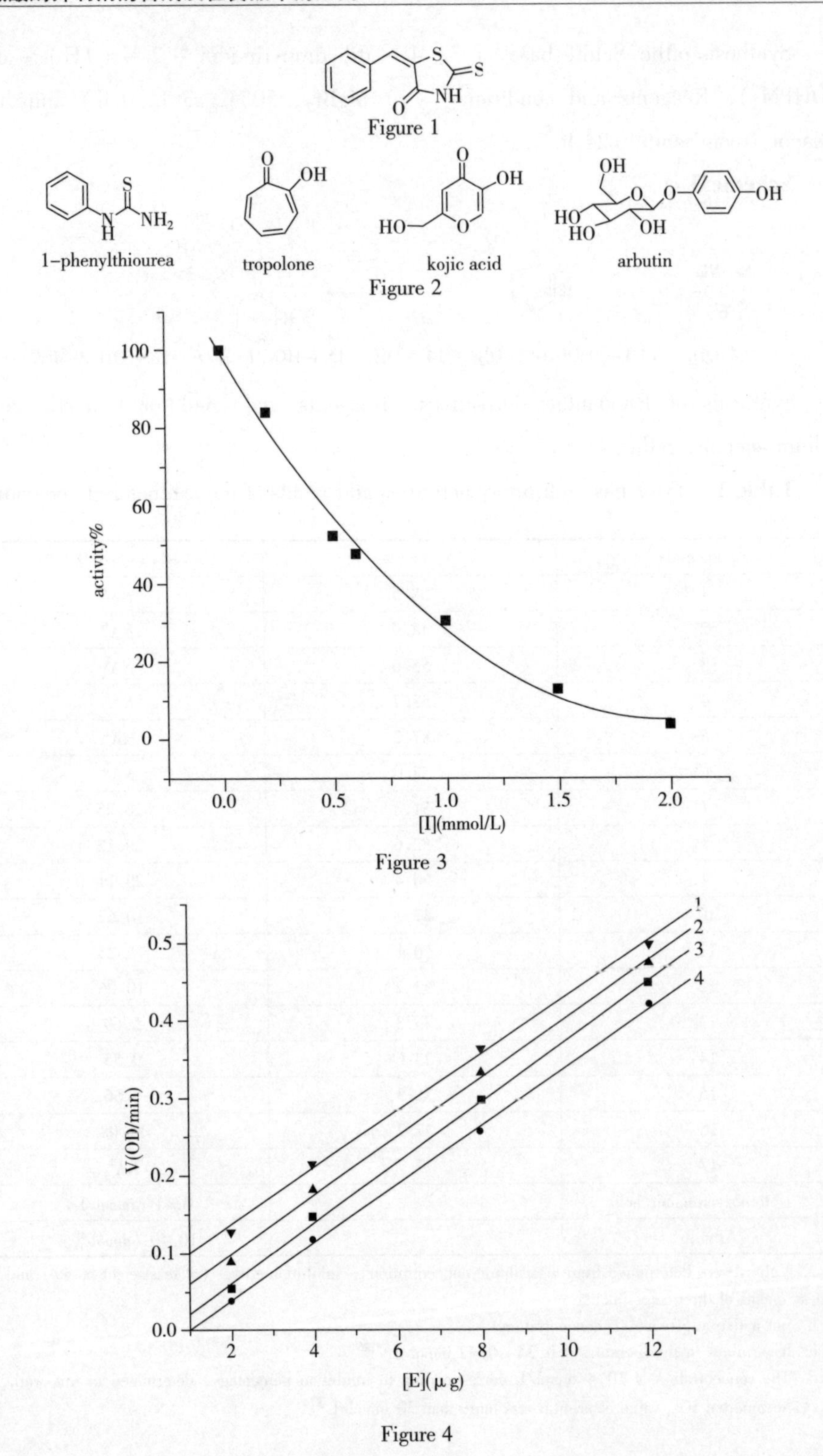

Figure 1

Figure 2

Figure 3

Figure 4

4.3.2 邵阳学院学报(自然科学版),2016,13(4):84-90.

噻二唑硫醚席夫碱酪氨酸酶抑制活性研究

刘进兵*,钟志坚,张宇,袁梦洁

(邵阳学院生物与化学工程系,湖南 邵阳 422000)

摘要 以二硫化碳和氨基硫脲为起始原料,缩合得到2-氨基-5-巯基-1,3,4噻二唑,再经硫醚化、缩合得到噻二唑硫醚席夫碱,目标化合物经核磁和质谱表征。对部分化合物进行了酪氨酸酶抑制活性考察,所测试的化合物都表现出一定的酪氨酸酶抑制活性,其中化合物8活性最好,强于阳性对照曲酸。优选化合物8进行了抑制机理探讨,结果表明其为不可逆抑制剂,同时对化合物8进行了分子对接研究。

关键词 噻二唑硫醚;席夫碱;酪氨酸酶抑制活性

Biological evaluation of 1,3,4-thiadiazole sulfide derivatives bearing Schiff base moieties as tyrosinase inhibitors

Zhijian Zhong, Yu Zhang, Mengjie Yuan, Jinbing Liu*

Department of Biology and Chemical Engineering, Shaoyang University, Shao Shui Xi Road, Shaoyang 422100, *PRC*

Abstract

1,3,4-thiadiazole sulfide derivatives bearing Schiff base moieties were designed, synthesized, and their tyrosinase inhibitory activities were evaluated. Some compounds displayed potent tyrosinase inhibitory activities, especially, compound 8 showed more potent inhibitory effect than the other compounds with the IC_{50} value of 1.03μmol/L. The structure-activity relationships (SARs) were preliminarily discussed. Docking study also was carried out in the paper. The inhibition mechanisms study demonstrated that the inhibitory effects of compound 8 on the tyrosinase were irreversible.

Keywords: thiadiazole sulfide derivatives; Schiff base; tyrosinase inhibitory activities.

作者简介:刘进兵,男,博士,副教授,从事药物分子设计与合成研究。

E-mail:syuliujb@163.com

酪氨酸酶是一种多酚氧化酶,两个铜离子活性中心与三个组氨酸残相连形成四面体结构[1,2]。酪氨酸酶是黑色素生物合成的关键酶,在动植黑色素合成过程中起着重要作用,与人的毛发、皮肤颜色密切相关,因此广泛存在于微生物和动植物[3,4]。酪氨酸酶在人体内过量表达会导致色素沉着性疾病,如雀斑、黄褐斑、老年斑等皮肤病,严重的会导致皮肤癌[5]。酪氨酸酶也和果蔬褐变有密切关系,酪氨酸酶抑制剂也可以用作食品保鲜剂[6]。同时酪氨酸酶与昆虫的蜕皮及伤口愈合,为开发生物杀虫剂指明方向[7]。由于酪氨酸酶抑制剂在医药、农业、食品工业、化妆品领域有着重要的用途,开发新型高效低毒的酪氨酸酶抑制剂并探讨其作用机制成为了科学界的研究热点。

杂环是一类具有生物活性多样性的物质,噻唑一类重要的含硫、氮杂环化合物,噻二唑是头孢唑啉、头孢西酮等头孢类药物的重要中间体。研究表明1,3,4-噻二唑类化合物是设计生物活性分子的重要先导化合物,具有抗肿瘤、抗菌、抗氧化、抗抑郁、抗菌消炎等生物活性[8,9]。本文以氨基硫脲和二硫化碳为起始原料合成2-氨基-5-巯基-1,3,4噻二唑,再经硫醚化、缩合得到噻二唑硫醚席夫碱,并考察了其酪氨酸酶抑制活性、抑制机理和抑制动力学特性,合成路线见Scheme 1。

Compound	R^1	R^2
1	Bz	—
2	烯丙基	—
3	Bz	4-羟基-3-甲氧基苯基
4	Bz	5-溴-2-羟基苯基
5	Bz	2-甲氧基苯基
6	Bz	5-氯-2-羟基苯基
7	烯丙基	4-羟基-3-甲氧基苯基
8	烯丙基	2,4-二羟基苯基
9	烯丙基	5-氯-2-羟基苯基
10	烯丙基	5-溴-2-羟基苯基
11	烯丙基	3-氯苯基
12	烯丙基	3-羟基苯基
13	烯丙基	2-羟基苯基
14	烯丙基	4-羟基苯基

图1 目标化合物的合成

1　实验部分

1.1　试剂与仪器

醛、酮购置于上海达瑞试剂有限公司，酪氨酸酶、L－多巴，曲酸从Sigma－Aldrich公司购得，其他试剂及溶剂为市售分析纯，未进一步纯化。熔点仪为SGW® X－4（上海精密科学仪器有限公司），未校正；核磁共振仪为Bruker 400 or 300型（德国Bruker公司）；质谱仪为LCMS－2010A型（日本岛津）；红外为VECTOR 22型（德国Bruker公司）；元素分析仪为Vario EL（德国Elementar公司）紫外分光光度计为UV－2100（北京莱伯泰科仪器有限公司）。

1.2　化合物合成

2－氨基－5－硫醚1，3，4噻二唑的合成

分别取0.1mol氨基硫脲9.114g、无水碳酸钠5g及无水乙醇80mL加入到三口烧瓶中，开启搅拌，于室温下滴加0.1mol（7.614g）二硫化碳，30min内滴完。滴加完毕后，继续在室温下搅拌0.5h，升温至80℃，直到反应体系不再产生硫化氢为止，冷至室温，加入100mL水，用浓盐酸调节PH至4，有大量淡黄色固体析出，抽滤，滤饼用无水乙醇重结晶得到2－氨基－5－巯基－1，3，4－噻二唑，收率为78%。

将0.03mol（3.99g）2－氨基－5－巯基－1，3，4－噻二唑和0.03mol（1.683g）氢氧化钾加入到50mL圆底烧瓶中，加入10mL水，用恒压滴液漏斗缓慢滴加0.03mol卤代烃，30min内滴完，滴加完毕后于RT下反应，TLC跟踪反应，反应瓶内产生大量淡黄色固体物质，抽滤，滤饼用水洗至中性，用乙醇重结晶，干燥得2－氨基－5－硫醚1，3，4噻二唑。

Compound 1. 收率98.71%. 淡黄色固体，熔点：156.5～159.1℃；^{1}H NMR（DMSO－d_6，500 MHz）δ 7.36－7.25（m，7H，NH_2，ph－H），4.30（s，2H，CH_2），；^{13}C NMR（DMSO－d_6，125 MHz）δ 170.3，149.9，137.4，129.5，129.0，127.9，38.9；MS（ESI）：m/z(100%) 224（M^+）.

Compound 2. 收率52.48%. 淡黄色固体，熔点：112.3～112.8 ℃；^{1}H NMR（DMSO－d_6，500 MHz）δ 7.32（s，2H，NH_2），5.91－5.84（m，1H，CH），5.16（dd，J=44.7，13.5Hz，2H），3.70（d，J=7.0Hz，2H，CH_2）；^{13}C NMR（DMSO－d_6，125 MHz）δ 170.4，149.8，133.9，119.1，37.8；MS（ESI）：m/z(100%) 174（M^+）.

2－氨基－5－硫醚1，3，4噻二唑席夫碱的合成

取5mmol 2－氨基－5－硫醚1，3，4噻二唑加入到25mL圆底烧瓶中，加入

10mL 无水乙醇,搅拌下加入不同取代基的芳香醛 6mmol 和催化量的 TBAB,回流反应,用 TLC 跟踪反应进程,反应完毕后冷却至室温,有固体析出,抽滤,滤饼用无水乙醇洗三次,用无水乙醇重结晶。如冷却至室温后无固体析出,采用柱层析分离得产品。

Compound 3. 收率 49.89%,淡黄色固体,熔点:178.1~180.1℃;^{1}H NMR (DMSO-d_6, 500 MHz) δ 10.33 (s, 1H, OH), 8.72 (s, 1H, CH), 7.58 (s, 1H, Ph-H), 7.50-7.46 (m, 2H, Ph-H), 7.37-7.28 (m, 4H, Ph-H), 6.96 (d, 1H, J=8.2Hz, Ph-H), 4.58 (s, 2H, CH_2), 3.86 (s, 3H, CH_3); ^{13}C NMR (DMSO-d_6, 125 MHz) δ 174.8, 169.0, 162.3, 153.2, 148.7, 136.9, 129.6, 129.1, 128.2, 126.5, 116.1, 111.5, 55.9, 37.5; MS (ESI): m/z (100%) 358 (M^+).

Compound 4. 收率 84.94%. 淡黄色固体,熔点:146.0~148.7℃;^{1}H NMR ($CDCl_3$, 500 MHz) δ 11.80 (s, 1H, OH), 8.93 (s, 1H, CH), 7.56-7.52 (m, 2H, Ph-H), 7.43 (d, J=7.2Hz, 1H, Ph-H), 7.36-7.30 (m, 3H, Ph-H), 6.95 (d, J=8.8Hz, 1H, Ph-H), 4.56 (s, 2H, CH_2); ^{13}C NMR ($CDCl_3$, 125 MHz) δ 174.8, 169.0, 162.3, 153.2, 148.7, 136.9, 129.6, 129.1, 128.2, 126.5, 116.1, 111.5, 55.9, 37.5; MS (ESI): m/z(100%) 405 (M^+).

Compound 5. 收率 43.06%,淡黄色固体,熔点:95.9~97.3℃;^{1}H NMR (DMSO-d_6, 500 MHz) δ 9.08 (s, 1H, CH), 8.59 (d, J=6.9Hz, 1H, Ph-H), 8.05 (d, J=7.6Hz, 1H, Ph-H), 7.65 (t, J=7.3Hz, 1H, Ph-H), 7.47 (d, J=7.4Hz, 1H, Ph-H), 7.42-7.25 (m, 4H, Ph-H), 7.10 (d, J=8.1Hz, 1H, Ph-H), 4.59 (s, 2H, CH_2), 3.92(s, 3H, CH_3); ^{13}C NMR (DMSO-d_6, 125 MIIz) δ 174.9, 167.7, 163.7, 160.9, 151.1, 137.4, 136.9, 136.2, 129.1, 128.0, 122.6, 121.5, 112.9, 56.4, 37.8; MS (ESI): m/z(100%) 342 (M^+).

Compound 6. 收率 49.92%. 淡黄色固体,熔点:148.2~150.7℃;^{1}H NMR ($CDCl_3$, 500 MHz) δ 11.79 (s, 1H, OH), 8.95 (s, 1H, CH), 7.44-7.30 (m, 7H, Ph-H), 7.00 (d, J=8.6Hz, 1H, Ph-H), 4.57 (s, 2H, CH_2); ^{13}C NMR ($CDCl_3$, 125 MHz) δ 170.4, 166.7, 164.8, 160.2, 135.6, 132.6, 132.3, 129.3, 128.9, 128.1, 124.6, 119.4, 118.9, 38.4; MS (ESI): m/z(100%) 362 (M^+).

Compound 7. 收率 26.64%. 淡黄色固体,熔点:65.5~67.4 ℃;^{1}H NMR

($CDCl_3$, 500 MHz) δ 8.65 (s, 1H, CH), 7.66 (s, 1H, Ph－H), 7.34 (d, J= 8.1Hz, 1H, Ph－H), 7.02 (d, J=8.1Hz, 1H, Ph－H), 6.01－5.97 (m, 1H, CH), 5.29 (dd, J=69.6, 13.5Hz, 2H), 3.97 (s, 3H, CH_3), 3.96 (d, J=7.0Hz, 2H, CH_2); ^{13}C NMR ($CDCl_3$, 125 MHz) δ 174.5, 166.6, 162.7, 151.4, 147.4, 132.0, 128.0, 127.3, 119.6, 114.6, 109.3, 56.2, 36.7; MS (ESI): m/z(100%) 308 (M^+).

Compound 8. 收率47.83%. 淡黄色固体, 熔点:152.7~154.9 ℃; ^1H NMR (DMSO－d_6, 500 MHz) δ 11.53 (s, 1H, OH), 10.73 (s, 1H, OH), 8.93 (s, 1H, CH), 7.69 (d, J=8.7Hz, 1H, Ph－H), 6.46 (d, J=8.7Hz, 1H, Ph－H), 6.38 (s, 1H, Ph－H), 6.00－5.95 (m, 1H, CH), 5.28 (dd, J=76.8, 13.5Hz, 2H), 3.96 (d, J=6.8Hz, 2H, CH_2); ^{13}C NMR (DMSO－d_6, 125 MHz) δ 174.1, 167.1, 165.5, 163.4, 162.0, 134.0, 133.1, 120.0, 112.6, 109.8, 102.8, 36.7; MS (ESI): m/z(100%) 294 (M^+).

Compound 9. 收率31.42%. 淡黄色固体, 熔点:125.9~127.5℃; ^1H NMR ($CDCl_3$, 500 MHz) δ 11.80 (s, 1H, OH), 8.97 (s, 1H, CH), 7.43 (s, 1H, Ph－H), 7.41 (d, J=8.7Hz, 1H, Ph－H), 7.01 (d, J=8.7Hz, 1H, Ph－H), 6.01－5.97 (m, 1H, CH), 5.31 (dd, J=68.7, 13.5Hz, 2H), 3.97 (d, J=6.9Hz, 2H, CH_2); ^{13}C NMR ($CDCl_3$, 125 MHz) δ 170.7, 166.7, 164.8, 160.3, 135.5, 132.6, 131.8, 124.6, 119.9, 119.4, 118.9, 36.6; MS (ESI): m/z (100%) 312 (M^+).

Compound 10. 收率28.41%. 淡黄色固体, 熔点:104.7~105.9℃; ^1H NMR ($CDCl_3$, 500 MHz) δ 11.82 (s, 1H, OH), 8.96 (s, 1H, CH), 7.58 (s, 1H, Ph－H), 7.55 (d, J=8.8Hz, 1H, Ph－H), 6.96 (d, J=8.8Hz, 1H, Ph－H), 6.01－5.97 (m, 1H, CH), 5.31 (dd, J=68.7, 13.5Hz, 2H), 3.97 (d, J=6.9Hz, 2H, CH_2); ^{13}C NMR ($CDCl_3$, 125 MHz) δ 170.7, 166.6, 164.8, 160.7, 138.3, 135.7, 131.8, 119.9, 119.6, 111.4, 36.6; MS (ESI): m/z(100%) 356 (M^+).

Compound 11. 收率 6.72%. 淡黄色固体, 熔点:79.3~81.2℃; ^1H NMR ($CDCl_3$, 500 MHz) δ 8.84 (s, 1H, CH), 7.97 (s, 1H, Ph－H), 7.81 (d, J= 7.6Hz, 1H, Ph－H), 7.54 (d, J=6.4Hz, 1H, Ph－H), 7.46－7.43 (m, 1H, Ph－H), 6.01－5.97 (m, 1H, CH), 5.37 (dd, J=68.7, 13.5Hz, 2H), 3.98 (d, J=6.9Hz, 2H, CH_2); ^{13}C NMR ($CDCl_3$, 125 MHz) δ 168.7, 165.0, 153.8,

139.7, 134.6, 131.9, 130.3, 129.5, 126.7, 119.8, 119.4, 36.6; MS (ESI): m/z(100%) 296 (M^+).

Compound 12. 收率 58.31%. 淡黄色固体, 熔点:156.2 ~ 158.1℃; ^{1}H NMR (DMSO - d_6, 500 MHz) δ 9.92 (s, 1H, OH), 8.84 (s, 1H, CH), 7.47 (s, 1H, Ph - H), 7.44 (d, *J* = 7.6Hz, 1H, Ph - H), 7.39 - 7.33 (m, 1H, Ph - H), 7.07 (d, *J* = 6.4Hz, 1H, Ph - H), 5.99 - 5.96 (m, 1H, CH), 5.37 (dd, *J* = 68.7, 13.5Hz, 2H), 3.99 (d, *J* = 6.9Hz, 2H, CH_2); ^{13}C NMR (DMSO - d_6, 125 MHz) δ 193.6, 174.3, 170.0, 163.5, 158.5, 136.2, 130.7, 122.3, 121.5, 120.0, 115.6, 36.6; MS (ESI): m/z(100%) 278 (M^+).

Compound 13. 收率 42.89%. 淡黄色固体, 熔点:102.2 ~ 103.4℃; ^{1}H NMR ((CDCl$_3$, 500 MHz) δ 11.82 (s, 1H, OH), 9.03 (s, 1H, CH), 7.50 - 7.45 (m, 2H, Ph - H), 7.04 (d, *J* = 8.3Hz, 1H, Ph - H), 6.99 (d, *J* = 7.4Hz, 1H, Ph - H), 6.03 - 5.97 (m, 1H, CH), 5.30 (dd, *J* = 68.8, 13.4Hz, 2H), 3.96 (d, *J* = 6.9Hz, 2H, CH_2); ^{13}C NMR (CDCl$_3$, 125 MHz) δ 171.4, 168.1, 164.1, 161.8, 135.8, 134.1, 131.9, 120.0, 119.8, 118.2, 117.8, 36.7; MS (ESI): m/z(100%) 278 (M^+).

Compound14. 收率 4.51%. 淡黄色固体, 熔点:96.9 ~ 100.5℃; ^{1}H NMR (DMSO - d_6, 500 MHz) δ 10.63 (s, 1H, OH), 8.76 (s, 1H, CH), 7.89 (d, *J* = 8.5Hz, 2H, Ph - H), 6.94 (d, *J* = 8.5Hz, 2H, Ph - H), 5.98 - 5.95 (m, 1H, CH), 5.28 (dd, *J* = 77.1, 13.5Hz, 2H), 3.96 (d, *J* = 6.8Hz, 2H, CH_2); ^{13}C NMR (DMSO - d_6, 125 MHz) δ 175.0, 169.0, 163.4, 162.3, 133.2, 133.1, 126.2, 119.9, 116.6, 36.6; MS (ESI): m/z(100%) 278 (M^+).

1.3　酪氨酸酶抑制活性测试

参照本课题组已经建立起的筛选方法[10],在总计 1000 μL 的测试体系中加入 940 ~ 950μL 0.1mol/L pH 值 6.8 磷酸盐缓冲液,10 μL 样品溶液(10 μL 样品作对照,以去除样品的颜色对实验的影响),10 - 20μL 酶液,25℃孵化 10min,加入 30μL DOPA 溶液,475nm 进行时间扫描,记录一分钟的 OD 值。改变样品溶液的剂量,分别测试酶催化活性。可以通过稀释的方式改变样品的浓度。以没有加入样品的酶催化活性 OD/min 值为 100%,用抑制百分数(指被抑制而失去活力的百分数)表示不同浓度样品对酶的抑制活性,做图求出抑制 50% 时样品的浓度(IC_{50})。抑制率的计算公式如下:

$$抑制率(\%) = [(B - S)/B] \times 100$$

式中 B 为空白吸光度，S 为样品吸光度。

1.4 抑制机理研究

适当选择3个不同样品浓度以及空白对照四个系列，保持底物浓度不变，改变酶浓度，分别测试酶催化的活性。以酶浓度[E]（μmol/L/mL）对酶催化活性 V（OD/min）的作用作图，根据四条线的交点判断抑制剂的类型。

1.5 分子对接

Auto Dock 是美国 Scripps 研究所的 Olson 实验室开发与维护的一款分子对接模拟软件，主要应用于配体—蛋白分子对接，Auto Dock Tools 1.5.4 从官网下载，编号 2y9x 酪氨酸酶晶体结构文件来自 Protein Data Bank，2y9x 为双孢菇酪氨酸酶的 X 衍射三维晶体结构，有 8 个亚基 A～H，对接位点选在 A 亚基活性中心小分子抑制剂 tropolone（OTR）所在区域。为充分暴露其活性中心，删去其余亚基及配体 OTR，得到用于对接的受体三维结构，保存为 pdb 格式，带对接化合物结构式用 Auto Dock Tools 1.5.4 软件处理受体分子：去水加氢，计算 Gasteiger 电荷，合并非极性氢原子，保存为 pdbqt 格式。同样，Auto Dock 可对配体进行处理，其非极性氢原子被合并，Gasteiger 电荷被加上，使能量最小化。Auto Dock Tools 1.5.4 软件中的 Ligand 子程序包可自动检测可旋转键的个数，在对接过程中，这些键可以灵活旋转与受体分子对接。Auto Dock4.2 可以进行柔性对接，但选取柔性基团的不确定因素较多，本实验采用刚性对接形式。

2 结果与讨论

2.1 目标化合物的合成

以二硫化碳和氨基硫脲为起始原料，无水乙醇做溶剂，在无水碳酸钠存在的情况下发生缩合的2－氨基－5－巯基1,3,4－噻二唑，通过改变反应温度得到反应体系在回流状态下收率最好，可以达到78%。将所得的2－氨基－5－巯基1,3,4－噻二唑分别于烯丙基溴及氯苄反应得到相应的硫醚化合物，其中和氯苄反应产物收率几乎是定量的，达到98.71%，然而和烯丙基溴反应的收率只有52.48%，说明氯苄反应活性更高。将所的相应的硫醚与不同的醛反应生成席夫碱，收率并不是很高，烯丙基取代的化合物收率偏低。

2.2 目标化合物的波谱分析

从核磁共振氢谱来看，苄基硫醚在 δ 4.3 处的单峰是苄基碳原子上的两个氢，烯丙基硫醚在 δ 5.16 处的 dd 峰为烯丙位碳原子上的两个氢原子，δ 5.91 处多重峰为双键中间碳原子上的氢，在 δ 3.70 处双峰为双末端碳原子上的氢。形成席夫碱后原来醛基碳上氢原子位置发生变化，约在 δ 8.50－9.00 处。在核磁共振碳谱中,形成席夫碱后处于席夫碱上的碳原子约在 δ 170。

2.3 化合物的酪氨酸酶抑制活性

为考察所合成的化合物对酪氨酸酶的抑制活性,采用本课题组已经建立起的帅选方法,以曲酸为阳性对照,对所合成的噻二唑硫醚席夫碱进行了酪氨酸酶抑制活性,其结果见表1。

表1 目标化合物酪氨酸酶抑制活性

Compounds	CLog P[a]	IC_{50} (μmol/L)[b]
3	3.438	112.52
4	4.545	276.84
5	3.688	103.16
6	4.395	247.35
7	2.644	228.28
8	2.388	1.03
9	3.601	—[c]
10	3.751	94.41
11	3.263	—
12	2.784	107.98
13	2.784	117.90
14	2.784	—
曲酸	—	23.6

[a] Log P 来源于 ChemBioDraw Ultra 12.0。
[b] 取三次实验平均值。
[c] 未测。

测试结果表明,目标化合物具有一定的酪氨酸酶抑制活性,其中化合物 8 抑制活性最强,其 IC_{50} 为 1.03μmol/L,强于阳性对照曲酸,其结构特征是含有两个羟基。苄基硫醚类化合物酪氨酸酶抑制活性最强的是化合物 5。对比苄基硫醚和烯丙基硫醚同种醛席夫碱的抑制活性,化合物 3 的抑制活性强于化合物 7,而化合物 10 的抑制活性强于化合物 4,这就说明该类化合物的酪氨酸酶抑制活性不仅与形成硫醚的基团有关,还与形成席夫碱的芳香醛上取代基的种类有关。从化合物 12,13 的活性数据可知,羟基在苯环上的位置可能对抑制活性有影响。

为了探讨这类化合物的抑制机理,优选化合物 8 进行抑制机理探讨,如图 2 所示,以酶的浓度为横坐标,V 为纵坐标作图得到一组平行线,因此化合物 8 对酪氨酸酶的抑制作用是不可逆的。

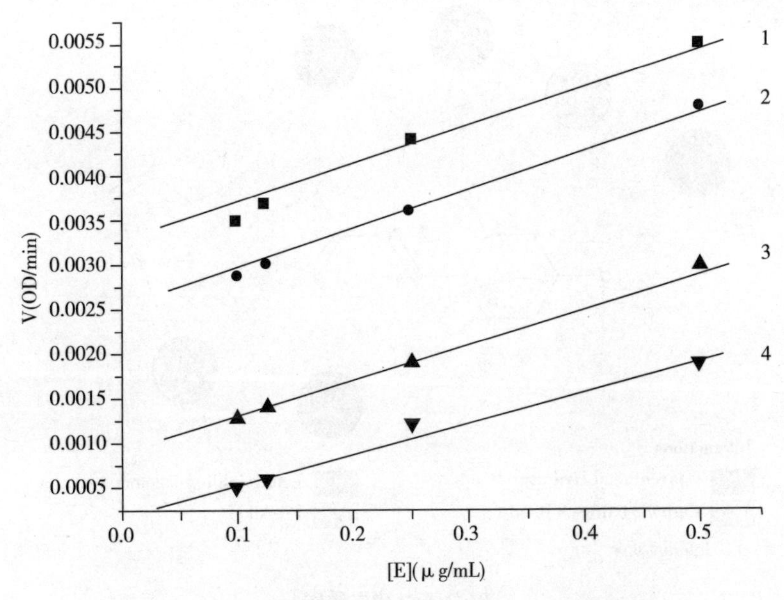

图2　化合物8对酪氨酸酶的抑制机理

化合物8对应直线1~4的浓度为0，0.5，1，1.25μmol/L

化合物8活性最好，选择这个化合物与酪氨酸酶进行分子对接，对接结果如图3所示。通过对接结果发现，化合物8与酪氨酸酶受体结合的总能量是-80.38 kcal/mol，这就说明化合物8能够与酪氨酸酶形成稳定的络合物。化合物8与酪氨酸酶可能发生相互作用的活性位点如图3所示。苯环上的两个酚羟基可以分别和GLU359，ILE13形成比较牢固氢键，其键长分别为2.85 Å和2.98 Å，噻二唑环上硫原子可以与LYS345形成氢键，其键长为2.56 Å。烯丙基的碳碳双键能够与TRP350形成π-π叠加。

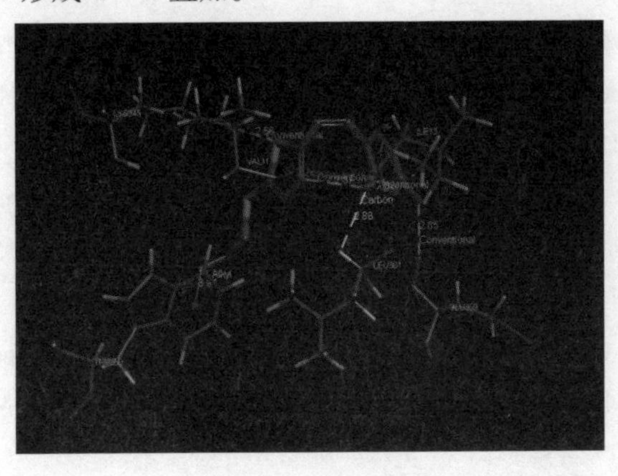

A. 分子对接三维图

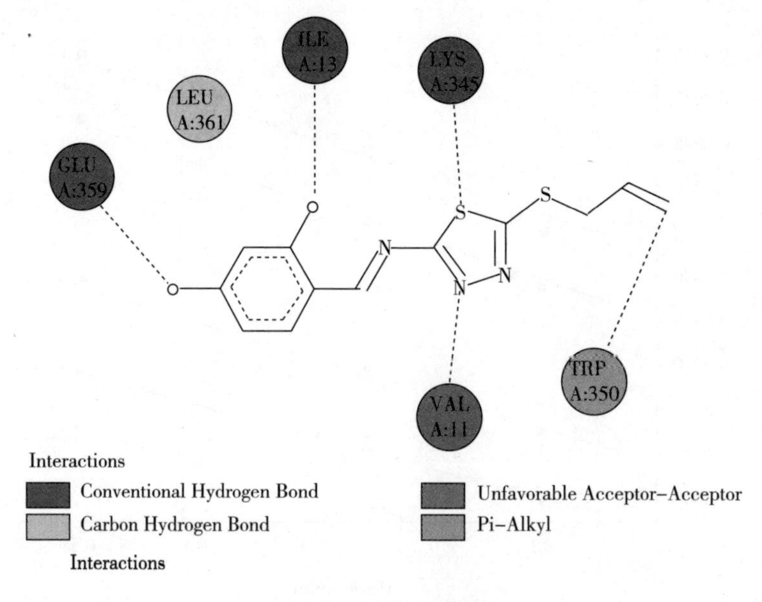

B. 分子对接二维图

图 3 分子对接图

3 结论

将 1,3,4 - 噻二唑和席夫碱拼接到一个分子中,考察其酪氨酸酶抑制活性,初步讨论了构效关系,化合物 8 的 IC_{50} 值为 1.03μmol/L,强于阳性对照曲酸,为该类化合物中活性最强的,并对该化合物进行了抑制机理探讨,结果表明其为不可逆抑制剂。同时对化合物 8 进行了分子对接研究,发现羟基在抑制酪氨酸酶的过程中起着重要作用。

参考文献

[1] MICHAEL R L S, CHRISTOPHER A R, PATRICK A R. Mechanistic studies of the inactivation of tyrosinase by resorcinol [J]. Bioorg Med Chem, 2013, 21(5): 1166 - 1173.

[2] LI Y C, WANG Y, JIANG H B, DENG J P. Crystal structure of *Manduca sexta* prophenoloxidase provides insights into the mechanism of type 3 copper enzymes [J]. P Natl Acad Sci USA,2009, 106(40): 17002 - 17006.

[3] CHEN Q X, LIU X D, HUANG H. Inactivation kinetics of mushroom tyrosinase in the dimethyl sulfoxide solution [J]. Biochemistry (Mosc), 2003, 68 (6): 644 - 649.

[4] DONG X, ZHU Q, DAI Y, HE J, PAN H, CHEN J, ZHENG Z - P.

Encapsulation artocarpanone and ascorbic acid in O/W microemulsions: Preparation, characterization, and antibrowning effects in apple juice [J]. Food Chem. 2016, 192: 1033 – 1040.

[5] TAN X, SONG Y H, PARK C, LEE K – W, KIM J Y, KIM D W, KIM K D, LEE K W. Highly potent tyrosinase inhibitor, neorauflavane from Campylotropis hirtella and inhibitory mechanism with molecular docking [J]. Bioorg Med Chem, 2016, 24(2): 153 – 159.

[6] ARTÉS F, CASTANER M, GIE M I. Revisión: EI pardeamiento enzimático en frutasy hortalizas mínimamente procesadas Review: Enzymatic browning in minimally processed fruit and vegetables [J]. Food Sci Technol Int, 1998, 4: 377.

[7] ORTIZ – URQUIZA A, KEYHANI N O. Action on the surface: Entomopathogenic fungi versus the insect cuticle [J]. Insects, 2013, 4(3): 357 – 374.

[8] YANG S – J, LEE S – H, KWAK H – J, GONG Y – D. Regioselective Synthesis of 2Amino – Substituted 1,3,4 – Oxadiazole and 1,3,4 – Thiadiazole Derivatives via Reagent – Based Cyclization of Thiosemicarbazide Intermediate [J]. J Org Chem, 2013, 78(2): 438 – 444.

[9] FARGHALY T A, ABDALLAH M A, MASARET G S, MUHAMMAD Z A. New and efficient approach for synthesis of novel bioactive [1,3,4] thiadiazoles incorporated with 1,3 – thiazole moiety [J]. Eur J Med Chem, 2015, 97: 320 – 333.

[10] LIU J B, YI W, WAN Y Q, MA L, SONG H C. 1 – (1 – Arylethylidene) – thiosemicarbazide derivatives: A new class of tyrosinase inhibitors [J]. Bioorg Med Chem, 2008, 16(3): 1096 – 1102.

(本论文得到邵阳学院大学生研究性学习和创新性实验计划项目资助)

4.3.3 Bioorganic Chemistry

Molecular docking studies and biological evaluation of 1,3,4 – thiadiazole derivatives bearing Schiff base moieties as tyrosinase inhibitors

Junyuan Tang, Jinbing Liu*, Fengyan Wu Department
of Biology and Chemical Engineering, Shaoyang University, Shao Shui Xi Road,

Shaoyang 422100, PR China

Abstract

1,3,4 – Thiadiazole derivatives bearing Schiff base moieties were designed, synthesized, and their tyrosinase inhibitory activities were evaluated. Some compounds displayed potent tyrosinase inhibitory activities, especially, 4 – (((5 – mercapto – 1, 3,4 – thiadiazol – 2 – yl) – imino) methyl) – 2 – methoxy – phenol (14) exhibited superior inhibitory effect to the other compounds with an IC_{50} value of 0.036μmol/L. The structure – activity relationships (SARs) were preliminarily discussed and docking studies showed compound14had strong binding affinity to mushroom tyrosinase. Hydroxy might be the active groups. The inhibition kinetics study revealed that compounds (13and14) inhibited tyrosinase by acting as uncompetitive inhibitors. The LD_{50} value of the compound14was 5000 mg/kg.

1. Introduction

Tyrosinase (monophenol or o – diphenol, oxygen oxidoreductase, EC 1.14.18.1, syn. polyphenol oxidase), as known as polyphenol oxidase (PPO) possesses an active site where two copper ions bind dioxygen in a form of the enzyme termed as oxy – tyrosinase[1]. The binuclear center together with three specific histidine residues forms a square – pyramidal structure [2]. Tyrosinase is a key enzyme in the biosynthesis of melanin pigment, and catalyzes two distinct reactions involving molecular oxygen in the hydroxylation of monophenols too – diphenols (monophenolase), as well as oxidation of o – diphenols too – quinones (diphenolase) [3]. Therefore, it plays a vital role in the processes of melanin formation in bacteria, fungi, plants and mammals [4]. Because of the high reactivity of tyrosinase, quinones could polymerize spontaneously to form high molecular weight brown – pigments (melanins), which widely spreads in skin, hair and eyes of mammals, or react with amino acids and proteins to enhance brown color of the pigment produced [5]. Although melanin is important in preventing skin cancer caused by overexposure to ultraviolent lights[6,7], excessive melanin production and accumulation may lead to hyperpigmentation such as melasma, freckles, ephelide, and solar lentigines [8]. Tyrosinase is also known to oxidize dopamine to form melanin in the brain, and thus tyrosinase is implicated in the pathogenesis of Parkinson's disease and related neurodegenerative disorders [9, 10]. In food industry, tyrosinase catalyzes the reactions resulting in undesirable browning in fruits, vegetables, and beverages, which impairs the color and sensory properties of these food products and may shorten shelf life, reduce market value, and lose nutritional value during the postharvest handling processes [11]. One way to avoid these negative effects resulted from tyrosinase is to use inhibitors of this enzyme. Furthermore, in insects, tyrosinase is associated with three biochemical processes, including sclerotization of cuticle, defensive encapsulation and melanization of foreign organism, and wound healing [12,13]. These processes provide potential targets for developing safer and effective tyrosinase inhibitors as insecticides and ultimately for insect control. In fact, tyrosinase inhibitors have been well appreciated in the fields of cosmetics, medicine, food sciences and agriculture[14,15]. Most of the recent researches have focused on below following fields: first, finding new naturally occurring inhibitors of tyrosinase; second, proceeding medicinal modifications or synthesis chemical analogs of target inhibitors for better inhibitory activities and lower side effects; third, investigating structure - activity and inhibitory mechanism of tyrosinase inhibitors. By far, some natural and synthetic tyrosinase inhibitors were reported, for example, hydroquinone, ascorbic acid, arbutin, kojic acid, aromatic aldehydes, aromatic acids, aromatic alcohol, tropolone, and polyphenols [16 - 19]. However, only handful molecules such as kojic acid, arbutin, tropolone, and 1 - phenyl - 2 - thiourea (PTU) are commercially useful [20]. Moreover, it has been reported that arbutin decomposes at room temperature (10% decomposition at 20℃ for 15 days), while arbutin and tropolone have been demonstrated not clinically efficacious when systematically analyzed in carefully controlled studies[21]. As a result, there is still an urgent need for novel tyrosinase inhibitors with higher activity, better pharmacology properties and lower side effect. Gratifyingly, it was reported that phenyl thioureas and alkyl thioureas could exhibit weak to moderate depigmenting activity [22], and the interaction with the hydrophobic protein pocket is close to the binuclear copper active site of tyrosinase. The sulfur atom of in the thioureas can bind to both copper ions in the active site. Similarly, thiosemicarbazide derivatives and curcumin have potential to coordinate the two copper ions in the active site of tyrosinase [23]. Inspired by these precedent literatures, our research provides latest progress in designing tyrosinase inhibitors.

We envisioned that thiadiazole, the hetero sulfur containing aromatic ring, could improve lipophilicity, and the mesoionic nature of thiadiazole makes this class of compounds to show good tissue permeability [24, 25]. The two - electron donor nitrogen system (- N = C - S) as well as the hydrogen binding domain, may help with pharmacodynamics of the molecule for better binding with the receptors[26]. Plus, thiadiazoles are bioisosteres of pyrimidines, oxadiazoles, oxazoles, benzene and the derivatives have been proved to display diverse range of biological and pharmacological properties[27]. 1,3,4 - thiadiazole, a unique structure, represents a key motif in heterocyclic chemistry and medicinal chemistry [28]. For instance, many drugs containing 1,3,4 - thiadiazole moieties are available on the market, e.g., acetazolamide, methazolamide and sulfamethazole[29]. Recently, derivatives of 1, 3, 4 - thiadiazole have been reported for their antibacterial, antifungal, inflammatory, antianxiety, and antitubercular activities[30]. Some scientists have studied extensively on 1, 3, 4 - thiadiazole derivatives as potential drugs to treat Alzheimer's disease (AD)[31]. 1,3,4 - Thiadiazole derivatives bearing Schiff base moieties even display anticancer, antibacterial, antidepressant, antidiabetic and antifungal activities [32]. Lately, our studies indicated that thiosemicarbazide derivatives exhibited potent inhibitory activities against mushroom tyrosinase. Inhibition mechanism analysis showed that this series of compounds could potentially bind to the binuclear active site of tyrosinase [33]. 1,3,4 - Thiadiazole - 2(3H) - thiones exhibited potent inhibitory activities against mushroom tyrosinase [34]. But to the best of our knowledge, the tyrosinase inhibitory activities of the 1, 3, 4 - thiadiazole Schiff base derivatives have never been reported in the literatures. Combining the biological significance of 1,3,4 - thiadiazole and Schiff base moiety, we designed and synthesized a library of 1,3,4 - thiadiazole derivatives bearing Schiff base moieties, and their inhibitory activities against mushroom tyrosinase were evaluated using kojic acid as positive control. The structure - activity relationship (SAR) was analyzed and discussed. To better elucidate the inhibition mechanism, the docking studies were carried out and discussed as well.

2. Materials and methods

2.1. Chemistry

Melting points (m. p.) were determined on SGW - X4 melting point apparatus and the thermometer was uncorrected. NMR spectra were recorded on Bruker 300 spectrometers at 25C in DMSOd6using TMS as internal standard for protons. Chemical shift values are mentioned ind(ppm) and coupling constants (J) are given in Hz. Mass – spectrometric (MS) data is reported in m/zusing the LCMS – 2010A. The reaction progress was monitored by TLC (Merck Kieselgel 60 F254). The chromatograms were visualized under UV 254 – 365 nm and iodine. Infrared (IR) spectra were recorded on VECTOR 22 spectrometer in KBr pellets and are reported in cm^{-1}. Tyrosinase, L – 3,4 – dhydroxyphenylalanine (L – DOPA) and kojic acid were purchased from Sigma – Aldrich Chemical Co (Shanghai, China). Other chemicals were purchased from commercial suppliers and were used without further purification.

2.1.1. Synthesis of 1,3,4 – thiadiazolederivatives bearing Schiff base moieties [30,35]

Thiosemicarbazide 18.228 g (0.2 mol) and anhydrous sodium carbonate 10.6 g (0.1 mol) were suspended in 80 mL anhydrous ethanol. The reaction mixture was stirred at RT and carbon disulphide 0.2 mol (15.228 g) was added dropwise in 1 h. The reaction mixture was stirred under reflux for 4 h and monitored by TLC until the reaction was completed. After completion of reaction, the mixture was allowed to cool to room temperature, water (100 mL) was then added to the mixture and neutralized by the 10% solution of hydrochloric acid, white precipitates of 2 – amino – 5 – mercapto – 1,3,4 – thiadiazole were generated. 2 – Amino – 5 – mercapto – 1,3,4 – thiadiazole (5 mmol) was suspended in 10 mL anhydrous ethanol and 6 mmol of different aromatic aldehyde was added, using three drops concentrated sulfuric acid as catalyst. The reaction mixture was stirred under reflux for 8 h and monitored by TLC until the reaction was completed.

After completion of reaction, the mixture was allowed to cool to room temperature, and then left overnight. Precipitates of 1,3,4 – thiadiazole derivatives bearing Schiff base moieties were generated. The solids were filtered, and washed with the mixture of ethanol and hot water (1 : 1); dried solid was purified by crystallization from anhydrous alcohol to afford the desired compound.

2.1.1.1. 2 – (Benzylideneamino) – 5 – mercapto – 1,3,4 – thiadiazole (1). Yield 51.4%. Yellow solid, mp: 237 ~ 239.2℃, IR (KBr) v_{max} 3310, 1567, 1476, 1376, 1312, 1121, 1051, 749, 711 cm^{-1}; ^{1}H NMR (DMSO – d_6, 300

MHz) δ 13.02 (s, 1H, SH), 8.62 (s, 1H, =CH), 7.91 (d, 2H, J = 8.7 Hz, ph - H), 7.60 (t, 1H, J = 7.5Hz, ph - H), 7.45 (t, 2H, J = 7.5 Hz, ph - H); ^{13}C NMR (DMSO - d_6, 75 MHz) δ 193.1, 180.9, 161.4, 136.1, 133.7, 129.9, 128.6; MS (ESI): m/z(100%) 222 (M^+); Anal. Calcd for $C_9H_7N_3S_2$: C, 48.85; H, 3.19; N, 18.99; found: C, 48.82; H, 3.18; N, 18.96.

2.1.1.2. 2 - ((2 - Fluorobenzylidene)amino) - 5 - mercapto - 1,3,4 - thiadiazole (2). Yield 16%. Yellow solid, mp: 242.6 ~ 244.1℃, IR (KBr) v_{max} 3223, 2995, 1562, 1491, 1382, 1313, 1267, 1208, 1108, 1051, 942, 754, 671 cm^{-1}; ^{1}H NMR (DMSO - d_6, 300 MHz) δ 13.18 (s, 1H, SH), 8.83 (s, 1H, = CH), 8.25 (d, 1H, J = 7.5 Hz, ph - H), 7.54 (t, 1H, J = 8.1 Hz, ph - H), 7.34 (d, 1H, J = 7.5 Hz, ph - H), 7.29 (t, 1H, J = 8.1 Hz, ph - H); ^{13}C NMR (DMSO - d_6, 75 MHz) δ 187.9, 180.8, 157.4, 136.9, 131.7, 129.4, 125.2, 122.1, 116.8; MS (ESI): m/z(100%) 240 (M^+); Anal. Calcd for $C_9H_6FN_3S_2$: C, 45.17; H, 2.53; N, 17.56; found: C, 45.15; H, 2.53; N, 17.52.

2.1.1.3. 2 - ((3 - Chlorobenzylidene)amino) - 5 - mercapto - 1,3,4 - thiadiazole (3). Yield 59.9%. Yellow solid, mp: 253.4 ~ 255.3℃, IR (KBr) v_{max} 3284, 1570, 1510, 1373, 1309, 1259, 1188, 1105, 1047, 738, 677 cm^{-1}; ^{1}H NMR (DMSO - d_6, 300 MHz) δ 13.12 (s, 1H, SH), 8.86 (s, 1H, =CH), 8.12 (s, 1H, ph - H), 7.96 (d, 1H, J = 7.8 Hz, ph - H), 7.80 (d, 1H, J = 6.9 Hz, ph - H), 7.45 (t, 1H, J = 8.7 Hz, ph - H); ^{13}C NMR (DMSO - d_6, 75 MHz) δ 192.1, 180.5, 161.4, 136.5, 134.2, 131.2, 130.4, 127.9, 127.0; MS (ESI): m/z(100%) 256 (M^+); Anal. Calcd for $C_9H_6ClN_3S_2$: C, 42.27; H, 2.36; N, 16.43; found: C, 42.25; H, 2.35; N, 16.42.

2.1.1.4. 2 - ((4 - Chlorobenzylidene)amino) - 5 - mercapto - 1,3,4 - thiadiazole (4). Yield 30.2%. Yellow solid, mp: 222.1 ~ 224.5℃, IR (KBr) v_{max} 3275, 1563, 1488, 1376, 1313, 1259, 1185, 1094, 1051, 1015, 783, 744 cm^{-1}; ^{1}H NMR (DMSO - d_6, 300 MHz) δ 13.18 (s, 1H, SH), 8.75 (s, 1H, =CH), 7.95 (d, 2H, J = 6.9 Hz, ph - H), 7.71 (d, 2H, J = 6.9 Hz, ph - H); ^{13}C NMR (DMSO - d6, 75 MHz) d192.1, 180.9, 161.4, 139.3, 134.8, 131.5, 129.3; MS (ESI): m/z(100%) 256 (M^+); Anal. Calcd for $C_9H_6ClN_3S_2$: C, 42.27; H, 2.36; N, 16.43; found: C, 42.28; H, 2.34; N, 16.45.

2.1.1.5. 2 – ((3,4 – Dichlorobenzylidene)amino) – 5 – mercapto – 1,3,4 – thiadiazole (5). Yield 67.8%. Yellow solid, mp: 223.3 ~ 225.7°C, IR (KBr) v_{max} 3404, 1566, 1469, 1376, 1309, 1264, 1190, 1103, 1053, 746, 667 cm^{-1}; ^1H NMR (DMSO – d_6, 300 MHz) δ 13.15 (s, 1H, SH), 8.82 (s, 1H, =CH), 7.97 (s, 1H, ph – H), 7.75 (d, 1H, J = 8.4 Hz, ph – H), 7.65 (d, 1H, J = 8.4 Hz, ph – H); ^{13}C NMR (DMSO – d_6, 75 MHz) δ 189.7, 181.3, 159.8, 136.8, 135.1, 134.5, 133.2, 131.3, 128.4; MS (ESI): m/z (100%) 291 (M^+); Anal. Calcd for $C_9H_5Cl_2N_3S_2$: C, 37.25; H, 1.74; N, 14.48; found: C, 37.28; H, 1.76; N, 14.49.

2.1.1.6. 4 – Chloro – 2 – (((5 – mercapto – 1,3,4 – thiadiazol – 2 – yl) imino)methyl)phenol (6). Yield 53%. Yellow solid, mp: 247.3 ~ 250.1°C, IR (KBr) v_{max} 3440, 3067, 2862, 1599, 1561, 1517, 1470, 1231, 1173, 1063, 804, 711 cm^{-1}; ^1H NMR (DMSO – d_6, 300 MHz) δ 14.58 (s, 1H, SH), 11.27 (s, 1H, OH), 8.80 (s, 1H, =CH), 7.84 (s, 1H, ph – H), 7.54 (d, 1H, J = 8.7 Hz, ph – H), 7.05 (d, 1H, J = 8.7 Hz, ph – H); ^{13}C NMR (DMSO – d_6, 75 MHz) δ 189.7, 180.7, 164.0, 158.7, 135.0, 127.8, 123.6, 121.1, 119.5; MS (ESI): m/z (100%) 272 (M^+); Anal. Calcd for $C_9H_6ClN_3OS_2$: C, 39.78; H, 2.23; N, 15.46; found: C, 39.76; H, 2.25; N, 15.48.

2.1.1.7. 4 – Bromo – 2 – (((5 – mercapto – 1,3,4 – thiadiazol – 2 – yl) imino)methyl)phenol (7). Yield 51.5%. Yellow solid, mp: 253.0 ~ 255.1°C, IR (KBr) v_{max} 3420, 3058, 2858, 1606, 1555, 1512, 1467, 1347, 1269, 1171, 1062, 965, 828, 729 cm^{-1}; ^1H NMR (DMSO – d_6, 300 MHz) δ 14.57 (s, 1H, SH), 11.28 (s, 1H, OH), 8.78 (s, 1H, =CH), 7.97 (s, 1H, ph – H), 7.61 (d, 1H, J = 8.7 Hz, ph – H), 7.00 (d, 1H, J = 8.7 Hz, ph – H); ^{13}C NMR (DMSO – d_6, 75 MHz) δ 189.6, 180.7, 163.9, 159.3, 137.7, 130.9, 124.2, 121.7, 110.9; MS (ESI): m/z (100%) 317 (M^+); Anal. Calcd for $C_9H_6BrN_3OS_2$: C, 34.19; H, 1.91; N, 13.29; found: C, 34.18; H, 1.93; N, 13.32.

2.1.1.8. 2 – ((4 – (Tert – butyl)benzylidene)amino) – 5 – mercapto – 1,3,4 – thiadiazole (8). Yield 66.1%. White solid, mp 214.2 ~ 216.5°C, IR (KBr) v_{max} 3395, 2960, 2868, 1689, 1607, 1567, 1499, 1379, 1315, 1264, 1185, 1105, 1052, 931, 835, 761 cm^{-1}; ^1H NMR (DMSO – d_6, 300 MHz) δ 13.25 (s, 1H,

SH), 8.83 (s, 1H, =CH), 7.66 (d, 2H, J = 8.4 Hz, ph – H),

7.48 (d, 2H, J = 8.4 Hz, ph – H), 1.35 (s, 9H, $3CH_3$); ^{13}C NMR ($DMSOd_6$, 75 MHz) δ 192.2, 180.7, 155.7, 152.9, 131.3, 129.7, 126.8, 34.8, 31.2; MS (ESI): m/z(100%) 278 (M^+); Anal. Calcd for $C_{13}H_{15}N_3S_2$: C, 56.28; H, 5.45; N, 15.15; found: C, 56.29; H, 5.44; N, 15.12.

2.1.1.9. 2 – ((2 – Nitrobenzylidene)amino) – 5 – mercapto – 1,3,4 – thiadiazole (9). Yield 22.7%. Yellow solid, mp: 229.7 ~ 232.2℃, IR (KBr) v_{max} 3221, 2987, 1610, 1562, 1519, 1352, 1309, 1265, 1215, 1188, 1110, 1055, 947, 856, 781, 733 cm^{-1}; ^1H NMR (DMSO – d_6, 300 MHz) δ 13.19 (s, 1H, SH), 9.36 (s, 1H, =CH), 8.16 (d, 1H, J = 8.1 Hz, ph – H), 7.92 (d, 1H, J = 7.5 Hz, ph – H), 7.79 (t, 1H, J = 8.4 Hz, ph – H), 7.67 (t, 1H, J = 7.8 Hz, ph – H); ^{13}C NMR (DMSO – d_6, 75 MHz) δ 189.9, 180.1, 158.6, 149.5, 134.2, 133.7, 130.8, 128.2, 124.9; MS (ESI): m/z (100%) 267 (M^+); Anal. Calcd for $C_9H_6N_4O_2S_2$: C, 40.59; H, 2.27; N, 21.04; found: C, 40.58; H, 2.28; N, 21.06.

2.1.1.10. 2 – ((2 – Methoxybenzylidene)amino) – 5 – mercapto – 1,3,4 – thiadiazole (10). Yield 38.3%. Yellow solid, mp: 148.2 ~ 151.3℃, IR (KBr) v_{max} 3328, 2934, 2835, 1564, 1491, 1373, 1312, 1252, 1175, 1116, 1051, 933, 802, 754 cm^{-1}; ^1H NMR (DMSO – d_6, 300 MHz) δ 13.17 (s, 1H, SH), 9.32 (s, 1H, =CH), 7.69 (d, 1H, J = 8.1 Hz, ph – H), 7.52 (t, 1H, J = 8.4 Hz, ph – H), 7.03 (d, 1H, J = 8.4 Hz, ph – H), 6.99 (t, 1H, J = 7.8 Hz, ph – H); ^{13}C NMR (DMSO – d_6, 75 MHz) d189.9, 181.2, 157.4, 155.5, 135.9, 131.0, 128.6, 122.4, 121.0, 56.0; MS (ESI): m/z(100%) 252 (M^+); Anal. Calcd for $C_{10}H_9N_3OS_2$: C, 47.79; H, 3.61; N, 16.72; found: C, 47.82; H, 3.63; N, 16.75.

2.1.1.11. 2 – ((3,5 – Dimethoxybenzylidene)amino) – 5 – mercapto – 1,3,4 – thiadiazole (11). Yield 44.8%. Yellow solid, mp: 227 ~ 229℃, IR (KBr) v_{max} 3339, 3082, 2926, 1609, 1557, 1463, 1325, 1264, 1200, 1155, 1059, 971, 845, 754 cm^{-1}; ^1H NMR (DMSO – d_6, 300 MHz) δ 13.19 (s, 1H, SH), 8.67 (s, 1H, =CH), 7.08 (s, 2H, ph – H), 6.61 (s, 1H, ph – H), 3.76 (s, 6H, 2CH3); ^{13}C NMR (DMSO – d_6, 75 MHz) δ 186.8, 180.9, 168.7, 161.4, 138.7,

107.5, 104.2, 55.58; MS (ESI): m/z(100%) 282 (M^+); Anal. Calcd for $C_{11}H_{11}N_3O_2S_2$: C, 46.96; H, 3.94; N, 14.93; found: C, 46.95; H, 3.96; N, 14.97.

2.1.1.12. 2 – (((5 – Mercapto – 1,3,4 – thiadiazol – 2 – yl)imino)methyl) phenol (12). Yield 51%. Yellow solid, mp: 252 ~ 254℃, IR (KBr) v_{max} 3434, 3072, 2875, 1607, 1571, 1528, 1482, 1374, 1276, 1199, 1146, 1067, 761 cm^{-1}; ^1H NMR (DMSO – d_6, 300 MHz) δ 14.53 (s, 1H, SH), 11.15 (s, 1H, OH), 8.88 (s, 1H, =CH), 7.87 (d, 1H, J = 7.8 Hz, ph – H), 7.53 (t, 1H, J = 7.2 Hz, ph – H), 7.02 – 6.96 (m, 2H, ph – H); ^{13}C NMR (DMSO – d_6, 75 MHz) δ 191.7, 186.5, 166.1, 160.8, 135.8, 130.1, 119.9, 119.5, 117.2; MS (ESI): m/z(100%) 238 (M^+); Anal. Calcd for $C_9H_7N_3OS_2$: C, 45.55; H, 2.97; N, 17.71; found: C, 45.57; H, 2.98; N, 17.73.

2.1.1.13. 4 – (((5 – Mercapto – 1,3,4 – thiadiazol – 2 – yl)imino)methyl) benzene – 1,3 – diol (13). Yield 57.6%. Yellow solid, mp: 240 ~ 243℃, IR (KBr) v_{max} 3305, 3139, 1635, 1587, 1516, 1438, 1319, 1241, 1213, 1171, 1113, 1059, 973, 864, 797 cm^{-1}; ^1H NMR (DMSO – d_6, 300 MHz) δ 14.37 (s, 1H, SH), 11.40 (s, 1H, OH), 10.73 (s, 1H, OH), 8.70 (s, 1H, =CH), 7.71 (d, 1H, J = 6.9 Hz, ph – H), 7.65 (s, 1H, ph – H), 6.34 (d, 1H, J = 6.9 Hz, ph – H); ^{13}C NMR (DMSO – d_6, 75 MHz) δ 190.9, 185.9, 165.7, 164.4, 163.2, 133.2, 119.9, 109.3, 102.3; MS (ESI): m/z(100%) 254 (M^+); Anal. Calcd for $C_9H_7N_3O_2S_2$: C, 42.68; H, 2.79; N, 16.59; found: C, 42.66; H, 2.82; N, 16.57.

2.1.1.14. 4 – (((5 – Mercapto – 1,3,4 – thiadiazol – 2 – yl)imino)methyl) – 2 – methoxyphenol (14). Yield 63.2%. Yellow solid, mp: 219.5 ~ 221.8℃, IR (KBr) v_{max} 3275, 1606, 1565, 1512, 1382, 1322, 1272, 1205, 1101, 1053, 928, 753 cm^{-1}; ^1H NMR (DMSO – d_6, 300 MHz) δ14.41 (s, 1H, SH), 10.34 (s, 1H, OH), 8.55 (s, 1H, =CH), 7.55 (s, 1H, ph – H), 7.42 (d, 1H, J = 8.7 Hz, ph – H), 6.99 (d, 1H, J = 8.7 Hz, ph – H), 3.86 (s, 3H, CH_3); ^{13}C NMR (DMSO – d_6, 75 MHz) δ 190.9, 186.2, 167.9, 152.9, 148.1, 128.6, 126.7, 115.6, 111.1, 55.5; MS (ESI): m/z (100%) 268 (M^+); Anal. Calcd for $C_{10}H_9N_3O_2S_2$: C, 44.93; H, 3.39; N, 15.72; found: C, 44.96; H, 3.38; N, 15.73.

2.1.1.15. 4 – (((5 – Mercapto – 1 ,3 ,4 – thiadiazol – 2 – yl)imino)methyl) – 2 ,6 – dimethoxyphenol (15). Yield 52.2%. Yellow solid, mp: 267 ~ 268.5℃, ^{1}H NMR (DMSO – d_6, 300 MHz) δ 14.42 (s, 1H, SH), 9.77(s, 1H, OH), 8.53 (s, 1H, = CH), 7.31 (s, 2H, ph – H), 3.84 (s, 6H, $2CH_3$); ^{13}C NMR (DMSO – d_6, 75 MHz) δ 191.1, 186.0, 168.1, 148.1, 142.0, 124.4, 107.6, 56.2; MS (ESI): m/z (100%) 298 (M^+); Anal. Calcd for $C_{11}H_{11}N_3O_3S_2$: C, 44.43; H, 3.73; N, 14.13; found: C, 44.45; H, 3.76; N, 14.13.

2.1.1.16. 2 – ((2 ,4 ,6 – Trimethoxybenzylidene)amino) – 5 – mercapto – 1 , 3 ,4 – thiadiazole (16). Yield 86%. Yellow solid, mp: 263 ~ 265℃, IR (KBr) v_{max} 3084, 2890, 1599, 1520, 1463, 1384, 1325, 1276, 1239, 1113, 1064, 767 cm^{-1}; ^{1}H NMR (DMSO – d_6, 300 MHz) δ 14.29 (s, 1H, SH), 8.69 (s, 1H, = CH), 6.31 (s, 2H, ph – H), 3.90 (s, 6H, $2CH_3$), 3.82 (s, 3H, CH_3); ^{13}C NMR (DMSO – d_6, 75 MHz) δ 190.8, 185.8, 167.2, 163.4, 161.3, 104.5, 91.7, 56.3, 55.7; MS (ESI): m/z(100%) 312 (M^+); Anal. Calcd for $C_{12}H_{13}N_3O_3S_2$: C, 46.29; H, 4.21; N, 13.49; found: C, 46.31; H, 4.23; N, 13.52.

2.1.1.17. 2 – (((5 – (Benzylthio) – 1 ,3 ,4 – thiadiazol – 2 – yl)imino) methyl) – 4 – chlorophenol (17). 2 – Amino – 5 – mercapto – 1,3,4 – thiadiazole 3.99 g (0.03 mol) was added to a 50 mL flask, and the solution of KOH (1.683 g, 0.03 mol) in 10 mL was added under stirring at room temperature. Halogenated hydrocarbon (0.1 mol) was added by dropwise in 0.5 h with vigorous stirring. The reaction mixture was stirred at room temperature and checked by TLC (petroleum ether: ethyl acetate 1:1). When reaction was complete, the yellow precipitate of crude thiadiazole sulfide was formed. This was filtered, washed with water to neutrality, and crystallized from 95% ethanol. 2 – Amino – 5 – sulfide – 1,3,4 – thiadiazole (5 mmol) was suspended in 10 mL anhydrous ethanol and 6 mmol of aromatic aldehyde was added, using tetrabutylammonium bromide (TBAB) as catalyst. The reaction mixture was stirred under reflux for 8 h and monitored by TLC (petroleum ether: ethyl acetate 4:1) until the reaction was complete. After completion of reaction, the mixture was allowed to cool to room temperature, and then left overnight. Precipitates of 1,3,4 – thiadiazole sulfide derivative bearing Schiff base moieties were produced. The solid that separated out was filtered, washed with the mixture of ethanol and hot water (1:1); dried solid was purified by crystallization from anhydrous alcohol to afford the desired compound. Yield 49.92%. Yellow solid, mp: 148.2 ~ 150.7℃, ^{1}H NMR ($CDCl_3$, 500 MHz) δ 11.79 (s, 1H, OH), 8.95 (s, 1H, CH), 7.44 - 7.30 (m, 7H, Ph - H), 7.00 (d, J = 8.6 Hz, 1H, Ph - H), 4.57 (s, 2H, CH2); ^{13}C NMR (CDC_{13}, 125 MHz) δ 170.4, 166.7, 164.8, 160.2, 135.6, 132.6, 132.3, 129.3, 128.9, 128.1, 124.6, 119.4, 118.9, 38.4; MS (ESI): m/z (100%) 362 (M^+). Anal. Calcd for $C_{16}H_{12}ClN_3OS_2$: C, 53.11; H, 3.34; N, 11.61; found: C, 53.09; H, 3.36; N, 11.63.

2.2. Tyrosinase activity assay

Tyrosinase inhibition assayswere performed according to the method reported by our group [36]. Briefly, 1,3,4 - thiadiazole derivatives were tested for diphenolase inhibitory activity of tyrosinase using L - DOPA (dihydroxyphenylalanine) as substrate. All the compounds were dissolved in DMSO and its final concentration in the reaction mixture was 2.0%. Thirty units of mushroom tyrosinase (0.5 mg/mL) were first pre - incubated with the compounds, in 50 mmol/L phosphate buffer (pH 6.8), for 10 min at 25℃. Then the L - DOPA (0.5 mmol/L) was added to the reaction mixture and the enzyme reaction was continuously monitored by measuring the change in absorbance at 475 nm of formation of the DOPA chrome for 1 min. The measurement was performed in triplicate for each concentration and averaged before further calculation. The activity was expressed as the sample concentration that gave a 50% inhibition in the enzyme activity (IC_{50}). The percent of inhibition of tyrosinase was calculated as follows:

$$\text{Inhibition rate}(\%) = [(B - S)/B] \times 100$$

Here, the B and S were the absorbances for the blank and samples. All the experiments were carried out at least in triplicate and averaged. Kojic acid was used as the positive control.

2.3. Kinetic analysis of the inhibition of tyrosinase

A series of experiments were performed to determine the inhibition kinetics by following method [37]. Inhibitor (13) with concentrations 0, 0.025, 0.05, 0.075, 0.1μmol/L, and (14) with concentrations 0, 0.005, 0.01, 0.015 and 0.02μmol/L, respectively were used. Substrate L - DOPA concentration was among 1014 - 304Mm in all kinetic study. Preincubation and measurement time was the same as discussed in mushroom tyrosinase inhibition assay protocol. The inhibition type on the enzyme was assayed by Lineweavere Burk plots of inverse of velocities (1/V) versus inverse of substrate concentration 1/[S] $mmol/L^{-1}$, and the inhibition constants were determined by the secondary plots of the apparent 1/Vm versus the concentrations of the inhibitors.

2.4. In silico docking simulation of tyrosinase with compounds 14 [38 - 40]

For docking simulations, we used the AutoDock4.2 program. Among the many tools available for in silico protein - ligand docking, AutoDock4.2 is the most commonly used due to its automated docking capability. The 3D structure of tyrosinase was used the crystal structure of Agaricus bisporus (PDB ID: 2Y9X) without homology modeling. We conducted simulations of the docking of tyrosinase to the compounds 14. To prepare the compound for the docking simulation, the following steps was performed: (1) conversion of 2D structures into 3D structures, (2) calculation of charges, and (3) addition of hydrogen atoms using the ChemOffice program (http://www.cambridgeoft.com).

2.5. Acute toxicity [41]

Acute toxicity experiments were carried out using KM mice of both sexes weighing 18 - 22 g. The food and water were provided according to EPA OPPTS Harmonized Test guideline. All animals were provided by Qinlongshan Laboratory animal center SCXK (Yu) 2005 - 0001. Various doses of the compound 14were given by intragastric administration to the healthy KM mice. After the intragastric administration of the compound, mice were observed continuously for any gross behavioral changes and deaths, and intermittently for 1 week, the LD_{50} values can be obtained.

3. Results and discussion

3.1. Chemistry

The synthesis of 1,3,4 - thiadiazole derivatives was shown in Scheme 1. As depicted in Scheme 1, 2 - amino - 5 - mercapto - 1,3,4 - thiadiazole can be synthesized by the condensation reaction of thiosemicarbazide with carbon disulphide in ethanol using anhydrous sodium carbonate as catalyst under reflux conditions. Thiadiazole sulfide was synthesized from 2 - amino - 5 - mercapto - 1,3,4 - thiadiazole and halogenated hydrocarbon. 1,3,4 - thiadiazole derivatives bearing

Schiff base moieties were prepared by condensing 2 – amino – 5 – mercapto – 1,3,4 – thiadiazole or 5 – (benzylamino) – 1,3,4 – thiadiazole – 2 – thiol with different aromatic aldehydes in ethanol using concentrated sulfuric acid as catalyst under reflux conditions. The resulting products were obtained in moderate yields. All the target compounds were characterized spectroscopic methods.

3.2. Biological activity

The 17 synthesized compounds were assessed using tyrosinase inhibition assay with kojic acid as a positive control, and the IC_{50} values against tyrosinase were recorded inTable 1. As we expected, 15compounds exhibited potent inhibition activities against mushroom tyrosinase, with IC_{50} values at single digit or sub – micromolar level, better than the reference inhibitor kojic acid. The other compounds showed some inhibitory activities against tyrosinase. Pleasingly, compound14exhibited the most potent tyrosinase inhibitory activity with IC_{50} value of 0.036μmol/L. The results showed that the substitution pattern of the benzene ring has a great influence in determining activities against tyrosinase. From the result of Table 1, if the hydrogens on benzene ring were substituted with halogens, all of the compounds showed better inhibitory activities than compound1 (49.246μmol/L). Compound3exhibited lower inhibitory activity than compound4. However, the dihalogen substituted compound 5 showed lower inhibitory activity than the compound 4. The results showed that the type, position and number of halogen atom at the benzene ring might modulate the activities. When the hydrogen of benzene ring was replaced by fluorine, better inhibitory activity was achieved than other halogen substituted compounds. The IC_{50} value of compound 2 was 0.765μmol/L, the IC_{50} values of the two compounds3 and 4 were 4.753μmol/L and 2.460μmol/L, respectively. This result may be because fluorine atom has smaller atomic radius and greater electronegativity. Moreover, comparing to the tyrosinase inhibitory activity of compounds 3, 4and6, when the hydroxyl group was introduced onto the benzene ring, an elevated inhibitory activity could be observed. These results showed that hydroxyl group may be beneficial to increasing the inhibitory activity of this kind of compounds with halogen atom at benzene ring. From Table 1, the IC_{50} value of compound 10 with a methoxyl substituted benzene ring (2 – position) was 0.733μmol/L. However, compound 11 with disubstituted methoxyl groups (3,5 – position) exhibited some tyrosinase inhibitory activity with an IC_{50} value of 5. 353μmol/L, it showed lower inhibitory activity than the compound 10. Interestingly, the IC_{50} of compound 16 with trisubstituted methoxyl groups（2,4,6 – position）was 0. 907μmol/L. This result might be explained that the inhibitory activity methoxyl substituted compound might be related to the position of methoxyl group. For compounds12 and 13, the inhibitory activity increased as the number of hydroxyl group increased. This result showed the hydroxyl group number might have some influences on the inhibitory activity. Compound 14was the best inhibitor among all the compounds, but in contrast, with the increased number of the methoxyl groups, the inhibitory activity of the compound 15was decreased. This result might be related to the steric hindrance. However, compound15exhibited better inhibitory activity than the compound 11. The result also showed that hydroxyl group may be beneficial to increasing the inhibitory activity. From the Table 1, based on the inhibitory activity of compounds 6 and 17, when the thiol group was etherified by the benzyl chloride, the inhibitory activity dropped drastically. Thus, free thiol group could be an active group in the inhibition. In addition, the o – dopaquinone can react with the thiol group of the inhibitor [42].

3. 3. Inhibitory mechanism

Among the tested target compounds, compound 13and 14 showed superior inhibitory activities to the other compounds, and therefore, we carried out the kinetic analysis of the two compounds for tyrosinase inhibition using L – DOPA as the substrate.

The Lineweaver – Burk plots for theinhibition of tyrosinase by the two compounds were obtained with various concentrations of the two compounds and the substrate (Figs. 1 and 2). The inhibition types on the enzyme by compounds13and14for oxidation of LDOPA were studied as shown inFigs. 1 and 2, respectively. The plots of 1/v versus 1/[S] gave a family of parallel straight lines with the same slopes. Meanwhile, as the inhibitor concentration increased, the values of both k_m and V_m decreased, but the ratio of k_m/V_m remained unchanged. The results indicated that both compounds 13 and 14 were uncompetitive inhibitors. The inhibition constants (K_I) of the two compounds13 and 14 were obtained from a plot of the vertical intercept (1/Vm) versus the concentration of the inhibitors as 0.

075μmol/L and 0.038μmol/L, respectively. The catalytic constants (k_{cat}) of the two compounds 13 and 14 were 0. 61 s^{-1} and 0. 83 s^{-1}. kcat/km values of the two compounds 13 and 14 were 1. 27 mmol/L^{-1} s^{-1} and 1. 08 mmol/L^{-1} s^{-1}. The inhibitory mechanism of the compound9with no hydroxyl group on mushroom tyrosinase for the oxidation of L - DOPA was determined (Fig. 3). The result displayed that the plots of V versus [E] gave a family of parallel straight lines with the same slopes, thus compound 9 was irreversible inhibitor.

3.4. In silico docking between tyrosinase and compounds 14

The binding affinities of the inhibitors to the receptor tyrosinase were calculated using computational docking studies. The lower value showed the more stable complex formed between the ligand and target protein. The binding energy of compound 14 was -5.94 kcal/mol, it proved that compound14formed stable drug receptor complex with tyrosinase. The phenolic hydroxyl present at position 4 in compound14can interact with the copper atom (Cu 401) of tyrosinase (Fig. 4). This result showed hydroxyl group might be activity group. Van der Waals force is also the interaction between compound14and tyrosinase. The 1,3,4 - thiadiazole ring formed pi - lone pair to GLY281 with bond length of 2. 91 Å. The p - pstacking present in the compound 14between phenyl ring and LYS180 having distance 1. 82 Å (Fig. 4). ALA286 and VAL283 can form p - alkyl interaction present phenyl ring having bond lengths 4. 36 Å and 4. 04 Å respectively. ASN260 and HIS259 can form hydrogen bonds with methoxy of the compound. There is no significant interaction between mercapto and tyrosinase.

3.5. Acute toxicity

To evaluate acute toxicities of the synthesized compounds, compound14was selected to test the acute toxicity. The LD_{50} value for acute toxicity of compound14in mice after intragastric administration was obtained. Test result indicated the LD_{50} value of compound14was 5000 mg/kg. It showed that compound 14 oral dose commonly used in clinical treatment should be safe. The result might be beneficial to search and develop novel tyrosinase inhibitor with better activity together with lower side effects.

4. Conclusion

In summary, we have developed a series of 1,3,4 - thiadiazole derivatives bearing Schiff base moieties as novel tyrosinase inhibitors. The results showed that many of them exhibited remarkable tyrosinase inhibitory activities with IC_{50} values at single digit or sub - micromolar level. Compound 14 was found to be the best tyrosinase inhibitor in this series with the IC_{50} value of 0. 036μmol/L. Structure - activity relationship (SAR) analysis indicated that (1) free thiol group and hydroxyl group could be active groups in the inhibition; (2) the steric hindrance could be an unfavorable factor; (3) substituted group was introduced onto the benzene ring, an elevated inhibitory activity could be observed; (4) the position and numbers of hydroxyl groups might play an important role in the increase of inhibitory activity. Moreover, the inhibition kinetics study revealed that compounds13and14exhibited such inhibitory effects on tyrosinase by acting as the uncompetitive inhibitors. Docking studies showed compound 14 combined strongly with mushroom tyrosinase. Hydroxyl group interact with the copper atom of tyrosinase, this result showed that the group was active groups. The acute toxicity of compound14was evaluated, LD_{50} value of the compound was 5000 mg/kg. It indicated that compound14oral dose commonly used in clinical treatment should be safe.

Acknowledgments

We thank the innovation team of Shaoyang University for funding this project. This work was also financially supported by the Foundation of Education Department of Hunan Province, China (15A172).

References

[1] R. L. S. Michael, A. R. Christopher, A. R. Patrick, Bioorg. Med. Chem. Lett. 21 (2013)1166 - 1173.

[2] Y. C. Li, Y. Wang, H. B. Jiang, J. P. Deng, Proc. Natl. Acad. Sci. U. S. A. 106 (2009)17002 - 17006.

[3] Q. X. Chen, X. D. Liu, H. Huang, Biochemistry (Mosc) 68 (2003) 644 - 649.

[4] X. Dong, Q. Zhu, Y. Dai, J. He, H. Pan, J. Chen, Z. - P. Zheng, Food Chem. 192(2016) 1033 - 1040.

[5] R. Matsuura, H. Ukeda, M. Sawamura, J. Agric. Food Chem. 54 (2006) 2309 - 2313.

[6] C. M. Kumar, U. V. Sathisha, S. Dharmesh, A. G. Rao, S. A. Singh,

Biochimie 93(2011) 562 - 569.

[7] A. Pinon, Y. Limami, L. Micallef, J. Cook - Moreau, B. Liagre, C. Delage, R. E. Duval, A. Simon, Exp. Cell Res. 317 (2011) 1669 - 1676.

[8] X. Tan, Y. H. Song, C. Park, K. - W. Lee, J. Y. Kim, D. W. Kim, K. D. Kim, K. W. Lee, Bioorg. Med. Chem. 24 (2016) 153 - 159.

[9] T. Pan, X. Li, J. Jankovic, Int. J. Cancer 128 (2011) 2251 - 2260.

[10] Y. J. Zhu, H. T. Zhou, Y. H. Hu, Food Chem. 124 (2011) 298 - 302.

[11] F. Artés, M. Castañer, M. I. Gil, Food Sci. Technol. Int. 4 (1998) 377 - 389.

[12] M. Ashida, P. Brey, Proc. Natl. Acad. Sci. U. S. A. 92 (1995) 10698 - 10702.

[13] A. Ortiz - Urquiza, N. O. Keyhani, Insects 4 (2013) 357 - 374.

[14] Y. Mu, L. Li, S. Q. Hu, Spectrochim. Acta A. Mol. Biomol. Spectrosc. 107 (2013) 235 - 240.

[15] P. Han, C. Chen, C. Zhang, K. Song, H. Zhou, Q. Chen, Food Chem. 107 (2008) 797 - 803.

[16] M. T. H. Khan, Curr. Med. Chem. 19 (2012) 2262 - 2272.

[17] I. E. Orhan, M. T. H. Khan, Curr. Top Med. Chem. 14 (2014) 1486 - 1493.

[18] L. Xia, A. Idhayadhulla, Y. R. Lee, Y. J. Wee, S. H. Kim, Eur. J. Med. Chem. 86 (2014) 605 - 612.

[19] A. You, J. Zhou, S. Song, G. Zhu, H. Song, W. Yi, Bioorg. Med. Chem. 23 (2015) 924 - 931.

[20] G. Battaini, E. Monzani, L. Casella, L. Santagostini, R. Pagliarin, J. Biol. Inorg. Chem. 5 (2000) 262 - 268.

[21] H. R. Kim, H. J. Lee, Y. J. Choi, Y. J. Park, Med. Chem. Commun. 5 (2014) 1410 - 1417.

[22] T. Klabunde, C. Eicken, J. C. Sacchettini, B. Krebs, Nat. Struct. Biol. 5 (1998) 1084 - 1090.

[23] A. You, J. Zhou, S. Song, G. Zhu, H. Song, W. Yi, Eur. J. Med. Chem. 93 (2015) 255 - 262.

[24] N. Grynberg, A. C. Santos, A. Echevarria, Anticancer Drugs 8 (1997) 88 - 91.

[25] A. Senff - Ribeiro, A. Echevarria, E. F. Silva, C. R. Franco, S. S. Veiga, M. B. Oliveira, Br. J. Cancer 91 (2004) 297 - 304.

[26] Z. Luo, B. Chen, S. He, Y. Shi, Y. Liu, C. Li, Bioorg. Med. Chem. Lett. 22 (2012) 3191 - 3193.

[27] N. Polkam, P. Rayam, J. S. Anireddy, S. Yennam, H. S. Anantaraju, S. Dharmarajan, Bioorg. Med. Chem. Lett. 25 (2015) 1398 - 1402.

[28] S. - J. Yang, S. - H. Lee, H. - J. Kwak, Y. - D. Gong, J. Org. Chem. 78 (2013) 438 - 444.

[29] T. A. Farghaly, M. A. Abdallah, G. S. Masaret, Z. A. Muhammad, Eur. J. Med. Chem. 97 (2015) 320 - 333.

[30] K. Zhang, P. Wang, L. - N. Xuan, X. - Y. Fu, F. Jing, S. Li, Y. - M. Liu, B. - Q. Chen, Bioorg. Med. Chem. Lett. 24 (2014) 5154 - 5156.

[31] A. Skrzypek, J. Matysiak, A. Niewiadomy, M. Bajda, P. Szymanski, Eur. J. Med. Chem. 62 (2013) 311 - 319.

[32] H. - C. Zhao, Y. - P. Shi, Y. - M. Liu, C. - W. Li, L. - N. Xuan, P. Wang, K. Zhang, B. - Q. Chen, Bioorg. Med. Chem. Lett. 23 (2013) 6577 - 6579.

[33] J. Xie, H. Dong, Y. Yu, S. Cao, Food Chem. 190 (2016) 709 - 716.

[34] U. Ghani, N. Ullah, Bioorg. Med. Chem. 18 (2010) 4042 - 4048.

[35] M. Yusuf, R. A. Khan, B. Ahmed, Bioorg. Med. Chem. 16 (2008) 8029 - 8034.

[36] J. B. Liu, W. Yi, Y. Q. Wan, L. Ma, H. C. Song, Bioorg. Med. Chem. 16 (2008) 1096 - 1102.

[37] Y. - H. Hu, X. Liu, Y. - L. Jia, Y. - J. Guo, Q. Wang, Q. - X. Chen, J. Biosci. Bioeng. 117(2014) 142 - 146.

[38] Z. Ashraf, M. Rafiq, S. - Y. Seo, K. S. Kwon, M. M. Babar, Eur. J. Med. Chem. 98(2015) 203 - 211.

[39] Y. Matoba, T. Kumagai, A. Yamamoto, H. Yoshitsu, M. Sugiyama, J. Biol. Chem. 281 (2006) 8981 - 8990.

[40] S. Radhakrishnan, R. Shimmon, C. Conn, A. Baker, Bioorg. Chem. 63 (2015)116 - 122.

[41] J. Liu, F. Wu, H. Song, H. Wang, L. Zhao, Med. Chem. Res. 22 (2013) 4228 - 4238.

[42] S. Li, Y. Xue, H. Zhang, H. Nie, L. Zhu, Fine Chem. 26 (2009) 889 - 893.

H_2N-C(=S)-NH-NH_2 + CS_2 —(Na_2CO3; C_2H_5OH,reflux)→ HS-(1,3,4-thiadiazole)-NH_2 —(R-CHO; H_2SO_4/C_2H_5OH,reflux)→ R-CH=N-(1,3,4-thiadiazole)-SH (1–16)

HS-(1,3,4-thiadiazole)-NH_2 + benzyl chloride (Cl) → benzyl-S-(1,3,4-thiadiazole)-NH_2 → benzyl-S-(1,3,4-thiadiazole)-N=CH-R

1. R = phenyl;　　2. R = 2 - fluorophenyl;

3. R = 3 - chlorophenyl;　　4. R = 4 - chlorophenyl;

5. R = 3,4 - dichlorophenyl;　　6. R = 5 - chloro - 2 - hydroxylphenyl;

7. R = 5 - bromo - 2 - hydroxyphenyl;　　8. R = 4 - t - Bu - phenyl;

9. R = 2 - nitrophenyl;　　10. R = 2 - methoxylphenyl;

11. R = 3,5 - dimethoxylphenyl;　　12. R = 2 - hydroxylphenyl;

13. R = 2,4 - dimethoxylphenyl;

14. R = 2 - hydroxyl - 3 - methoxylphenyl;

15. R = 3,5 - dimethoxyl - 4 - hydroxylphenyl;

16. R = 2,4,6 - trimethoxylphenyl;

17. R = 5 - chloro - 2 - hydroxylphenyl.

Scheme 1　The synthesis of 1,3,4 - thiadiazole derivatives bearing Schiff base moieties.

Table 1 Tyrosinase inhibitory activities of the synthesized compounds.

Compounds	R	IC_{50} (μmol/L)^2 ± SD
1	Phenyl	49. 246 ±0. 184
2	2 - Fluorophenyl	0. 765 ±0. 077
3	3 - Chorophenyl	4. 753 ±0. 109
4	4 - Chorophenyl	2. 460 ±0. 155
5	3,4 - Dichlorophenly	3. 621 ±0. 142
6	5 - Chloro - 2 - hydroxylphenyl	1. 718 ±0. 135
7	5 Bromo 2 hydroxylphcnyl	1. 971 ±0. 131
8	4 - Bu - phenyl	1. 390 ±0. 118
9	2 - Nitrophenyl	0. 671 ±0. 093
10	2 - Methoxylphenyl	0. 7331 ±0. 083
11	3,5 - Dimethoxylphenyl	5. 353 ±0. 136
12	2 - Hydroxylphenyl	0. 785 ±0. 064
13	2,4 - Dihydroxylphenyl	0. 255 ±0. 065
14	4 - Hydroxyl - 3 - methoxylphenyl	0. 036 ±0. 002
15	3,5 - Dimethxoyl - 4 - hydroxylphenyl	0. 473 ±0. 096
16	2,4,6 - Trimethoxylphenyl	0. 907 ±0. 142
17	5 - Chloro - 2 - hydroxylphenyl	247. 350 ±3. 782
Kojic acid	—	23. 6

[a]Inhibitor concentration (mean of three independent experiments) required for 50% inactivation of tyrosinase.

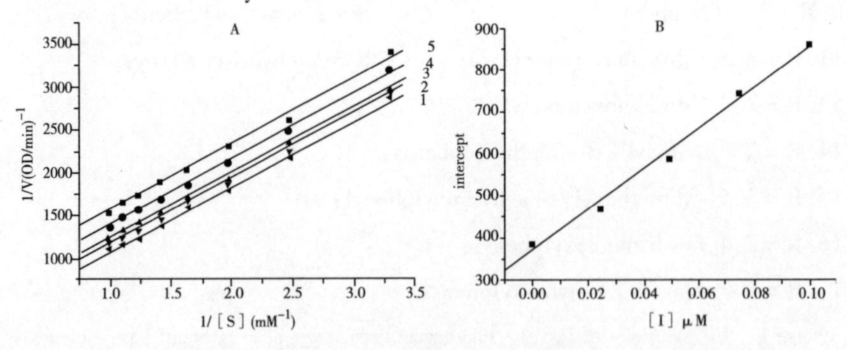

Fig. 1 (A) Kinetic inhibition analysis of compound 13 against the oxidation of L - DOPA by tyrosinase. The mode of inhibition type exhibited through Lineweavere - Burk plot with different concentrations of compound13. The inhibitor concentrations of compound 13 for Curves 1 - 5 were 0, 0.025, 0.05, 0.075 and 0.1μmol/L. (B) The plot of intercept versus concentration of compound 13 was used to determine the inhibition constant K_I.

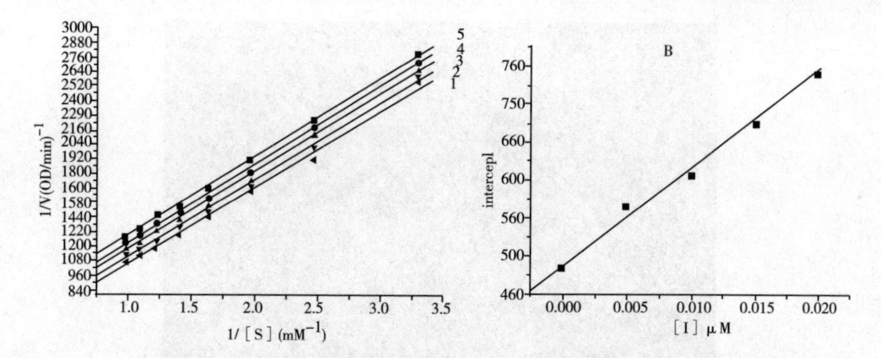

Fig. 2 (A) Kinetic inhibition analysis of compound 14 against the oxidation of L - DOPA by tyrosinase. The mode of inhibition type exhibited through Lineweavere - Burk plot with different concentrations of compound 14. The inhibitor concentrations of compound 14 for Curves 1 - 5 were 0, 0.005, 0.01, 0.015 and 0.02μmol/L. (B) The plot of intercept versus concentration of compound 14 was used to determine the inhibition constant K_I.

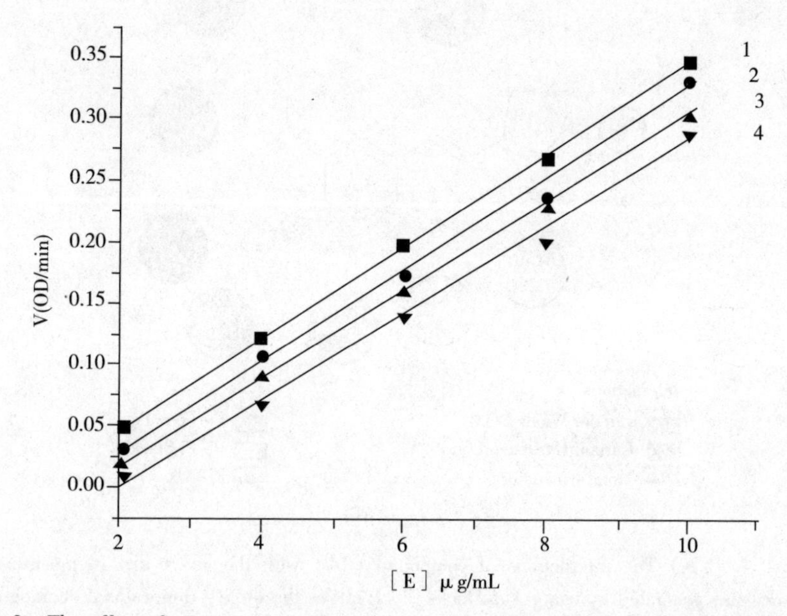

Fig. 3 The effect of concentrations of tyrosinase on its activity for the catalysis of LDOPA at different concentration of compound 9. The concentrations of compound 9 for curves 1 - 4 are 0, 0.25, 0.5 and 1μmol/L, respectively.

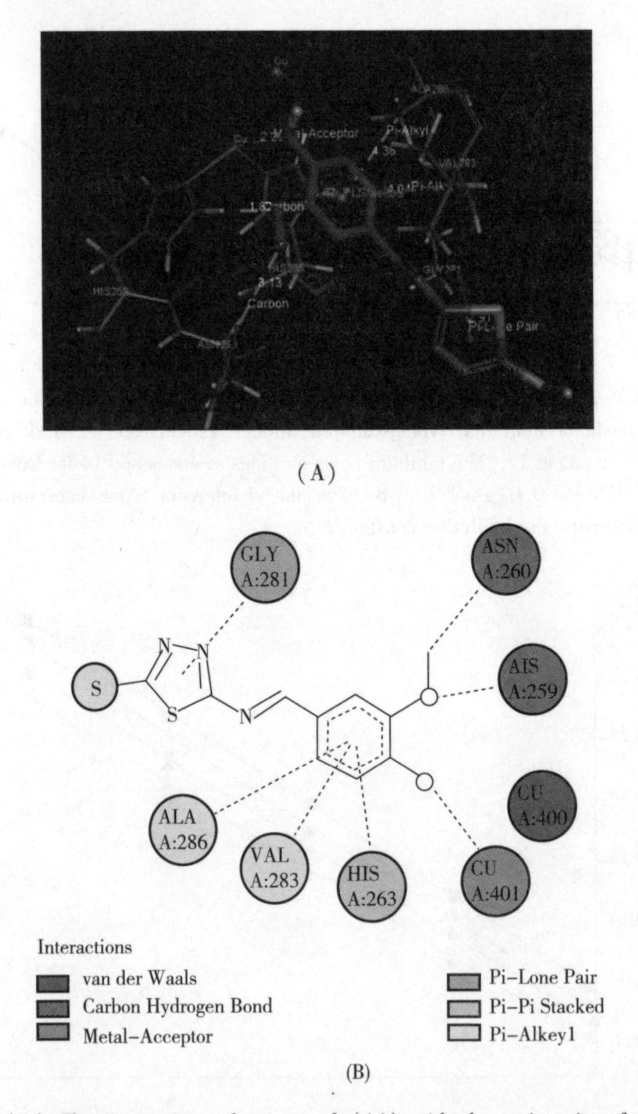

Fig. 4 (A) The interactions of compound (14) with the active site of mushroom tyrosinase generated by using AutoDock4. 2. It shows the three - dimensional docking of the compound in the binding pocket. Dashed lines indicate bond distances between interacting functionalities of the ligand and receptor. (B) The interactions of compound (14) with the active site of mushroom tyrosinase generated by using Auto Dock4. 2. It shows the two dimensional interaction patterns. The legend inset represents the type of interaction between the ligand atoms and the amino acid residues of the protein.

4.3.4 唐君源硕士论文部分内容

摘 要

酪氨酸酶是一种含铜离子的氧化还原酶,又称多酚氧化酶。酪氨酸酶广泛存在于动植物、微生物及人体中,是导致果蔬褐变的关键酶,严重影响了果蔬制品的营养、风味及外观品质和商品价值,使产品难以进入市场,抑制其活性保证延长果蔬保鲜期成为了研究热点。

研究表明1,3,4-噻二唑类化合物是设计生物活性分子的重要先导化合物,具有抗肿瘤、抗菌、抗氧化、抗抑郁、抗菌消炎及抑制酪氨酸酶等生物活性。席夫碱同样具有多种生物活性,本课题组已经证明这类化合物具有较好的酪氨酸酶抑制活性。为了找到高效低毒的酪氨酸酶抑制剂,本论文采用活性片段组合原理,将噻二唑和席夫碱组合到一个分子中,考察其酪氨酸酶抑制活性。优选化合物进行抑制机理、抑制动力学及分子对接研究,在此基础上选择高效低毒的化合物进行褐变抑制效果考察。为进一步深入开展噻二唑衍生物酪氨酸酶抑制剂奠定基础。

本文合成的34个噻二唑衍生物中,有13个新化合物,21个已知化合物,其中有16个化合物的酪氨酸酶抑制活性高于曲酸,T1系列样品除T1-16 IC_{50}为49.246 μmol/L外,其余均低于曲酸,样品T1-13和T1-4 IC_{50}最低分别为0.036 μmol/L和0.255 μmol/L;T2及T3系列化合物除T3-2 IC_{50}为1.03 μmol/L小于曲酸外,其余均大于曲酸。为了考察该类化合物的抑制动力学,优选活性较好的化合物T1-4和T1-13进行试验,结果表明样品T1-13和T1-4为非竞争性可逆抑制剂。

为了进一步考察这类化合物与酪氨酸酶的作用方式,优选化合物T1-4、T1-13与2Y9X进行了分子对接研究,结果表明自由结合能分别为-5.94 Kcal/mol和-5.65 Kcal/mol。T1-4与氨基酸残基ALA 286、VAL 283、HIS 263、GLY 281、MET 280和VAL 283有作用力。T1-13与氨基酸残基ALA 286、VAL 283、HIS 263、GLY 281、ASN 260和HIS 259有作用力。T1-4、T1-13苯环4号位上的羟基均与酪氨酸酶中的铜离子(401)以Metal-Acceptor的方式结合,键长分别为2.32 Å和2.28 Å。

为了初步考察这类化合物的毒性,优选化合物T1-13和T1-4进行急毒性试验,结果表明T1-13和T1-4的LD_{50}分别为6364 mg/Kg和5710 mg/Kg,均大于曲酸LD_{50}2700 mg/Kg。因此,T1-13和T1-4可能能用于食品工业、临床及化妆品。

在急毒性试验基础上,优选化合物 T1 - 13 进行苹果酪氨酸酶的抑制效果研究,经过单因素和响应面试验确定最佳的浓度、作用时间和温度为 0.27 μmol/L、1901.98 s 和 20.81 ℃。在此条件下,经过 3 次验证试验得出酪氨酸酶抑制率分别为:78.15%、77.32% 及 73.16%,平均值为 76.21%。

关键词:1,3,4 - 噻二唑衍生物;酪氨酸酶抑制剂;酶动力学;响应面法;分子对接

Abstract

Tyrosinase is a copper - containing oxidoreductase, as known as polyph - enol oxidase, is widelydistributed in animals, plants, microorganisms and human body. It is a key enzyme leading to browning of fruits and vegetables, a serious impact on the nutrition, flavor, appearance, quality and value of fruit and vegetable products, so that the product is difficult to enter the market. Inhibiting tyrosinase activity has become a wide range of research topic in food industry and Pharmaceutical industry. 1,3,4 - thiadiazole was an important leader compound in the design of biologically active molecules. This compound exhibited some important biological activities, such as ant - itumor, antibacterial, antioxidation, antidepressant, antibacterial and tyros - inase inhibition. In order to find safer and effective tyrosinase inhibitors, based on the 1,3,4 - thiadiazole and Schiff base moieties, a series of 1,3,4 - th - iadiazole derivatives bearing Schiff base moieties were synthsisized and their their inhibitory activities against mushroom tyrosinase were also eval - uated. In addition, inhibitory mechanism, molecular docking and the kinetic analysis were also carried out. Anti - browning effect was further investigat - ed. These results suggested that such compounds might be served as lead compounds for further designing new potential tyrosinase inhibitors.

All of the synthesized compounds displayed potent tyrosinase inhibito - ry activities, especially, 16 compounds exhibited superior inhibitory effect to the kojic acid. Comparing the inhibitory activities of T1 series compoun - ds, except for theIC_{50} value of T1 - 16 was 49.246μmol/L, and the other com - pounds showed higher inhibirory activities than kojic acid. Pleasingly, compound T1 - 3 exhibited the most potent tyrosinase inhibitory activity with IC_{50} value of 0.036μmol/L. Comparing to the tyrosinase inhibitory activity of T3 and T2 series compounds, IC_{50} value of T3 - 2 was lower than that of kojic acid, but all of the other compounds showed higher IC_{50} value than the kojic acid. Among the tested target compounds, compound T1 – 13 and T1 – 4 showed superior inhibitory activities to the other compounds, and therefore, we carried out the kinetic analysis of the two compounds for tyrosinase inhibition using L – DOPA as the substrate. The results indicated that both compounds T1 – 13 and T1 – 4 were uncompetitive inhibitors.

In order to further investigate the interaction of these compounds with tyrosinase, the molecular docking studies of compounds T1 – 4, T1 – 13 with 2Y9X were carried out. The results showed that the free binding energies were – 5.94Kcal / mol and – 5.65Kcal / mol, respectively. T1 – 4 is reactive with the amino acid residues ALA286, VAL283, HIS263, GLY281, MET280 and VAL283. T1 – 13 is reactive with the amino acid residues ALA286, VAL283, HIS263, GLY281, ASN260 and HIS259. The phenolic hydroxyl present at position 4 of compounds T1 – 4 and T 1 – 13 can interact with the copper atom (Cu 401) of tyrosinase. The bond lengths were 2.32 Å and 2.28 Å. This result showed hydroxyl group might be activity group.

In order tovaluate the acute toxicities of the synthesized compounds, T1 – 4 and T 1 – 13 were selected to test the acute toxicity. The LD_{50} of T1 – 13 was 6364mg/Kg, and the LD_{50} of T1 – 4 was 5710mg/Kg. It showed that compounds T1 – 4 and T1 – 13 used in food industry, clinical treatment and cosmetics might be safe.

Based on urgent toxicity test, Anti – browning effect of compound T1 – 13 was further investigated. The optimum conditions of concentration, reaction time and temperature was definite by the single test and response surface experiment. The optimum conditions as following: The concentration of compound T1 – 13 is 0.27μmol/L, the reaction time is 1901.98s and the reaction temperature is 20.81℃. Under the conditions, the result was obtained by three validation experiments. The inhibition rates of tyrosinase were 78.15%, 77.32% and 73.16%, respectively. The average value was 76.21%.

KeyWords: 1, 3, 4 – Thiadiazole derivatives; Tyrosinase inhibitor; Enzme kinetics; Response surface methodology; Molecular docking studies

第1章　绪 论(节选)

1.4　研究意义和内容

1.4.1　课题来源

横向科研项目

1.4.2 研究意义

果蔬在贮藏、加工及销售过程中易发生褐变,严重降低了果蔬的内在品质及商业价值。当前广大研究人员认为酪氨酸酶是酶促褐变相关的重要酶,而酪氨酸酶广泛存在于果蔬中,因此通过抑制酪氨酸酶活性从而有效地遏制果蔬的褐变。

本课题在前期研究的基础上,设计、合成新型酪氨酸酶抑制剂,测定其对酪氨酸酶的抑制活性,研究其抑制机理和抑制动力学,探讨分子作用机制,总结构效关系,优选化合物进行毒性评价和防褐变效果实验,其研究意义主要体现在以下两个方面:(一)研究新型酪氨酸酶抑制剂的抑制机理和抑制动力学,利于推动我国食品工业的发展;(二)研究新型酪氨酸酶抑制剂的抑制机理和抑制动力学,总结化合物的构效关系并建立新型酪氨酸酶抑制剂的筛选平台,为新型酪氨酸酶抑制剂的研发提供扎实的理论和技术依据。

1.4.3 研究内容

基于以上几点,本论文主要研究内容如下:

(1)化合物设计与合成工作是基于活性片段组合原理,因为噻二唑和席夫碱都具有酪氨酸酶抑制活性,所以本文以氨基硫脲和二硫化碳为原料合成2-氨基-5-巯基-1,3,4-噻二唑后,再与不同的取代芳香醛反应得到一系列的噻二唑衍生物,并对巯基部位进行结构修饰改造。

(2)研究噻二唑衍生物对蘑菇酪氨酸酶的抑制活性和抑制动力学,讨论构效关系,为进一步开展噻二唑席夫碱酪氨酸酶抑制剂研究工作奠定基础。

(3)优选1-2种对蘑菇酪氨酸酶抑制活性较强的抑制剂进行毒理实验及分子对接研究。

(4)优选毒性较低的酪氨酸酶抑制剂进行鲜切果蔬保藏效果评价

第2章 噻二唑衍生物的合成

本文以氨基硫脲和二硫化碳为起始原料合成2-氨基-5-巯基-1,3,4噻二唑,再与不同的芳香醛得到噻二唑席夫碱,并对部分化合物的巯基部位进行结构修饰。

2.1 实验材料和仪器

2.1.1 实验材料

实验中使用的材料如表2.1所示。

表2.1　实验材料

名称	规格(纯度)	生产厂家
氨基硫脲	分析纯	上海达瑞精细化学品有限公司
无水乙醇	分析纯(99%)	成都金山化学试剂有限公司
无水碳酸钠	分析纯	天津市恒兴化学试剂制造有限公司
二硫化碳	分析纯	天津市恒兴化学试剂制造有限公司
浓盐酸	分析纯	天津市富宇精细化工有限公司
氢氧化钾	分析纯	天津市永大化学试剂有限公司
浓硫酸	分析纯(98%)	邵阳市方华化工有限公司
4-氯苯甲醛	分析纯	上海达瑞精细化学品有限公司
5-溴-2-羟基苯甲醛	分析纯	上海达瑞精细化学品有限公司
2-甲氧基苯甲醛	分析纯	上海达瑞精细化学品有限公司
2,4-二羟基苯甲醛	分析纯	上海达瑞精细化学品有限公司
5-氯-2-羟基苯甲醛	分析纯	上海达瑞精细化学品有限公司
2-羟基苯甲醛	分析纯	上海达瑞精细化学品有限公司
4-叔丁基-苯甲醛	分析纯	上海达瑞精细化学品有限公司
3,5-二甲氧基苯甲醛	分析纯	上海达瑞精细化学品有限公司
3,4-二氯苯甲醛	分析纯	上海达瑞精细化学品有限公司
3-氯苯甲醛	分析纯	上海达瑞精细化学品有限公司
2-氟苯甲醛	分析纯	上海达瑞精细化学品有限公司
2,4,6-三甲氧基苯甲醛	分析纯	上海达瑞精细化学品有限公司
4-羟基-3-甲氧基苯甲醛	分析纯	上海达瑞精细化学品有限公司
2-硝基苯甲醛	分析纯	上海达瑞精细化学品有限公司
4-羟基-3,5-二甲氧基苯甲醛	分析纯	上海达瑞精细化学品有限公司
苯甲醛	分析纯	上海达瑞精细化学品有限公司
烯丙基溴	分析纯	上海达瑞精细化学品有限公司
苄氯	分析纯	上海达瑞精细化学品有限公司
四丁基溴化铵	分析纯	上海达瑞精细化学品有限公司
石油醚	分析纯	天津市富宇精细化工有限公司
乙酸乙酯	分析纯	天津市富宇精细化工有限公司
3,4-二羟基苯甲醛	分析纯	上海达瑞精细化学品有限公司
3-羟基4-甲氧基苯甲醛	分析纯	上海达瑞精细化学品有限公司
3-羟基苯甲醛	分析纯	上海达瑞精细化学品有限公司

续表

名称	规格(纯度)	生产厂家
4-羟基苯甲醛	分析纯	上海达瑞精细化学品有限公司
2,4-二甲氧基苯甲醛	分析纯	上海达瑞精细化学品有限公司

2.1.2 实验仪器

实验中所需实验仪器如表2.2所示。

表2.2 实验仪器

名称	型号	生产厂家
集热式恒温加热磁力搅拌器	DF-101S	巩义市予华仪器有限责任公司
电子天平	CP213	上海越磁电子科技有限公司
电热鼓风干燥机	101-1AB	天津市泰斯仪器有限公司
电热恒温干燥箱	202-0A	天津泰斯特仪器有限公司
显微熔点仪(未校正)	SGW X-4	上海精密科学仪器有限公司
玻璃仪器气流烘干器	RQ-C	巩义市予华仪器有限责任公司
低温冷却循环泵	DLSB-5/25	巩义市予华仪器有限责任公司
循环水式真空泵	SHZ-D(Ⅲ)	巩义市宇华仪器有限公司
手持紫外分光光度计	ZF-7	巩义市予华仪器有限责任公司
旋转蒸发仪	IKARHB10	德国IKA公司
红外光谱分析仪	Nicolet-iS5	赛默飞世尔科技公司
核磁共振分析仪	Bruker300/400	德国Bruker公司
质谱仪	LCMS-2010A	日本岛津公司
海尔柜式冰箱	海尔	上海市安亭电子仪器厂

2.2 噻二唑衍生物的合成

2.2.1 2-氨基-5-巯基-1,3,4-噻二唑的合成

合成路线：

$$H_2N-C(=S)-NH-NH_2 + CS_2 \xrightarrow[C_2H_5OH,\ reflux]{Na_2CO_3} \text{HS-(1,3,4-thiadiazole)-}NH_2$$

91.02　　75.94　　132.98

取氨基硫脲19.228 g(0.2 mol)、无水乙醇80 mL及无水碳酸钠10 g加入到250 mL三口反应烧瓶中,室温下(25 ℃)开启搅拌,使用恒压滴液漏斗缓慢加入二硫化碳15.228 g(0.2mol),一小时后缓慢升温至75 ℃后保温回流,产生大量

二硫化氢气体(使用氢氧化钠溶液尾气吸收)。反应至无硫化氢气体产生后停止反应,冷却至室温后加入 100 mL 水,用浓盐酸调 pH 值至 4,产生大量淡黄色沉淀。放置冰箱 15 分钟后抽滤,滤饼用无水乙醇冲洗 3 遍,再用无水乙醇重结晶得淡黄色晶体 20.744 g,摩尔收率[(产物的物质的量/投料较小的原料的物质的量)× 100%]为 78%,熔点:212.2℃ ~213.8 ℃。

2.2.1.1 5-(4-氯苯亚甲基)氨基-1,3,4-噻二唑-2-硫醇的合成

合成路线:

$$\text{HS-噻二唑-NH}_2\ (132.98) + \text{4-Cl-C}_6\text{H}_4\text{CHO}\ (140.00) \xrightarrow{H_2SO_4/C_2H_5OH,\ reflux} \text{4-Cl-C}_6\text{H}_4\text{CH=N-噻二唑-SH}\ (254.97)$$

将 0.66 g(5 mmol)2-氨基-5-巯基-1,3,4-噻二唑、0.84 g(6 mmol)对氯苯甲醛加入到 25 mL 反应瓶中。加入 10mL 无水乙醇后在搅拌下加入 3 滴浓硫酸,升温至 80 ℃后保温回流反应,通过薄层色谱(TLC)检测于 1.5 h 后停止加热并冷却至室温,静置过夜后有固体析出,抽滤,滤饼用乙醇和热水($V_{乙醇}:V_{热水}=1:1$)的混合溶剂淋洗 3 次,再用无水乙醇重结晶干燥得黄色固体 0.385 g,摩尔收率为 30.2%,产物标记为 T1-1。IR(cm^{-1},KBr):3275,1563,1488,1376,1313,1259,1185,1094,1051,1015,783,744;^1H NMR(DMSO-d_6,300MHz):δ13.18(s,1H,S-H),8.75(s,1H,=CH),7.95(d,2H,J=6.9Hz,Ph-H),7.71(d,2H,J=6.9Hz,Ph-H);^{13}C NMR(DMSO-d_6,75MHz):δ192.1,180.9,161.4,139.3,134.8,131.5,129.3;MS(ESI):m/z(100%)256(M^+)。

2.2.1.2 4-溴-2-{[(5-硫烷基-1,3,4-噻二唑-2-基)亚氨基]甲基}苯酚的合成

合成路线:

$$\text{HS-噻二唑-NH}_2\ (132.98) + \text{5-溴水杨醛}\ (199.95) \xrightarrow{H_2SO_4/C_2H_5OH,\ reflux} \text{产物}\ (314.91)$$

将 0.66 g(5 mmol)2-氨基-5-巯基-1,3,4-噻二唑、1.20 g(6 mmol)5-溴水杨醛加入到 25 mL 反应瓶中。加入 10mL 无水乙醇后在搅拌下加入 3 滴浓硫酸,升温至 80℃后保温回流反应,通过 TLC 检测于 0.5h 后停止加热并冷却至室温,静置过夜后有固体析出,抽滤,滤饼用乙醇和热水($V_{乙醇}:V_{热水}=1:1$)的混

合溶剂淋洗 3 次，再用无水乙醇重结晶干燥得黄色固体 0.811 g，摩尔收率为 51.5%，产物标记为 T1－2。IR(cm^{-1}，KBr)：3420，3058，2858，1606，1555，1512，1467，1347，1269，1171，1062，965，828，729；^{1}H NMR(DMSO－d_6，300MHz)：δ14.57(s，1H，S－H)，11.28(s，1H，O－H)，8.78(s，1H，＝CH)，7.97(s，1H，Ph－H)，7.71(d，1H，J＝8.7Hz，Ph－H)，7.00(d，1H，J＝8.7Hz，Ph－H)；^{13}C NMR(DMSO－d_6，75MHz)：δ189.6，180.7，163.9，159.3，137.7，130.9，124.2，121.7，110.9；MS(ESI)：m/z(100%)317(M^+)。

2.2.1.3　5－[(2－甲氧苯亚甲基)氨基]－1，3，4－噻二唑－2－硫醇的合成

合成路线：

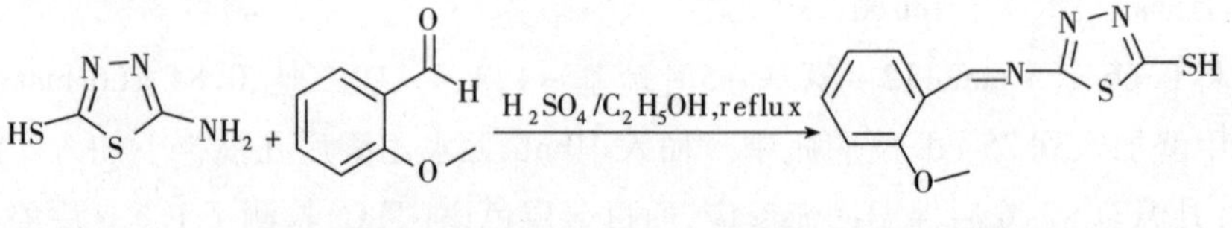

H_2SO_4/C_2H_5OH, reflux

132.98　　136.05　　251.02

将 0.66 g(5 mmol)2－氨基－5－巯基－1，3，4－噻二唑、0.816 g(6 mmol)邻甲氧基苯甲醛加入到 25 mL 反应瓶中。加入 10mL 无水乙醇后搅拌下加入 3 滴浓硫酸。升温至 80℃保温回流反应，通过 TLC 检测反应。2h 后停止加热并冷却至室温，静置过夜后有固体析出，抽滤，滤饼用乙醇和热水($V_{乙醇}$：$V_{热水}$＝1：1)的混合溶剂淋洗 3 次，用无水乙醇重结晶，干燥得黄色固体 0.483 g，摩尔收率为 38.3%，产物标记为 T1－3。熔点：148.2－151.3℃。IR(cm^{-1}，KBr)：3328，2934，2835，1564，1491，1373，1312，1252，1175，1116，1051，933，802，754；^{1}H NMR(DMSO－d_6，300 MHz)：δ13.17(s，1H，S－H)，9.32(s，1H，＝CH)，7.69(d，1H，J＝8.1Hz，Ph－H)，7.52(t，1H，J＝8.4Hz，Ph－H)，7.03(d，1H，J＝8.4Hz，Ph－H)，6.99(t，1H，J＝7.8Hz，Ph－H)；^{13}C NMR(DMSO－d_6，75 MHz)：δ189.9，181.2，157.4，155.5，135.9，131.0，128.6，122.4，121.0，56.0；MS(ESI)：m/z(100%)252(M^+)。

2.2.1.4　4－{[(5－硫烷基－1，3，4－噻二唑－2－基)亚氨基]甲基}苯－1，3－二酚的合成

合成路线：

将 0.66 g(5 mmol)2－氨基－5－巯基－1，3，4－噻二唑、0.828 g(6 mmol)2，4－二羟基苯甲醛加入到 25mL 反应瓶中。加入 10mL 无水乙醇后搅拌下加入 3

H_2SO_4/C_2H_5OH,reflux

132.98　138.03　253.00

滴浓硫酸。升温至 80℃保温回流反应,通过 TLC 检测反应。2h 后停止加热并冷却至室温,静置过夜后有固体析出,抽滤,滤饼用乙醇和热水($V_{乙醇}:V_{热水}=1:1$)的混合溶剂淋洗 3 次,用无水乙醇重结晶,干燥得黄色固体 0.732 g,摩尔收率为 57.8%,产物标记为 T1 - 4。熔点:240 ~ 243℃。IR(cm^{-1},KBr):3305,3139,1635,1587,1516,1438,1319,1241,1213,1171,1113,1059,973,864,797;^1H NMR(DMSO - d6,300 MHz):δ 14.37(s,1H,S - H),11.40(s,1H,O - H),10.73(s,1H,O - H),8.70(s,1H, = CH),7.71(d,1H,J = 6.9Hz ,Ph - H),7.65(s,1H,Ph - H),6.34(d,1H,J = 6.9Hz,Ph - H);^{13}C NMR(DMSO - d_6,75 MHz):δ 190.9,185.9,165.7,164.4,163.2,133.2,119.9,109.3,102.3;MS(ESI):m/z(100%)254(M +)。

2.2.1.5　4 - 溴 - 2 - {[(5 - 硫烷基 - 1,3,4 - 噻二唑 - 2 - 基)亚氨基]甲基}苯酚的合成

合成路线:

H_2SO_4/C_2H_5OH,reflux

132.08　156.00　270.96

将 0.66 g(5 mmol)2 - 氨基 - 5 - 巯基 - 1,3,4 - 噻二唑 0.939 g(6 mmol)5 - 氯水杨醛加入到 25 mL 反应瓶中。加入 10mL 无水乙醇后搅拌下加入 3 滴浓硫酸。升温至 80℃保温回流反应,通过 TLC 检测反应。0.5h 后停止加热并冷却至室温,静置过夜后有固体析出,抽滤,滤饼用乙醇和热水($V_{乙醇}:V_{热水}=1:1$)的混合溶剂淋洗 3 次,用无水乙醇重结晶,干燥得黄色固体 0.732 g,摩尔收率为 53.0%,产物标记为 T1 - 5。熔点:247.3 ~ 250.1℃。IR(cm^{-1},KBr):3440,3067,2862,1599,1561,1517,1470,1231,1173,1063,804,711;^1H NMR(DMSO - d_6,300 MHz):δ 14.58(s,1H,S - H),11.27(s,1H,O - H),8.80(s,1H, = CH),7.84(s,1H,Ph - H),7.54(s,1H,J = 8.7Hz,Ph - H),7.05(d,1H,J = 8.7Hz,Ph - H);^{13}C NMR(DMSO - d_6,75 MHz):δ189.7,180.7,164.0,158.7,135.0,127.8,123.6,121.1,119.6;MS(ESI):m/z(100%)272(M^+)。

2.2.1.6 2-{[(5-硫烷基-1,3,4-噻二唑-2-基)亚氨基]甲基}苯酚的合成

合成路线:

HS–(1,3,4-噻二唑)–NH_2 + 水杨醛 $\xrightarrow{H_2SO_4/C_2H_5OH, reflux}$ 产物(–SH)

132.98 122.04 237.00

将0.66 g(5 mmol)2-氨基-5-巯基-1,3,4-噻二唑0.733 g(6 mmol)水杨醛加入到25 mL反应瓶中。加入10 mL无水乙醇后搅拌下加入3滴浓硫酸。升温至80℃保温回流反应,通过TLC检测反应。1.5h后停止加热并冷却至室温,静置过夜后有固体析出,抽滤,滤饼用乙醇和热水($V_{乙醇}$: $V_{热水}$ =1 : 1)的混合溶剂淋洗3次,用无水乙醇重结晶,干燥得黄色固体0.607 g,摩尔收率为51.0%,产物标记为T1-6。熔点:252~254℃。IR(cm^{-1},KBr):3434,3072,2875,1607,1571,1528,1482,1374,1276,1199,1146,1067,761;^{1}H NMR(DMSO-d_6,300 MHz):δ14.53(s,1H,S-H),11.15(s,1H,O-H),8.88(s,1H,=CH),7.87(d,1H,J=7.8 Hz,Ph-H),7.53(t,1H,J=7.2Hz,Ph-H),7.02-6.96(m,2H,Ph-H);^{13}C NMR(DMSO-d6,75 MHz):δ191.7,186.5,166.1,160.8,135.8,130.1,119.9,119.5,117.2;MS(ESI):m/z(100%)238 (M+)。

2.2.1.7 5-((4-叔-丁基苯亚甲基)氨基)-1,3,4-噻二唑-2-硫醇的合成

合成路线:

HS–(1,3,4-噻二唑)–NH_2 + 对叔丁基苯甲醛 $\xrightarrow{H_2SO_4/C_2H_5OH, reflux}$ 产物(–SH)

132.98 162.10 277.07

将0.66 g(5 mmol)2-氨基-5-巯基-1,3,4-噻二唑0.973 g(6 mmol)对叔丁基苯甲醛加入到25 mL反应瓶中。加入10 mL无水乙醇后搅拌下加入3滴浓硫酸。升温至80℃保温回流反应,通过TLC检测反应。2.5h后停止加热并冷却至室温,静置过夜后有固体析出,抽滤,滤饼用乙醇和热水($V_{乙醇}$: $V_{热水}$ =1 : 1)的混合溶剂淋洗3次,用无水乙醇重结晶,干燥得白色固体0.919 g,摩尔收率为66.1%,产物标记为T1-7。熔点:214.2-216.5℃。IR(cm^{-1},KBr):3395,2960,2868,1689,1607,1567,1499,1379,1315,1264,1185,1105,1052,931,835,

761;^{1}H NMR(DMSO－d_6,300 MHz):δ13.25(s,1H,S－H),8.83(s,1H,＝CH),7.66(d,2H,J＝8.4 Hz,Ph－H),7.48(d,2H,J＝8.4Hz,Ph－H),1.35(s,9H,$3CH_3$);^{13}C NMR(DMSO－d_6,75 MHz):δ 192.2,180.7,155.7,152.9,131.3,129.7,126.8,34.8,31.2;MS(ESI):m/z(100%)278 (M^+)。

2.2.1.8 5－[(3,5－二甲氧苯亚甲基)氨基]－1,3,4－噻二唑－2－硫醇的合成

$$\text{HS-thiadiazole-NH}_2 + \text{3,5-(H}_3\text{CO)}_2\text{C}_6\text{H}_3\text{CHO} \xrightarrow{H_2SO_4/C_2H_5OH,\text{reflux}} \text{3,5-(H}_3\text{CO)}_2\text{C}_6\text{H}_3\text{CH=N-thiadiazole-SH}$$

合成路线:

将0.66 g(5 mmol)2－氨基－5－巯基－1,3,4－噻二唑0.997 g(6 mmol)3,5－二甲氧基苯甲醛加入到25 mL反应瓶中。加入10 mL无水乙醇后搅拌下加入3滴浓硫酸。升温至80℃保温回流反应,通过TLC检测反应。3h后停止加热并冷却至室温,静置过夜后有固体析出,抽滤,滤饼用乙醇和热水($V_{乙醇}:V_{热水}=1:1$)的混合溶剂淋洗3次,用无水乙醇重结晶,干燥得黄色固体0.632 g,摩尔收率为44.8%,产物标记为T1－8。熔点:227～229℃。IR(cm^{-1},KBr):3339,3082,2926,1609,1557,1463,1325,1264,1200,1155,1059,971,845,754;^{1}H NMR(DMSO－d_6,300 MHz):δ13.19(s,1H,S－H),8.67(s,1H,＝CH),7.08(s,2H,Ph－H),6.61(s,1H,Ph－H),3.76(s,6H,$2CH_3$);^{13}C NMR(DMSO－d_6,75 MHz):δ186.8,180.9,168.7,161.4,138.7,107.5,104.2,55.58;MS(ESI):m/z(100%)282 (M^+)。

2.2.1.9 5－[(3,4－二氯苯亚甲基)氨基]－1,3,4－噻二唑－2－硫醇的合成

合成路线:

$$\underset{132.98}{\text{HS-thiadiazole-NH}_2} + \underset{173.96}{\text{3,4-Cl}_2\text{C}_6\text{H}_3\text{CHO}} \xrightarrow{H_2SO_4/C_2H_5OH,\text{reflux}} \underset{288.93}{\text{3,4-Cl}_2\text{C}_6\text{H}_3\text{CH=N-thiadiazole-SH}}$$

将0.66 g(5 mmol)2－氨基－5－巯基－1,3,4－噻二唑1.050 g(6 mmol)3,4－二氯苯甲醛及四丁基溴化铵0.16 g(0.5 mmol)加入到25 mL反应瓶中。加入10 mL无水乙醇后搅拌,升温至80℃保温回流反应,通过TLC检测反应。1.5h

后停止加热并冷却至室温，静置过夜后有固体析出，抽滤，滤饼用乙醇和热水（$V_{乙醇}$：$V_{热水}$ =1：1）的混合溶剂淋洗3次，用无水乙醇重结晶，干燥得黄色固体0.986 g，摩尔收率为67.8%，产物标记为T1－9。熔点：223.3～225.7℃。IR（cm^{-1}，KBr）：3404，1566，1469，1376，1309，1264，1190，1103，1053，746，667；^{1}H NMR（DMSO－d_6，300 MHz）：δ 13.15（s，1H，S－H），8.82（s，1H，＝CH），7.97（s，1H，Ph－H），7.75（d，1H，J＝8.4 Hz，Ph－H），7.65（d，1H，J＝8.4 Hz，Ph－H）；^{13}C NMR（DMSO－d_6，75 MHz）：δ 189.7，181.3，159.8，136.8，135.1，134.5，133.2，131.3，128.4；MS（ESI）：m/z（100%）291（M^+）。

2.2.1.10　5－［（3－氯苯亚甲基）氨基］－1，3，4－噻二唑－2－硫醇的合成

合成路线：

H_2SO_4/C_2H_5OH, reflux

132.98　　140.00　　254.97

将0.66 g（5 mmol）2－氨基－5－巯基－1，3，4－噻二唑、0.843 g（6 mmol）3－氯苯甲醛及四丁基溴化铵0.16 g（0.5 mmol）加入到25 mL反应瓶中。加入10 mL无水乙醇后搅拌，升温至80℃保温回流反应，通过TLC检测反应。2h后停止加热并冷却至室温，静置过夜后有固体析出，抽滤，滤饼用乙醇和热水（$V_{乙醇}$：$V_{热水}$ =1：1）的混合溶剂淋洗3次，用无水乙醇重结晶，干燥得黄色固体0.767 g，摩尔收率为59.9%，产物标记为T1－10。熔点：253.4～255.3℃。IR（cm^{-1}，KBr）：3404，1566，1469，1376，1309，1264，1190，1103，1053，746，667；^{1}H NMR（DMSO－d_6，300 MHz）：δ13.12（s，1H，S－H），8.86（s，1H，＝CH），8.12（s，1H，Ph－H），7.96（d，1H，J＝7.8 Hz，Ph－H），7.80（d，1H，J＝6.9 Hz，Ph－H），7.45（t，1H，J＝8.7 Hz，ph－H）；^{13}C NMR（DMSO－d_6，75 MHz）：δ 192.1，180.5，161.4，136.5，134.2，131.2，130.4，127.9，127.0；MS（ESI）：m/z（100%）256（M^+）。

2.2.1.11　5－（（2－氟苯亚甲基）氨基）－1，3，4－噻二唑－2－硫醇的合成

合成路线：

H_2SO_4/C_2H_5OH, reflux

132.98　　124.03　　239.00

将0.66 g(5 mmol)2-氨基-5-巯基-1,3,4-噻二唑、0.745 g(6 mmol)邻氟苯甲醛加入到25 mL反应瓶中。加入10 mL无水乙醇后搅拌下加入3滴浓硫酸,升温至80℃保温回流反应,通过TLC检测反应。4h后停止加热并冷却至室温,静置过夜后有固体析出,抽滤,滤饼用乙醇和热水($V_{乙醇}:V_{热水}=1:1$)的混合溶剂淋洗3次,用无水乙醇重结晶,干燥得黄色固体0.192 g,摩尔收率为16.0%,产物标记为T1-11。熔点:242.6~244.1℃。IR(cm^{-1},KBr):3233,2995,1562,1491,1382,1313,1267,1208,1108,1051,942,754,671;^{1}H NMR(DMSO-d_6,300 MHz):δ 13.18(s,1H,S-H),8.83(s,1H,=CH),8.25(d,1H,J=7.5 Hz,Ph-H),7.54(t,1H,J=8.1 Hz,Ph-H),7.34(d,1H,J=7.5 Hz,Ph-H),7.29(t,1H,J=8.1 Hz,Ph-H);^{13}C NMR(DMSO-d_6,75 MHz):δ 187.9,180.8,157.4,136.9,131.7,129.4,125.2,122.1,116.8;MS(ESI):m/z(100%)240(M^+)。

2.2.1.12 5-((2,4,6-三甲氧苯亚甲基)氨基)-1,3,4-噻二唑-2-硫醇的合成

合成路线:

将0.66 g(5 mmol)2-氨基-5-巯基-1,3,4-噻二唑1.177 g(6 mmol)邻氟苯甲醛加入到25 mL反应瓶中。加入10 mL无水乙醇后搅拌下加入3滴浓硫酸,升温至80℃保温回流反应,通过TLC检测反应。2h后停止加热并冷却至室温,静置过夜后有固体析出,抽滤,滤饼用乙醇和热水($V_{乙醇}:V_{热水}=1:1$)的混合溶剂淋洗3次,用无水乙醇重结晶,干燥得黄色固体1.342 g,摩尔收率为86.0%,产物标记为T1-12。熔点:263~265℃。IR(cm^{-1},KBr):3084,2890,1599,1520,1463,1384,1325,1276,1239,1113,1064,767;^{1}H NMR(DMSO-d_6,300 MHz):δ 14.29(s,1H,S-H),8.69(s,1H,=CH),6.31(s,2H,Ph-H),3.90(s,6H,2CH_3),3.82(s,3H,CH_3);^{13}C NMR(DMSO-d_6,75 MHz):δ 190.8,185.8,167.2,163.4,161.3,104.5,91.7,56.3,55.7;MS(ESI):m/z(100%)312(M^+)。

2.2.1.13 4-{[(5-巯基-1,3,4-噻二唑-2-基)亚氨基]甲基}-2-甲氧基苯酚的合成

合成路线:

N－N　HS　S　NH_2　＋　H_3CO　O　H　HO　H_2SO_4/C_2H_5OH,reflux　H_3CO　HO　＝N－　N－N　S　SH

132.98　　152.05　　267.01

将0.66 g(5 mmol)2－氨基－5－巯基－1,3,4－噻二唑0.913 g(6 mmol)4－羟基－3－甲氧基苯甲醛加入到25 mL反应瓶中。加入10 mL无水乙醇后搅拌下加入3滴浓硫酸,升温至80℃保温回流反应,通过TLC检测反应。1.5h后停止加热并冷却至室温,静置过夜后有固体析出,抽滤,滤饼用乙醇和热水($V_{乙醇}$: $V_{热水}$ =1 : 1)的混合溶剂淋洗3次,用无水乙醇重结晶,干燥得黄色固体0.847 g,摩尔收率为63.2%,产物标记为T1－13。熔点:219.5～221.8℃。IR(cm^{-1},KBr):3275,1606,1565,1512,1382,1322,1272,1205,1101,1053,928,753;^1H NMR(DMSO－d_6,300 MHz):δ 14.41(s,1H,S－H),10.34(s,1H,－OH),8.55(s,1H,＝CH),7.55(s,1H,Ph－H),7.42(d,1H,J＝8.7 Hz,Ph－H),6.99(d,1H,J＝8.7Hz,Ph－H),3.86(s,3H,CH3);^{13}C NMR(DMSO－d_6,75 MHz):δ 190.9,186.2,167.9,152.9,148.1,128.6,126.7,115.6,111.1,55.5;MS(ESI):m/z(100%)268(M^+)。

2.2.1.14　5－[(2－硝基苯亚甲基)氨基]－1,3,4－噻二唑－2－硫醇的合成

合成路线:

N－N　HS　S　NH_2　＋　O　H　NO_2　H_2SO_4/C_2H_5OH,reflux　＝N－　N－N　S　SH　NO_2

132.98　　151.03　　265.99

将0.66 g(5 mmol)2－氨基－5－巯基－1,3,4－噻二唑0.907 g(6 mmol)邻硝基苯甲醛及四丁基溴化铵0.16 g(0.5 mmol)加入到25 mL反应瓶中。加入10 mL无水乙醇后搅拌,升温至80℃保温回流反应,通过TLC检测反应。2h后停止加热并冷却至室温,静置过夜后有固体析出,抽滤,滤饼用乙醇和热水($V_{乙醇}$: $V_{热水}$ =1 : 1)的混合溶剂淋洗3次,用无水乙醇重结晶,干燥得黄色固体0.303 g,摩尔收率为22.7%,产物标记为T1－14。熔点:229.7～232.2℃。IR(cm^{-1},KBr):3221,2987,1610,1562,1519,1352,1309,1265,1215,1188,1110,1055,947,856,781,733;^1H NMR(DMSO－d_6,300 MHz):δ 13.19(s,1H,S－H),9.36(s,1H,＝CH),8.16(d,1H,J＝8.1 Hz,Ph－H),7.92(d,1H,J＝7.5 Hz,Ph－H),

7.79(t,1H,J = 8.4 Hz,Ph - H),7.67(t,1H,J = 7.8 Hz,Ph - H);^{13}C NMR (DMSO - d_6,75 MHz):δ 189.9,180.1,158.6,149.5,134.2,133.7,130.8,128.2,124.9;MS(ESI):m/z(100%)267(M^+)。

2.2.1.15　4 - {[(5 - 巯基 - 1,3,4 - 噻二唑 - 2 - 基)亚氨基]甲基} - 2,6 - 二甲氧基苯酚的合成

合成路线：

H_2SO_4/C_2H_5OH,reflux

132.98　182.06　297.02

将0.66 g(5 mmol)2 - 氨基 - 5 - 巯基 - 1,3,4 - 噻二唑 1.093 g(6 mmol)4 - 羟基 - 3,5 - 二甲氧基苯甲醛加入到25 mL反应瓶中。加入10 mL无水乙醇后搅拌下加入3滴浓硫酸,升温至80℃保温回流反应,通过TLC检测反应。1.5h后停止加热并冷却至室温,静置过夜后有固体析出,抽滤,滤饼用乙醇和热水($V_{乙醇}$: $V_{热水}$ = 1 : 1)的混合溶剂淋洗3次,用无水乙醇重结晶,干燥得黄色固体0.778 g,摩尔收率为52.2%,产物记为T1 - 15。熔点:267 ~ 268.5℃。IR(cm^{-1},KBr):3275,1606,1565,1512,1382,1322,1272,1205,1101,1053,928,753;^{1}H NMR(DMSO - d_6,300 MHz):δ 14.42(s,1H,S - H),9.77(s,1H,- OH),8.53(s,1H,= CH),7.31(s,2H,Ph - H),3.84(s,6H,2CH_3);^{13}C NMR(DMSO - d_6,75 MHz):δ 191.1,186.0,168.1,148.1,142.0,124.4,107.6,56.2;MS(ESI):m/z (100%)298(M^+)。

2.2.1.16　5 - [(吡啶 - 2 - 基甲亚基)氨基] - 1,3,4 - 噻二唑 - 2 - 硫醇的合成

合成路线：

H_2SO_4/C_2H_5OH,reflux

132.98　107.04　222.00

将0.66 g(5 mmol)2 - 氨基 - 5 - 巯基 - 1,3,4 - 噻二唑 1.093 g(6 mmol)吡啶 - 2 - 甲醛加入到25 mL反应瓶中。加入10 mL无水乙醇后搅拌下加入3滴浓硫酸,升温至80℃保温回流反应,通过TLC检测反应。1.5h后停止加热并冷

却至室温，静置过夜后有固体析出，抽滤，滤饼用乙醇和热水($V_{乙醇}:V_{热水}=1:1$)的混合溶剂淋洗3次，用无水乙醇重结晶，干燥得黄色固体0.778 g，摩尔收率为52.2%，产物标记为T1－16，熔点：267～268.5℃。IR(cm^{-1}，KBr)：3275，1606，1565，1512，1382，1322，1272，1205，1101，1053，928，753；^{1}H NMR(DMSO－d_6，300 MHz)：δ 14.42(s，1H，S－H)，9.77(s，1H，－OH)，8.53(s，1H，＝CH)，7.31(s，2H，Ph－H)，3.84(s，6H，2CH_3)；^{13}C NMR(DMSO－d_6，75 MHz)：δ 191.1，186.0，168.1，148.1，142.0，124.4，107.6，56.2；MS(ESI)：m/z(100%)298(M^+)。

2.2.2　2－氨基－5－苯甲基醚基－1，3，4－噻二唑的合成

合成路线：

HS–(1,3,4-噻二唑)–NH_2 + 苄氯(Cl) $\xrightarrow[RT]{KOH/H_2O}$ H_2N–(1,3,4-噻二唑)–S–CH_2–Ph

132.98　　126.02　　223.02

取2－氨基－5－巯基－1，3，4－噻二唑0.03 mol(3.99 g)、KOH 0.03 mol(1.683 g)及水20 mL于50 mL圆底烧瓶中，用恒压滴液漏斗缓慢滴加苄氯0.03 mol(30min内滴完)，立即有乳化物出现，滴加完毕后于RT下反应一段时间(通过TLC检测)，反应瓶中产生大量淡黄色固体物质，过滤，滤饼用水洗至中性，用无水乙醇重结晶干燥黄色得黄色固体物质6.604 g，摩尔收率为98.71%，熔点：156.5～159.1℃。IR(cm^{-1}，KBr)：3275，1606，1565，1512，1382，1322，1272，1205，1101，1053，928，753；^{1}H NMR(DMSO－d_6，500 MHz)δ 7.36－7.25(m，5H，ph－H)，4.30(s，2H，CH_2)；^{13}C NMR(DMSO－d_6，125 MHz)δ 170.3，149.9，137.4，129.5，129.0，127.9，38.9；MS(ESI)：m/z(100%)224(M^+)。

2.2.2.1　4－{[(5－苯甲基醚基－1，3，4－噻二唑－2－基)亚氨基]甲基}－2－甲氧基苯酚的合成

合成路线：

H_2N–(1,3,4-噻二唑)–S–CH_2–Ph + 4-羟基-3-甲氧基苯甲醛(HO，OCH_3，CHO) $\xrightarrow[C_2H_5OH,reflux]{TBAB}$ HO(OCH_3)–Ph–CH＝N–(1,3,4-噻二唑)–S–CH_2–Ph

223.02　　152.05　　357.06

将1.115 g(5 mmol)2－氨基－5－苯甲基醚基－1，3，4－噻二唑0.913 g(6mmol)4－羟基－3－甲氧基苯甲醛及四丁基溴化铵0.16 g(0.5 mmol)加入到25 mL反应瓶中。加入10 mL无水乙醇后搅拌下并升温至80℃保温回流反应，

通过 TLC 检测反应。9h 后停止加热并冷却至室温，静置过夜后有固体析出，抽滤，滤饼用乙醇和热水（$V_{乙醇}:V_{热水}=1:1$）的混合溶剂淋洗 3 次，用无水乙醇重结晶，干燥得黄色固体 0.892 g，摩尔收率为 49.92%，产物标记为 T2－1。熔点：148.2～150.7℃。IR（cm^{-1}，KBr）：3275，1606，1565，1512，1382，1322，1272，1205，1101，1053，928，753；^{1}H NMR（$CDCl_3$，500 MHz）：δ 11.79（s，1H，－OH），8.95（s，1H，＝CH），7.44－7.30（m，7H，Ph－H），7.00（d，J＝8.6 Hz，1H，Ph－H），4.57（s，2H，CH_2）；^{13}C NMR（$CDCl_3$，125 MHz）：δ 170.4，166.7，164.8，160.2，135.6，132.6，132.3，129.3，128.9，128.1，118.9，38.4；MS（ESI）：m/z（100%）362（M^+）。

2.2.2.2　2－{[（5－苯甲基醚基－1，3，4－噻二唑－2－基）亚氨基]甲基}－4－溴苯酚的合成

合成路线：

TBAB
C_2H_5OH,reflux
223.02　199.95　404.96

将 1.115 g（5 mmol）2－氨基－5－苯甲基醚基－1，3，4－噻二唑 1.206 g（6 mmol）5－溴－水杨醛及四丁基溴化铵 0.16 g（0.5 mmol）加入到 25 mL 反应瓶中。加入 10 mL 无水乙醇后搅拌下并升温至 80℃保温回流反应，通过 TLC 检测反应。8h 后停止加热并冷却至室温，静置过夜后有固体析出，抽滤，滤饼用乙醇和热水（$V_{乙醇}:V_{热水}=1:1$）的混合溶剂淋洗 3 次，用无水乙醇重结晶，干燥得黄色固体 1.728 g，摩尔收率为 84.94%，产物标记为 T2－2。熔点：178.1～180.1℃。IR（cm^{-1}，KBr）：3275，1606，1565，1512，1382，1322，1272，1205，1101，1053，928，753；^{1}H NMR（$CDCl_3$，500 MHz）：δ 11.80（s，1H，－OH），8.93（s，1H，＝CH），7.56－7.52（m，2H，Ph－H），7.43（d，J＝8.6 Hz，1H，Ph－H），7.36－7.30（m，3H，Ph－H），6.95（d，J＝8.6 Hz，1H，Ph－H），4.56（s，2H，CH_2）；^{13}C NMR（$CDCl_3$，125 MHz）：δ 174.8，169.0，162.3，153.2，148.7，136.9，129.6，129.1，128.2，126.5，116.1，111.5，55.9，37.5；MS（ESI）：m/z（100%）405（M^+）。

2.2.2.3　N－[（5－苯甲基醚基）－1，3，4－噻二唑－2－基]－1－（2－甲氧苯基）甲亚胺的合成

合成路线：

TBAB
C_2H_5OH,reflux
223.02 136.05 341.07

将1.115 g(5 mmol)2－氨基－5－苯甲基醚基－1,3,4－噻二唑0.817 g(6 mmol)2－甲氧基苯甲醛及四丁基溴化铵0.16 g(0.5 mmol)加入到25 mL反应瓶中。加入10 mL无水乙醇后搅拌下并升温至80℃保温回流反应,通过TLC检测反应。9h后停止加热并冷却至室温,静置过夜后有固体析出,抽滤,滤饼用乙醇和热水($V_{乙醇}$∶$V_{热水}$=1∶1)的混合溶剂淋洗3次,用无水乙醇重结晶,干燥得白色固体0.735 g,摩尔收率为43.06%,产物标记为T2－3。熔点:95.9～97.3℃。IR(cm^{-1},KBr):3275,1606,1565,1512,1382,1322,1272,1205,1101,1053,928,753;^{1}H NMR(DMSO－d_6,500 MHz):δ 9.08(s,1H,＝CH),8.59(d,J=6.9Hz,1H,Ph－H),8.05(d, J=7.6Hz,1H,Ph－H),7.65(t,J=7.3Hz,1H,Ph－H),7.47(d,J=7.4Hz,1H,Ph－H),7.42－7.25(m,4H,Ph－H),7.10(d,J=8.1Hz,1H,Ph－H),4.59(s,2H,CH_2),3.92(s,3H,CH_3);^{13}C NMR(DMSO－d_6,125 MHz):δ 174.9,167.7,163.7,160.9,151.1,137.4,136.9,136.2,129.1,128.0,122.6,121.5,112.9,56.4,37.8;MS(ESI):m/z(100%)342(M^+)。

2.2.2.4 2－{[(5－苯甲基醚基－1,3,4－噻二唑－2－基)亚氨基]甲基}－4－氯苯酚的合成

合成路线:

TBAB
C_2H_5OH,reflux
223.02 156.00 361.01

将1.115 g(5 mmol)2－氨基－5－苯甲基醚基－1,3,4－噻二唑0.936 g(6 mmol)5－氯－2－羟基苯甲醛及四丁基溴化铵0.16 g(0.5 mmol)加入到25 mL反应瓶中。加入10 mL无水乙醇后搅拌下并升温至80℃保温回流反应,通过TLC检测反应。8h后停止加热并冷却至室温,静置过夜后有固体析出,抽滤,滤饼用乙醇和热水($V_{乙醇}$∶$V_{热水}$=1∶1)的混合溶剂淋洗3次,用无水乙醇重结晶,干燥得黄色固体0.903 g,摩尔收率为49.92%,产物标记为T2－4。熔点:148.2～150.7℃。IR(cm^{-1},KBr):3275,1606,1565,1512,1382,1322,1272,1205,1101,1053,928,753;^{1}H NMR(DMSO－d_6,500 MHz):δ 11.79(s,1H,－OH),

8.95(s,1H,=CH),7.44–7.30(m,7H,Ph–H),7.00(d,J=8.6Hz,1H,Ph–H),4.57(s,2H,CH_2);^{13}C NMR(DMSO–d_6,125 MHz):δ 170.4,166.7,164.8,160.2,135.6,132.6,132.3,129.3,128.9,128.1,124.6,119.4,118.9,38.4;MS(ESI):m/z(100%)362(M^+)。

2.2.2.5　2–{[(5–苯甲基醚基–1,3,4–噻二唑–2–基)亚氨基]甲基}–苯酚的合成

合成路线：

H_2N-噻二唑-S-苄基 (223.02) + 2-羟基苯甲醛 (122.04) $\xrightarrow[C_2H_5OH,reflux]{TBAB}$ 产物 (327.05)

将1.115 g(5 mmol)2–氨基–5–苯甲基醚基–1,3,4–噻二唑0.732 g(6 mmol)2–羟基苯甲醛及四丁基溴化铵0.16 g(0.5 mmol)加入到25 mL反应瓶中。加入10 mL无水乙醇后搅拌下并升温至80℃保温回流反应，通过TLC检测反应。10h后停止加热并冷却至室温，静置过夜后有固体析出，抽滤，滤饼用乙醇和热水($V_{乙醇}$∶$V_{热水}$=1∶1)的混合溶剂淋洗3次，用无水乙醇重结晶，干燥得黄色固体0.791 g，摩尔收率为48.32%，产物标记为T2–5。熔点：136.9～138.6℃。IR(cm^{-1},KBr):3275,1606,1565,1512,1382,1322,1272,1205,1101,1053,928,753;^{1}H NMR(DMSO–d_6,500 MHz):δ 11.22(s,1H,–OH),9.09(s,1H,=CH),7.90(d,J=7.8Hz,1H,Ph–H),7.68(d,J=7.8Hz,1H,Ph–H),7.55–7.27(m,6H,Ph–H),6.99(d,J=7.8Hz,1H,Ph–H),;^{13}C NMR(DMSO–d_6,125 MHz):δ 173.4,169.8,166.9,160.7,137.1,136.4,129.1,128.9,128.4,127.7,127.4,119.4,117.2,37.3;MS(ESI):m/z(100%)328(M^+)。

2.2.2.6　5–{[(5–苯甲基醚基–1,3,4–噻二唑–2–基)亚氨基]甲基}–2–甲氧基苯酚的合成

合成路线：

H_2N-噻二唑-S-苄基 (223.02) + 3-羟基-4-甲氧基苯甲醛 (H_3CO, OH; 152.05) $\xrightarrow[C_2H_5OH,reflux]{TBAB}$ 产物 (H_3CO, OH; 357.06)

将1.115 g(5 mmol)2–氨基–5–苯甲基醚基–1,3,4–噻二唑0.913 g(6 mmol)3–羟基–4–甲氧基–苯甲醛及四丁基溴化铵0.16 g(0.5 mmol)加入

到25 mL反应瓶中。加入10 mL无水乙醇后搅拌下并升温至80℃保温回流反应,通过TLC检测反应。8.5h后停止加热并冷却至室温,静置过夜后有固体析出,抽滤,滤饼用乙醇和热水($V_{乙醇}:V_{热水}=1:1$)的混合溶剂淋洗3次,用无水乙醇重结晶,干燥得黄色固体1.049 g,摩尔收率为58.69%,产物标记为T2-6。熔点:185.1~186.7℃。IR(cm^{-1},KBr):3275,1606,1565,1512,1382,1322,1272,1205,1101,1053,928,753;^{1}H NMR(DMSO-d_6,500 MHz):δ 9.77(s,1H,-OH),8.71(s,1H,=CH),7.47(d,J=7.8Hz,2H,Ph-H),7.40-7.10(m,6H,Ph-H),4.58(s,2H,-CH_2),3.87(s,2H,-CH_2);^{13}C NMR(DMSO-d_6,125 MHz):δ 174.2,169.8,162.1,153.3,149.4,147.0,137.1,129.8,128.9,127.7,125.1,114.0,111.8,55.8,37.2;MS(ESI):m/z(100%)358(M^+)。

2.2.2.7 4-{[(5-苯甲基醚基-1,3,4-噻二唑-2-基)亚氨基]甲基}苯-1,2-二酚的合成

合成路线:

H_2N S S N-N 223.02 + HO OH O H 138.03 —TBAB / C_2H_5OH,reflux→ HO OH N N-N S S 343.04

将1.115 g(5 mmol)2-氨基-5-苯甲基醚基-1,3,4-噻二唑0.829 g(6 mmol)3,4-二羟基苯甲醛及四丁基溴化铵0.16 g(0.5 mmol)加入到25 mL反应瓶中。加入10 mL无水乙醇后搅拌下并升温至80℃保温回流反应,通过TLC检测反应。11h后停止加热并冷却至室温,静置过夜后有固体析出,抽滤,滤饼用乙醇和热水($V_{乙醇}:V_{热水}=1:1$)的混合溶剂淋洗3次,用无水乙醇重结晶,干燥得黄色固体1.204 g,摩尔收率为70.12%,产物标记为T2-7。熔点:141.0~143.5℃。IR(cm^{-1},KBr):3275,1606,1565,1512,1382,1322,1272,1205,1101,1053,928,753;^{1}H NMR(DMSO-d_6,500 MHz):δ 10.18(s,1H,-OH),9.61(s,1H,-OH),8.64(s,1H,=CH),7.48(d,J=7.8Hz,2H,Ph-H),7.35-7.26(m,5H,Ph-H),6.94(d,J=7.8Hz,1H,Ph-H),4.57(s,2H,-CH_2);^{13}C NMR(DMSO-d_6,125 MHz):δ 174.4,168.6,152.1,149.4,146.0,137.1,129.8,128.6,127.6,126.1,115.8,114.3,37.2;MS(ESI):m/z(100%)344(M^+)。

2.2.3 2-氨基-5-烯丙基醚基-1,3,4-噻二唑的合成

合成路线:

132.98　　119.96　　173.01

取2－氨基－5－巯基－1,3,4－噻二唑0.03 mol(3.99 g)、KOH 0.03 mol(1.683 g)及水20 mL于50 mL圆底烧瓶中,用恒压滴液漏斗缓慢滴加烯丙基溴0.03 mol(30min内滴完),立即有乳化物出现,滴加完毕后于RT下反应一段时间(通过TLC检测),反应瓶中产生大量淡黄色固体物质,过滤,滤饼用水洗至中性,用无水乙醇重结晶干燥黄色得黄色固体物质2.724 g,摩尔收率为52.48%,产物标记为T3,熔点:112.3～112.8℃。IR(cm^{-1},KBr):3275,1606,1565,1512,1382,1322,1272,1205,1101,1053,928,753;^{1}H NMR(DMSO－d_6,500 MHz)δ 7.32(s,2H,NH_2),5.91－5.84(m,1H,HC＝),5.16(d,J＝13.5Hz,2H,＝CH_2),3.70(d,J＝7.0Hz,2H,CH_2);^{13}C NMR(DMSO－d_6,125 MHz)δ 170.4,149.8,133.9,119.1,37.8;MS(ESI):m/z(100%)174(M^+)。

2.2.3.1　4－{[(5－烯丙基醚基－1,3,4－噻二唑－2－基)亚氨基]甲基}－2－甲氧基苯酚的合成

合成路线:

173.01　　152.05　　307.04

将0.865 g(5 mmol)2－氨基－5－烯丙基醚基－1,3,4－噻二唑、0.837g(5.5 mmol)4－羟基－3－甲氧基苯甲醛及四丁基溴化铵0.16 g(0.5 mmol)加入到25 mL反应瓶中。加入10 mL无水乙醇后搅拌下并升温至90℃保温回流反应,通过TLC检测反应。9h后停止加热并冷却至室温,用旋转蒸发仪将体系溶剂蒸干,用$V_{石油醚}:V_{乙酸乙酯}=3:1$的洗脱剂进行层析柱,将收集到的黄色液体用旋转蒸发仪蒸干得黄色固体0.412 g,摩尔收率为28.57%,产物标记为T3－1,熔点:65.5～67.4℃。IR(cm^{-1},KBr):3275,1606,1565,1512,1382,1322,1272,1205,1101,1053,928,753;^{1}H NMR(DMSO－d_6,500 MHz):δ 11.18(s,1H,－OH),8.65(s,1H,CH),7.66(s,1H,Ph－H),7.34(d,J＝8.1Hz,1H,Ph－H),7.02(d,J＝8.1Hz,1H,Ph－H),6.01－5.97(m,1H,CH＝),5.29(d,J＝13.5Hz,2H,＝CH_2),3.97(s,3H,－CH_3),3.96(s,J＝7.0Hz,2H,－CH_2);^{13}C NMR(DMSO－d_6,125 MHz):δ 174.5,166.6,162.7,151.4,147.4,132.0,128.0,127.3,119.6,

114.6,1109.3,56.2,36.7;MS(ESI):m/z(100%)308(M^+)。

2.2.3.2　4-{[(5-烯丙基醚基-1,3,4-噻二唑-2-基)亚氨基]甲基}苯-1,3-二酚的合成

合成路线:

H₂N-噻二唑-S-烯丙基 (173.01) + 2,4-二羟基苯甲醛 (138.03) —TBAB, reflux→ 产物 (293.03)

将0.865 g(5 mmol)2-氨基-5-烯丙基醚基-1,3,4-噻二唑、0.760 g(5.5 mmol)3,4-二羟基苯甲醛及四丁基溴化铵0.16 g(0.5 mmol)加入到25 mL反应瓶中。加入10 mL无水乙醇后搅拌下并升温至90℃保温回流反应,通过TLC检测反应。12h后停止加热并冷却至室温,静置过夜后有固体析出,抽滤,滤饼用乙醇和热水($V_{乙醇}:V_{热水}=1:1$)的混合溶剂淋洗3次,用无水乙醇重结晶,干燥得黄色固体0.702 g,摩尔收率为47.83%,产物标记为T3-2,熔点:152.7~154.9℃。IR(cm^{-1},KBr):3275,1606,1565,1512,1382,1322,1272,1205,1101,1053,928,753;^1H NMR(DMSO-d_6,500 MHz):δ 11.53(s,1H,-OH),10.73(s,1H,-OH),8.93(s,1H,CH),7.69(s,1H,Ph-H),6.46(d,J=8.6Hz,1H,Ph-H),6.38(s,1H,Ph-H),6.00-5.95(m,1H,CH=),5.28(d,J=13.5Hz,2H,=CH_2),3.96(d,J=6.8Hz,2H,-CH_2);^{13}C NMR(DMSO-d_6,125 MHz):δ 174.1,167.1,165.5,163.4,162.0,134.0,133.1,120.0,112.6,109.8,102.8,36.7;MS(ESI):m/z(100%)294(M^+)。

2.2.3.3　2-{[(5-烯丙基醚基-1,3,4-噻二唑-2-基)亚氨基]甲基}-4-氯苯酚的合成

合成路线:

H₂N-噻二唑-S-烯丙基 (173.01) + 5-氯-2-羟基苯甲醛 (156.00) —TBAB, reflux→ 产物 (311.00)

将1.038 g(6 mmol)2-氨基-5-烯丙基醚基-1,3,4-噻二唑、1.017 g(6.5 mmol)3,4-二羟基苯甲醛加入到25 mL反应瓶中。加入15 mL无水乙醇后搅拌下加入3滴浓硫酸,升温至90℃保温回流反应,通过TLC检测反应。6h后停止加热并冷却至室温,用旋转蒸发仪将体系溶剂蒸干,用$V_{石油醚}:V_{乙酸乙酯}=6:1$

的洗脱剂进行层析柱,将收集到的黄色液体用旋转蒸发仪蒸干得黄色固体0.553 g,摩尔收率为31.42%,产物标记为T3-3,熔点:125.9~127.5℃。IR(cm^{-1},KBr):3275,1606,1565,1512,1382,1322,1272,1205,1101,1053,928,753;^1H NMR($DMSO-d_6$,500 MHz):δ 11.80(s,1H,-OH),8.97(s,1H,CH),7.43(s,1H,Ph-H),7.41(d,J=8.6Hz,1H,Ph-H),7.01(d,J=8.6Hz,1H,Ph-H),6.00-5.97(m,1H,CH=),5.31(d,J=13.5Hz,2H,=CH_2),3.97(d,J=6.8Hz,2H,-CH_2);^{13}C NMR($DMSO-d_6$,125 MHz):δ 170.7,166.7,164.8,160.3,135.5,132.6,131.8,124.6,119.9,119.4,118.9,36.6;MS(ESI):m/z(100%)312(M^+)。

2.2.3.4 2-{[(5-烯丙基醚基-1,3,4-噻二唑-2-基)亚氨基]甲基}-4-溴苯酚的合成

合成路线:

H_2N(噻二唑)S-烯丙基 173.01 + 5-溴-2-羟基苯甲醛 199.95 —TBAB, reflux→ 产物 354.94

将1.038 g(6 mmol)2-氨基-5-烯丙基醚基-1,3,4-噻二唑、1.307 g(6.5 mmol)5-溴-2-羟基苯甲醛加入到25 mL反应瓶中。加入15 mL无水乙醇后搅拌下加入3滴浓硫酸,升温至90℃保温回流反应,通过TLC检测反应。6h后停止加热并冷却至室温,用旋转蒸发仪将体系溶剂蒸干,用$V_{石油醚}:V_{乙酸乙酯}$=6:1的洗脱剂进行层析柱,将收集到的黄色液体用旋转蒸发仪蒸干得黄色固体0.312 g,摩尔收率为28.41%,产物标记为T3-4,熔点:104.7~105.9℃。IR(cm^{-1},KBr):3275,1606,1565,1512,1382,1322,1272,1205,1101,1053,928,753;^1H NMR($DMSO-d_6$,500 MHz):δ 11.82(s,1H,-OH),8.96(s,1H,CH),7.58(s,1H,Ph-H),7.55(d,J=8.8Hz,1H,Ph-H),6.96(d,J=8.8Hz,1H,Ph-H),6.01-5.97(m,1H,CH=),5.31(d,J=13.5Hz,2H,=CH_2),3.97(d,J=6.8Hz,2H,-CH_2);^{13}C NMR($DMSO-d_6$,125 MHz):δ 170.7,166.7,164.8,160.7,138.3,135.7,131.8,119.9,119.6,111.4,36.6;MS(ESI):m/z(100%)356(M^+)。

2.2.3.5 1-(3-氯苯基)-N-(5-烯丙基醚基-1,3,4-噻二唑-2-基)甲亚胺的合成

合成路线:

$$\text{H}_2\text{N-thiadiazole-S-allyl (173.01)} + \text{3-Cl-C}_6\text{H}_4\text{CHO (140.00)} \xrightarrow[\text{reflux}]{\text{TBAB}} \text{product (295.00)}$$

将0.865 g(5 mmol)2－氨基－5－烯丙基醚基－1,3,4－噻二唑、0.773 g(5.5 mmol)3－氯－苯甲醛及四丁基溴化铵0.16 g(0.5 mmol)加入到25 mL反应瓶中。加入15 mL无水乙醇后搅拌并升温至90℃保温回流反应,通过TLC检测反应。9h后停止加热并冷却至室温,用旋转蒸发仪将体系溶剂蒸干,用$V_{石油醚}$:$V_{乙酸乙酯}$=6∶1的洗脱剂进行层析柱,将收集到的黄色液体用旋转蒸发仪蒸干得黄色固体0.093 g,摩尔收率为6.72%,产物标记为T3－5,熔点:104.7～105.9℃。IR(cm^{-1},KBr):3275,1606,1565,1512,1382,1322,1272,1205,1101,1053,928,753;^{1}H NMR(DMSO－d_6,500 MHz):δ 8.84(s,1H,CH),7.97(s,1H,Ph－H),7.81(d,J=7.6Hz,1H,Ph－H),7.54(d,J=6.4Hz,1H,Ph－H),7.46－7.43(m,1H,Ph－H),6.01－5.97(m,1H,CH=),5.37(d,J=13.5Hz,2H,=CH_2),3.97(d,J=6.8Hz,2H,－CH_2);^{13}C NMR(DMSO－d_6,125 MHz):δ 168.7,165.0,153.8,139.7,134.6,131.9,130.3,129.6,126.,119.8,119.4,36.6;MS(ESI):m/z(100%),296(M^+)。

2.2.3.6 3－{[(5－烯丙基醚基－1,3,4－噻二唑－2－基)亚氨基]甲基}苯酚的合成

合成路线:

$$\text{H}_2\text{N-thiadiazole-S-allyl (173.01)} + \text{3-HO-C}_6\text{H}_4\text{CHO (122.04)} \xrightarrow[\text{reflux}]{\text{TBAB}} \text{product (277.03)}$$

将1.038 g(6 mmol)2－氨基－5－烯丙基醚基－1,3,4－噻二唑、0.794 g(6.5 mmol)3－羟基苯甲醛加入到25 mL反应瓶中。加入10 mL无水乙醇后搅拌下加入浓硫酸3滴,升温至90℃保温回流反应,通过TLC检测反应。8h后停止加热并冷却至室温,静置过夜后有固体析出,抽滤,滤饼用乙醇和热水($V_{乙醇}$:$V_{热水}$=1∶1)的混合溶剂淋洗3次,用无水乙醇重结晶,干燥得黄色固体0.904 g,摩尔收率为58.31%,产物标记为T3－6,熔点:156.2～158.1℃。IR(cm^{-1},KBr):3275,1606,1565,1512,1382,1322,1272,1205,1101,1053,928,753;^{1}H NMR(DMSO－d_6,500 MHz):δ 9.92(s,1H,－OH),8.48(s,1H,CH),7.07(d,J=6.4Hz,1H,Ph－H),5.99－5.96(m,1H,CH=),5.37(d,J=13.5Hz,2H,=

CH_2),3.99(d,J=6.8Hz,2H,$-CH_2$);^{13}C NMR(DMSO-d_6,125 MHz):δ 193.6,174.3,170.0,163.5,158.5,136.2,130.7,122.3,121.5,120.0,115.6,36.6;MS(ESI):m/z(100%)278(M^+)。

2.2.3.7 2-{[(5-烯丙基醚基-1,3,4-噻二唑-2-基)亚氨基]甲基}苯酚的合成

合成路线：

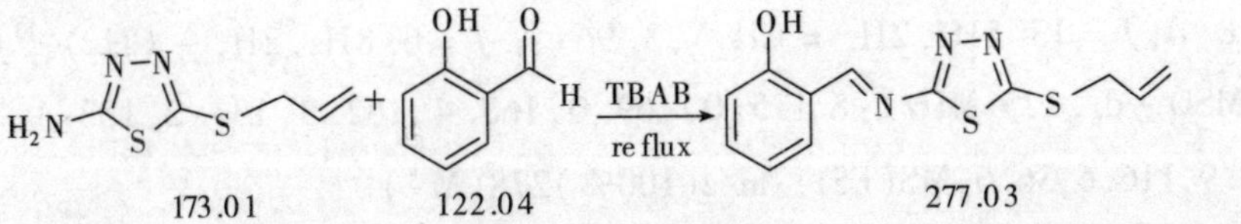

173.01　　122.04　　TBAB reflux　　277.03

将1.038 g(6 mmol)2-氨基-5-烯丙基醚基-1,3,4-噻二唑、0.794 g(6.5 mmol)3-羟基苯甲醛加入到25 mL反应瓶中。加入10 mL无水乙醇后搅拌下加入浓硫酸3滴,升温至90℃保温回流反应,通过TLC检测反应。7h后停止加热并冷却至室温,静置过夜后有固体析出,抽滤,滤饼用乙醇和热水($V_{乙醇}$:$V_{热水}$=1:1)的混合溶剂淋洗3次,用无水乙醇重结晶,干燥得黄色固体0.715 g,摩尔收率为42.89%,产物标记为T3-7,熔点:102.2~103.4℃。IR(cm^{-1},KBr):3275,1606,1565,1512,1382,1322,1272,1205,1101,1053,928,753;^{1}H NMR(DMSO-d_6,500 MHz):δ 11.82(s,1H,-OH),9.03(s,1H,CH),7.50-7.45(m,2H,Ph-H),7.04(d,J=8.3Hz,1H,Ph-H),6.99(d,J=6.4Hz,1H,Ph-H),6.03-5.97(m,1H,CH=),5.30(d,J=13.5Hz,2H,=CH_2),3.96(d,J=6.8Hz,2H,$-CH_2$);^{13}C NMR(DMSO-d_6,125 MHz):δ 171.4,168.1,164.1,161.8,135.8,134.1,131.9,120.0,119.8,118.2,117.8,36.7;MS(ESI):m/z(100%)278(M^+)。

2.2.3.8 4-{[(5-烯丙基醚基-1,3,4-噻二唑-2-基)亚氨基]甲基}苯酚的合成

合成路线：

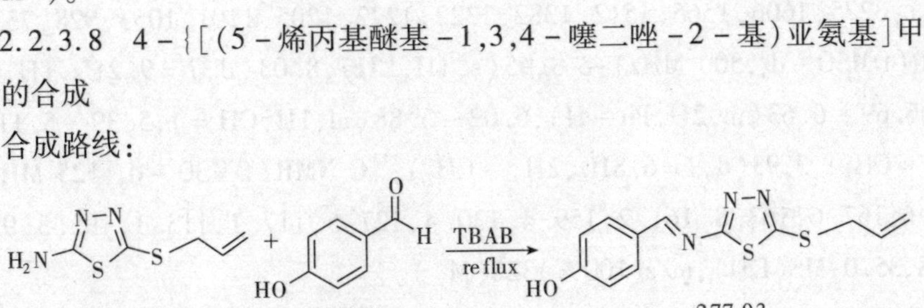

173.01　　122.04　　TBAB reflux　　277.03

将1.038 g(6 mmol)2-氨基-5-烯丙基醚基-1,3,4-噻二唑、0.794 g(6.5 mmol)3-羟基苯甲醛加入到25 mL反应瓶中。加入10 mL无水乙醇后搅拌下加入浓硫酸3滴,升温至90℃保温回流反应,通过TLC检测反应。8h后停

止加热并冷却至室温，静置过夜后有固体析出，抽滤，滤饼用乙醇和热水($V_{乙醇}$: $V_{热水}$ = 1 : 1)的混合溶剂淋洗3次，用无水乙醇重结晶，干燥得黄色固体0.075 g，摩尔收率为4.51%，产物标记为T3 - 8，熔点：96.9 ~ 100.5℃。IR(cm^{-1}，KBr)：3275，1606，1565，1512，1382，1322，1272，1205，1101，1053，928，753；^{1}H NMR(DMSO - d_6，500 MHz)：δ 10.63(s，1H，- OH)，8.76(s，1H，CH)，7.89(d，J = 8.5Hz，2H，Ph - H)，6.94(d，J = 8.5Hz，2H，Ph - H)，5.98 - 5.95(m，1H，CH =)，5.28(d，J = 13.5Hz，2H，= CH_2)，3.96(d，J = 6.8Hz，2H，- CH_2)；^{13}C NMR(DMSO - d_6，125 MHz)：δ 175.0，169.0，163.4，162.3，133.2，133.1，126.2，119.9，116.6，36.6；MS(ESI)：m/z(100%)278(M^+)。

2.2.3.9　N - (5 - 烯丙基醚基 - 1，3，4 - 噻二唑 - 2 - 基) - 1 - (2，4 - 二甲氧苯基)甲亚胺的合成

合成路线：

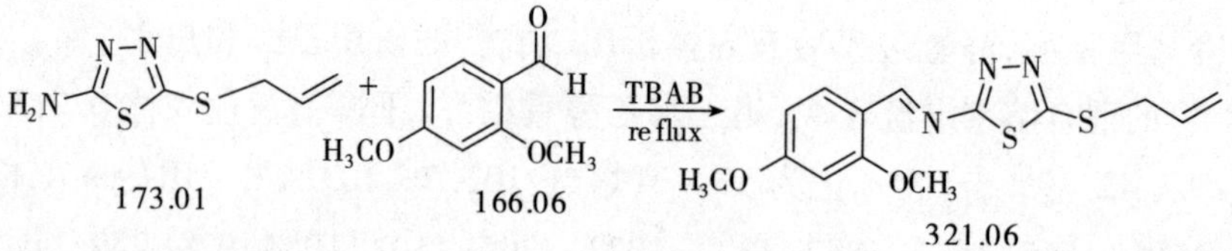

H_2N ... S + O H H_3CO OCH_3 —TBAB / reflux→ N ... H_3CO OCH_3

173.01　　166.06　　321.06

将0.865 g(5 mmol)2 - 氨基 - 5 - 烯丙基醚基 - 1，3，4 - 噻二唑、0.914 g(6.5 mmol)2，4 - 二甲氧基苯甲醛加入到25 mL反应瓶中。加入10 mL无水乙醇后搅拌下加入浓硫酸3滴，升温至90℃保温回流反应，通过TLC检测反应。9h后停止加热并冷却至室温，静置过夜后有固体析出，抽滤，滤饼用乙醇和热水($V_{乙醇}$: $V_{热水}$ = 1 : 1)的混合溶剂淋洗3次，用无水乙醇重结晶，干燥得黄色固体0.775 g，摩尔收率为51.24%，产物标记为T3 - 9，熔点：113.8 ~ 115.1℃。IR(cm^{-1}，KBr)：3275，1606，1565，1512，1382，1322，1272，1205，1101，1053，928，753；^{1}H NMR(DMSO - d_6，500 MHz)：δ 8.95(s，1H，CH)，8.03(d，J = 9.2Hz，1H，Ph - H)，6.69 - 6.63(m，2H，Ph - H)，6.02 - 5.88(m，1H，CH =)，5.39 - 5.11(m，2H，= CH_2)，3.93(d，J = 6.8Hz，2H，- CH_2)；^{13}C NMR(DMSO - d_6，125 MHz)：δ 172.6，167.6，163.8，161.2，159.4，130.4，127.5，117.1，113.1，105.3，95.8，53.8，35.0；MS(ESI)：m/z(100%)322(M^+)。

2.2.3.10　N - (5 - 烯丙基醚基 - 1，3，4 - 噻二唑 - 2 - 基) - 1 - (4 - 叔丁基)甲亚胺的合成

合成路线：

将0.865 g(5 mmol)2 - 氨基 - 5 - 烯丙基醚基 - 1，3，4 - 噻二唑、0.811 g

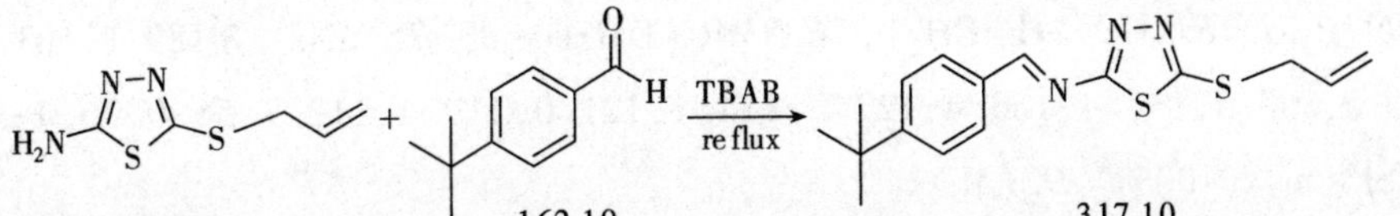

173.01 162.10 317.10

(5.5 mmol)4-叔丁基苯甲醛加入到25 mL反应瓶中。加入10 mL无水乙醇后搅拌下加入浓硫酸3滴,升温至90℃保温回流反应,通过TLC检测反应。11h后停止加热并冷却至室温,用旋转蒸发仪将体系溶剂蒸干,用$V_{石油醚}:V_{乙酸乙酯}$ = 17:1的洗脱剂进行层析柱,将收集到的黄色液体用旋转蒸发仪蒸干得黄色固体0.378 g,摩尔收率为26.64%,产物标记为T3-10,熔点:65.5~67.4℃。IR(cm^{-1}, KBr):3275,1606,1565,1512,1382,1322,1272,1205,1101,1053,928,753;^1H NMR(DMSO-d_6,500 MHz):δ 8.90(s,1H,CH),7.99(d,J=8.1Hz,2H,Ph-H),7.63(d,J=8.1Hz,2H,Ph-H),5.98-5.84(m,1H,CH=),5.40-5.10(m,2H,=CH_2),4.00(d,J=6.8Hz,2H,-CH_2),1.33(s,9H,3CH_3);^{13}C NMR(DMSO-d_6,125 MHz):δ 192.6,169.1,163.5,157.8,133.9,132.6,129.4,126.0,119.4,37.3,35.0,31.1;MS(ESI):m/z(100%)318(M^+)。

2.2.3.11 N-((5-烯丙基醚基)-1,3,4-噻二唑-2-基)-1-(2-甲氧苯基)甲亚胺的合成

合成路线:

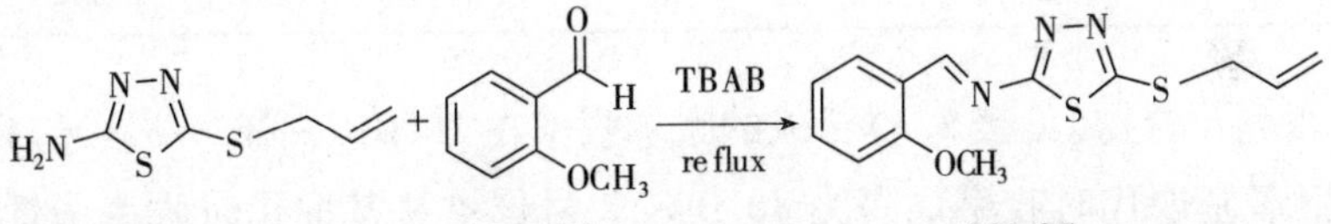

173.01 136.05 291.05

将0.865 g(5 mmol)2-氨基-5-烯丙基醚基-1,3,4-噻二唑、0.749 g(5.5 mmol)2-甲氧基苯甲醛加入到25 mL反应瓶中。加入10 mL无水乙醇后搅拌下加入浓硫酸3滴,升温至90℃保温回流反应,通过TLC检测反应。11h后停止加热并冷却至室温,用旋转蒸发仪将体系溶剂蒸干,用$V_{石油醚}:V_{乙酸乙酯}$=9:1的洗脱剂进行层析柱,将收集到的黄色液体用旋转蒸发仪蒸干得黄色固体0.335 g,摩尔收率为24.59%,产物标记为T3-10,熔点:70.2~72.1℃。IR(cm^{-1},KBr):3275,1606,1565,1512,1382,1322,1272,1205,1101,1053,928,753;^1H NMR(DMSO-d_6,300 MHz):δ 9.10(s, 1H, CH),8.08(d, J=7.8Hz, 1H, Ph-H),7.67-7.64(m, 1H, Ph-H),7.25(d, J=8.4Hz, 1H, Ph-H),7.11-7.06(m, 1H, Ph-H),5.99-5.84(m, 1H, HC=),5.39-5.13(m, 2H, =CH_2),

3.99(d, J=6.9Hz, 2H, CH_2);^{13}C NMR (DMSO - d_6, 75 MHz):δ 189.1,163.4,161.5,160.5,136.4,133.4,127.7,124.1,121.0,119.4,112.7,56.0,36.1;MS (ESI):m/z(100%)292(M^+)。

2.3　合成结果分析及讨论

合成了34个1,3,4-噻二唑衍生物,得到了这些化合物相关的合成工艺参数和理化性质。所有化合物用溴化钾为载体进行了红外表征,也用氘代氯仿或者氘代DMSO为溶剂进行核磁共振氢谱检测及碳谱检测。

表2.3　目标化合物收率

化合物	摩尔收率(%)	化合物	摩尔收率(%)
T1-1	30.2	T1-9	67.8
T1-2	51.5	T1-10	59.9
T1-3	38.3	T1-11	16.0
T1-4	57.5	T1-12	86.0
T1-5	53.0	T1-13	63.2
T1-6	51.0	T1-14	22.7
T1-7	66.1	T1-15	52.2
T1-8	44.8	T1-16	52.2

T1系列化合物(见表2.3)合成的摩尔收率在16.0%~86.0%。苯环2,4,6号位被甲氧基取代时收率最高为86.0%,2号位被氟基取代时收率最低为16.0%。由苯环2号位被硝基取代的T1-14收率为22.7%,由此推测,吸电子基团可能不利于反应的进行。当苯环被羟基取代时(T1-2,T1-4,T1-6,T1-13,T1-15),收率在51.0%~63.2%,而当羟基上的氢被甲基取代后(T1-3)其收率下降至38.3%,对反应的进行不利。

从红外光谱图上看,芳香醛及2-氨基-5-巯基-1,3,4-噻二唑在1600cm^{-1}及3400 cm^{-1}左右有羰基及氨基的特征峰,当形成希夫碱后均消失。从核磁氢谱上来看,T1系列化合物巯基在δ(13.12-14.58)ppm范围内的单峰为噻唑环上巯基上的氢,当苯环上被羟基取代或者苯环上被多取代基取代时(T1-2、T1-4、T1-5、T1-6、T1-12、T1-13、T1-15),由于氢键效应及位阻效应使氢的位移向低场移动,处于(14.29-14.58)ppm。其余均在(13.12-13.25)ppm,较苯环上无取代时(δ 13.02 ppm)向低场移动了(0.1-0.2)ppm。亚氨基上(HC

=N)氢为该系列化合物另一特征峰,在δ(8.67 - 9.77)ppm。当苯环上被硝基、甲氧基(T1 - 3、T1 - 14、T1 - 15)等取代时,其位移向低场移动,其余均在δ(8.67 - 8.88)ppm,较苯环上无取代时(δ 8.62 ppm)向低场移动了0.05 - 0.26。

2.4 本章小结

通过实验合成了34个噻二唑类化合物,其中13个新化合物,21个已知化合物,得到了这些化合物的合成工艺条件和工艺参数以及基本的物理化学性质。并且通过^{1}H NMR、^{13}C NMR、IR表征,确定了其化学结构是目标化合物。

第3章 酪氨酸酶抑制剂的筛选及分子对接的研究

3.1 酪氨酸酶抑制剂的筛选

3.1.1 实验材料与实验仪器

实验中使用的材料如表3.1所示。

表3.1 实验材料

名称	规格(纯度)	生产厂家
蘑菇酪氨酸酶	2687U/mg	Sigma化学公司
磷酸氢二钾	分析纯(95%)	天津市恒兴化学试剂制造有限公司
磷酸二氢钾	分析纯(95%)	天津市恒兴化学试剂制造有限公司
L - DOPA(左旋多巴)	分析纯(99%)	Johnson Matthey化学公司
二甲基亚砜(DMSO)	分析纯(95%)	Sigma化学公司
去离子水		自制

实验中所需实验仪器如表3.2所示。

表3.2 实验仪器

名称	型号	生产厂家
可见紫外分光光度计	UV - 4802 BCD - 256KFA	UNICO公司
冰箱	FE20型	海尔公司
pH计	BD - 198E	梅特勒 - 托利多仪器公司
低温冰柜	CP213	海尔公司
电子天平		奥豪斯仪器(上海)有限公司

3.1.2 试验方法

3.1.2.1 配制溶液方法

(1) 0.05 mol/L pH 6.8 磷酸盐缓冲溶液:

准确称取磷酸二氢钾和磷酸氢二钾 5.7055 g 和 3.4023 g,分别将其配置成 500 mL 0.05 mol/L 的溶液。将适量的0.05 mol/L 的 KH_2PO_4溶液加入到500 mL 0.05 mol/L K_2HPO_4中,搅拌均匀,用 pH 计将混合溶液的 pH 值调至 6.8 即可。

(2) 样品溶液:

以 DMSO 为溶剂配制 1.0 mmol/L 的样品溶液 1 mL 备用。

(3) 蘑菇酪氨酸酶溶液(现用现配):

用0.05 mol/L pH 6.8 的磷酸盐缓冲溶液将 1mg 蘑菇酪氨酸酶溶解并定容至 1mL。

(4) L-DOPA 溶液(现用现配):

用0.05 mol/L pH 6.8 的磷酸盐缓冲液将 15 mg L-DOPA,溶解,定容至 10 mL,配制成 7.6 mmol/L 的 L-DOPA 溶液备用。

3.1.2.2 酪氨酸酶抑制剂半抑制率(IC_{50})测定方法

酪氨酸酶是形成黑色素的限速酶,因此可通过多巴或者多巴醌的生成速度来表示酪氨酸酶的活性。以 L-多巴(L-DOPA)为底物,在 475 nm 波长下测定一定时间内吸光度变化值,此值相当于多巴醌的生成速度。IC_{50}是指酪氨酸酶活性被抑制一半时酪氨酸酶抑制剂的浓度,IC_{50}值越小则可说明该抑制剂的酪氨酸酶抑制效果越好。

参照本课题组已建立起的筛选方法的方法,在 1.5 mL 的离心管中配制 1000 μL 的测试体系,包括 pH 6.8 浓度为0.05 mol/L 的磷酸盐缓冲液 950 μL,待测浓度为 1.0 mmol/L 的样品溶液 10 μL(10 μL 样品作对照,以去除样品的颜色对实验的影响),10 μL 的酪氨酸酶溶液,在 25℃ 的条件下孵化 10 min 后,加入 30 μL 浓度为7.6 mmol/L 的 L-DOPA 溶液,在475 nm 波长下进行动力学测试,记录一分钟的 OD 值的改变量。再通过改变样品溶液的浓度(通过稀释的方式改变),分别测试酪氨酸酶的催化活性。以空白对照时的酶催化活性 OD 值为 100%,而不同浓度的样品对酶的抑制活性则用抑制百分数来表达,然后再通过 Origin7.0 做图得到 IC_{50}值。抑制率的计算公式如下:

抑制率 = $[(B-S)/B] \times 100\%$ ——公式 3.1

式中 B 为空白吸光度值,S 为样品吸光度值。

3.1.2.3 酪氨酸酶抑制剂动力学的测定方法

同样按照本文 3.1.2.2 的试验方法,选择 2 个不同浓度(在样品 IC_{50}附近选取)的样品组以及空白对照组,在不改变底物浓度的基础上,通过改变酪氨酸酶溶液的浓度$[E]$(μmol/mL)来测试其酪氨酸酶催化活性 V(OD/min),并通过

Origin7.0 以［E］对 V 的作用作图，通过三条线交点的位置来确定抑制剂的类型。

当抑制剂表现为可逆抑制时，再按照本文 3.1.2.2 的试验方法，选择 2 个不同浓度（在样品 IC_{50} 附近选取）的样品组以及空白对照组，通过改变底物 L－DOPA 的浓度［S］及酪氨酸酶的浓度［E］（μmol/mL）来测试其酪氨酸酶的催化活性，并通过 Origin7.0 以 1/［S］对酶催化活性 OD 样的倒数 1/Vss 作用作图，通过三条线有无交点或者交点的位置来确定抑制剂是竞争性抑制剂、反竞争性抑制剂还是非竞争性抑制剂。

3.1.3　实验结果与分析

3.1.3.1　酪氨酸酶抑制剂 IC_{50} 测试结果及分析

根据 3.1.2.2 的方法筛选到的所有样品的 IC_{50} 列在表 3.3 中。

表 3.3　IC_{50} 测定结果

化合物	IC_{50}（μmol/L）	化合物	IC_{50}（μmol/L）
T1－1	2.460	T2－2	276.84
T1－2	1.971	T2－3	103.16
T1－3	0.733	T2－4	247.35
T1－4	0.255	T2－5	125.47
T1－5	1.718	T2－6	511.73
T1－6	0.785	T2－7	42.50
T1－7	1.390	T3－1	228.24
T1－8	5.353	T3－2	1.03
T1－9	3.621	T3－3	87.22
T1－10	4.753	T3－4	94.41
T1－11	0.765	T3－5	128.75
T1－12	0.907	T3－6	135.08
T1－13	0.036	T3－7	117.90
T1－14	0.671	T3－8	108.11
T1－15	0.473	T3－9	337.73
T1－16	49.246	T3－10	188.12
T2－1	112.52	T3－11	91.33
曲酸	23.6		

由表3.4可知，T1系列样品除T1－16 IC_{50}为49.246 μmol/L大于阳性对照曲酸外，其余均小于曲酸，说明其酪氨酸酶抑制效果均优于曲酸，样品T1－13和T1－4IC_{50}最低分别为0.036 μmol/L和0.255 μmol/L；T2系列化合物IC_{50}均大于曲酸，说明其酪氨酸酶抑制效果均比曲酸差；T3系列化合物除T3－2 IC_{50}为1.03 μmol/L小于曲酸外，其余均大于曲酸，说明其酪氨酸酶抑制效果均比曲酸差。由T1、T2、T3系列化合物IC_{50}可知噻唑环上的游离巯基为酪氨酸酶抑制活性的必需基团，当巯基上的氢被取代时，酪氨酸酶抑制活性下降甚至几乎没有活性；通过比较化合物T1－4、T1－13、T3－2与其他化合物的IC_{50}说明苯环4号位上的羟基可能为活性必需基团。比较化合物与T1系列其他化合物的活性可以发现苯环上有取代基存在时其酪氨酸酶抑制活性升高。

3.1.3.3　抑制动力学探讨

选择T1－4和T1－13作为研究对象，进行抑制剂动力学的探讨，并通过Origin7.0以［E］(μmol/mL)对V(OD/min)的作用作图，根据三条线交点的位置或有无交点来确定抑制剂的作用方式为何种类型。由图3.1和图3.2可知T1－13和T1－4为反竞争性抑制剂。

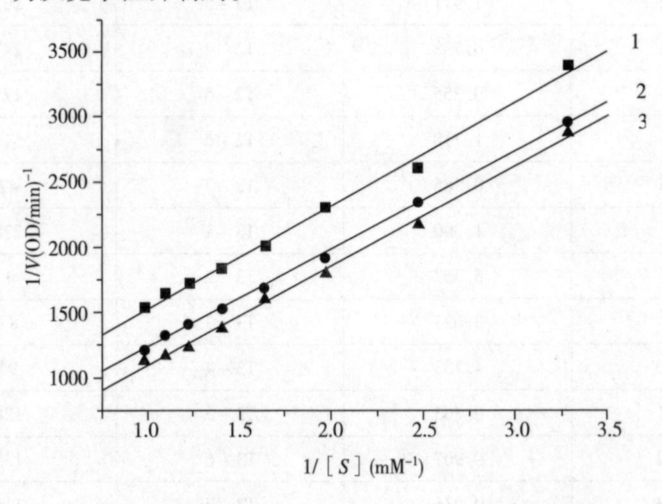

图3.1　化合物T1－13的Lineweaver－Burk曲线

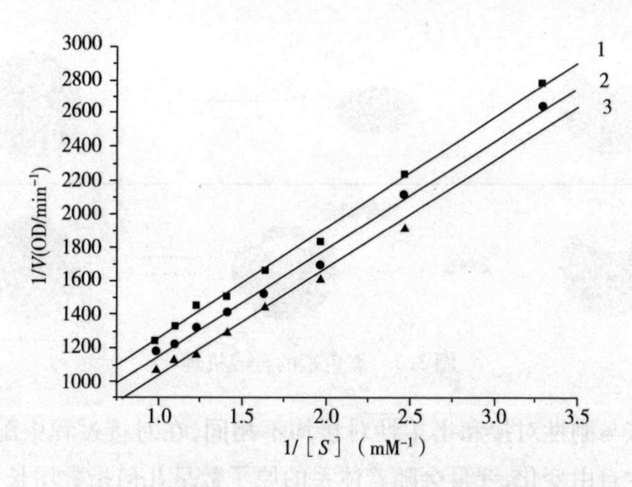

图 3.2　化合物 T1 - 4 的 Lineweaver - Burk 曲线

3.2　分子对接的研究

分子对接方法最初是科学家在分子水平上研究生物体系中的化学问题时提出的，随着计算机辅助药物设计的深入，分子对接的应用越来越广，通过分子对接软件来研究这些抑制剂与酪氨酸酶的相互作用，并且酪氨酸酶抑制剂与酪氨酸酶的活性空间部位和电性互补匹配特征等也可通过此方法得到确定，为了解这些化合物的酶活性抑制机理提供了一定的参考。

Fisher. E 提出的"锁 - 钥模型"原理为分子对接的研发提供了最初的理论依据，配体进入受体的方式类似于锁和钥匙，此时视受体和配体均为刚性结构，空间构象不发生变化。在配体与受体的几何与能量匹配识别的过程中，要求配体与受体在空间结构、氢键作用、静电作疏水作用等方面相互匹配。由于"锁 - 钥模型"原理的局限性，D. E. Koshland 在 1958 年提出了"诱导契合学说"，指出酶分子活性中心的结构原来并非和底物的结构互相吻合，但酶的活性中心并不是刚性的，而是柔软的，酶的活性中心的构象会在底物与酶相遇发生相应的变化。该原理说明在分子对接的过程中，配体和受体均被视为柔性结构。

刚性对接、半柔性对接和柔性对接是分子对接的三种方法。刚性对接时，参与对接的分子分子的空间位置和形态发生变化，但对接的分子构象并不发生变化，适用于处理结构比较大的分子用于模型建立后的初步分析。有文献报道采用刚性对接法成功实现了麦芽糖和蛋白质的对接。

半柔性对接过程中，只是受体的构象固定不变，而配体的构象如某些非关键部位的键角和键长等是可以在一定范围内发生变化的，此方法被广泛应用于小分子和大分子（酶或核酸）之间的对接。

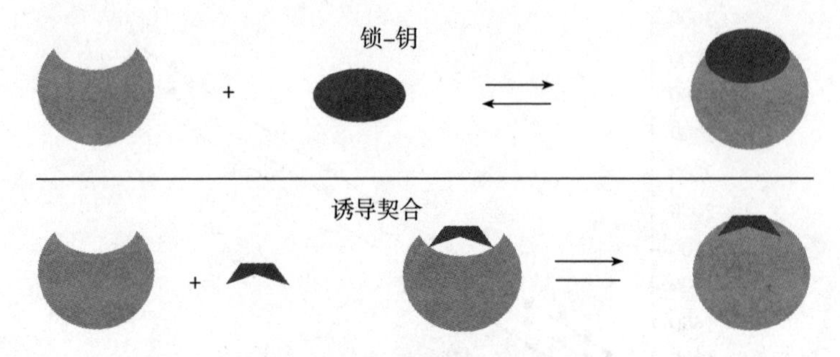

图 3.3　黑色素的合成机理

柔性对接与刚性对接和半柔性对接均不相同,在对接过程中配体与受体的构象均可发生自由变化,变量会随着体系的原子数呈几何指数增长,因此柔性对接方法不仅对计算机软硬件及计系统要求高,还要求分子构象的精度高,虽然计算量相当大,耗时也较久,但得到的对接精度相当高,适合用于精确考察分子间的识别情况。Mangoni 等用柔性的配体对柔性的受体进行了对接。

目前配体与蛋白质分子之间的对接主要采用由斯克利普斯研究所 Olson 课题组研发的 AutoDock 软件。AutoDock 软件通过以下几点来实现分子对接:①高标准的打分函数;②基于网格的刚性分子对接以及柔性对接受体活性中心;③采用拉马克遗传算法来寻找亲和性结合最好时配体的位置;④采用取样与次好策略相结合的方法来解决潜在的全局最优化问题等。

3.2.1　材料与方法

3.2.1.1　材料

实验中使用的材料如表 3.4 所示。

表 3.4　实验材料

名称	规格(纯度)	生产厂家
分子对接软件	Auto dock4.2	Auto dock
Auto dock tools	1.5.4 版	Auto dock tools
蘑菇酪氨酸酶晶体	2Y9X	Protein data bank

3.2.1.2　方法

(1) 2Y9X 的处理:

编号 2y9x 蘑菇酪氨酸酶晶体结构从 PDB 官方网站下载(下载地址:http://www.rcsb.org/pdb/home/home.do),含有 8 个亚基 A ~ H,选取 A 亚基活性中心

(tropolone(OTR))所在区域为对接位点。删除其余亚基及A亚基活性中心配体OTR,得到可用于分子对接的受体三维结构,将其保存为pdb格式文件。再用auto dock软件对其处理,删除水分子后加氢,并计算其Gasteiger电荷,再将其非极性氢原子合并后保存为pdbqt格式文件备用。

(2) 配体分子的处理:

用autodock将T1-4和T1-13打开,将其非极性氢原子合并,并加上其Gasteiger电荷,在能量最小化上再在Ligand子程序包中自动检测可旋转键的个数,将其保存为pdbqt格式文件备用。

(3) 对接参数的设置:

在Grid Box中,格点间距设置为0.375 Å,大小设置为60×60×60 Å,默认格子的中心为受体活性的中心。在搜索配体构象过程采用半经验打分函数和拉马克遗传算法,Number of GA Runs修改为100,Max Number of evals修改为2700000,Rate of Crossover修改为0.8,其他参数保持不变。另外由于酪氨酸酶为含铜的氧化还原酶,其电荷数的大小将对活性对接有一定的影响,然而铜离子并不是Auto Dock软件的默认原子类型,在对接前需参考auto dock官网数据自行添加铜离子的相关参数。

3.2.2 实验结果与分析

3.2.2.1 化合物T1-13与2Y9X对接模型实验结果与分析

化合物T1-13与2Y9X的自由结合能为-5.94 Kcal/mol,说明该化合物与酪氨酸酶有稳定的结合构象。图3.4为化合物T1-13与2Y9X氨基酸残基对接图,由图可知苯环对位上的羟基可与酪氨酸酶铜离子(401)以Metal-Acceptor的方式结合,其键长为2.28 Å,由此可知对位上的羟基可能为酪氨酸酶抑制活性的活性基团;另外苯环可与ALA 286和VAL 283以π-Alkyl方式结合,与HIS 263以Pi-Pi Stacked形式结合,键长分别为4.36 Å,4.04 Å和3.63 Å;1,3,4-噻唑环可与GLY 281以Pi-Lone Pair形式结合,键长为2.91 Å;苯环3号位上的甲氧基可与ASN 260和HIS 259以Carbon Hydrogen Bond形式结合,键长分别为3.13 Å和1.82 Å。

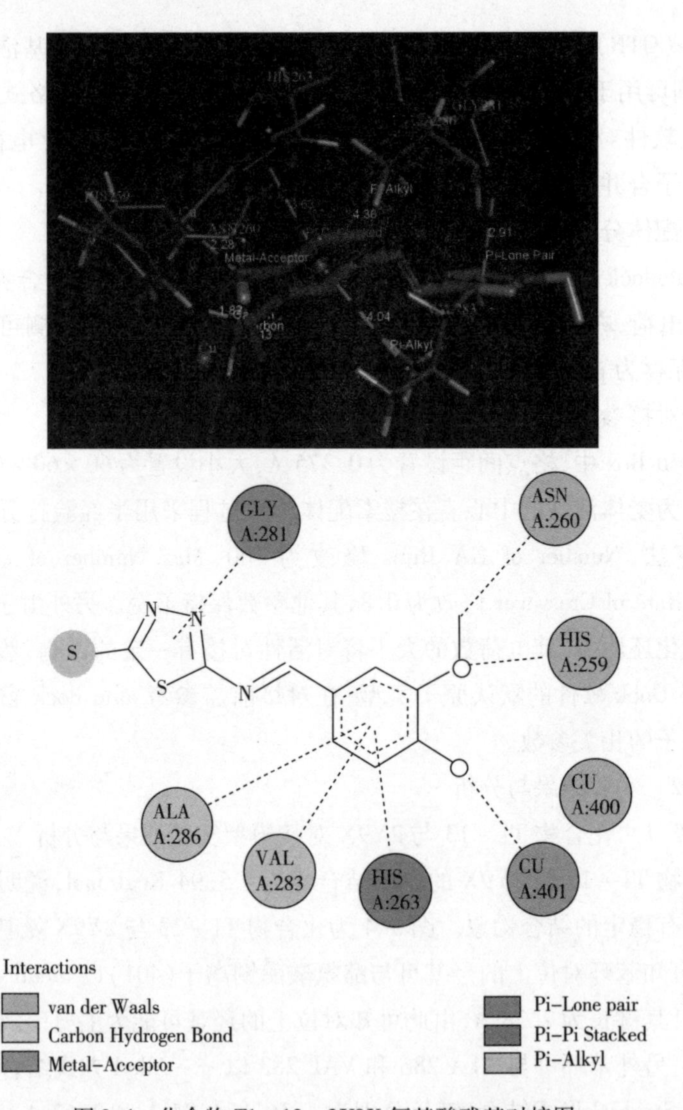

图 3.4 化合物 T1 -13 -2Y9X 氨基酸残基对接图

3.2.2.1 化合物 T1 -4 与 2Y9X 对接模型实验结果与分析

化合物 T1 -4 与 2Y9X 的自由结合能为 -5.65 Kcal/mol，说明该化合物与酪氨酸酶有稳定的结合构象。图 3.5 为化合物 T1 -4 与 2Y9X 氨基酸残基对接图，由图可知苯环对位上的羟基可与酪氨酸酶铜离子(401)以 Metal - Acceptor 的方式结合，其键长为 2.32 Å，由此可知对位上的羟基可能为酪氨酸酶抑制活性的活性基团；另外苯环可与 ALA 286 以 π - Alkyl 方式结合，与 VAL 283 以 Pi - Sigma 形式结合，与 HIS 263 以 Pi - Pi Stacked 形式结合，键长分别为 4.65 Å，2.77 Å 和

3.69 Å;1,3,4-噻唑环可与 GLY 281 以 Pi-Lone Pair 形式结合,键长为 2.99 Å;苯环 2 号位上的羟基氢可与 MET 280 以氧氢键的形式结合,其键长为 2.10 Å;亚胺上的 N 可以和 VAL 283 形成氢键,其键长为 2.42 Å。

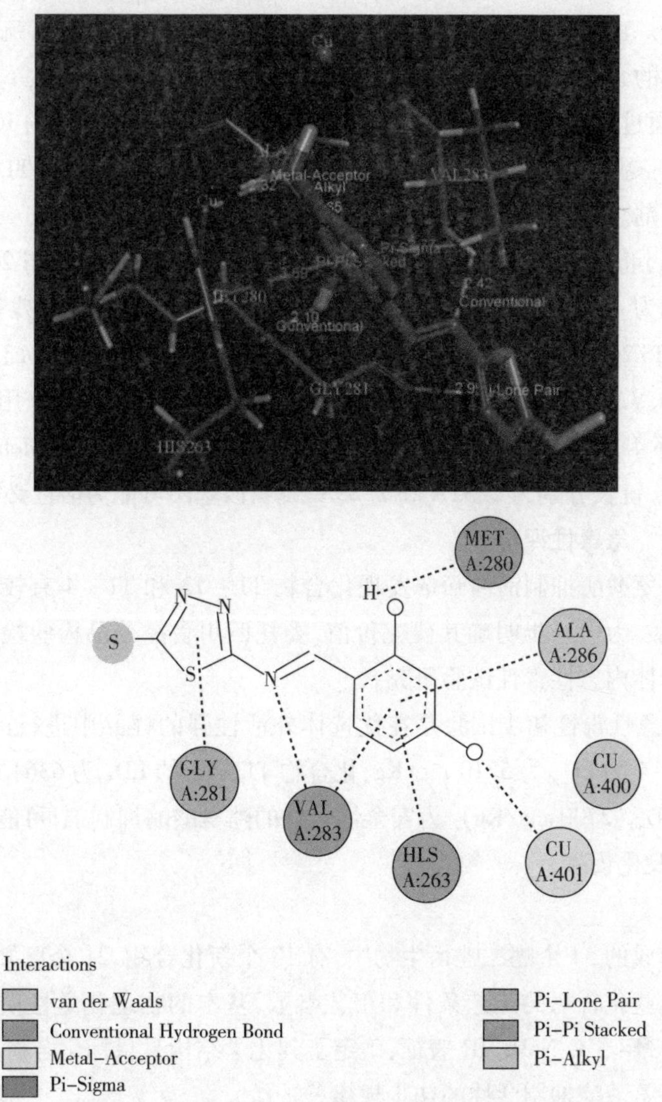

图 3.5 化合物 T1-4-2Y9X 氨基酸残基对接图

3.3 本章小结

(1)34 个合成化合物中,有 16 个化合物的活性比曲酸要高。T1 系列样品除 T1-16 IC_{50} 为 49.246 μmol/L 大于阳性对照曲酸外,其余均小于曲酸,说明其酪氨酸酶抑制效果均优于曲酸。样品 T1－13 和 T1－4 IC_{50}最低，分别为 0.036 μmol/L 和 0.255 μmol/L；T2 系列化合物 IC_{50}均大于曲酸，说明其酪氨酸酶抑制效果均比曲酸差；T3 系列化合物除 T3－2 IC_{50}为 1.03 μmol/L 小于曲酸外，其余均大于曲酸。由 T1、T2、T3 系列化合物 IC_{50}可知噻唑环上的游离巯基为酪氨酸酶抑制活性的必需基团，当巯基上的氢被取代，酪氨酸酶抑制活性下降甚至几乎没有活性；通过比较化合物 T1－4、T1－13、T3－2 与其他化合物的 IC_{50}说明苯环 4 号位上的羟基可能为活性必需基团。动力学研究表明 T1－4 和 T1－13 均为反竞争性抑制剂。

（2）通过将化合物和酪氨酸酶分子对接发现 T1－4、T1－13 与 2Y9X 的自由结合能分别为－24.80kJ/mol 和－23.65kJ/mol。T1－4 与氨基酸残基 ALA 286、VAL 283、HIS 263、GLY 281、MET 280 和 VAL 283 有作用力。T1－13 与氨基酸残基 ALA 286、VAL 283、HIS 263、GLY 281、ASN 260 和 HIS 259 有作用力。T1－4、T1－13 苯环 4 号位上的羟基均与酪氨酸酶中的铜离子（401）以 Metal－Acceptor 的方式结合，键长分别为 2.32Å 和 2.28Å，说明该基团可能为活性必需基团。

第 4 章　急毒性评价

通过酪氨酸酶抑制活性筛选发现化合物 T1－13 和 T1－4 有较好的抑制活性，高于曲酸，为进一步明确其研究价值，委托四川食品药品检验检测院安评中心对其进行体内急性毒性试验研究。

LD_{50}的急性毒性初步试验是在优选体外活性强的样品中进行的，结果表明化合物 T1－4 的 LD_{50}为 5710 mg/Kg，化合物 T1－13 的 LD_{50}为 6364 mg/Kg，均大于曲酸的 LD_{50}（2700 mg/Kg），为安全性较好的酪氨酸酶抑制剂，可能能用于食品工业、临床及化妆品。

结论

本文合成的 34 个噻二唑衍生物中，有 13 个新化合物，21 个已知化合物，得到了合成这些化合物的工艺条件和工艺参数，基本的物理化学性质。这些化合物通过^{1}H NMR、^{13}C NMR、IR 表征，确定了其化学结构是目标化合物。

1. 对酪氨酸酶抑制活性的初步规律。

（1）含游离巯基化合物的酪氨酸酶抑制活性高于巯基被醚化的化合物，说明游离巯基可能为该类化合物的活性中心。

（2）苯环四号位上被羟基取代的化合物（T1－4、T1－13、T3－2）的酪氨酸酶抑制活性比无取代或其他取代基取代时要高，同时分子对接研究结果也表明苯环四号位上的羟基能与酪氨酸酶活性中心铜离子发生作用，说明苯环四号位上的羟基可能为酪氨酸酶抑制活性所必须基团。

(3)苯环上有取代基存在时比无取代基时的酪氨酸酶抑制活性要好。

2. 抑制动力学研究结果表明化合物 T1 - 13 和 T1 - 4 为反竞争性抑制剂。

3. 急性毒性初步试验结果表明 T1 - 13 和 T1 - 4 的 LD_{50} 为 6364 mg/Kg 和 5710 mg/Kg,大于曲酸 LD_{50} 2700 mg/Kg。因此,这类化合物毒性较低。鉴于其毒理实验仍在继续,将进一步确定该化合物是否是无毒的酪氨酸酶抑制剂,该化合物是否适合大量用于鲜切果蔬的加工、临床及化妆品等相关领域有待进一步的求证。

第5章 酪氨酸酶抑制剂在食品领域的应用

5.1 三唑类化合物对果蔬的作用(陈玲娟硕士论文部分内容)

新鲜果品蔬菜中含有大量丰富的维生素和矿物质等营养物质,是保障人体健康需求的物质基础。采收后的果蔬仍为活体,仍进行着复杂的生命活动。由于采后果蔬处于与母体隔离状态,缺乏水分和营养,不仅会促进其成熟和衰老,而且其外观色泽、营养特性和食用价值等也将发生显著的变化。特别是切开后的果蔬,切口破坏了果体细胞中细胞膜、线粒体结构,释放出酪氨酸酶等多酚类氧化酶,加速果蔬褐变、腐烂,影响色泽和口感。酶促褐变是在氧化酶催化下的多酚类物质发生氧化和抗坏血酸氧化下的褐变,是鲜切果蔬主要的问题之一。对于马铃薯、苹果等含有大量多酚及多酚氧化酶的果蔬,通常在切分后马上采用物理、化学等方法进行灭酶处理,并且在加工生产线上通常保持原料在水中隔绝氧气进行加工,以最大限度降低氧化褐变,持果蔬原有品质目的,否则加工过程中发生的果肉黄色、褐色或黑色变化使制品失去天然色泽,不仅影响了美观、破坏产品风味和营养,而且成为不能食用的标志。

现今我国的果蔬产量居世界第一,据相关统计,现阶段新鲜果蔬的腐烂损耗率较高,因此对果蔬进行护色保鲜及加工是必不可少的。为了解决这一问题,人们对果蔬产品的褐变原理和防止方法进行了许多深入的研究。其中化学方法在食品工业上一直是最为广泛地用来抑制酶促褐变的方法。在试验研究中,已有数百种化合物被证实可抑制酶促褐变。如作用于酪氨酸酶辅基的抑制剂有抗坏血酸、氰化物、氟化物、CO、二乙基二硫氨基甲酸钠(DIECA)、巯基苯噻唑、二巯基丙醇、叠氮化物、甲基黄原酸钾等,还有作用于酶蛋白的抑制剂,如 SO_2 或亚硫酸盐。但许多化合物存在取之不易,剂量太大,或是本身毒性等问题,如硫氢化合物、亚硫酸盐等。4-己基间苯二酚在对虾褐变的抑制上,与亚硫酸盐相比,不仅稳定性好,而且对人体无害。但是此类的酪氨酸酶抑制剂极少,且价格居高位。所以当前寻找食品工业上的安全、无毒、无硫、高效的抗褐变剂是一个活跃的研究领域,也是一个函待解决的问题。

三氮唑化合物名称为3-苯基-4-氨基-5-硫酮-1,2,4-三(4-amino

-3-phenyl-1H-1,2,4-triazole-5(4H)-thione,APTT)。

1.1 仪器与材料

材料

雪莲果、红富士苹果、牛蒡、葛根、香蕉、龙牙百合

湖南省邵阳市市售新鲜、颜色浅黄、无机械损伤、无腐烂、无虫害的

磷酸二氢钾　　天津大茂化学试剂厂

磷酸氢二钾　　天津博迪化工有限公司

二甲基亚砜(DMSO)　　北京鼎国昌盛生物技术有限责任公司

3-苯基-4-氨基-5-硫酮-1,2,4-三唑　　邵阳学院生化系提供

仪器

紫外分光光度计　UV-4802 型　UNICO 公司

台式高速冷冻离心机 3-18R　TOMOS 公司

恒温水浴器　DK-98-ⅡA　天津市泰斯特仪器有限公司

移液枪　1000μL、20μL　芬兰 Dragon 公司

200μL、100μL　德国 Eppendorf 公司

电子天平　AUY2220　Shimadzu 公司

pH 计　FE20 型　梅特勒-托利多仪器(上海)有限公司

冰箱　BCD-256KF A　海尔公司

1.2 实验方法

溶液配制

(1) 0.05mol/L pH 6.8 磷酸盐缓冲溶液:

向 500mL 0.05 mol/L K_2HPO_4 中加入适量 0.05 mol/L 的 KH_2PO_4 溶液混匀,用 pH 计调节溶液 pH 值到 6.8。

(2)样品溶液:

配制 1mL 浓度为 1.0mmol/L 的样品溶液,以 DMSO 为溶剂。

(3)蘑菇酪氨酸酶溶液(现用现配):

称取 1mg 的蘑菇酪氨酸酶,用 0.05mol/L 磷酸盐缓冲溶液溶解,定容至 1mL。

(4) L-DOPA 溶液(现用现配):

称取 1.5mg 的 L-DOPA,用 pH 6.8 的磷酸盐缓冲液溶解,定容至 1mL,配制成 7.6mmol/L 的多巴溶液。

测量波长的选择

取新鲜无机械损伤、无腐烂的红富士苹果、香蕉、雪莲果、牛蒡、葛根、龙牙百合，去外皮，取少量果肉，迅速置于加有磷酸盐缓冲溶液的研钵中研磨成糜，然后用纱布过滤，放入离心机中在 25℃ 时离心 13000 r/min，离心 1min，取上清液制成酪氨酸酶粗提液。取酪氨酸酶粗提液置于紫外分光光度计中进行全波长 190 ~ 1100nm 扫描，以磷酸盐缓冲溶液为空白对照，根据 OD 的变化值，每间隔 5min 测定 1 次，重复 3 次，作图选择合适的测量波长，确定波长 *W*。

1.3　单因素实验

APTT 浓度的选择

以 10mmol/L 的 APTT 母液，配制不同浓度 APTT 的磷酸盐缓冲溶液溶液，如表 1。

表 1　APTT 浓度配制表(μmol/L)

样品	1	2	3	4	5
红富士苹果	24	28	32	36	40
香蕉	34	38	42	46	50
雪莲果	16	20	24	28	32
牛蒡	300	350	400	450	500
葛根	500	550	600	650	700
龙牙百合	10	20	30	40	50

参照前述波长选择方法配制的样品粗提液在于最适波长 *W* 下动力学测定，记录 $OD_{标}$/min。将 APTT 的磷酸盐缓冲溶液溶液代替磷酸盐缓冲溶液制成样品粗提液，将样品粗提液至于最适波长 *W* 下进行动力学测定，记录 $OD_{样}$/min。根据公式 1 计算出抑制率(%)。

抑制率 = $[1-(OD_{样}/\text{min})/(OD_{标}/\text{min})]\times 100\%$ ……………公式 1

根据抑制率判断最佳测定浓度 *C*(μmol/L)。

APTT 作用时间的选择

参照前述波长选择方法将适量红富士苹果、香蕉、雪莲果、牛蒡、葛根、龙牙百合置于 APTT 最佳测定浓度 *C* 中制成粗提液，然后置于紫外分光光度计中进行动力学扫描，根据 $OD_{样}$变化值结合公式 1 计算出抑制率。

APTT 作用温度的选择

参照前述波长选择方法将适量红富士苹果、香蕉、雪莲果、牛蒡、葛根、龙牙百合置于不同温度保温的 APTT 最佳测定浓度 C 溶液中制成粗提液，然后置于紫外分光光度计中进行动力学扫描，根据 $OD_{样}$/min 结合公式 1 计算出抑制率。

正交实验

根据单因素试验确定的 APTT 浓度、作用时间、作用温度作为考察的3因素，每因素3水平，以 $OD_{样}$/min 为指标，用 $L_9(3^4)$ 正交表安排试验，分析得到结果。表2到表7即 APTT 对红富士苹果、香蕉、雪莲果、牛蒡、葛根、龙牙百合酪氨酸酶抑制最佳浓度、时间和温度测定的因素水平表。然后根据公式1计算出抑制率(%)。

表2　APTT 抑制苹果效果 $L_9(3^4)$ 正交试验因素水平表

水平/因素	(*A*)抑制剂浓度/(μmol/L)	(*B*)温度/℃	(*C*)时间/s
1	28	17	1715
2	32	20	1895
3	36	23	2075

表3　APTT 抑制香蕉正交实验因素水平

水平/因素	(*A*)抑制剂浓度/(μmol/L)	(*B*)温度/℃	(*C*)时间/s
1	46	15	360
2	48	20	420
3	50	25	480

表4　APTT 抑制雪莲果正交实验因素水平

水平/因素	(*A*)抑制剂浓度/(μmol/L)	(*B*)温度/℃	(*C*)时间/s
1	24	21	394
2	28	26	454
3	32	31	514

表5　APTT 抑制牛蒡正交实验因素水平

水平/因素	(*A*)抑制剂浓度/(μmol/L)	(*B*)温度/℃	(*C*)时间/s
1	400	20	420
2	450	25	480
3	500	30	540

表6 APTT抑制葛根正交实验因素水平

水平/因素	(*A*)抑制剂浓度/(μmol/L)	(*B*)温度/℃	(*C*)时间/s
1	500	20	407
2	550	25	467
3	600	30	527

表7 APTT抑制龙牙百合正交实验因素水平

水平/因素	(*A*)抑制剂浓度/(μmol/L)	(*B*)温度/℃	(*C*)时间/s
1	40	15	600
2	50	20	700
3	60	25	800

根据因素水平表设计的$L_9(3^4)$如表8。

表8 $L_9(3^4)$正交试验表

序号	APTT浓度/(μmol/L)	作用时间/s	作用温度/℃	空列
1	1	1	1	1
2	1	2	2	2
3	1	3	3	3
4	2	1	2	3
5	2	2	3	1
6	2	3	1	2
7	3	1	3	2
8	3	2	1	3
9	3	3	2	1

1.4 结果与分析

测定波长结果

根据*OD*的变化值,每间隔5min测定一次,重复3次,作图选择合适的测量波长,确定波长*W*。如图1为牛蒡粗提液的波长扫描图,根据*OD*变化值,确定牛蒡粗提液的最佳扫描波长是755nm。根据已有的测试方法,测定出各样品的最佳测定波长红富士苹果475nm、香蕉518.5nm、雪莲果759nm、牛蒡755nm、葛根592nm、龙牙百合442nm。

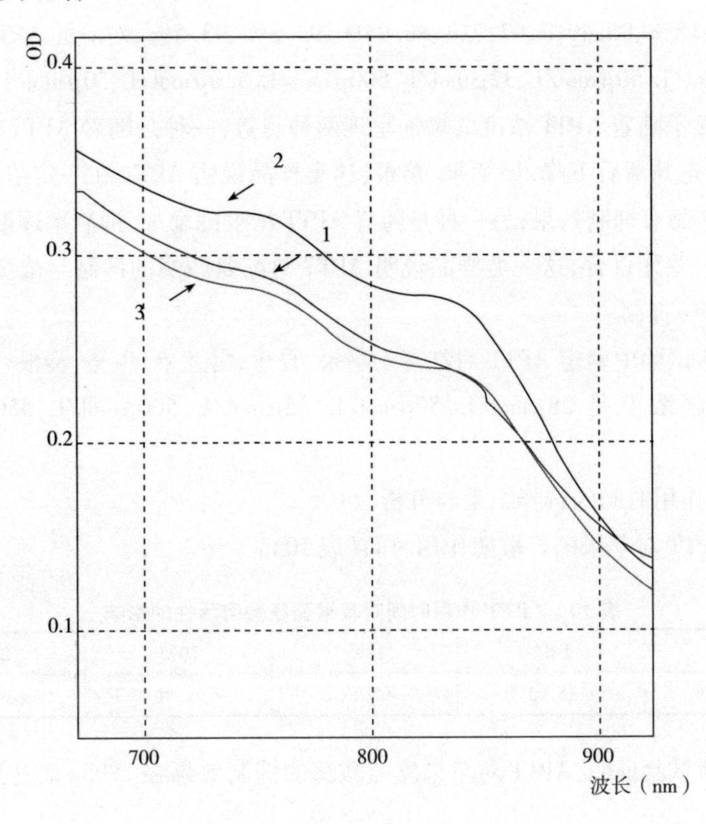

图1 牛蒡粗提液波长扫描图

单因子实验结果

APTT浓度选择结果与分析

表9 不同APTT浓度对不同样品的抑制率(%)

序号	1	2	3	4	5
红富士苹果	47.42	88.18	78.90	71.98	61.07
香蕉	22.01	28.41	71.44	75.11	86.49
雪莲果	35.3	72.3	79.0	88.3	97.7
牛蒡	31.61	50.21	58.89	75.92	88.94
葛根	84.1	94.3	91.2	65.4	67.9
龙牙百合	5.6	67.6	84.6	87.1	93.5

根据试验结果表9，随着APTT的浓度变化，样品酪氨酸酶的活性受到不同层度的抑制，其中红富士苹果、香蕉、雪莲果、牛蒡、葛根、龙牙百合的最高抑制率分别为88.18%、86.49%、97.7%、88.94%、94.3%、93.5%，对应的APTT浓度分别为28μmol/L、50μmol/L、32μmol/L、500μmol/L、550μmol/L、50μmol/L。

表9显示随着APTT浓度的增加呈现两种趋势：一种是随着APTT浓度的增加，抑制率先升高后下降，如苹果、葛根，这类样品说明APTT在一定浓度下对样品中酪氨酸酶有抑制效果；另一种是随着APTT浓度的增加，抑制率逐渐增高，如香蕉、牛蒡、龙牙百合，这一类样品说明APTT对酪氨酸酶的抑制与浓度有关，但是不能100%抑制。

单因素试验中确定APTT对红富士苹果、香蕉、雪莲果、牛蒡、葛根、龙牙百合的最佳抑制浓度是28μmol/L、50μmol/L、32μmol/L、500μmol/L、550μmol/L、50μmol/L。

APTT作用时间的选择结果与分析

（1）APTT对苹果酪氨酸酶作用时间（表10）：

表10 APTT作用时间对苹果酪氨酸酶活性的影响

时间/s	1715	1895	2075	2250
抑制率/%	75.42	92.18	91.90	89.98

单因素试验证明，APTT对苹果酪氨酸酶的抑制效果在1895s时达到最高值92.18%。

（2）APTT对香蕉酪氨酸酶作用时间（表11）：

表11 APTT作用时间对香蕉酪氨酸酶活性的影响

时间/s	60	120	180	240	300	360
抑制率/%	41.89	91.41	89.63	90.38	78.03	89.15

单因素试验证明，APTT对香蕉酪氨酸酶的抑制效果在120s时达到最高值91.41%。

（3）APTT对雪莲果酪氨酸酶作用时间（表12）：

表12 APTT作用时间对雪莲果酪氨酸酶活性的影响

时间/s	450	455	460	465
抑制率/%	46.08	60.42	44.61	27.45

单因素试验证明,APTT 对雪莲果酪氨酸酶的抑制效果在 455s 时达到最高值 60.42%。

(4)APTT 对牛蒡酪氨酸酶作用时间(表 13):

表 13　APTT 作用时间对牛蒡酪氨酸酶活性的影响

时间/s	180	240	300	360	420	480	540
抑制率/%	56.89	68.23	69.38	72.59	79.57	81.32	78.53

单因素试验证明,APTT 对牛蒡酪氨酸酶的抑制效果在 480s 时达到最高值 81.32%。

(5)APTT 对葛根酪氨酸酶作用时间(表 14):

表 14　APTT 作用时间对葛根酪氨酸酶活性的影响

时间/s	347	407	467	527
抑制率/%	78.38	81.42	92.41	74.58

单因素试验证明,APTT 对牛蒡酪氨酸酶的抑制效果在 1895s 时达到最高值 92.18%。

(6)APTT 对龙牙百合酪氨酸酶作用时间(表 15):

表 15　APTT 作用时间对龙牙百合酪氨酸酶活性的影响

时间/s	100	200	300	400	500	600
抑制率/%	77.48	74.58	84.48	98.30	92.62	85.18

单因素试验证明,APTT 对龙牙百合酪氨酸酶的抑制效果在 400s 时达到最高值 98.30%。

从 APTT 对果蔬酪氨酸酶作用时间试验中,APTT 对苹果、香蕉和葛根的酪氨酸酶抑制效果较好,可以达到 90% 以上,相比对牛蒡的抑制效果没有如此理想,APTT 对不同果蔬的抑制效果不一致,说明果蔬自身的酪氨酸酶含量及活性与 APTT 的抑制效果有密切的有关。

APTT 作用温度的选择结果与分析

如图 2,样品粗提液的抑制率随温度的变化而变化,抑制率越大,说明 APTT 抑制效果越好,反之则 APTT 的抑制效果越差。根据图示,APTT 对香蕉酪氨酸酶的抑制效果较高,其在各个温度下的抑制率普遍高于其他的样品,而 APTT 对苹果酪氨酸酶的抑制效果在各个温度下偏差,其在20℃时抑制率最低,抑制效果最差,仅38.15%。

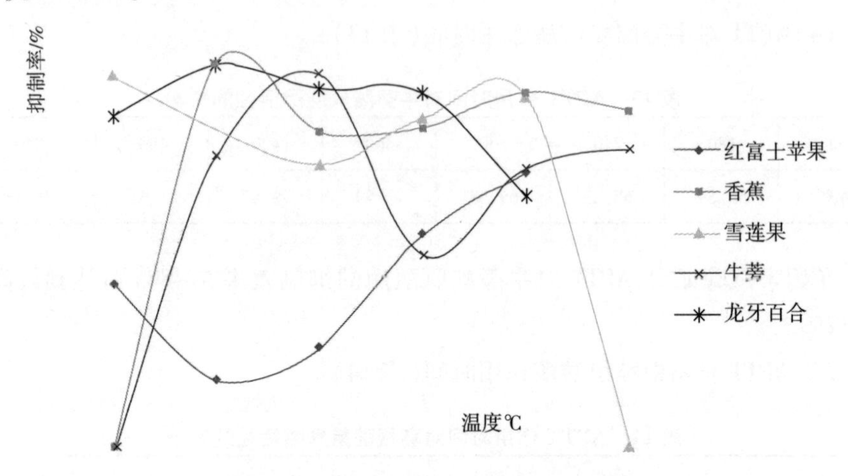

图2　APTT对样品酪氨酸酶的作用温度

根据图形趋势判断红富士苹果中APTT的抑制效果随着温增加,抑制效果先降低后增强,在35℃时抑制效果最佳,说明APTT抑制苹果酪氨酸酶受温度影响较大。在香蕉中APTT的抑制效果随着温增加,抑制效果先降低后增强,但趋势较苹果平缓,在30℃后趋于平衡。说明APTT抑制香蕉酪氨酸酶受温度影响较小。在雪莲果中,情况与香蕉相似。在牛蒡中APTT的抑制效果与温度有很大关系,在25℃时抑制效果最佳,温度升高或者降低都会影响APTT的抑制效果。APTT对葛根酪氨酸酶抑制的最佳抑制温度是25℃,APTT抑制效果受温度影响,温度升高或者降低都会影响APTT的抑制效果。龙牙百合的最佳抑制温度是20℃。APTT对红富士苹果、香蕉、雪莲果、牛蒡、葛根、龙牙百合的抑制效果根据样品的不同会有不同影响,有的与温度有莫大关系,有的关系不大,需要通过正交试验来验证温度对各样品的影响大小。

根据图2判断各样品的最佳抑制温度是苹果35℃,香蕉15℃,雪莲果35℃,牛蒡25℃,葛根25℃,龙牙百合20℃。

正交试验结果

APTT对红富士苹果中酪氨酸酶抑制作用的研究

APTT对红富士苹果酪氨酸酶抑制的效果如表16。

表 16　三因素四水平的正交试验结果分析表

序号	因素				抑制率/%
	A 温度/℃	B 时间/s	C 浓度/(μmol/L)	空列	
1	1(17)	1(1715)	1(28)	1	68.75
2	1(17)	2(1895)	2(32)	2	77.33
3	1(17)	3(2075)	3(36)	3	69.77
4	2(20)	1(1895)	2(32)	3	74.56
5	2(20)	2(1895)	3(36)	1	81.36
6	2(20)	3(2075)	1(28)	2	90.18
7	3(23)	1(1715)	3(36)	2	71.54
8	3(23)	2(1895)	1(28)	3	74.81
9	3(23)	3(2075)	2(32)	1	96.73
k_1	71.947	71.617	77.913	82.280	
k_2	82.033	77.830	82.870	79.680	
k_3	81.027	85.560	74.223	73.047	
R	10.086	13.943	8.647	9.233	

由表 16 可知，$R_B > R_A > R_C$，则时间对 APTT 抑制苹果酪氨酸酶的作用效果影响最大，温度次之，浓度影响最小。由于最优组合 $A_2B_3C_2$ 不在正交试验组中，故通过验证试验来验证其抑制效果，结果表明确定 APTT 浓度为 32μmol/L、作用时间为 2095s、温度为 20℃时，抑制效果最佳，其抑制率高于正交组中第 9 组的 96.73%，达到 98.51%。

APTT 对香蕉中酪氨酸酶抑制作用的研究

APTT 对香蕉酪氨酸酶抑制的效果如表 17。

表 17　三因素四水平的正交试验结果分析表

序号	因素				抑制率/%
	A 浓度/(μmol/L)	B 温度/℃	C 时间/s	D 空白	
1	1(46)	1(15)	1(360)	1	66.20
2	1(46)	2(20)	2(420)	2	89.20
3	1(46)	3(25)	3(480)	3	93.90
4	2(48)	1(15)	2(420)	3	87.79
5	2(48)	2(20)	3(480)	1	87.79
6	2(48)	3(25)	1(360)	2	87.79

续表

序号	因素				抑制率/%
	A 浓度/(μmol/L)	B 温度/℃	C 时间/s	D 空白	
7	3(50)	1(15)	3(480)	2	99.06
8	3(50)	2(20)	1(360)	3	82.63
9	3(50)	3(25)	2(420)	1	18.12
k_1	83.100	84.350	78.873	57.370	
k_2	87.790	86.540	65.037	92.017	
k_3	66.603	66.603	93.583	88.107	
R	21.187	19.937	28.546	34.647	

由表17得知,$R_C > R_A > R_B$,则在APTT抑制香蕉酪氨酸酶的作用效果中时间影响最大,浓度次之,温度对其的影响最小。正交实验确定最优组合为$A_2B_2C_3$,其抑制率为87.79%,但实际试验中$A_3B_1C_3$组合的抑制率比前一组合的抑制率高,达到99.06%,考虑抑制率直接反应酶被抑制的效果,低温、高浓度抑制剂和长时间作用都会对酶抑制效果,故确定最佳的抑制条件为APTT浓度为50μmol/L、作用时间为480s、温度为15℃,抑制率为99.06%。

APTT对雪莲果中酪氨酸酶抑制作用的研究

APTT对雪莲果酪氨酸酶抑制的效果如表18。

表4.18　三因素四水平的正交试验

序号	因素				抑制率/%
	A 浓度/(μmol/L)	B 温度/℃	C 时间/s	D 空白	
1	1(24)	1(21)	1(394)	1	10.58
2	1(24)	2(26)	2(454)	2	90.21
3	1(24)	3(31)	3(514)	3	74.34
4	2(28)	1(21)	2(454)	3	58.20
5	2(28)	2(26)	3(514)	1	80.16
6	2(28)	3(31)	1(394)	2	96.30
7	3(32)	1(21)	3(514)	2	80.42
8	3(32)	2(26)	1(394)	3	95.24
9	3(32)	3(31)	2(454)	1	89.42
k_1	58.377	49.733	67.373	60.053	
k_2	78.220	88.537	79.277	88.977	
k_3	88.360	86.687	78.307	75.927	
R	29.983	38.804	11.904	28.924	

由表18得知，$R_B > R_A > R_C$，则在APTT抑制雪莲果酪氨酸酶的作用效果中温度对其反应的影响最大，浓度次之，时间则最小。正交试验确定最优组合为$A_3B_2C_2$，根据验证试验最优组合的抑制率为96.82%，略高于试验组中的最高抑制率组合为$A_2B_3C_1$，综合各个因素水平考虑，最终确定APTT抑制雪莲果酪氨酸酶的最佳工艺条件为APTT浓度为32μmol/L、作用时间为454s、温度为26℃，抑制率为96.82%。

APTT对牛蒡中酪氨酸酶抑制作用的研究

APTT对牛蒡酪氨酸酶抑制的效果如表19。根据试验结果分析得到，$R_C > R_B > R_A$，则在APTT抑制牛蒡酪氨酸酶的作用效果中时间对其反应的影响最大，温度次之，浓度则最小。根据正交试验结果分析可知最优组合为$A_3B_1C_2$，通过验证试验证实此工艺对APTT抑制牛蒡酪氨酸酶抑制剂的抑制效果优于正交试验组中抑制率最高的第9组，达到93.41%。因此确认最优的工艺条件为APTT浓度为500μmol/L、作用时间为480s、温度为20℃时，抑制率为93.41%。

表19　三因素四水平的正交试验结果分析表

序号	因素				抑制率/%
	*A*浓度/(μmol/L)	*B*温度/℃	*C*时间/s	*D*空白	
1	1(400)	1(20)	1(420)	1	73.55
2	1(400)	2(25)	2(480)	2	87.22
3	1(400)	3(30)	3(540)	3	20.48
4	2(450)	1(20)	2(480)	3	91.32
5	2(450)	2(25)	3(540)	1	72.73
6	2(450)	3(30)	1(420)	2	70.19
7	3(500)	1(20)	3(540)	2	89.52
8	3(500)	2(25)	1(420)	3	81.57
9	3(500)	3(30)	2(480)	1	92.71
k_1	60.417	84.797	75.103	79.633	
k_2	78.080	80.507	90.417	82.310	
k_3	87.933	61.127	60.910	64.457	
R	27.516	23.670	29.507	17.853	

APTT对葛根中酪氨酸酶抑制作用的研究

APTT对葛根牛蒡酪氨酸酶抑制的效果如表20。

表 20　三因素四水平的正交试验结果分析表

序号	因　素				抑制率/%
	A 浓度/(μmol/L)	*B* 温度/℃	*C* 时间/s	*D* 空白	
1	1(500)	1(20)	1(467)	1	94.59
2	1(500)	2(25)	2(407)	2	78.04
3	1(500)	3(30)	3(527)	3	93.92
4	2(550)	1(20)	2(407)	3	89.86
5	2(550)	2(25)	3(527)	1	94.26
6	2(550)	3(30)	1(467)	2	88.85
7	3(600)	1(20)	3(527)	2	58.45
8	3(600)	2(25)	1(467)	3	86.82
9	3(600)	3(30)	2(407)	1	95.95
k_1	88.850	80.967	90.087	94.933	
k_2	90.990	86.373	87.950	75.113	
k_3	80.407	92.907	82.210	90.200	
R	10.583	11.940	7.877	19.820	

根据表 20 确定 $R_B > R_A > R_C$，在 APTT 抑制牛蒡酪氨酸酶的作用效果中温度对其反应的影响最大，浓度次之，时间对其影响最小。根据试验结果分析确定最优组合为 $A_2B_3C_1$，其抑制率为 88.85%，但实际正交试验组中抑制率最高的组合为 $A_3B_3C_2$，其抑制率达 95.95%，两组相比在 APTT 浓度和作用时间上都有区别，抑制率最高组合比最优组合的 APTT 浓度和时间都有所提高，但是从抑制效果来看，工艺条件最优组合为 APTT 浓度为 600μmol/L 、作用时间为 407s、温度为 30℃，抑制率为 95.95%。

APTT 对龙牙百合中酪氨酸酶抑制作用的研究

APTT 对龙牙百合酪氨酸酶抑制的效果如表 21。

表 21　三因素四水平的正交试验结果分析表

序号	因　素				抑制率/%
	A 浓度/(μmol/L)	*B* 温度/℃	*C* 时间/s	*D* 空白	
1	1(40)	1(20)	1(500)	1	75.33
2	1(40)	2(25)	2(600)	2	75.94
3	1(40)	3(30)	3(700)	3	80.53
4	2(50)	1(20)	2(600)	3	97.57
5	2(50)	2(25)	3(700)	1	84.38
6	2(50)	3(30)	1(500)	2	87.55
7	3(60)	1(20)	3(700)	2	86.51

续表

序号	因素				抑制率/%
	A 浓度/(μmol/L)	B 温度/℃	C 时间/s	D 空白	
8	3(60)	2(25)	1(500)	3	74.57
9	3(60)	3(30)	2(600)	1	82.47
k_1	77.267	86.470	79.150	80.727	
k_2	89.833	78.297	85.327	83.333	
k_3	82.183	83.517	83.807	84.223	
R	12.566	8.173	6.177	3.496	

根据表21确定 $R_A > R_B > R_C$，在APTT抑制牛蒡酪氨酸酶的作用效果中浓度对其反应的影响最大，温度次之，时间对其影响最小。根据试验结果分析确定最优组合为 $A_2B_1C_3$，根据验证试验结果，此工艺抑制率为98.12%略高于试验组最高抑制率97.57%的第4组。因此工艺条件最优组合为APTT浓度为50μmol/L、作用时间为700s、温度为20℃，抑制率为98.12%。

1.5　结论

APTT对红富士苹果中酪氨酸酶抑制作用的最优工艺是APTT浓度为32μmol/L、作用时间为2075s、温度为20℃，抑制率为98.51%；APTT对香蕉中酪氨酸酶抑制的最优工艺是APTT浓度为50μmol/L、作用时间为480s、温度为15℃，抑制率为99.06%；APTT对雪莲果中酪氨酸酶抑制的最优工艺是APTT浓度为28μmol/L、作用时间为394s、温度为31℃，抑制率为96.82%；APTT对牛蒡中酪氨酸酶抑制的最优工艺是APTT浓度为500μmol/L、作用时间为480s、温度为20℃，抑制率为93.41%；APTT对葛根中酪氨酸酶抑制的最优工艺是APTT浓度为550μmol/L、作用时间为467s、温度为30℃，抑制率为95.95%；APTT对龙牙百合中酪氨酸酶抑制的最优工艺是APTT浓度为50μmol/L、作用时间为700s、温度为20℃，抑制率为98.12%。从上述试验结果来看，APTT对果蔬具有较高的抑制效果，与传统的VC、柠檬酸相比，APTT的抑制效果更好。

5.2　噻二唑席夫碱对果蔬的作用(唐君源硕士论文部分内容)

新鲜果品蔬菜中含有大量丰富的维生素和矿物质等营养物质，是保障人体健康需求的物质基础。采收后的果蔬仍为活体，仍进行着复杂的生命活动。由于采后果蔬处于与母体隔离状态，缺乏水分和营养，不仅会促进其成熟和衰老，而且其外观色泽、营养特性和食用价值等也将发生显著的变化，特别是切开后的果蔬,切口破坏了果体细胞中细胞膜、线粒体结构,释放出酪氨酸酶等多酚类氧化酶,加速果蔬褐变、腐烂,影响色泽和口感。现今我国的果蔬产量居世界第一,据相关统计,现阶段新鲜果蔬的腐烂损耗率较高,因此对果蔬进行护色保鲜及加工是必不可少的。为了解决这一问题,人们对果蔬产品的褐变原理和防止方法进行了许多深入的研究。其中化学方法在食品工业上一直是最为广泛地用来抑制酶促褐变的方法,寻找安全、无毒、高效的抗褐变剂是一个活跃的研究领域,也是一个函待解决的问题。

5.2.1 噻二唑席夫碱对果蔬中酪氨酸酶抑制作用

材料与仪器

实验中使用的材料如表1所示。

表1 实验材料

名称	规格(纯度)	生产厂家
红富士苹果	新鲜无损伤	邵阳市步步高超市
磷酸氢二钾	分析纯(95%)	天津市恒兴化学试剂制造有限公司
磷酸二氢钾	分析纯(95%)	天津市恒兴化学试剂制造有限公司
二甲基亚砜(DMSO)	分析纯(95%)	Sigma 化学公司
T1 - 13		自制

实验中所需实验仪器如表2所示。

表2 实验仪器

名称	规格(纯度)	生产厂家
可见紫外分光光度计	UV - 4802 FE20 型	UNICO 公司
pH 计	BD - 198E	梅特勒 - 托利多仪器公司
低温冰柜	CP213	海尔公司
电子天平	3 - 18R	奥豪斯仪器(上海)有限公司 TOMOS 公司
高速冷冻离心机		

实验方法

酪氨酸酶是形成黑色素的限速酶,因此可通过多巴或者多巴醌的生成速度来表示酪氨酸酶的活性。以L - 多巴(L - DOPA)为底物,在475 nm波长下测定一定时间内吸光度变化值,此值相当于多巴醌的生成速度。IC_{50}是指酪氨酸酶活性被抑制一半时酪氨酸酶抑制剂的浓度,IC_{50}值越小则可说明该抑制剂的酪氨酸酶抑制效果越好。

参照本课题组已建立起的筛选方法的方法,在1.5 mL的离心管中配制1000

μL 的测试体系，包括 pH 6.8 浓度为 0.05 mol/L 的磷酸盐缓冲液 950 μL，待测浓度为 1.0 mmol/L 的样品溶液 10 μL（10 μL 样品作对照，以去除样品的颜色对实验的影响），10 μL 的酪氨酸酶溶液，在 25℃的条件下孵化 10 min 后，加入 30 μL 浓度为 7.6 mmol/L 的 L－DOPA 溶液，在 475 nm 波长下进行动力学测试，记录一分钟的 OD 值的改变量。再通过改变样品溶液的浓度（通过稀释的方式改变），分别测试酪氨酸酶的催化活性。以空白对照时的酶催化活性 OD 值为 100%，而不同浓度的样品对酶的抑制活性则用抑制百分数来表达，然后再通过 Origin7.0 做图得到 IC_{50}值。抑制率的计算公式如下：

抑制率 = $[(B-S)/B]\times100\%$ ……………………………………公式1

式中 B 为空白吸光度值，S 为样品吸光度值。

以酪氨酸酶粗提液代替蘑菇酪氨酸酶溶液，以抑制率为评定指标采用单因素和响应面的实验方法确定噻二唑席夫碱对苹果抑制作用的最佳浓度、最佳作用时间和最佳作用温度。

溶液配制

（1）0.05 mol/L pH 6.8 磷酸盐缓冲溶液：

准确称取磷酸二氢钾和磷酸氢二钾 5.7055 g 和 3.4023 g，分别将其配置成 500 mL 0.05 mol/L 的溶液。将适量的 0.05 mol/L 的 KH_2PO_4溶液加入到 500 mL 0.05 mol/L K_2HPO_4中，搅拌均匀，用 pH 计将混合溶液的 pH 值调至 6.8 即可。

（2）样品溶液：

以 DMSO 为溶剂配制 1.0 mmol/L 的样品溶液 1 mL 备用。

（3）蘑菇酪氨酸酶溶液（现用现配）：

用0.05 mol/L pH 6.8 的磷酸盐缓冲溶液将 1mg 蘑菇酪氨酸酶溶解并定容至 1mL。

（4）L－DOPA 溶液（现用现配）：

用 0.05 mol/L pH 6.8 的磷酸盐缓冲液将 15 mg L－DOPA，溶解，定容至 10 mL，配制成 7.6 mmol/L 的 L－DOPA 溶液备用。

配制浓度分别为 0.025 μmol/L、0.050 μmol/L、0.100 μmol/L、0.200 μmol/L 和 0.400 μmol/L 的 T1－13 溶液。

苹果酪氨酸酶粗提取液方法

取新鲜、无机械损伤、无腐烂的红富士苹果，去外皮，取少量果肉，迅速置于加有磷酸盐缓冲溶液的研钵中研磨成糜，然后用纱布过滤，放入离心机中，在 25℃及离心速度为 4000 r/min 的条件下离心 10 min，取上清液制成酪氨酸酶粗提液。

5.2.2 实验结果与分析

单因素实验结果与分析

(1) T1 -13 最佳浓度的确定

由图 1 知随噻二唑席夫碱浓度的增加,对苹果酪氨酸酶的抑制率逐渐增大,说明噻二唑席夫碱对酪氨酸酶的抑制与浓度有关,最佳浓度为 0.400 μmol/L。

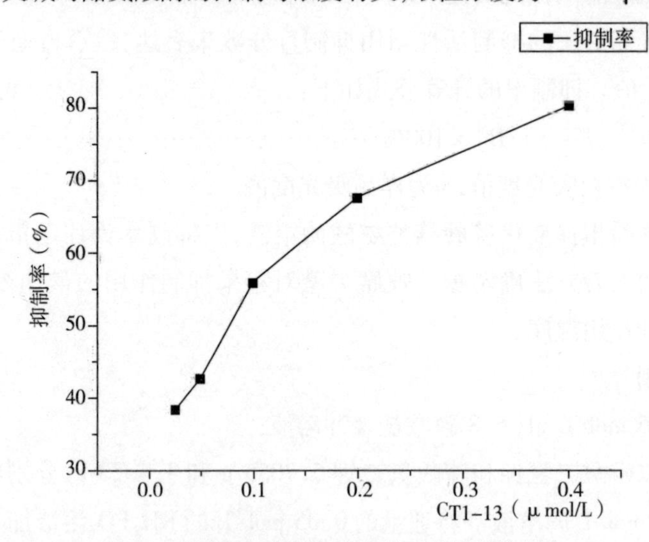

图 1 不同噻二唑席夫碱浓度对苹果的抑制率

(2) T1 -13 最佳作用时间的确定

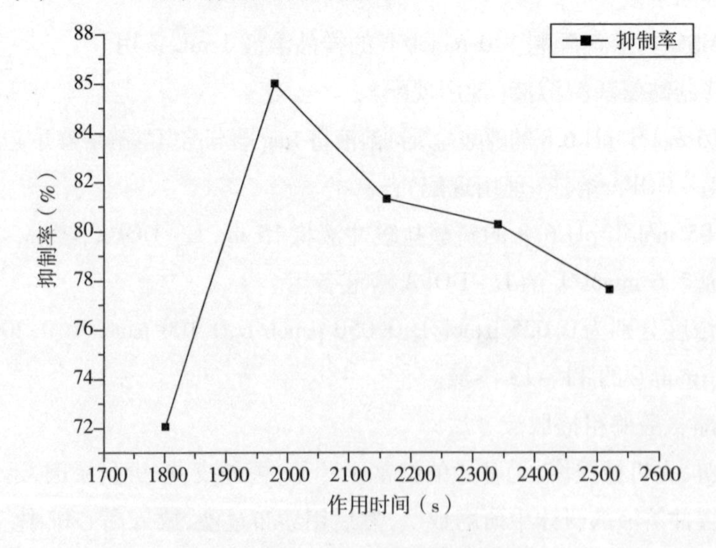

图 2 不同作用时间对苹果的抑制率

由图 2 可知随着作用时间的增加,对苹果酪氨酸酶的抑制率呈现先升高后降低的趋势,在 1980s 时达到最大抑制率 81.34%。

(3) T1 - 13 最佳作用温度的确定

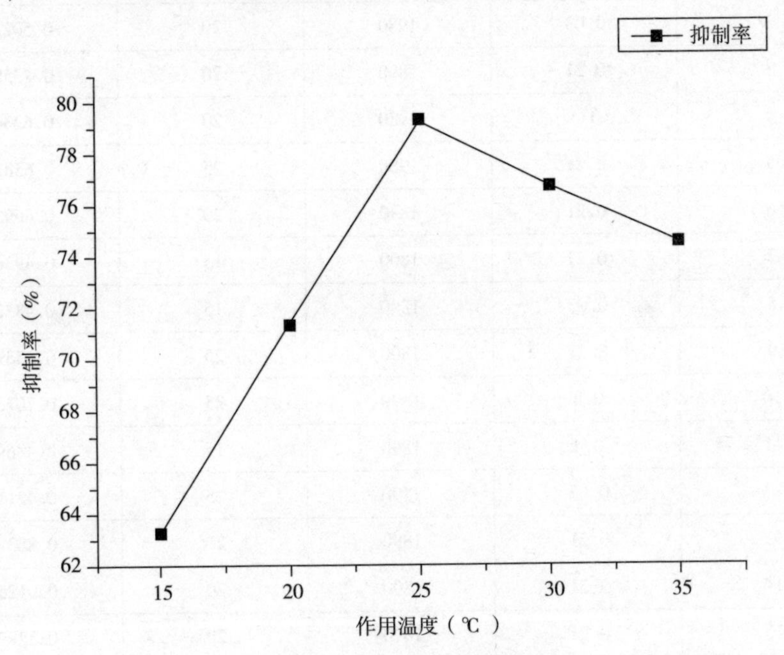

图 3　不同作用温度对苹果的抑制率

由图 3 可知随着作用温度的升高,对苹果酪氨酸酶的抑制率呈现先升高后降低的趋势,在 25℃时达到最大抑制率 79.28%。

响应面实验结果与分析

(1)响应面实验设计及结果

在单因素实验的基础上,根据 Box - Behnken 试验设计原理,以抑制率(Y)为响应值,选取浓度(*A*)、作用时间(*B*)、温度(*C*)进行响应面优化实验。Box - Behnken 因素水平编码表见表 3,响应面试验设计及结果见表 4。

表 3　Box - Behnken 试验因素水平表

编码水平	因素		
	A 浓度(μmol/L)	*B* 作用时间(s)	*C* 作用温度(℃)
-1	0.025	1800	15
0	0.212	1890	20
1	0.400	1980	25

表 4　响应面实验结果

实验号	*A* 浓度(μmol/L)	*B* 作用时间(s)	*C* 温度(℃)	R(抑制率)
1	0.21	1980	15	0.4837
2	0.03	1980	20	0.5022
3	0.21	1890	20	0.7358
4	0.4	1980	20	0.6234
5	0.21	1980	25	0.6587
6	0.21	1890	20	0.6698
7	0.21	1800	15	0.4936
8	0.03	1890	15	0.4832
9	0.21	1800	25	0.5539
10	0.4	1890	25	0.7023
11	0.4	1890	15	0.4869
12	0.03	1890	25	0.4211
13	0.21	1890	20	0.8837
14	0.21	1890	20	0.9125
15	0.21	1890	20	0.8299
16	0.4	1800	20	0.5554
17	0.03	1800	20	0.5219

(2)回归模型的建立与显著性分析

回归模型的方差分析见表 5,运用 Design - Expert 8.0.6 对表 4 进行多元回归拟合,得到抑制率(Y)对自变量浓度(A)、作用时间(B)、温度(C)的多元回归方程:

$$Y = 0.81 + 0.067A + 0.018B + 0.021C + 0.022AB + 0.044AC + 0.059BC - 0.13A^2 - 0.13B^2 - 0.13C^2$$

表 5　回归与方差结果分析

Source	Sum of Squares	df	Mean Square	*F* Value	p - value Prob > F	
Model	0.3001	9	0.0333	4.7841	0.0255	significant
A - 浓度	0.0364	1	0.0364	5.2212	0.0462	*
B - 时间	0.0026	1	0.0026	0.3677	0.5634	

续表

Source	Sum of Squares	df	Mean Square	*F* Value	p - value Prob > F	
C - 温度	0.0035	1	0.0035	0.5061	0.4998	*
AB	0.0019	1	0.0019	0.2758	0.6157	
AC	0.0079	1	0.0079	1.1299	0.3231	*
BC	0.0138	1	0.0138	1.9857	0.2016	*
A^2	0.0683	1	0.0683	9.7973	0.0166	*
B^2	0.0693	1	0.0693	9.9362	0.0161	*
C^2	0.0718	1	0.0718	10.3037	0.0149	*
Residual	0.0488	7	0.0070			
Lack of Fit	0.0074	3	0.0025	0.2371	0.8666	not significant
Pure Error	0.0414	4	0.0104			
Cor Total	0.3489	16				

注：*：差异显著，$P<0.05$；* *：差异极显著，$P<0.01$。回归系数 $R^2=0.8601$，$R^2_{AdJ}=0.6803$

用 Box - Behnken Design 响应面分析法对试验结果拟合的模型进行方差分析和显著性检验，各个变量对抑制率 Y 影响的显著性则是通过回归方程方差分析中的概率 P 值来判定，由表 7 可以看出，该二次多项式模型 P 值 <0.05，模型显著，失拟项 P 值 >0.05，失拟项不显著，表明该回归方程拟合度较好，误差小，与实际预测值能较好的拟合；该模型的复相关系数为 $R^2=0.8671$，校正决定系数 $R^2_{adj}=0.6803$，说明建立的模型能够解释 86.71% 的响应值变化，可用来进行酪氨酸酶抑制率 Y（响应值）的预测；由显著性检验可知，噻二唑席夫碱浓度及温度对酪氨酸酶抑制率影响显著，而一次项 B、交互项 AB 对酪氨酸酶抑制率影响不显著。因此，各试验因素对酪氨酸酶抑制率的影响并不是呈现简单的线性关系；另外，通过 F 值大小，可判定各因素对感官评分影响的重要性，F 值越大，重要性越大，所以各因素对感官评价总分的影响大小为：$A>C>B$（即 T1 - 13 浓度 > 温度 > 作用时间）。

（3）响应面分析

响应值（抑制率 Y）对各个试验因素所围成的三维的曲面图构成了响应面图形，并且最佳参数以及各个参数相互间的作用也可从响应面图形中得出。等高线图可判定各因素交互作用的显著性，等高图趋向椭圆，各因素间的交互作用显著，反之，则不显著；等高线的疏密程度可判定各因素对酪氨酸酶抑制率的影响大小，等高线越密，影响越大，反之则越小；而响应面图形是一个开口向下且向上凸起的曲面，因此抑制率 Y 在响应面图形的最高点存在极值，各试验因素的最佳作用点都处于试验设计值范围内。

由图 4 ~ 图 9 可知，AC、BC 及 AB 交互作用的等高图均呈椭圆形，但 AB 交互作用的等高图趋向于圆形，说明 AC、BC 之间的交互作用较 AB 显著。故 A 的影响比 B、C 的影响大，C 的影响比 B 的影响大，这与方差分析的结果是一致的。综上所述，噻二唑席夫碱浓度对酪氨酸酶抑制率影响最为突出，温度、作用时间的影响次之。

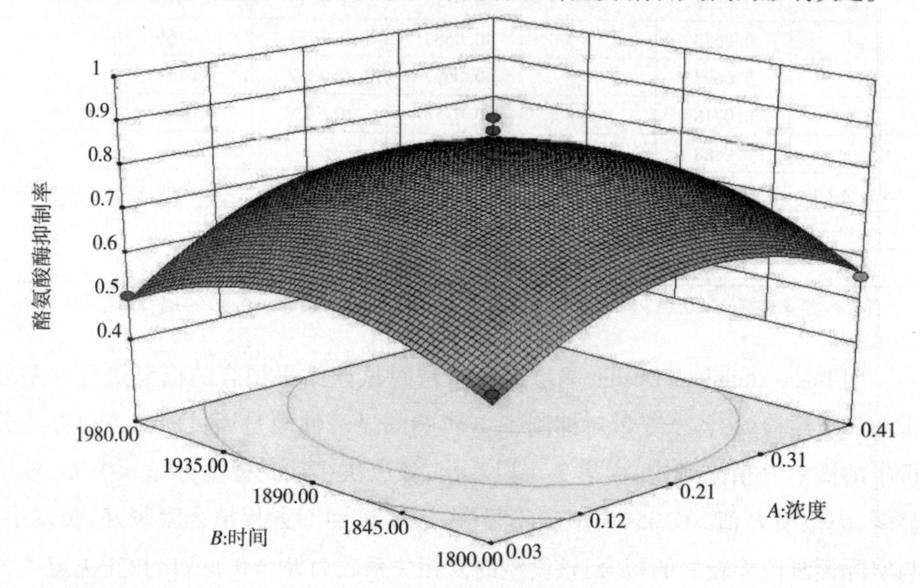

图 4　$Y = f(A,B)$ 的响应面

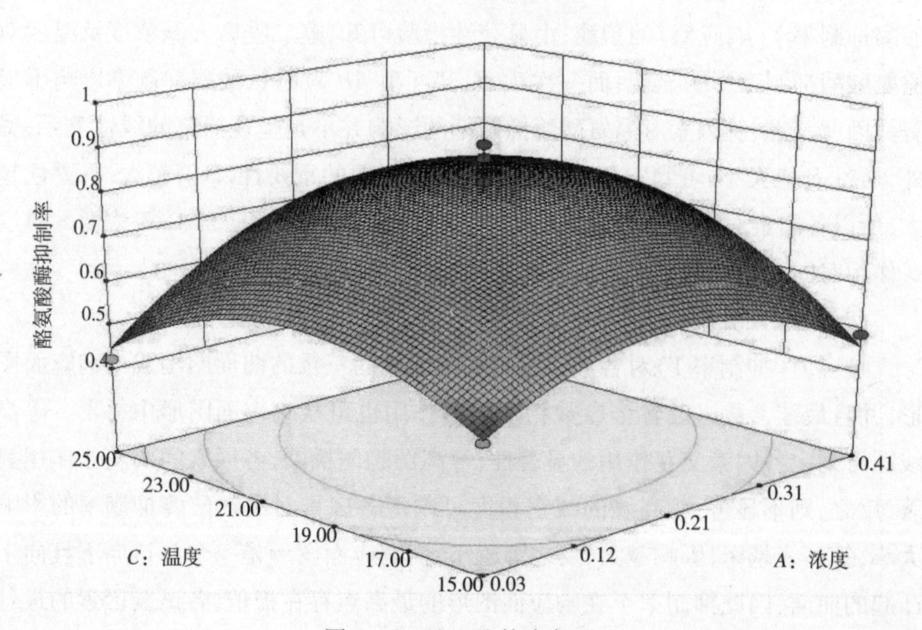

图 5　$Y = f(A,C)$ 的响应面

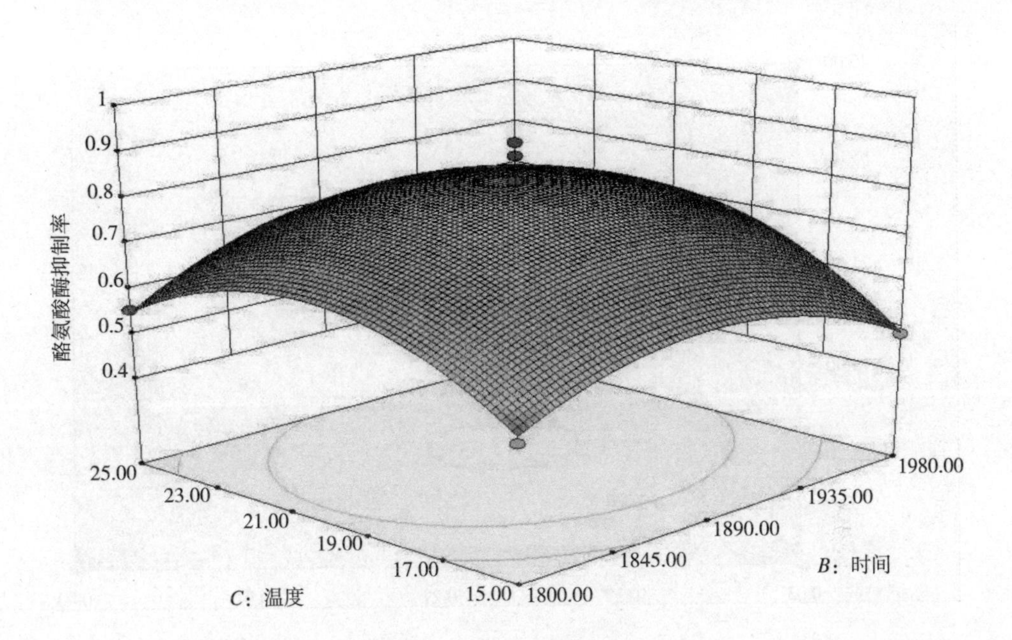

图6　$Y = f(B, C)$的响应面

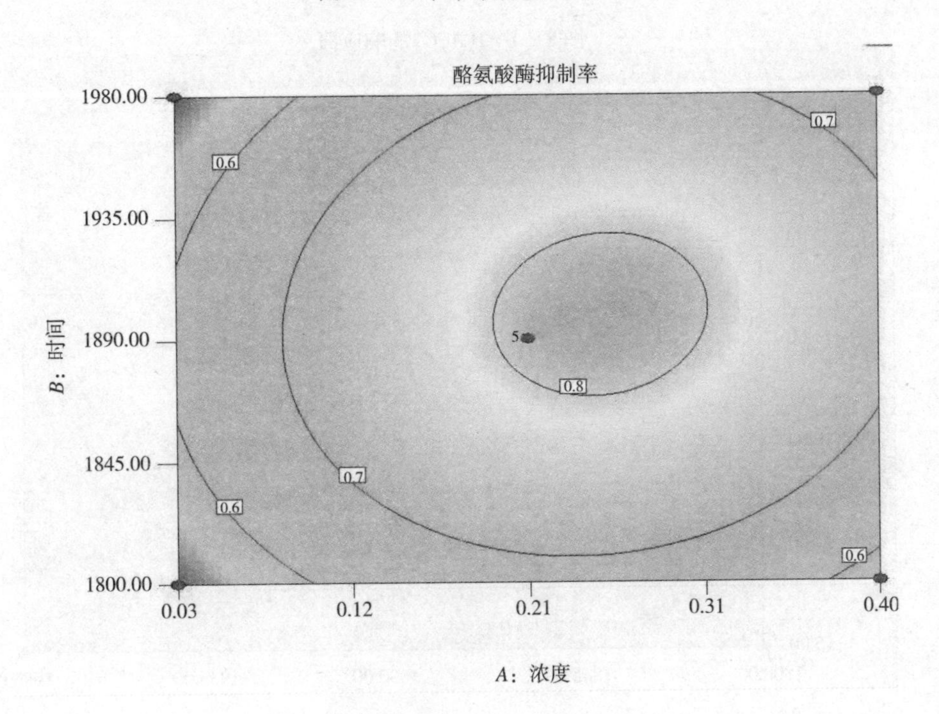

图7　$Y = f(A, B)$的响应面

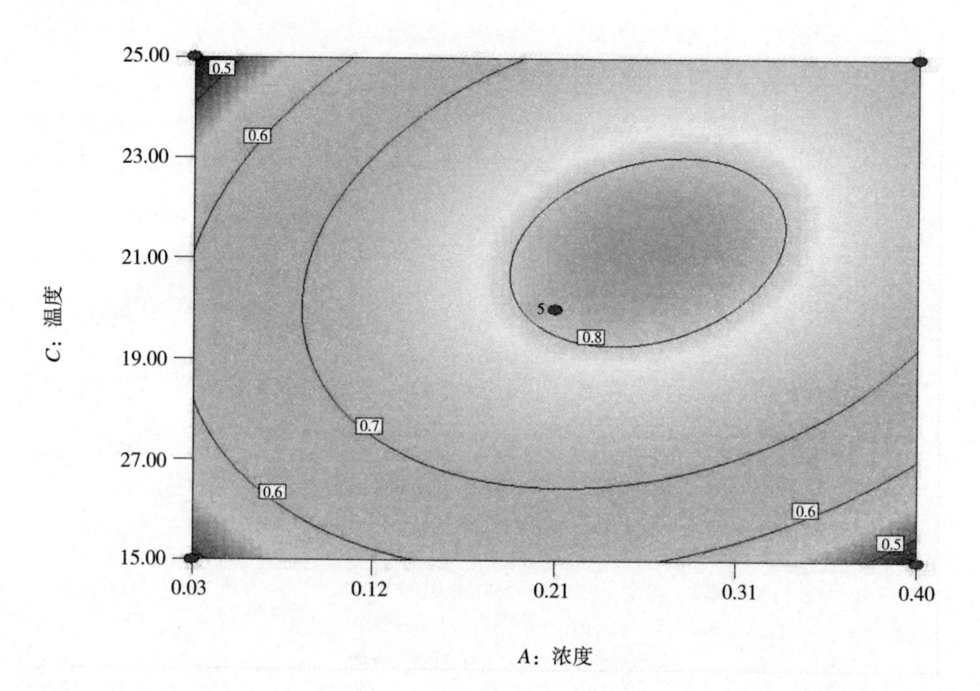

图 8　$Y = f(A,C)$的响应面

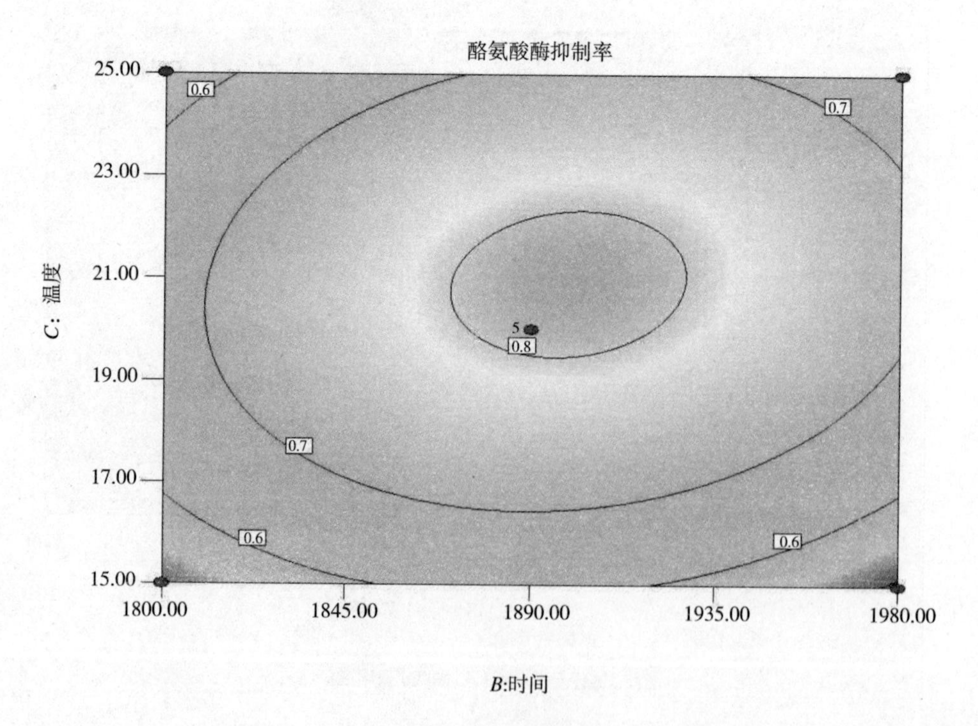

图 9　$Y = f(B,C)$的响应面

(4)噻二唑席夫碱对苹果酪氨酸酶抑制最佳条件的确定

通过软件 Design - Expert 8.0.6 分析,预测出酪氨酸酶抑制率最佳条件为:浓度为0.27 μmol/L,温度为20.81℃,作用时间为1901.98 s为1.37 g,此时模型预测最佳工艺条件的抑制率为81.53%。经过3次验证试验得出在浓度为0.27 μmol/L,温度为21℃,作用时间为1902 s条件下酪氨酸酶抑制率分别为:78.15%、77.32%及73.16%,平均值为76.21%,与预测值较接近,可用于酪氨酸酶抑制率最佳工艺条件感官评价总分的理论预测。

5.2.3 小结

通过软件 Design - Expert 8.0.6 分析 T1 - 13 对红富士苹果酪氨酸酶抑制率最佳条件为:浓度为0.27 mol/L,温度为20.81℃,作用时间为1901.98 s为1.37 g,此时模型预测最佳工艺条件的抑制率为81.53%。经过3次验证试验得出在浓度为0.27 umol/L,温度为21℃,作用时间为1902 s条件下酪氨酸酶抑制率分别为:78.15%、77.32%及73.16%,平均值为76.21%。

5.3 3-苯基-4-氨基-5-硫酮-1,2,4-三唑对牛蒡中酪氨酸酶的抑制作用。

赵良忠,凌晶晶,陈玲娟,刘进兵

(1. 邵阳学院生物与化学工程系,湖南邵阳 422000;2. 中南林业科技大学食品科学与工程学院,湖南长沙 410004)

摘 要:以3-苯基-4-氨基-5-硫酮-1,2,4-三唑(APTT)为效应物,通过SPSS设计正交试验,研究不同APTT浓度、时间以及温度对牛蒡酪氨酸酶提取液的护色效果的影响,并探讨p H值对抑制剂及反应体系作用效果的影响,数据用SPSS软件处理,在25℃条件下,牛蒡与APTT比为0.5∶3(g/mL)时,500μmol/L抑制剂对底物作用420s为最佳抑制反应条件,其 ΔA 为0.0082 ∶ pH7.0时,抑制效果最佳。APTT对牛蒡中酪氨酸酶有良好的抑制作用。

关键词:牛蒡;酪氨酸酶;正交试验;3-苯基- 4 -氨基- 5-硫酮- 1 ,2 ,4 -三唑

Inhibitory Effect of 4 - Amino - 3 - Phenyl - 1H - 1,2,4 - Triazole - 5(4H) - Thione on Tyrosinase in Burdock

ZHAO Liang - zhong, LING Jing - jing, CHEN Ling - juan, LIU Jin - bing

(1. Department of ChemicalBiotechnology and Engineering, Shaoyang University, Shaoyang 422000, China ;2. College of Food Science and Engineering, Central South University of Forestry and Technology, Changsha 410004, China)

Abstract: APTT, 4 - amino - 3 - phenyl - 1H - 1,2,4 - triazole - 5(4H) - thione, was used as tyrosinase inhibitor to suppress the browning of burdock. The inhibitory effects of APTT concentration, reaction time, reaction temperature and reaction pH on tyrosinase in burdock extract were determined by orthogonal array design. The optimal reaction conditions of APTT were substrate /APTT ratio of 1:6 (g/mL), APTT concentration of 500μmol/L, reaction temperature of 25 ℃, reaction time of 420 s and reaction pH of 7.0. Therefore, APTT is a highly effective inhibitor for tyrosinase from burdock.

Key words: burdock; tyrosinase; orthogonal test; 4 - amino - 3 - phenyl - 1H - 1,2,4 - triazole - 5(4H) - thione

牛蒡(Arctium lappa L.)又名白肌人参、蒡翁菜,属菊科牛蒡属2年生草本植物,其果实牛蒡子用于治疗感冒、麻疹、咽喉肿痛、丹毒等症[1]。牛蒡根含有各种氨基酸,含量高且具有特殊药理作用,还含有多酚类物质,具有抗癌、抗突变的作用[2]。牛蒡茎叶含挥发油、鞣质、黏液质、咖啡酸、绿原酸、异绿原酸等。牛蒡子含牛蒡苷、脂肪油、甾醇、硫胺素、牛蒡酚等多种化学成分[3],Li 等[4]采用高压微波法提取牛蒡苷达9.01%。根据尹丹丹等[5]研究表明,牛蒡子的不同溶剂提取物均具有抗氧化活性,其中乙醇和丙酮提取物的抗氧化活性较2,6 - 二叔丁基 - 4 - 甲基苯酚高。因此牛蒡具有较高的营养与药用价值。但是,牛蒡在贮藏、加工过程中发生褐变反应,影响产品的感官、味道、营养等,导致品质降低,产品销售降低,造成巨大的浪费和经济损失。

研究表明,牛蒡的褐变与其丰富的酪氨酸酶有着密切的关系[6]。牛蒡含丰富酚类物质,切分后极易发生褐变,影响牛蒡品质和制约其加工产业发展[7]。周志才等[8]采用钾明矾和0.02% SSA 的混合液,或0.58% 磷酸和0.02% SSA 的混合液对牛蒡中多酚氧化酶进行抑制取得了一定效果,但工业应用仍存许多问题。

酪氨酸酶(Tyrosinase,EC1.14.18.1)是一种多酚氧化酶,与果蔬的褐变、黑色素合成、昆虫的发育等有密切关系[9-11]。研究酪氨酸酶抑制剂对它的抑制作用,可在化妆品的开发,色素代谢相关疾病的治疗[12-13]与杀虫农药[14]开发方面发挥重大的作用。目前,氢醌、果酸、维生素 C[15]、曲酸和熊果苷等酪氨酸酶抑制剂已作为美白剂应用于化妆品中,但氢醌和果酸刺激性大,而 VC 和曲酸易氧化,因保质期短已逐渐被淘汰。

3 – 苯基 – 4 – 氨基 – 5 – 硫酮 – 1,2,4 – 三唑(4 – amino – 3 – phenyl – 1H – 1, 2,4 – triazole – 5(4H) – thione,APTT)的 IC_{50} 为 1.76μmol/L,其结构式如图 1 所示,理论分析是一种低毒性化合物。APTT 对牛蒡酪氨酸酶的抑制效果尚未有研究,但因 APTT 对酪氨酸酶具有高效抑制效果,其可对牛蒡保鲜贮藏和加工具有一定经济价值。

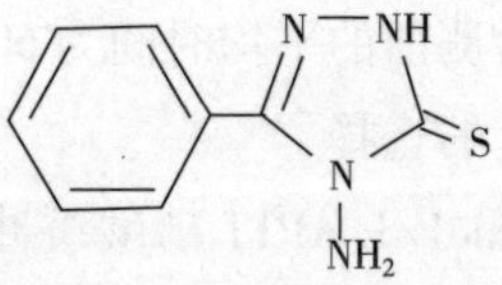

N—NH
S
N
NH_2

图1　3 – 苯基 – 4 – 氨基 – 5 – 硫酮 – 1, 2, 4 – 三唑化学结构式

1　材料与方法

1.1　材料、试剂与仪器

牛蒡(新鲜、颜色浅黄、无机械损伤、无腐烂、无虫害) 市售。磷酸二氢钾、磷酸氢二钾(分析纯) 天津市大茂化学试剂厂;二甲基亚砜(DMSO) 北京鼎国昌盛生物技术有限责任公司;AP TT 邵阳学院药物化学实验室。UV – 4802 型紫外分光光度计　美国尤尼柯仪器有限公司;TOMOS3 – 18R 台式高速冷冻离心机 上海托莫斯科学仪器有限公司;pH 计 梅特勒 – 托利多仪器(上海) 有限公司;电子天平 杭州汇尔仪器设备有限公司。

1.2　方法

1.2.1　测量波长选择

取新鲜无机械损伤、无腐烂的牛蒡,去皮,用不锈钢刀切取 0. 5g,迅速置于研钵中加 3. 0mL　pH6. 8 的磷酸盐缓冲溶液(PBS)研磨成糜,立即用 6 层纱布加压过滤,然后取上清液用紫外分光光度仪在波长(190 ~1100) nm 扫描吸光度(A),每间隔 5min 测定 1 次[16],重复 3 次,作图选择合适的测量波长。

1.2.2　牛蒡对照样吸光度的测定

取牛蒡酪氨酸酶粗提液 1. 0mL 于比色皿(以 PBS 溶液作为空白对照),迅速在 755nm 波长处进行动力学扫描,确定样品的最高值与最低值,计算两者之差为 ΔA 标。同样测定添加 APTT 的牛蒡酪氨酸酶粗提液的 A 变化值,记为 ΔA。

1.2.3　单因素试验

1.2.3.1　抑制剂 APTT 浓度的选择

参考 1. 2. 1 节配制含 300、350、400、450、500μmol/L(高于 500 μmol/L,

PBS 中的 APTT 有部分析出)APTT 的牛蒡粗提液,在实验测量波长条件下测定 A 最高值与最低值,计算两者之差计为 ΔA 抑。同一 APTT 浓度平行测定次,以 ΔA 平最低的组别所对应的 APTT 浓度为最佳浓度。

1.2.3.2　APTT 作用时间的选择

参考 1.2.1 节方法制得牛蒡粗提液,测定 0 ~ 300s 中每 60 s 吸光度的变化值,以不加抑制剂的作为对照,在试验测量波长条件下测定 A 值,探究 APTT 最佳作用时间。从开始切牛蒡至检测前的平均时间为 60s。

1.2.3.3　APTT 作用温度的选择

参考 1.2.1 节方法,450μmo1/L APTT 磷酸缓冲溶液 3.0 mL 分别在 15、20、25、30、35 ℃水浴中保温 10min,制得牛蒡粗提液在实验测量波长条件下进行动力学检测,记录 A 最高值与最低值,两者之差计为 ΔA_T,同一温度平行测定 3 次。

1.2.4　正交试验

根据单因素试验确定的 APTT 浓度、作用时间、作用温度作为考察的 3 因素,每因素 3 水平,以 A 变化值(ΔA)为指标,用 $L_9(3^4)$ 正交表安排试验,直接采用 SPSS18 软件进行正交设计[17],得到数据直接输入设计的表格进行计算。所有试验重复 3 次,得 ΔA_1。按照 SPSS 18 软件对三因素三水平试验的分析要求,再重复一次正交试验,得 ΔA_2。输出试验结果,采用方差分析法[18],确 APTT 的最佳作用条件。

1.2.5　pH 值对抑制剂及反应体系的影响

根据正交试验确定的最佳反应条件,在 pH 值分别为 6.0、6. 5、7.0、7. 5、8.0 条件下测定添加 APTT 和不添加 APTT 的牛蒡酪氨酸酶粗提液的 A 值,重复 3 次,然后按下式考察抑制效果。$R = A_1 - A_0$ 式中:A_1 为添加 APTT 的 3 次平均值;A_0 为不添加 APTT 的 3 次平均值;R 为抑制效果对比值。

2　结果与分析

2.1　波长测量

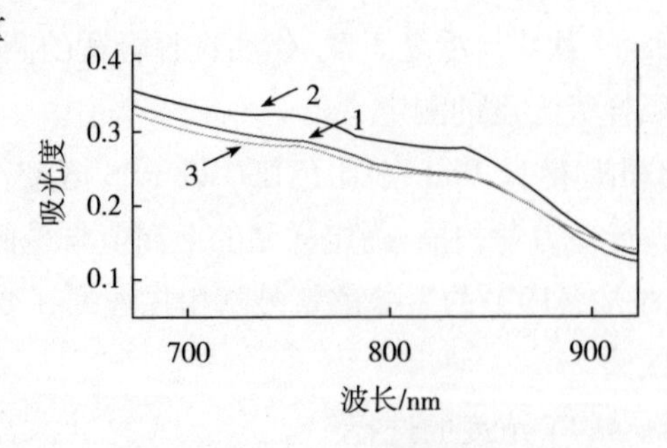

图 2　波长扫描图

由图 2 可知,最佳测量波长为 755nm。

2.2　单因素试验

2.2.1　抑制 APTT 浓度

从表 1 可知,$\Delta A_{平}$随 APTT 浓度的增加而减少,由于 APTT 浓度超过 500μmo1/L 时有晶体析出,故取最佳抑制 APTT 浓度为 500μmo1 / L。

表 1　APTT 浓度的选择记录表

吸光度	APTT 浓度/(μmol/L)					
	0	300	350	400	450	500
$\Delta A_{抑1}$	0.1302	0.0736	0.0595	0.0469	0.0202	0.0103
$\Delta A_{抑2}$	0.1083	0.0923	0.0627	0.0504	0.0275	0.0152
$\Delta A_{抑3}$	0.1279	0.0847	0.0602	0.0534	0.0329	0.0151
$\Delta A_{平}$	0.1221	0.0835	0.0608	0.0502	0.0294	0.0135

2.2.2　作用时间

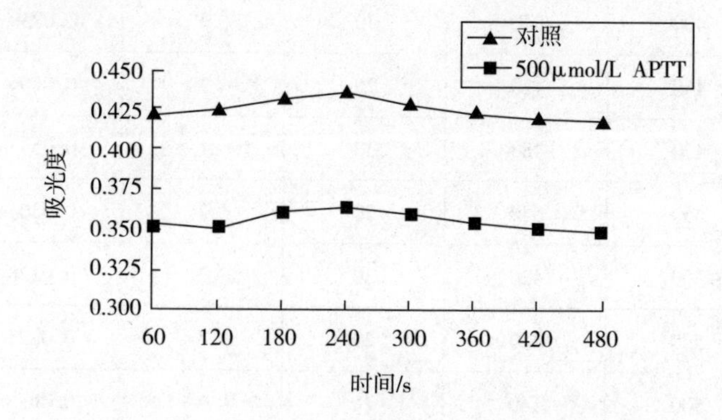

图 3　500μmol/L APTT 与标准品在不同时间段的抑制效果

图 3 表明,APTT 的最佳作用时间为 480s。

2.2.3　作用温度

表 2　APTT 作用温度的选择记录表

吸光度	APTT 温度/℃				
	15	20	25	30	35
ΔA_{T1}	0.0250	0.0244	0.0094	0.0216	0.0160
ΔA_{T2}	0.0211	0.0310	0.0155	0.0261	0.0173
ΔA_{T3}	0.0187	0.0356	0.0112	0.0302	0.0200
$\Delta A_{平}$	0.0216	0.0303	0.0120	0.0260	0.0178

如表2所示,ΔA平最低为0.0120,所对应的APTT温度25℃,因此设定25℃为APTT最佳作用温度。

2.3　正交试验

根据单因素试验确定的APTT浓度、时间、温度的结果,设计正交试验,同时根据SPSS 18软件计算要求,正交试验结果平行一次,试验设计及结果见表3 。

表3　APTT对牛蒡中酪氨酸酶抑制效果的正交试验设计及结果

编号	因素				ΔA_1^*	ΔA_2^*
	A APTT浓度/(μmol/L)	*B* 时间/s	*C* 温度/℃	*D* 空列		
1	400	420	20	1	0.0323	0.0498
2	400	480	25	2	0.0156	0.0167
3	400	540	30	3	0.0250	0.0335
4	450	420	25	3	0.0106	0.0134
5	450	480	30	1	0.0333	0.0276
6	450	540	20	2	0.0364	0.0436
7	500	420	30	2	0.0128	0.0113
8	500	480	20	3	0.0225	0.0254
9	500	540	25	1	0.0089	0.0085
k_1	0.029	0.022	0.035			
k_2	0.027	0.024	0.012			
k_3	0.015	0.026	0.024			
R	0.014	0.002	0.023			

注:*.两次正交试验数据分别为ΔA_1和ΔA_2,每组数据平行测定3次。

由表3可知,第9组试验,即APTT浓度500μmol/L、时间540 s、温度25℃时,A值最低,抑制效果最好。表3中R代表极差,极差越大,该因素对结果越显著,表3极差结果表明,$R_C > R_A > R_B$,因此,影响因素主次顺序为:$C > A > B$,即温度>APTT浓度>时间。

2.4　方差分析

表4　正交试验结果方差分析表

方差来源	Ⅲ型平方和	自由度	均方	*F* 值	*P* 值
APTT 浓度	0.00100000	2	0.00000000	13.193	0.002
时间	0.00005546	2	0.00002773	1.0340	0.394
温差	0.00200000	2	0.00100000	28.871	0.000
误差	0.00000000	9	0.00002682		

由表4可知，APTT浓度（*A*）、作用时间（*B*）、作用温度（*C*）的 *P* 值分别为0.002、0.394、0.000，表明租用温度（C）对APTT的抑制效果影响极显著（$P<0.001$），APTT浓度（A）对APTT的抑制效果有显著影响（$0.001<P<0.005$），作用时间（B）对APTT的抑制效果不明显（$P>0.001$）。根据影响因素的显著性，判断试验 *C*、*A* 为重要因素，按照各因素的最佳水平选取为 C_2A_3，即温度25℃、APTT浓度500μmol/L形成最优组合。而时间为次要因素，选取时间最短时间为420s，因此，最佳抑制条件为 $C_2A_3B_1$，即最佳抑制条件为APTT浓度500μmo1/L、时间420s、温度25℃。优化后的条件经实验验证，Δ*A* 为0.0082，比正交试验9组的结果更佳。因此，正交优化最佳条件为 $C_2A_3B_1$，即在25℃条件下，牛蒡以APTT浓度500μmo1/L为抑制剂对底物作用420s。

2.5　不同pH值条件对APTT抑酶效果的影响

表5　不同pH值条件下的A值

吸光度	APTT 作用 pH				
	6.0	6.5	7.0	7.5	8.0
A_1	0.0541	0.0392	0.0151	0.013	0.0119
A_0	0.0801	0.0674	0.0588	0.0422	0.0367
R	−0.026	−0.0282	−0.0437	−0.0292	−0.0248

由表5可知，在pH值6.0~8.0范围，*R* 值均为负值，说明APTT均存在抑制效果。*R* 值在pH值7.0时达到−0.0437，说明APTT在pH值7.0时抑制效果最佳。

3　结论

正交试验结果显示APTT对牛蒡酪氨酸酶有良好的抑制作用，其最佳作用条件为 $C_2A_3B_1$，即牛蒡∶PBS为0.5∶3（g/mL）、作用温度25℃、抑制剂APTT浓度

500μmol / L、作用时间 420s。在 pH6.0 ~ 8.0 范围内,APTT 对酪氨酸酶存在抑制性,在 p H7.0 时抑制效果最佳。

参考文献:

[1]陈世雄,陈靠山. 牛蒡根化学成分及活性研究进展[J]. 食品与药品, 2010, 12(7): 281 -285.

[2]蒋淑敏. 牛蒡化学成分和药理作用的研究现状[J] . 时珍国医国药, 2001, 12(10): 941 -942.

[3]赵占英,冯玉斌,何先波. 牛蒡的价值及加工利用[J]. 北方园艺, 1994 (6): 3.

[4] LI Chao, WANG Weidong, TANG Shirong, et al. High pressure microwave - assisted extraction of arctiin from Fructus Arctii [J]. Food Science, 2010, 31(16): 128 -134.

[5]尹丹丹,温新宝,袁保刚, 等. 牛蒡子提取物的抗氧化活性研究[J]. 西北农林科技大学学报: 自然科学版, 2011, 39(4): 201 -210.

[6]王学堂. 牛蒡开发技术简介[J]. 食品工业, 2003(2): 51 -52.

[7]刘玲,高维道,张建惠. 抑制牛蒡褐变的工艺研究[J]. 食品与机械, 1997(1): 27 -28.

[8]周志才,王美兰,王鲁敏. 牛蒡中多酚氧化酶的活性及其影响因素研究[J]. 烟台大学学报: 自然科学与工程版, 1998, 11(1): 62 -65.

[9] YAN Qin, CAO Rihui, YI Wei, et al. Inhibitory effects of 5 - benzylidene barbiturate derivatives on mushroom tyrosinase and their antibacterial activities[J]. European Journal of Medicinal Chemistry, 2009, 44(10): 4235 -4243.

[10] CHIARIM E, JORAY M B, RUIZ G, et al . Tyrosinase inhibitory activity of native plants from central argentina: isolation of anactive principle from *Lithrea molleoides* [J]. Food Chemistry, 2010, 120(1):10 -14.

[11] YI Wei, CAO Ri - hui, WEN Huan, et al. Discovery of 4 - functi onalized phenyl - O - β - D - glycosides as a new class of mushroom tyrosinase inhibitors [J]. Bioorganic and Medicinal Chemistry Letters, 2009, 19(21): 6157 - 6160.

[12] KUBO I, YOKOKAWA Y, KINST - HORI I. Tyrosinase inhibitors from

Belivian medicinal plants [J]. Journal of Natural Porducts, 1995, 58(5): 739 -743.

[13]邹先伟,蒋志胜. 植物源酪氨酸酶抑制剂研究进展[J]. 中草药, 2004, 35(6): 702 -705.

[14]张宗炳,冷欣夫. 杀虫药剂毒理及应用[M]. 北京: 化学工业出社, 1993(4): 541 -552.

[15] YI Wei, WU Xiaoqin, CAO Rihui, et al. Biological evaluations of novel vitamin Cesters as mushroom tyrosinase inhibitors and antioxidants [J]. Food Chemistry, 2009, 117(3): 381 -386.

[16]陈学红,秦卫东,秦杰,等. 鲜切牛蒡丝护色和制备工艺优化[J]. 食品科学, 2010, 31(4): 95 -98.

[17]郑卫星,苏秀榕,吴芝岳,等. SPSS正交设计提取林蛙油多糖[J]. 中国生化药物杂志, 2008, 29(1): 43 -45.

[18]邓振伟,于萍,陈玲. SPSS软件在正交试验设计、结果分析中的应用[J]. 电脑学习, 2009(5): 15 -17.

5.4　5-[4-(2-甲氧二氧)苯烯]-2-硫代噻唑-4-酮对雪莲果中酪氨酸酶的抑制作用

陈玲娟[1],唐雪娟[2],赵良忠[2],刘进兵[2]

(1.中南林业科技大学食品科学与工程学院,湖南长沙 410004;2.邵阳学院生物与化学工程系,湖南邵阳 422000)

摘　要:通过分光光度法探究5-[4-(2-甲氧二氧)苯烯]-2-硫代噻唑-4-(MBTO)对雪莲果中酪氨酸酶的抑制作用。结果表明:MBTO对雪莲果中酪氨酸酶有良好的抑制作用。单因素试验最佳抑制剂浓度为30.0μmol/L、最佳作用时间为210s、最佳作用温度为30℃。通过三因素三水平正交试验,分析得最佳反应条件为:pH7.5、抑制剂MBTO浓度30.0μmol/L、作用时间330s、作用温度30℃,可达到最佳抑制效果,在此条件下MBTO对雪莲果中酪氨酸酶的抑制率达84.24%。

关键词:5-[4-(2-甲氧二氧)苯烯]-2-硫代噻唑-4-酮;雪莲果;酪氨酸酶;正交试验

Inhibitory Effect of (Z)-5-[4-(2-Methoxyethoxy)benzylidene]-2-

thioxothiazolidin - 4 - oneon Tyrosinase in Yacon

CHEN Ling - juan[1], TANG Xue - juan[2], ZHAO Liang - zhong[2], * , LIU Jin - bing[2]

(1. College of Food Science and Engineering, Central South University of Forestry and Technology, Changsha 410004, China; 2. Department of Chemical Biotechnology and Engineering, Shaoyang University, Shaoyang 422000, China)

Abstract: The inhibitory effect of (Z) - 5 - (4 - (2 - methoxyethoxy) benzylidene) - 2 - thio - xothiazolidin - 4 - one (MBTO) on tyrosinase in yacon was studied by spectrometrically. The results showed that MBTO had excellent inhibitory effect on tyrosinase in yacon. The optimal inhibitory concentration, time and temperature, as determined by one - factor - at - a - time design, were 30 μmol/L, 210 s, and 30℃, respectively. The results of 3 - variable, 3 - level orthogonal array experiments showed that the best inhibition results were obtained by treatment with 30.0 μmol/L MBTO at pH 7.5 and 30℃ for 330 s, yielding an inhibitory rate as high as 84.24%.

Key words: (Z) - 5 - [4 - (2 - methoxyethoxy) benzylidene] - 2 - thioxothiazolidin - 4 - one; yacon; tyrosinase; orthogonal array design

雪莲果(Smallanthus sonchifolius),英文名为 Yacon ,别名亚贡、菊薯、地参果等,为菊科向日葵属双子叶草本植物,因其特殊的保健功能及药用价值被广泛种植[1-3]。雪莲果形似番薯[4],其肉质晶莹如玉,脆甜多汁,属低热量食品[5],雪莲果中含有大量水溶性植物纤维和丰富的低聚果糖,其低聚果糖含量为所有已知植物块茎中最高,被称为“低聚果糖之王”[6]。雪莲果以鲜食为主,但是鲜切雪莲果由于机械损伤或在加工过程中果肉直接暴露在空气中,果肉中的多酚类物质在酪氨酸酶的作用下氧化为邻醌,转而又迅速地通过聚合作用形成褐色素或黑色素,严重影响产品货架期间的感官质量和营养品质[7]。酪氨酸酶(tyrosinase , EC 1.14.18.1)是一种活性中心具有双铜离子的氧化还原酶,参与黑色素合成的前两步反应,是黑色素合成的限速酶,它广泛存在于微生物、动物和人体中,是生物体合成黑色素等产生色斑的关键酶[8-10]。国内外对酪氨酸酶抑制剂的研究广泛,但尚未找到一种高效、无毒的酪氨酸酶抑制剂来治疗诸如白化病、黑色素瘤等疾病的药物[11-13]。

目前,传统的果蔬护色剂产品主要是曲酸、半胱氨酸、维生素 C 衍生物[14]及一些中药提取物[15]、抗坏血酸、异抗坏血酸钠、4 - 乙基间苯二酚、查尔酮等[16-17],还有一些是化学药剂。

5 - [4 - (2 - 甲氧二氧)苯烯] - 2 - 硫代噻唑 - 4 - 酮{(Z) - 5 - [4 - (2

- methoxyethoxy) benzylidene] -2-thioxothiazolidin-4-one, MBTO}[18]是邵阳学院药物化学实验室合成的一种酪氨酸酶不可逆抑制剂,结构式见图1,它的 IC_{50} 值为0.56mmol/L时抑制效果良好,且理论上对人体无毒害性。本实验以雪莲果为研究对象,考察MBTO对雪莲中酪氨酸酶的抑制效果,以确定其是否具有开发潜质。

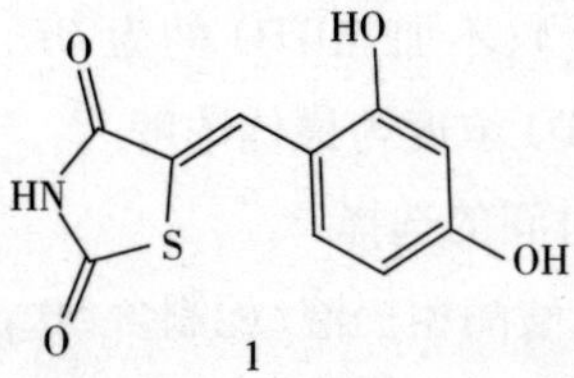

图1 5-(4-(2-甲氧二氧)苯烯)-2-硫代噻唑-4-酮结构式

1 材料与方法

1.1 材料、试剂与仪器

选取新鲜、无机械损伤、无腐烂、无虫害雪莲果。磷酸二氢钾、磷酸氢二钾 天津市大茂化学试剂厂;二甲基亚砜(DMSO) 北京鼎国昌盛生物技术有限责任公司;MBTO 邵阳学院药物化学实验室合成;其他试剂均为分析纯。UV-2100型紫外-可见分光光度计 上海尤尼柯仪器有限公司;3-18R台式高速冷冻离心机 上海托莫斯科学仪器有限公司;移液枪 德国艾本德股份公司;pH计 梅特勒-托利多仪器(上海)有限公司;电子天平 杭州汇尔仪器设备有限公司。

1.2 方法

1.2.1 雪莲果酪氨酸酶的粗提

准确称取新鲜去皮雪莲果1.0g,加入2.0mL未加MBTO、pH6.8的磷酸盐缓冲液(PBS),研磨,用4层纱布加压过滤,13000r/min迅速离心1min,上清液即为雪莲果酪氨酸酶粗提液,取1mL置于比色皿中(以PBS为空白对照),于25℃在700~800nm波长范围内扫描,并每隔5min重新扫描[19],观测光密度(*OD*)值的变化,重复3次,以选择合适的测量波长。

1.2.2 MBTO对酪氨酸酶抑制率的测定

取1.0mL雪莲果酪氨酸酶粗提液置于比色皿中(以PBS为空白对照),迅速于25℃条件下在最适波长处进行动力学扫描,测定样品的 *OD* 值,并记录每分钟光密度的变化值($OD_{标准}$)。测定MBTO溶液的雪莲果酪氨酸酶粗提液的光密度值,并同时记录每分钟光密度的变化值($OD_{样品}$)。抑制率的计算如下:

抑制率 = $(1 - OD_{样品}/OD_{标准}) \times 100\%$

1.2.3　单因素试验

1.2.3.1　MBTO 浓度的选择

配制含 20.0、30.0、35.0、40.0、50.0 μmol/L MBTO 的雪莲果酪氨酸酶粗提液，迅速于 25℃条件下在最适波长处进行动力学扫描，测定样品(240～300)s 范围内光密度的变化值($OD_{样品}$)，不加 MBTO 的为 $OD_{标准}$，重复 3 次。计算抑制率，以抑制率最高所对应的 MBTO 浓度为最佳浓度。

1.2.3.2　MBTO 作用时间的选择

取一定量的雪莲果酪氨酸酶粗提液，配制含 50 μmol/LMBTO 迅速于 25℃条件下在最适波长处进行动力学扫描，测得(0～360)s 内每 60s 的光密度记为 $OD_{样品}$，以不加 MBTO 的 $OD_{标准}$ 作对照，计算抑制率，探究 MBTO 最佳作用时间。加上雪莲果粗提液的加工时间，最佳时间为开始切雪莲果至检测前的平均加工时间 150s 加上抑制率最高所对应的时间。

1.2.3.3　MBTO 作用温度的选择

将 25、30、35、40℃制得雪莲果酪氨酸酶粗提液，配制含 50μmol/L MBTO 迅速于 25℃条件下在最适波长处进行动力学扫描，测定样品(240～300)s 范围内光密度的变化值($OD_{样品}$)，不加 MBTO 的为 $OD_{标准}$。计算抑制率，则抑制率最高的为最佳作用温度。

1.2.4　正交试验

根据单因素试验确定的 MBTO 浓度、作用时间、作用温度作为考察因素，采用 $L_9(3^4)$ 正交表[20]，每组重复 3 次取平均值，记录 $OD_{样品}$ 的平均值，计算出抑制率。数据处理采用 SPSS 软件作分析[21]，得出最佳护色条件。

1.2.5　pH 值对抑制剂及反应体系的影响

根据正交试验确定的最佳反应条件，在 pH 值分别为 5.5、6.0、6.5、7.0、7.5、8.0 的条件下参考 1.2.1 节的方法进行动力学扫描，不同的 pH 值添加 MBTO 和不添加 MBTO 的粗提液每分钟光密度变化值，分别记为 $OD_{样品}$ 和 $OD_{标准}$，重复 3 次取其平均值，计算抑制率。

2　结果与分析

2.1　测量波长结果

由图 2 可知，在(700～800)nm 波长处扫描得到一个曲线图，最高点 755nm 即为最适波长，因此后续实验都采用 755nm 波长进行检测。

2.2　单因素试验结果

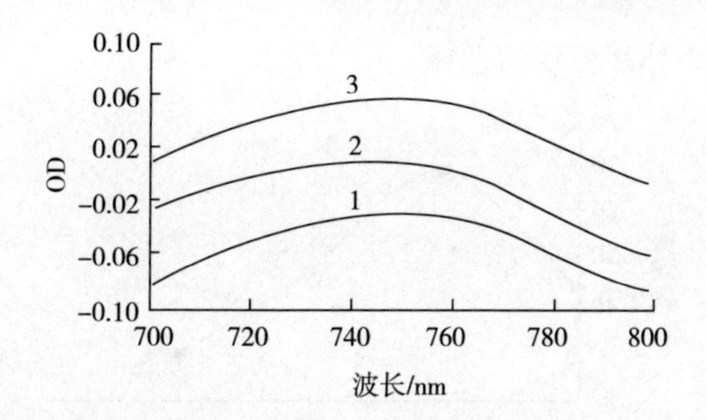

图 2　雪莲果酪氨酸酶粗提液在 700 ~800 nm 波长范围内的全波长扫描图

2.2.1　MBTO 抑制浓度的选择

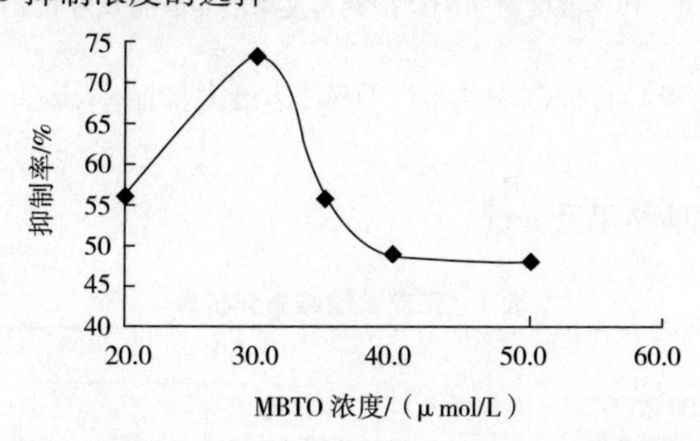

图 3　MBTO 浓度对雪莲果酪氨酸酶抑制效果的影响

由图 3 可知，当 MBTO 浓度为 30.0μmol/L 时达到最佳抑制效果，MBTO 对雪莲果酪氨酸酶的抑制率为 73.06%。

2.2.2　MBTO 作用时间的选择

由图 4 可知，60s 时 MBTO 的抑制率最高，加上从开始切雪莲果到检测前的平均加工时间 150s，确定 MBTO 的筛选作用时间为 210s，此条件下 MBTO 对雪莲果酪氨酸酶抑制率达 81.51%。

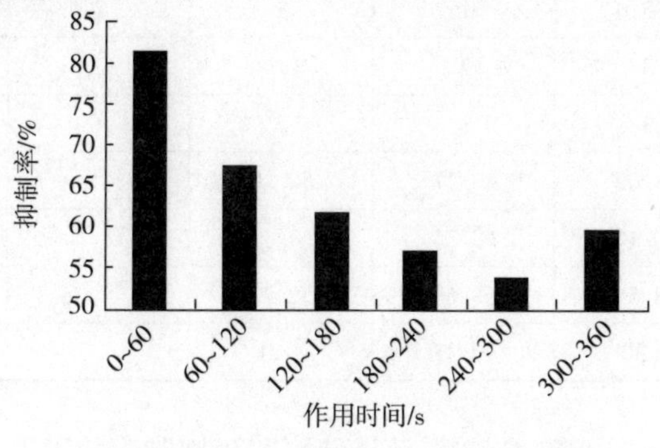

图 4　30.0μmol/ L MBTO 不同时间段对雪莲果酪氨酸酶抑制效果的影响

2.2.3 MBTO 作用温度的选择

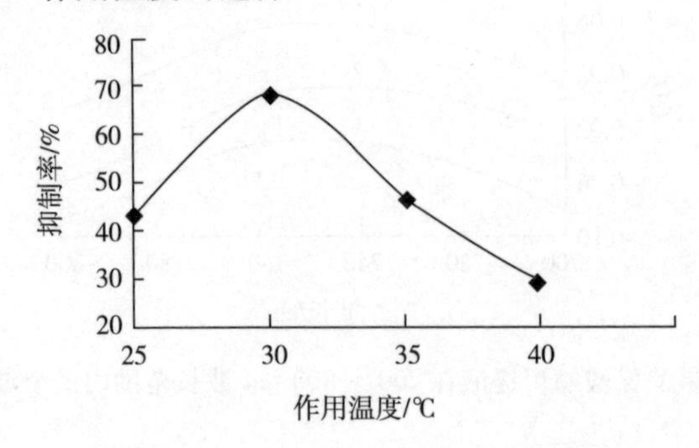

图 5 作用温度对 MBTO 抑制雪莲果酪氨酸酶效果的影响

由图 5 可知,30℃时抑制率为 67.97%,达最大抑制率,因此选择 30℃ MBTO 的筛选作用温度。

2.3 正交试验结果及分析

表 1 正交试验结果分析表

试验号	因素				抑制率/%
	*A*MBTO 浓度/(μmol/L)	*B* 作用时间/s	*C* 作用温度/℃	*D* 空列	
1	1(20.0)	1(210)	1(25)	1	46.10
2	1	2(270)	2(30)	2	48.09
3	1	3(330)	3(35)	3	80.03
4	2(30.0)	1	2	3	66.29
5	2	2	3	1	56.27
6	2	3	1	2	58.81
7	3(35.0)	1	3	2	32.41
8	3	2	1	3	76.29
9	3	3	2	1	66.96
k_1	58.07	48.27	60.40		
k_3	60.46	60.21	60.44		
k_3	58.55	68.60	56.23		
R	2.39	20.33	4.21		

由表 1 极差分析可知,影响雪莲果酪氨酸酶抑制率的因素主次顺序为:*B* > *C* > *A*,即作用时间 > 作用温度 > MBTO 浓度。根据直观分析,确定最佳方案是

$B_3C_3A_3$，最佳工艺条件为：MBTO 浓度 30. 0μmol/L、作用时间 330s、作用温度 30℃。

采用 SPSS18 软件对正交试验结果进行方差分析，结果见表 2。MBTO 浓度(A)、作用时间(B)、作用温度(C) 的 P 值分别为 0. 000、0. 000、0. 004，表明这 3 个因素对雪莲果酪氨酸酶的抑制效果影响均极显著($P < 0.01$)。再根据试验要求对最优方案进行验证，MBTO 对雪莲果酪氨酸酶的抑制率为 84. 24%，高于正交试验最高抑制率(即第 3 组抑制率 80. 03%)。因此，确定 MBTO 对雪莲果酪氨酸酶的最佳工艺条件是：MBTO 浓度 30. 0μmol/L、作用时间 330s、作用温度 30℃。

表 2　正交试验结果方差分析表

方差来源	Ⅲ型平方和	自由度	均方	F 值	P 值
A	6.679×10^{-5}	2	3.340×10^{-5}	29.525	0.000
B	0.001	2	0.000	359.485	0.000
C	2.490×10^{-5}	2	1.245×10^{-5}	11.007	0.004
误差	1.018×10^{-5}	9	1.131×10^{-6}		

2.4　pH 值对 MBTO 抑制效果的影响

表 3　不同 pH 值对 MBTO 抑制效果的影响

项目	作用 pH					
	5.5	6.0	6.5	7.0	7.5	8.0
$OD_{样品}$	0.0328	0.0437	0.0587	0.0660	0.0439	0.0298
$OD_{标准}$	0.0516	0.0768	0.1225	0.0802	0.0940	0.0103
抑制率/%	36.43	43.10	52.08	17.71	53.30	

由表 3 可知，在 pH 值为 5.5 ~7.5 时，MBTO 抑制效果随 pH 值变化而变化。其中 pH 值为 8.0 时，未检测出 MBTO 有抑制效果。故选择 MBTO 抑制效果的最佳 pH 值为 7.5，此时抑制率最大，为 53. 30% 。

3　结论

正交试验结果显示 MBTO 对雪莲果酪氨酸酶有良好的抑制作用，其最佳作用条件为：$B_3C_2A_2$，即作用时间为 330s、MBTO 浓度为 30. 0μmol/L、作用温度为 30℃时抑制效果最好，抑制率达 84. 24% 。王琼波等[22] 认为橘子汁可作为雪莲果酪氨酸酶抑制剂，运用于雪莲果深加工产品，而姚昕等[23] 也采用了混合液质量分数 0. 04% 耐晒通用型护色剂、0. 06% 植酸钠和 0. 15% 柠檬酸，同样可抑制雪莲果酪氨酸酶。在雪莲果深加工过程中,橘子汁的成本过高,柠檬酸、植酸钠等常用的护色剂对雪莲果的护色效果并不理想,这也就制约着雪莲果的深加工。本实验采用的 MBTO 与传统的雪莲果护色剂相比,MBTO 具有高效抑制酶促褐变的作用,且用量远低等特点,但是 MBTO 尚未进行毒理实验,所

以目前不能直接添加至食品中,其毒理学评价有待进一步研究。

参考文献:

[1]林若冰. 新的饮料植物:雪莲果[J]. 广西热带农业, 2005, 96(1): 31 -32.

[2]严和平,马学海. 雪莲果中绿原酸的提取工艺[J]. 食品研究与开发, 2009, 30(10): 89 -92.

[3]蒲海燕. 雪莲果的综合利用和开发[J]. 河南工业大学学报:自然科学版, 2010, 31(3): 86 -90.

[4]史云东,贾琳,李祥,等. 雪莲果多酚氧化酶活性及褐变控制的研究[J]. 粮油食品, 2011, 19(2): 46 -49.

[5]王文光,袁唯. 雪莲果加工过程中绿褐变原因初探[J]. 中国食品添加剂, 2008(2): 92 -95.

[6]唐松梅,杨翠香,陶鲜娇,等. 雪莲果片处方筛选及制备工艺研究[J]. 大理学院学报, 2010, 9(6): 93 -95.

[7]刘雪莲,杨希娟,郝学宁. 雪莲果的护色及其糖制工艺优化[J]. 安徽农业科学, 2011, 39(4): 2294 -2296.

[8]宋康康. 抑制剂对酪氨酸酶的效应及其对黑色素生成调控的研究[D]. 厦门:厦门大学, 2007.

[9] PARK Y D, KIM S Y, LYOU Y J, etal. A new type of uncompetitive inhibition of tyrosinase induced by Cl^- binding[J]. Biochime, 2005, 87(11): 931 -937.

[10] LIU Jinbing, CAO Rihui, YI Wei, et al. A class of potent tyrosinase inhibitors: alkylidenethiosemicarbazide compounds [J]. European Journal of Medicinal Chemistry, 2009, 44(4): 1773 -1778.

[11]陈清西,林建峰,宋康康. 酪氨酸酶抑制剂的研究进展[J]. 厦门大学学报:自然科学版, 2007, 46(2): 274 -282.

[12]宋康康,邱凌,黄璜,等. 熊果苷作为化妆品添加剂对酪氨酸酶抑制作用[J]. 厦门大学学报：自然科学版, 2003, 42(6)：791 - 794.

[13]王白强,曾晓军. 酪氨酸酶活性的抑制研究及皮肤美白化妆品的研制[J]. 福建轻纺, 2005, 158(7)：1 - 6.

[14]韩强,林惠芬,朱玲莉. 几种中药提取物对酪氨酸酶活性的抑制[J]. 香料香精化妆品, 1998, 12(4)：22 - 24.

[15] YI Wei, WU Xiaoqin, CAO Rihui, et al. Biological evaluationsof novel vitamin C esters as mushroom tyrosinase inhibitors and antioxidants [J]. Food Chemistry, 2009, 117(3)：381 - 386.

[16] KHATIB L, NERYA O, MUSA R, et al. Chalcones as potent tyrosinase inhibitors：the importance of a 2,4 - substituted resorcinol moiety[J]. Bioorganic & Medicinal Chemistry, 2005, 13(2)：433 - 441.

[17] NERYA O, MUSA R, KHATIB S, et al. Chalcones as potent tyrosinase inhibitors：the effect of hydroxyl positions and numbers [J]. Phytochemistry, 2004, 65(10)：1389 - 1395.

[18] LIU Jinbing, WU Fengyan, CHEN Lingjuan, et al. Evaluation of dihydropyrimidin - (2H) - one analogues and rhodanine derivatives as tyrosinase inhibitors[J]. Bioorganic & Medicinal Chemistry Letters, 2011, 21(8)：2376 - 2379.

[19]杨波,杨光,陈佳雯. 鲜切雪莲果护色保鲜工艺的研究[J]. 食品工业科技, 2009,(4)：239 - 242.

[20]邓振伟,于萍,陈玲. SPSS 软件在正交试验设计、结果分析中的应用[J]. 电脑学习, 2009,(5)：15 - 17.

[21]郑卫星,苏秀榕,吴芝岳,等. SPSS 正交设计提取林蛙油多糖[J]. 中国生化药物杂志, 2008, 29(1)：43 - 45.

[22]王琼波,魏永义,王玲玲. 雪莲果多酚氧化酶性质及其抑制方法研究[J]. 北方园艺, 2011(8)：49 - 51.

[23]姚昕,涂勇. 正交试验优化雪莲果打浆护色条件研究[J]. 西昌学院学报：自然科学版, 2010, 23(3)：34 - 35.

[illegible]

[illegible] 2005 [illegible]

[illegible]

[illegible] et al. [illegible] novel [illegible] J. Food Chem [illegible]

[illegible] Bioorganic & Medicinal Chemistry, 2005 [illegible]

[illegible] et al. [illegible] potent [illegible] Phytochemistry, 2004 [illegible] 1389-1395.

[illegible] Bioorganic & Medicinal Chemistry Letters [illegible]

[illegible] 2004 [illegible]

[illegible]

[illegible]

[illegible]

[illegible]